AF454143

Elements of Farm Power and Machinery

THE AUTHOR

Dr. S K Gupta, Born (1949), INAE Distinguished Visiting Professor, ICAR-NDRI, Karnal, is an Engineer with specialization in Soil and Water Resources Engineering and Civil Engineering. He obtained B. Tech (Agril. Engg.) from the Punjab Agricultural University (PAU), Ludhiana, in 1970 and Master of Agricultural Engineering from the same University in 1976. He obtained Ph.D. in Civil Engineering from the Jawahar Lal Nehru Technological University, Hyderabad, Telangana, India in 1984. Since his joining CSSRI in 1971, besides holding scientific positions, he has been Head, Division of Drainage and Water Management, Head, Indo-Dutch Network Project, Head, Division of Irrigation and Drainage Engineering and Project Coordinator (AICRP on Management of Salt Affected Soils and Use of Saline Water in Agriculture). After superannuation, Dr. Gupta worked as Emeritus Scientist (ICAR) and INAE Distinguished Professor at the same institute. Currently, he is professor of Agricultural Engineering (Guest Faculty) at Khalsa College, Amritsar. Dr. Gupta has been engaged in conducting researches in surface and subsurface drainage for land reclamation, irrigation water management, hydrology of salt affected soils, leaching and use of saline water in agriculture. He has been teaching at ICAR-NDRI, Karnal, Karnal Institute of Engineering and Technology, Karnal and Khalsa College, Amritsar. Dr. Gupta is Fellow of the National Academy of Agricultural Sciences, Indian National Academy of Engineering and Indian Society of Agricultural Engineering. He has published more than 150 research papers in high impact factor journals and published 15 books. Besides teaching on-line, he is pursuing his own interests in research and writing books. Dr. Gupta has been consultant to WAPCOS-Louis Berger, MoWR, MoRD, UP-Dutch Tube Well Project, ActionAid International, Synergics Hydro India-Oromia Water Works Enterprise, Ethiopia and others. He has been bestowed with Rafi Ahmad Kidwai Award by ICAR. Besides he is recipient of several other awards from NAAS, ISAE, IE (India), CBIP and MoWR among other organizations. He is the Chief Editor of Journal of Water Management of the Indian Society of Water Management.

Elements of Farm Power and Machinery

— *Author* —

S.K. Gupta, *Ph.D.*

Formerly–Professor of Agricultural Engineering (Guest Faculty),
Khalsa College, Amritsar
INAE Distinguished Visiting Professor
ICAR-National Dairy Research Institute, Karnal
Head, Division of Irrigation and Drainage Engineering
Project Coordinator, Use of Saline Water in Agriculture
ICAR-Central Soil Salinity Research Institute, Karnal

2023

Daya Publishing House®
A Division of

Astral International Pvt. Ltd.
New Delhi – 110 002

Published by   :   **Daya Publishing House®**
                             A Division of
                             Astral International Pvt. Ltd.
                             – ISO 9001:2015 Certified Company –
                             4736/23, Ansari Road, Darya Ganj
                             New Delhi-110 002
                             Ph. 011-43549197, 23278134
                             E-mail: info@astralint.com
                             Website: www.astralint.com

Preface

Increasing use of farm power and machines in crop production programs has been one of the most outstanding developments in agriculture India has seen since the first green revolution. The results are quite spectacular in many respects especially in increasing efficiency and the output per worker, timeliness of operations, increasing input use efficiencies, and reducing the drudgery of farm work. All these translated in greatly increasing production and productivity at the farms. Realizing the role of farm mechanization, tremendous amount of research and development efforts are being made at academic and industry level to realize the dream of ushering in an era of evergreen revolution. Innumerable research papers, thousands of technical reports, new and innovative machine designs are few indicators of these efforts. The book titled 'Elements of Farm Power and Machinery' is the latest addition to the literature on the subject intended primarily as a textbook for various degree courses leading to B.Sc. (Hon.) in agriculture, horticulture, forestry and B. Tech. in agricultural engineering. The textbook addresses the need of an introductory book on this vital subject. The purpose is to make the students aware of the functional requirements, principles, and performance evaluation of tractors, power tillers and farm machines for agriculture, gardening and forestry. After completing this course students will have the capability of applying the relevant knowledge in the selection and operation of machines at the farms. Design-related material has been left out of this basic course. Nonetheless a list of design related BIS standards on farm power and machines is included for those interested in design aspects of tractor components and farm machines.

The book, divided into well-defined 17 chapters, begins with the Introduction to farm power and basic machines, a basic pre-requisite of the course. The other chapters include Internal combustion engines, Tractor systems, Farm tractors and power tillers, Introduction to electric motors, Farm mechanization, Primary tillage and related equipment, Secondary tillage and related equipment, Field capacity power requirement and costing, Machinery for land leveling and land forming, Sowing and planting equipment, Irrigation pumps, Plant protection equipment, Harvesting and threshing equipment, Garden tools, Tools and machines for forestry and Miscellaneous equipment. Contents of each chapter dovetail with the requirement of the agricultural/horticulture/agriculture engineering/forestry students who compulsorily take this subject. The contents are as per the Dean's committee recommendations made to ICAR from time to time. Methods for testing or evaluating the performance of certain types of field machinery are included in the appropriate chapters. The following features of this book add to its value for the students.

- ☆ The book is highly illustrated with relevant diagrams and photographs for easy to understand reading
- ☆ Theoretical questions appear at the end of each chapter while all objective type questions and their answers appear in one section towards the end following the glossary of terms
- ☆ Special emphasis is placed on solved numerical problems at relevant places throughout the book
- ☆ Related major books and papers are referenced in the references. It provides a handy guide to the students or field practitioner who would like a more detailed study of a particular subject
- ☆ Glossary of terms and a subject index is included

While teaching this subject, I could understand that there is no better way to learn a subject than to teach it. The present form of the book owes its existence to rich contributions from the hundreds of students whom I've had the good fortune to teach at Khalsa College, Amritsar in online mode. The students have patiently sat through and asked searching questions that helped me to improve over the study material. This book bears the stamp of these students.

The book has heavily drawn from direct and indirect contributions of many national and international organizations and individuals. The author wishes to express his appreciation to all the individuals and organizations whose material I have freely used in preparing this manuscript. I am particularly thankful to some of the farm machinery manufacturers and others that have been most cooperative in supplying the required material. E-courses of farm power and machinery brought out by ICAR have been used as teaching material and it might reflect in the text book as well. I have tried to include several important contributions by referring the works in the references. Nonetheless, it is in no way a complete list. I place on record my appreciation for all the contributors who have painstakingly researched and developed the relevant processes, technologies and teaching material. I take this opportunity to whole heartedly thank the Principal, Khalsa College Amritsar who has been kind enough to give me an opportunity to work as guest professor to teach the students. I am personally thankful to Dr. Kanwaljit Singh Sandhu, Assistance Professor Agronomy, Er. Harwinder Singh and Er. Prashant Kumar, Assistant

Professors Agricultural Engineering at Khalsa College Amritsar for providing logical support and resolving day to day issues whenever cropped up. They have also painstakingly gone through the teaching material and offered constructive suggestions. I find it difficult to name each and everyone who have contributed or supported me in one way or the other while writing this book. Finally, I would like to thank my family and friends for their support. My wife and children have seen me spent days together on this book. There have been occasions when I sacrificed their quality time to pursue this work. They have never complained and have remained supportive and encouraging to the end.

As already stated, this book has been written to serve as a textbook to familiarize students on the basic practical approach to farm power selection, construction and operation of farm machines, and to provide necessary theoretical and practical skills for effective farm management. I believe that this book will prove to be a good resource to the teachers and students to have a better teaching/learning experience. I am sure that the book will be a cherished possession of the faculty and students of agricultural/horticulture/agricultural engineering/forestry and practicing agricultural graduates and agricultural engineers. Besides, it should be a valuable academic asset for libraries of colleges and universities worldwide.

S.K. Gupta

Contents

of Tillage Operations, Conservation Tillage, Ploughing, Methods of Ploughing a Land, Headland Pattern, Ploughs, Country or indigenous Plough, Animal Drawn Mould-Board Plough, Tractor Operated MB Plough, Kinds of MB Plough Bottoms, Components, Plough Accessories, Materials of Construction, Adjustments of Mould-Board Plough, Terms Used in Ploughing, Disc Plough, Commonly Used Terms, Types of Disc Plough, Plough Adjustments, Trouble Shooting, Special Ploughs, Rotavator, Reversible MB Plough, Ridge Plough, Chisel and Sub-soil Ploughs, Basin Lister, Questions (Theory)

Harrows, Disc Harrow, Components, Adjustments to Increase Penetration, Care and Maintenance of Disc Harrow, Other Kinds of Harrows, Power Harrow-Tractor Drawn, cultivators, Sweeps/Shovels and Shanks, Animal Drawn Cultivators, Tractor Drawn Cultivators, Rotary Tiller or Rotavator, Hoes, Puddlers, Secondary Tillage Implements for Rice Cultivation, Cage Wheels, Paddy Weeder, Land Rollers or Pulverizers, Cage Roller (Clod Crusher), Tractor drawn Spiked Clod Crusher, Improved Bakhar, V-shaped Roller Pulverizer, Clod Breaker, Sheep Foot Roller, Soil Packer, Good Agricultural Practice in Tillage, Questions (Theory)

Field Capacity, Theoretical Field Capacity, Actual/Effective Field Capacity, Field Efficiency, Material Capacity, Power Required, Estimation of Drawbar Power, PTO Power, Hydraulic Power, Wheel Slip, Selection of Tractor Power, Cost of Farm Machinery, Cost of Tractor and Machine, Tractor and Implement Selection, Selection of Farm Machinery, Ownership of Farm Machinery, Points to be Considered while Making Actual Purchase, Questions (Theory)

Land Leveling, Benefits of Land Leveling, Phases of Land Leveling Operation, Machinery for Land Leveling, Bulldozer/Dozer for Land Clearing, Methods of Land Leveling, Manual Leveling, Animal or Mechanical Power, Animal drawn Soil Scoop, Animal drawn Leveling Board, Patella Harrow, Animal drawn Buck Scraper, Tractor Drawn Implements, Tractor Mounted Dozer, Tractor Drawn Leveler, Tractor Drawn Land Plane, Box and Grading Scrapers, Laser Land Leveling, Laser Land Leveling Equipment, Steps in Land Leveling Program, Benefits and Limitations of Laser Land Leveling, Land Forming Equipment, Bund Former, Ridger, Irrigation Channel Former, Tractor Drawn Ditcher, Questions (Theory)

Methods of Sowing, Broadcasting, Dibbling, Drilling, Transplanting, Limitations of Traditional Methods, Seed Drill, Classification of Seed Drills, Components of Seed Drill, Seed Metering Mechanisms, Fertilizer Metering Systems, Furrow

Openers and their Types, Coulters and Covering Devices, Sowing Implements, The Seed-cum-fertilizer Drill, Tractor Mounted Ridger Seeder, Zero Till Drill, Turbo Happy Seeder, Pneumatic Multi-crop Drill/Planter, Calibration of Seed Drill, Farm Shed Calibration Procedure, Calibration in the Field, Rice Transplanting, Walk Behind/Manual Rice Planter, Engine Operated Paddy Transplanter, Preparation of Mat-type Nursery, Preparation of Field for Transplanting, Direct Seeded Rice, Tractor Mounted Direct Rice Seeder, Planter, Tractor Mounted Modular Planter, Bed Planter, Tractor Mounted Ridge Planter for Winter Maize, Potato Planter, General Precautions during Sowing, Questions (Theory)

Chapter 1

Introduction to Farm Power

Whenever a force moves a body from one place to another, it is said that some work has been done. On the other hand, some energy is required to do the said work. This energy comes from various sources notably food, fuels, electricity, sun or wind. The amount of energy consumed is equal to the work done. This chapter deals with the basics of work, power and energy, and their relationships. It is followed by a brief introduction to farm power and to some basic machines, a precursor to the main subject of farm power and machinery.

FORCE (F)

A force is *'an invisible agent that always tries to change the state of the body i.e. a change in speed, direction or shape'*. The SI unit of force is Newton (N) and in MKS system it is kg-f

WORK

Work is described as transfer of energy. When a force is applied to a body, and the body moves in the direction of the applied force, then it is said to have done some work. Thus, for a work to be done a force must be applied and there must be some displacement in the direction of the force. Thus, we can define work as *'the component of the force in the direction of the displacement times the magnitude of the displacement in the direction of force'*.

$$W = F \times d \tag{1.1}$$

Here W is the work done, N-m, F is the force, N, and d is the distance, m. The SI unit of work is the joule (J) and 1 N-m = 1 joule. For example, if a force of 20 Newton (N) pushes an object 2.5 m in the same direction as that of the force, it will accomplish 50 joules (J) or 50 N-m of work. The work done by the force is positive, if the force acting on a body has a component in the direction of displacement. For example, when a body falls freely under gravity, then the work done by the gravity is positive. On the other hand if the force acting on a body has a component in the opposite direction of displacement, the work done is negative. As an example, the work done is negative when a body sliding on a slanting surface moves against the friction force. A good example of work is the lifting a weight from the ground and putting it on a shelf (Figure 1.1 left). In this case, the force is equal to the weight of the object, F (weight = mass x acceleration due to gravity) and the distance is equal to the height of the shelf, d. Note that if the force is applied at an angle of the actual direction of displacement, the force component in the direction of displacement is used to calculate the work. For example, when the force of 100 N is applied at an angle of 30° (Figure 1.1 middle), then the force in the direction of motion is:

F = 100 x cos 30 = 100 x 0.866 = 86.6 N

Thus, general relationship of work done can be written as:

$$F = F \cos \theta \times d \tag{1.2}$$

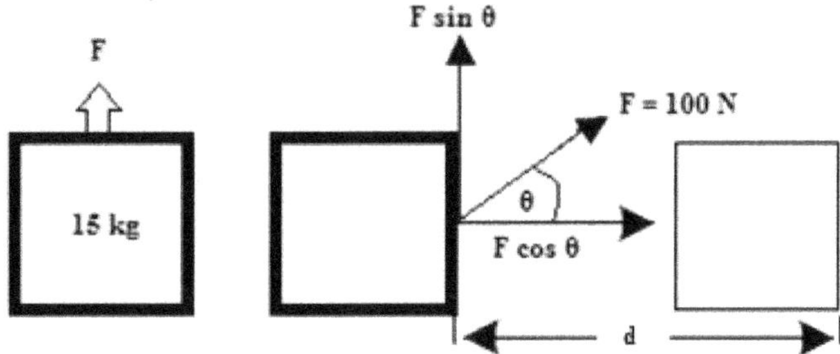

Figure 1.1. Illustration of the Concept of Work.

The work done is zero, if

☆ If there is no displacement, *i.e.* d = 0. For example, a person pushing a wall does not do any work

☆ Body is displaced perpendicular to the direction of force, such that cos 90 = 0

Example 1.1

A 100 N force acting parallel to the surface horizontally drags a body. What is the amount of work done by the force in moving the object through a distance of 8 m?

We know that,

F = 100 N, d = 8 m

Since F and d are in the same direction, we use eq. (1.1) to calculate the work done.

W = 100 x 8 = 800 J

Example 1.2

A person is holding a bucket by applying a force of 10 N. He moves a horizontal distance of 5 m and then climbs up a vertical distance of 10 m. Calculate the total work done by him?

F = 10 N

d = 5 m

θ = 90° (When moving in horizontal direction)

Work done, W_1=F x d x cos θ = 10 x 5 x cos 90° = 0

For vertical motion, the angle between force and displacement is 0°.

Work done, W_2=10 x 10 x cos θ = 10 x 10 x cos 0 = 100 J

Total work done = W_1+W_2 = 100 J

Example 1.3

A block of steel is pushed through a distance of 20 m. Find the work done if force applied is 20 kg weight in a direction inclined at 60° to the ground.

Let us illustrate the question with the line diagram as follows:

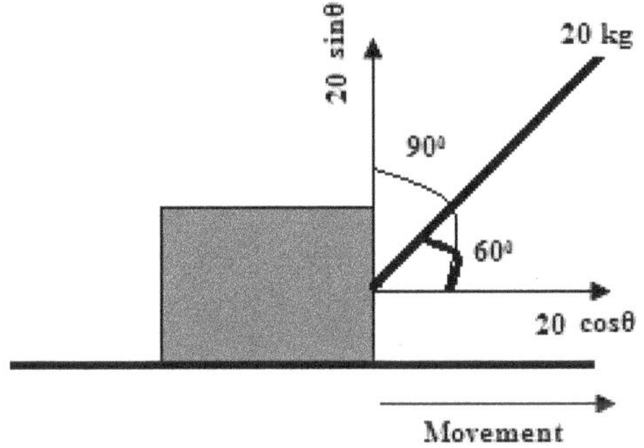

F = 20 kg = 20 x 9.8 = 196.2 N

d = 20 m

θ = 60°

Using eq. (1.2), work done is given as

W = F x d x cos θ = 196.2 x 20 x cos 60° = 196.2 x 20 x 0.5

W = 1962 J

Note that the work is done only by the force in the direction of motion. Therefore, work done by the component of the force normal to the motion is zero.

Example 1.4

Calculate the work done for the following 3 cases shown in the figure.

1. A 100 N force is applied to move a 12 kg object to 5 m.

2. A 100 N force is applied at an angle of 30° with the horizontal to move 12 kg body to a distance of 5 m

3. An upward force is applied to lift 12 kg object to a height of 5 m

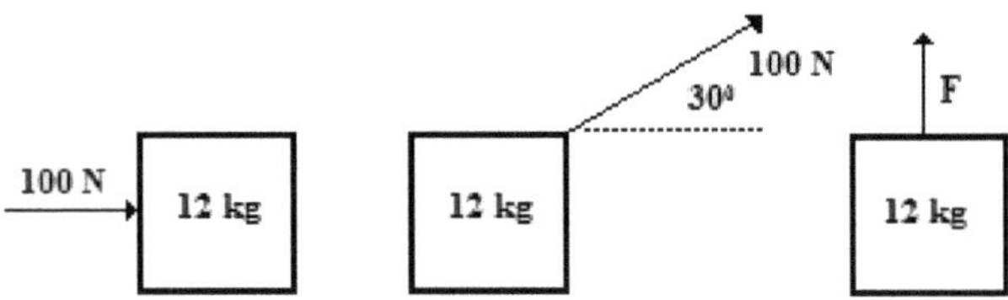

Work done = 100 x 5 = 500 J

Work done = 100 x 5 x cos 30 = 433 J

Work done = 12 x 9.81 x 5 = 588.6 J

Note that cos θ in first and third cases is equal to 1.0 and therefore omitted in the equations to calculate the work done.

POWER

Power is defined *'as the rate of doing work'. In other words, it means that power is a measure of how quickly the work is done. Therefore, the work done per unit time is called power'* and is expressed in N-m/s

Power = Rate of doing work = Work done/Time   (1.3)

When work is done, an equal amount of energy is consumed. Thus, power can also be defined *'as the rate at which energy is consumed or utilized'*.

Power = Energy consumed/Time

If under a constant force F a body is displaced through a distance d in time t, the power, p is given as

$p = W/t = F \times d/t = F \times v$ (1.4)

Since d/t = v, uniform velocity with which body is displaced. This new equation for power reveals that a powerful machine is both strong (have higher force) and fast (higher velocity). As an example, a powerful car engine is both strong and fast. The work in SI unit is expressed in Joule/s, and is also expressed in N-m/s. Other units are Watts, W. A power of 1W means that work is being done at the rate of 1 J/s. The unit of Watt is dedicated to James Watt, the developer of the steam engine. The bigger unit is kilo Watt, kW. It is also given in horsepower, the force needed to move 75 kg force through a distance of one meter in one second. Power is a scalar quantity as the power doesn't have any direction but only magnitude.

Example 1.4

A hoist lifts a truck up 1.5 m above the ground in 12 seconds. Find the power delivered to the truck. Assume the mass of the truck as 1500 kg.

F = mg = 1500 x 9.81 = 14715 N.

W = F x d = 14715 x 1.5 = 22072.5 N-m

P = W/t = 22072.5/12 = 1308 J/s = 1839.3 W

Horsepower

Engine power is commonly measured in terms of horsepower. The term had its origin from the time when engines and horses were competing as sources of power. *'Horsepower is that amount of force which is capable of displacing 75 kg force through a distance of one meter in one second'*. One HP is approximately equal to 746 Watts (W) or 0.746 kilowatts (kW). A dynamometer is used to measure horsepower. Although not in common use, a term metric horsepower denoted as PS is also reported in the literature. It is equal to 735.5 watts such that 1 HP = 1.01 PS.

ENERGY

'Energy of a body is its capacity of doing work'. Since, energy can neither be created nor destroyed; it is only transformed from one form of energy to another form. There are several types of energies such as kinetic energy (KE) and potential energy (PE) *etc.* While the energy in motion is known as KE, the energy stored in an object due to its position is PE. These energies respectively are given as:

$$KE = (1/2) \, m \, V^2 \qquad\qquad (1.5)$$

$$PE = m \, g \, h \qquad\qquad (1.6)$$

Here m is the mass, V is the velocity, g is the acceleration due to gravity and h is the position or height. Because energy is the capacity to do work, energy and work have the same units. If the mass has units of kilograms and the velocity in m/s, the kinetic energy has units of $kg\text{-}m^2/s^2$ or N-m. The SI unit of energy is joules (J), the name assigned in honour of James Prescott Joule. Other units are Calories or kilo Watt hr. Note that all the 3 quantities namely work, power and energy are scalar quantities. It means they have magnitude but no direction. It may not be out of place to mention that the term vector quantities is used for quantities, which have both magnitude and direction *e.g.* velocity. Some other types of energy are as follows:

- ☆ Mechanical energy
- ☆ Mechanical wave energy
- ☆ Chemical energy
- ☆ Electric energy
- ☆ Magnetic energy
- ☆ Radiant energy
- ☆ Nuclear energy
- ☆ Ionization energy
- ☆ Elastic energy
- ☆ Gravitational energy
- ☆ Thermal energy
- ☆ Heat Energy

CONVERSION FACTORS

One horsepower (HP) = 746 Watts or 0.746 kW

One Watt = 1 Joule/s

One joule = 0.2389 Calories

One kW = 102 kg-m/s

1 Calorie = 4.2 Joules

1 kg = 9.81 Newton (N)

1 N-m = 1 Joule

1 N-m/s = 1 Joule/s = 1 Watt

1 kg/cm^2 = 9.8 x 10^4 N/m^2 = 9.8 x 10^4 Pascal (Pa) = 98 kPa

1 Pa = 1N/m^2

1 kg-m/s = 9.8 Joules/s = 9.8 Watt

1 J/s = 0.10198 kg-m/s

1 kW-h = 860 k Calories = 3.6 E+6 Joules

1 kW = 1000 W = 102 kg-m/s

SOURCES OF FARM POWER

Various sources of power for doing different mobile and stationary operations at the farm are as under (Srinivasan *et al.*, 2011).

Mobile Power

The mobile sources of power are mainly used for tractive work such as seedbed preparation, cultivation, harvesting and transportation *etc.* Some sources of mobile power are:

Human: Men, women, children

Draught animals: Bullocks, buffaloes, camels, horses and ponies, mules and donkeys, elephants (mostly in forestry)

Mechanical power: Tractors, power tillers, self-propelled machines such as combines, dozers, reapers, sprayers *etc.*

Stationary Power

These are mainly used for lifting of irrigation water, stationary work like silage cutting, feed grinding, threshing and winnowing. Some sources of stationary power are:

Diesel/oil engines: Used to run pump sets, threshers, sprayers and other stationary operations

Electric motors: Used to run pump sets, threshers, sprayers and other stationary operations

On the basis of sources of farm power available in India, the farm power is categorized as follows:

- ☆ Human power
- ☆ Animal power
- ☆ Mechanical power (Tractors + Power tillers + Oil engines)
- ☆ Electrical power
- ☆ Renewable (Solar energy + Wind energy + Biogas)

Human Power

Human power has been used in agriculture probably from the time agriculture started on the earth. All operations such as land preparation, cultivation, harvesting, and processing of final product were performed more or less by human power. It has been the main source of power for operating small implements and tools at the farm. Manual labour used to perform all stationary work like chaff cutting, lifting water, threshing and winnowing *etc.* An average man can develop maximum power of about 0.1 HP for doing farm work (Sahay, 2006). A woman labour can develop about 0.05 HP power (www.Agrimoon.com, 2016). On an average human power is taken as 0.05 kW per human (The Working Group Report, 2018). All reports reveal that there is decline in number of labours available or employed in agriculture. Advantages and disadvantages of human power are listed as follows:

Advantages	Disadvantages
Traditional skill no knowledge required	Costly
It is easily available	Very low efficiency, only about 25 per cent of the energy consumed is converted to physical work when handling relatively easy tasks, which may reduce to as low as 5 per cent or less for heavier tasks
Can be used for all types of work	Needs full maintenance even when not in use
Decision making process can be performed on the spot saving time and cost	Weather condition and seasons affect its use and efficiency

Animal Power

Development and inventions of heavier and more effective field tools and machines such as heavier and larger ploughs and harrows, mechanical planters, cultivators, and harvesting devices required more powerful and better sources of power. As a result, the bullocks, the oxen, the horses, and the mules entered the farm power scene. Because of their availability for use in most stringent conditions, they are often referred to as the beast of burden (Bello, 2012). Although, Indian agriculture mostly revolves around bullocks, other animals like camels, buffaloes, horses, donkeys, mules and elephants are also used in some parts of India. Animal power is still being used all over the world and more so in India and other developing and underdeveloped countries. Power developed by an average pair of bullocks is about 1 HP for usual farm work or 0.38 kW per animal. At an operating speed of 0.7 m/s *i.e.* 2.5 km/h, push exerted by a pair of bullocks is 107.5 kgf (Richard, 2014). It can be shown as follows:

Push (*kgf*) = HP x 75/Speed   (1.7)

Push (*kgf*) = 1.0 x 75/0.7=107.5

Assuming pull makes an angle of 45° with horizontal, draft (D) is given by,

D= P cos θ   (1.8)

D = 107.5 x cos 45° = 76.0 kg

Average force exerted by an animal is nearly one-tenth of its body weight, although it may vary from one place to another depending upon the kind and breed of animals, species, sex, age, temperament and other environmental factors. Some of the advantages and disadvantages of animal power are listed in Table 1.1.

Table 1.1. Advantages and Disadvantages of Animal Power

Advantages	Disadvantages
It is easily available	Less efficient
Can be used for all types of work	Slow and cannot work continuously
Initial investment is less	Needs full maintenance even when not in use, health problems
Serves as a source of manure and fuel	Weather condition and seasons affect its use and efficiency
Part of upkeep comes from farm produce	Creates unhealthy and unhygienic atmosphere near the residence

Mechanical Power

Mechanical power is mostly derived from fossil fuels (petrochemicals). Fossil fuels are produced from plant and animal remains through natural processes over millions of years into a form which we can use as a source of energy. Fossil fuels such as coal, oil and natural gas are the main sources of energy used to run power stations and vehicles. Mechanical power at the farm includes stationary oil engines, tractors, power tillers and self-propelled machines. The internal combustion engine in these machines is used to convert liquid fuel into mechanical power. These engines are of two types namely spark ignition engines (petrol or kerosene engine) and compression ignition engines (diesel engines). The efficiencies of petrol engines are in the range of 25 and 32 per cent. On the other hand, the efficiencies of diesel engines vary between 32 and 38 per cent. Some of the issues related to these engines are discussed in Chapter 2. Currently almost all the tractors and power tillers are operated by diesel engines. The diesel engines are also used for stationery operations like water lifting, threshing, winnowing, chaff cutting *etc.* and in agricultural industry such as flour mills, oil expellers, vegetable washers, gardeners, mulberry processing equipment. Some advantages and disadvantages of mechanical power are as follows:

Advantages	Disadvantages
High efficiency	High initial investments
Not affected by weather or seasonal conditions	Fuel used to run is costly and scarce
Can run at a stretch for longer period	Requires technical knowledge for repairs and maintenance
Require less space and is a cheaper source of power	They produce noise and exhaust fumes, both are not environmentally friendly
They have multi-purpose use, operate in harder conditions than animals or humans	High operator fatigue in walk-behind power sources

A number of terms are widely used to describe power of the tractor. They are drawbar horsepower, brake horsepower and power take off *i.e.* power at the tractor's power take off shaft. These terms are described in the next chapter of the book. To calculate the average power on agricultural lands, tractor power on an average is taken as 26.1 kW while that of a power tiller as 5.6 kW (The Working Group Report, 2018). The diesel engine on an average is assumed to develop power of 5.6 kW. A comparative statement of tractor and animal power is made in Table 1.2.

Table 1.2. Comparison of Tractor and Animal Power

Parameter	Tractor	Animal
Availability	Various sizes of tractors and power tillers are available	Available in plenty but reducing over time
Acceptability	Quite acceptable	Limited acceptability only by small farmers
Tractive work	Suitable for any traction job	Suitable for all kinds of farm work
Stationary work	All kinds of stationary works can be performed	Limited use for such works
Transport work	Quick means of medium distance transport	Good for short and medium distances
Initial investment	Though cost per horsepower is low but overall investment is high	Cost per horsepower is high but overall investment is less
Maintenance cost	Reasonable	Very high
Cost of operation	Cheaper per horsepower hour	Costlier than tractor
Limitations	Because of limited technical knowhow, initially people were discouraged, but it is no longer valid now	Traditional method, no technical knowhow is required. Constant care required to keep the animal in good health
Output	High	Low
Cost during lean periods	No fuel or lubricant is consumed while not in use	Feed and fodder is required for all times, but provides manure for the crops

Electrical Power

Electric power is used in various forms. One of the forms is fuel cells or batteries. Batteries are electrochemical devices that convert the chemical energy of a fuel directly and very efficiently into electricity (DC) and heat, thus doing away with combustion. Now-a-day's electricity derived from hydro, thermal and nuclear sources of energy has become a very important source of power globally and on farms in all the states of India. Electrical power is a clean and efficient source of power. It is mostly used for operating electrical motors for water pumping, threshing, diary, cold storage, farm product processing, fruit industry and many similar operations and sites (Desai and Sivakumar, 2017). In calculating the average power used on Indian farms, the power developed per electric motor is taken as 3.7 kW. The advantages and disadvantage of electrical power are reported in Table 1.3.

Table 1.3. Advantages and Disadvantages of Electrical Power

Advantages	Disadvantages
The cheapest form of power	Capital investment is high
Highly efficient	Requires good amount of technical knowledge
Can work continuously and is not affected by weather or seasonal conditions	If handled carelessly, it may prove to be dangerous
Operational cost is low, and it remains almost constant throughout its life	
Needs less attention and care for operation and maintenance	

Renewable Sources of Power

Because of the depleting non-renewable resources and environmental concerns, people are switching over to renewable sources of energy like sun, wind, biomass (biogas, producer gas, ethanol and biodiesel) *etc.* These kinds of energy are used in agriculture including domestic purposes with suitable devices. Renewable energy is now commonly used for lighting, cooking, water heating, space heating, water distillation, food processing, water pumping, and electric generation. This type of energy is inexhaustible in nature and is available on a regular basis. Besides, these sources are considered to be environmental friendly. A brief description of these power sources is included in the following sections.

Solar Power

An enormous amount of energy from the sun reaches the earth each day. It is estimated that all the energy stored in earth's reserves of coal, oil, and natural gas is just equal to the energy from only 20 days of sunshine (www.AgriMoon.com, 2016). Most states and regions receive enough sunshine to make solar energy practical. The energy conversion process involves three major steps:

- ☆ Absorption of the sunlight by solar cell (heat source)
- ☆ Heating up of the thermocouple junction or black surface results in a temperature difference
- ☆ The transfer of the thermoelectric potentials or temperature difference for outside application in the form of electric current or heat

The solar power can be used in agriculture in a number of ways for which some equipment are available. To name a few are: solar pumps, solar dryers, lantern, cooker, solar still, solar refrigeration, solar lighting *etc.* Photovoltaic solar panels that convert solar energy to electrical energy are capable of powering most farm operations such as lifting water, run pumps, lights, and even electric fences at remote places. At home also solar heat collectors can be used to dry crops and warm livestock buildings, and greenhouses. Water heaters that run on solar power can provide hot water for dairy operations, pen cleaning, and homes. Solar power in the long-run is often less expensive than electrical lines, although initial investments are high. It can cut a farm's electricity and heating bills while reducing pollution. Government of India provides subsidy for switching over to solar energy.

Wind Power

Wind power has long been used to pump water, generate electricity and run flour mills. It is inexhaustible in nature. It is most suited in areas where strong winds blow throughout the year. Windmills take advantage of the faster and less turbulent wind at 30 m or more above ground. They can be used for lifting water where the wind velocity is more than 32 kmph. Average capacity of a windmill may be about 0.5 HP but several of them can be combined to increase the power. The main components of a wind turbine system include the rotor, generator, tower, and storage devices. Usually, two or three blades are mounted on a shaft to form a rotor. The blade can spin in either horizontal or vertical axis to the ground. The rotor consists of a hub that connects the rotor to the turbine system. The usefulness of wind energy for farm work is quite limited, the main limitation being the uncertainty. Although it is the cheapest source of farm power, it is yet to become popular in India. Wind turbines are now commercially available and can be installed either individually or on cooperative basis.

Biomass Energy

Biomass energy is also produced from agricultural crops including forest crop residues, organic wastes and food grains. These can be directly used as solid fuels, or can be converted to charcoal or briquettes. Solid wood used as fire wood is a fine example of direct use of biomass for heating. The briquettes are composed of fine charcoal plus a binder/filler. The filler may be wood ash, coal or petroleum solids, sawdust, or calcium carbonate. The fillers control the burning rate, but have to be significantly cheaper than charcoal to minimize the cost. Besides, organic waste can be converted to manure or some form of energy. It can be done either on the farm itself or sold to composting companies. Some industries use the biomass in gasifiers to produce liquid fuels, and biogas/producer gas *etc.* through the process of pyrolysis that involves thermal decomposition of wastes. These fuels can be used in automobiles including tractors besides having other applications including heating. Some industries use the biomass to produce electricity that can be directly supplied to homes and businesses.

Example 1.5

A farmer employs 5 labours, 2 of them being men and 3 women. On the same day he makes use of a pair of bullocks. Calculate the farm power in kW used by the farmer on that day.

Human power = 0.1 x 2 + 0.05 x 3 = 0.35 HP

Animal power = 1 x 1 = 1.0

Total = 1+0.35 = 1.35 HP

Since 1 HP = 746 W, we have

Power in kilo-Watt = 1.35 x 0.746 = 1 kW

FARM POWER IN INDIA AND PUNJAB

Farm power is the most important factor in agriculture as it helps to operate different farm equipment to undertake various farm operations. Modern agriculture largely depends upon power for operations like tilling, transplanting, weeding, harvesting of fruits and vegetables. Besides stationary operations like irrigation, threshing/shelling/cleaning/grading and other post harvest operations also require power. Even for increasing productivity of dry land agriculture, which constitute about 58 per cent of the cultivated area in India as reported in Vision 2030 of ICAR-CRIDA, Hyderabad, timeliness in farm operations especially for seedbed preparation and sowing operations for establishing good crop stand is of paramount importance. Indirectly, power is also required to produce fertilizers and other chemical inputs, a part and parcel of modern agriculture. Modern agriculture gives much higher yields than traditional methods but is much more energy intensive. Various sources of farm power includes human, animal, tractors, power tillers, diesel engine and electric motor, which are run by mechanical, electrical or renewable sources of energy. During the last 65 years or so, the average farm power availability in India has increased from about 0.25 kW/ha in 1951 to 2.76 kW/ha in 2020-21. Farm power availability has been defined as the total power consumed in kW per ha of cultivated land in a year. It is considered as one of the parameters of expressing the level of mechanization. Increase in productivity has been observed to be well correlated with the increasing farm power on Indian farms. It has been assessed that to feed the increasing population of India, it is necessary to increase the availability of farm power from 2.76 kW/ha to 4.0 kW/ha by the end of 2030. Another notable feature of power use globally and particularly in India is a clear shift towards the use of mechanical and electrical sources of power over animate sources derived from human beings and draught animals. In 1960-61 about 92.31 per cent of farm power was coming from animate sources. The contribution of animate sources of power reduced to about 13.0 per cent in 2009-10 while that of mechanical and electrical sources of power increased from 7.70 per cent to about 87 per cent during the same period. Net sown area reduced from 2162 ha/tractor in 1965-66 to 27 ha/tractor in 2013-14. According to some projections, the tractor population by 2050 is likely to stabilize at around 7 million units. The available farm power will then stabilize at around 4.5 kW/ha. There has been a sharp growth in electricity use in the agriculture sector. The consumption rose from 8 per cent of total consumption in 1969 to 17 per cent in 2016. Higher on-farm power use has translated in higher cropping intensity, which increased from 114 per cent with power availability of 0.32 kW/ha during 1965-66 to about 142 per cent with increase in power availability of 2.02 kW/ha in 2013-14. For the same period, average food productivity increased from 0.64 t/ha to 2.11 t/ha.

Punjab is one of the most fertile regions suitable for growing multiple crops such as cereal, vegetable, fruit and cash crops. It is also called as the "Granary of India" or India's bread-basket. On the farm power front, the scenario in Punjab is much better than India. In Punjab agriculture, the human and animal power has substantially reduced from 7.5 to 0.69 per cent and 73 to 0.61 per cent respectively during the period 1960-61 to 2012-13. It is because of the increase in the use of mechanical power from 17 to 76 per cent and electrical power from 1.7 to 23.5 per cent during the same period. Out of 1.476 million tube wells in Punjab in 2018-19, nearly 91 per cent were electrically operated. The percentage contribution of animate power reduced from 80.5 per cent to only 1.3 per cent and mechanical power increased from 19.5 to 98.7 per cent of the total power consumed in the respective years (Figure 1.2). The intensity of farm power availability has increased from 0.37 to over 3.5 kW/ha in 2001 and further to 5.68 kW/ha in 2012-13. Number of tractors during the same period increased from 0.2/1000 ha to 115/1000 ha. Presently one tractor for every 9 ha of net cultivated land of the state is the highest and constitutes 11 per cent of tractors in the country (https://agri.punjab.gov.in/). The same is true for electrical motors, which increased from 2/1000 ha to 287/1000 ha. As a result cropping intensity increased from 112 to 196 per cent, and total food grains productivity from 668 to 3638 kg/ha. Operation-wise energy input in Punjab agriculture is highest in irrigation about 51 per cent followed by harvesting and threshing at 19 per cent. Since the production is related to the farm power input, Punjab has always remained ahead in agriculture productivity. Since mechanization of agriculture, be it in India or Punjab, holds the key for sustainable development of agriculture, more power will be required in times to come. High power use on the farm benefits the farming community in the following manner.

- ✰ Higher farm productivity and income
- ✰ Increased cropping intensity
- ✰ Timeliness of operations
- ✰ Reduced drudgery in farming
- ✰ Increased farm mechanization
- ✰ Enabling the operation of large farm holdings and intensified cropping patterns

A more detailed discussion including other benefits resulting from farm mechanization is included in Chapter 6 of the book.

BASIC MACHINES

Machines are commonly used to perform various kinds of work. The use of the machines makes these tasks easier to perform so that the workers find the works interesting and pleasant to do. Use of agricultural machines and equipment also help to perform the farm works faster than without machines. In principle, a machine gives some mechanical advantage which eases the operation of doing the work. Although the name machine strikes well with a big tool/equipment but most machines are made up of many simple called basic machines. *'A simple/basic machine is a device with few or no moving parts that are used to modify motion and the magnitude of a force in order to perform work'*. In fact, most complex machines are a complex combination of simple machine units and use mechanical and/or electrical energy to do the job. Six commonly used basic machines in assembling big machines are:

☆ The lever

☆ The pulley system

☆ The wheel and axle

☆ The inclined plane

☆ The screw and gears

☆ The wedge

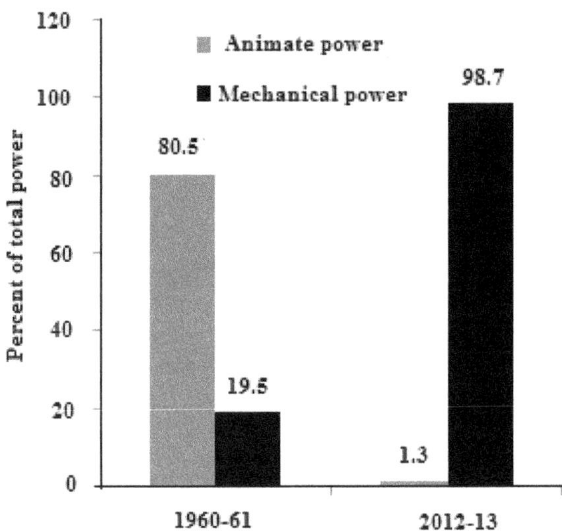

Figure 1.2. Relative Contribution of Animate and Mechanical Power in Punjab Agriculture.

As already defined, work is the force acting on an object in the direction of motion times the magnitude of the displacement. A machine makes the work easier to perform by accomplishing one or more of the following functions.

☆ Transferring a force from one place to another

☆ Changing the direction of a force

☆ Increasing the magnitude of a force

☆ Increasing the distance or speed of a force

Three commonly used terms in relation to basic machines are mechanical advantage, velocity ratio and efficiency.

Mechanical Advantage: Mechanical advantage (MA) of a machine is the *'ratio of the force delivered by the machine to the force applied'*. If the force delivered by the machine is the load (W) and the force which operates the machine is the applied force (effort, F), then

$$MA = Load/Effort = W/F \tag{1.9}$$

Being a ratio, mechanical advantage has no units.

Velocity Ratio (VR): *'It is defined as the distance moved by the effort to the distance moved by load'*. If d_e represents the distance moved by effort and d_l represents the distance moved by load, then

$$VR = d_e/d_l \tag{1.10}$$

Efficiency: *'The efficiency of the machine is the ratio of the work done by the machine to the work done on it'*.

Efficiency = Work done on the load/Work done by the effort

It is also given as the ratio of mechanical advantage to velocity ratio. Thus, the efficiency in per cent can be written as:

$$Efficiency = 100 \, MA/VR \tag{1.11}$$

The efficiency of the machines is always less than 100 per cent because some of the effort is lost in overcoming the friction. If friction is neglected, the work done by the machine (through the load) must be equal to the work done on it (through the effort) resulting in 100 per cent efficiency.

The Lever

The lever is one of the simplest forms of machine. Levers rely on torque for their operation. Torque is the amount of force required to cause an object to rotate around its axis (or pivot point). For example, a long pole or a rigid bar pivoted on a fulcrum is used to raise heavy loads (Figure 1.3). The fixed point at which the straight long pole or rigid bar rotates is the fulcrum. A load of mass 300 kg can be easily lifted by a man with the lever, which without its help may be nearly impossible. In workshops, the lever principle is commonly applied while pulling nails using a crowbar. In this case, fulcrum is the point upon which a crowbar rests. The effort is applied at the far end of the fulcrum. The effort required with a crowbar to pull a nail is much less than when the work is performed without a crowbar. Other examples of application of levers are: wheelbarrows, fishing rods, shovels, brooms, arms, legs, boat oars, and bottle openers. Depending upon the location of the fulcrum, load and effort, levers are categorized as first class levers – the fulcrum is in the middle of the effort and the load, second class levers – the load is in the middle between the fulcrum and the effort and third class levers – the effort is in the middle between the fulcrum and the load. For example, a shovel works as a third-order lever. The hand closest to the end acts as the fulcrum, the second hand provides effort and the shovel end lifts and moves the load.

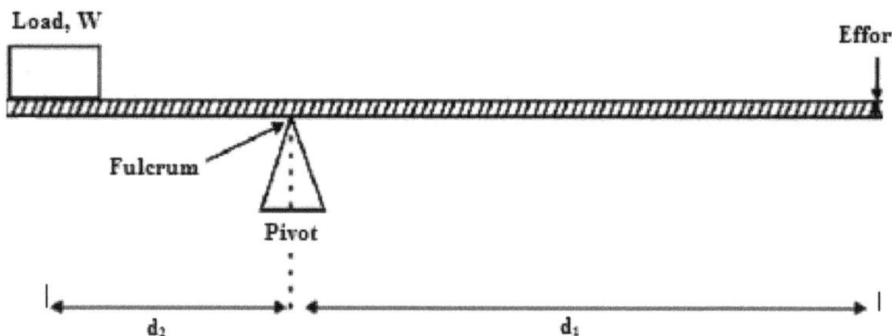

Figure 1.3. A First Class Lever to Lift Heavy Load.

Mechanical advantage of the lever shown in Figure 1.3 can be calculated using the formula:

$$MA = Effort/Load = d_1/d_2 \tag{1.12}$$

Here d_1 is the distance from the effort to fulcrum, and d_2 is the distance from load to fulcrum. Using eq. (1.12), it should be possible to calculate the effort required to lift a particular load or vice-versa.

The Pulley System

A simple pulley is made from a wheel that carries a flexible rope, cord, cable, chain, or belt on its rim. The rim of the wheel is grooved such that the rope, or cord, fits well into that groove. The distance moved by the effort downwards is equal to the distance moved upwards by the load. Hence its mechanical advantage is equal to 1 because the force with which load is pulled down must be equal to the load being lifted. But, it makes moving the object easier, because pulling a rope downward is easier than lifting a heavy object upward. To improve the mechanical advantage, pulleys are usually paired or combined to form systems of pulleys to transmit energy and motion. A compound pulley system uses a combination of fixed and movable pulleys (Figure 1.4). The effort needed to move a load rigged to a compound pulley system is always less than half of the original load. The disadvantage of the compound pulley system is that with each addition of pulley, both the length of rope required and the distance that the rope must travel increase. In belt drive, pulleys are affixed to shafts at their axes, and power is transmitted between the shafts by means of endless belts running over the pulleys. It can be used to increase or decrease the speed of rotation depending upon the sizes of drive and driven pulleys.

Let us consider the case of block and tackle system shown in Figure 1.4. It has three pulleys, two on the fixed upper block and one on the movable lower block. The effort E =F is to raise a load L. The pull or the tension in all parts of the rope is equal to F. Then the total upward force on the lower block is 3 F while the downward force on the lower block is L. In order to raise the load, the following equality holds good.

$$W = 3 F$$

$$MA = Load/Effort = 3F/F = 3 \tag{1.13}$$

For an ideal pulley system, VR will also be equal to 3.

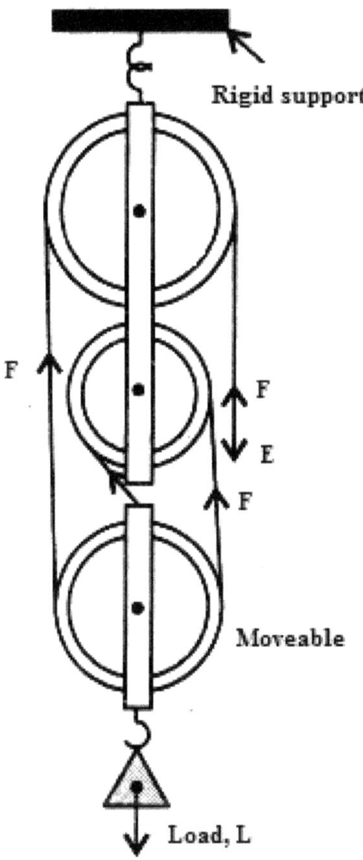

Figure 1.4. Block and Tackle System.

The Wheel and Axle

The simple machine, a wheel and axle, refers to the assembly of two disks, or cylinders of different diameters mounted on the same axle so that they rotate together. Wheel and axle is used to amplify force. The principle of this machine is that a small force applied to the rim of a wheel exerts a larger force on an object attached to the axle. The wheel and axle is used in steering wheels, doorknobs, windmills, and bicycle wheels. The wheel and axle is also used in lifting water from deep wells and in raising heavy loads such as anchors of ships. Let us consider that the wheel and axle are made up of a wheel of radius R placed or mounted on an axle whose radius is r, such that R > r. The load is borne by the axle while the effort is applied on the wheel (Figure 1.5). Let the load on the axle be W while the effort applied on the wheel is F. The mechanical advantage is the ratio of the radius of the wheel to that of the axle as shown in the following derivation.

MA = Load/Effort = W/F

When the wheel rotates once, the distance covered in one revolution is equal to the circumference of the wheel.

$D = 2\pi R$

Similarly, the distance covered in one revolution of the axle of radius r, $d = 2\pi r$

Thus the velocity ratio, VR = D/d = R/r   (1.14)

Since Work done = Force x distance

Then work done on load = W x $2\pi r$ = $2\pi r W$

Work done by effort = F x $2\pi R$

Efficiency = Work done on load/Work done by effort

= $2\pi r W / 2\pi R F$ = Wr/FR   (1.15)

It can also be given as: MA/VR or

Efficiency = (W/F)/(R/r) = W r/F R   (1.16)

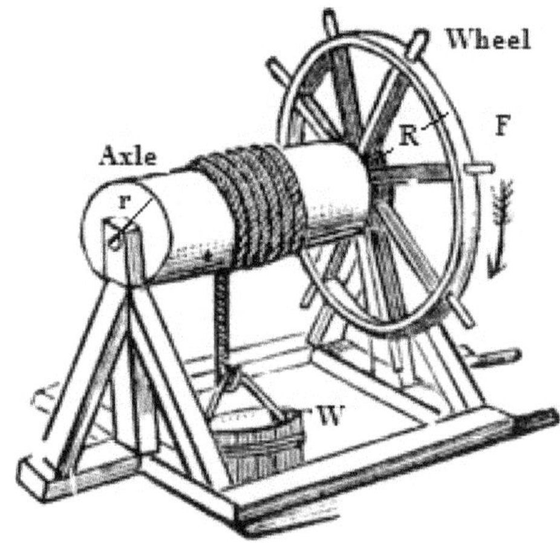

Figure 1.5. A Schematic View of a Simple Wheel and Axle.

The Inclined Plane

An inclined plane is a flat surface with one end higher than the other such that it forms an angle with the horizontal plane. This allows for heavy objects to slide up to a higher point rather than be lifted. It is generally easier to slide something than to lift it. It is also a simple machine since it modifies the intensity and the direction of the force needed to move an object. Examples of inclined planes are ramps, sloping roads and hills, ploughs, chisels, hatchets, carpenter's planes, and wedges, ramps being the simplest example. Let us take the case of a load, W, which is lifted into the trailer with the use of a plank placed in an inclined plane at an angle θ to the horizontal (Figure 1.6). The inclined plane is therefore serving as a machine to accomplish the rolling of the load into the trailer. There is a trade-off with this simple machine. If the slope is gentle, a person has to push or pull the object over a longer distance, but with little effort. If the slope is steep, a person has to push or pull the object over a very short distance, but with more effort. Thus, there is a greater mechanical advantage if the slope is gentle because less force will be needed to move the object up or down the slope.

Let L be the length of the inclined plane AC

θ = angle of inclination of the plane to the horizontal

H = the height through which the load is lifted

W = mg = Load

F = Effort applied to roll the load on the inclined plane

VR = Distance moved by effort/Distance moved by load = L/H

From triangle ABC, $\sin \theta = H/L$

Thus VR = $1/\sin \theta$ (1.17)

If friction is neglected, MA = VR = $1/\sin \theta$ (1.18)

Also MA = mg/F = mg/mg $\sin \theta$ = $1/\sin \theta$ (1.19)

Thus, the mechanical advantage for a ramp can be expressed in terms of the distance traveled. The ideal mechanical advantage for an inclined plane is simply L/H.

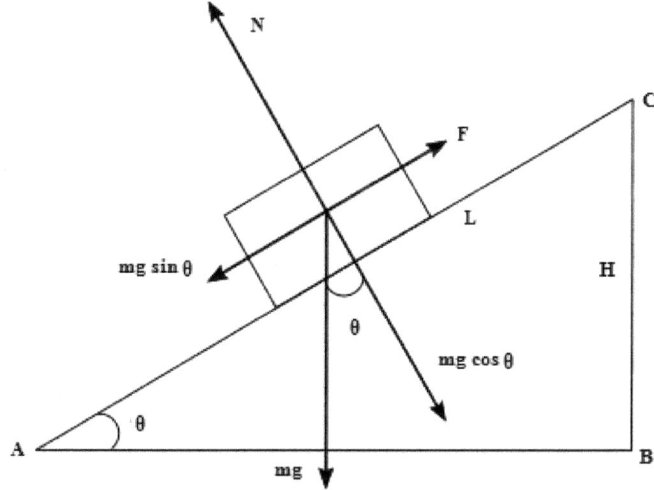

Figure 1.6. Inclined Plane for Loading Heavy Loads.

The Wedge

An inclined plane and a wedge are two forms of the same simple machine. *A wedge is a device that is thick at one end and tapers to a thin edge at the other end.* A wedge may be a single wedge or double wedge, each doing a slightly different job. We discuss herein only a double wedge, which is simply two inclined planes placed back to back to form a sharp edge with an angle θ (Figure 1.7). It is normally driven between two objects to force them apart. Force is applied to the thick end of the wedge. The sloping sides of the wedge change the direction of the force, which acts on the object at a 90° angle to its slope (Figure 1.7 right). It results in cutting or splitting it apart. The mechanical advantage of a wedge is more than 1. Some examples of wedges are a shovel, a knife, an axe, a pick axe, a saw, a needle, scissors, or an ice pick. But wedges can also hold things together as in the case of a staple, push pins, tack, nail, doorstop, or a shim. Force multiplication varies inversely with the size of the wedge angle; a sharp wedge (small inclined angle) yields a large force. The ideal mechanical advantage of a wedge is determined by dividing the length of the wedge by its width, such that

$$MA = L/t \qquad (1.20)$$

Clearly, the longer and thinner a wedge is, the greater will be its mechanical advantage.

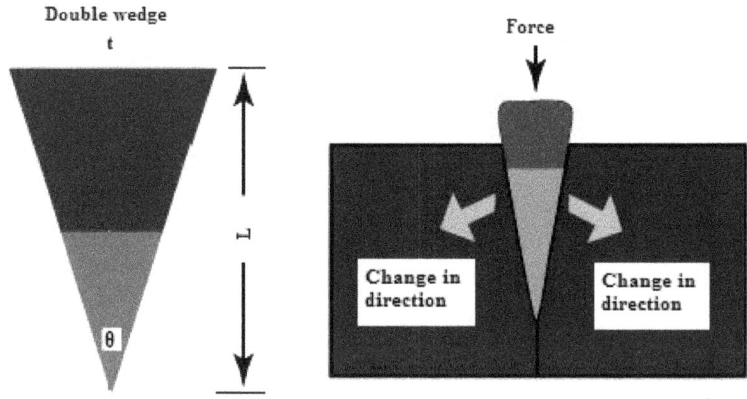

Figure 1.7. A Double Wedge (Left) and Change in the Direction of Force (Right).

The Screw and Gears

Screw

A screw is a simple machine that can be used to fasten two or more objects together or it can be used to lift up heavy objects/loads. In most applications, a lever is used to turn the screw. A good example of this is a screw driver or a screw jack used to lift cars or tractors (Figure 1.8 left). During usage, as the screw head is made to turn one complete revolution, the screw (load) travels forward and covers a distance equal to the pitch of the screw. The pitch of a screw is the distance between two adjacent threads on the screw. The circumference of the lever or screw driver and the pitch of the screw determine the mechanical advantage of the screw. As another example, let us look at the functioning of a screw jack shown in Figure 1.8 left. A screw jack has a long vertical screw which threads into a heavy base. It also has a long rod called tommy bar. The ideal mechanical advantage of a jack is found approximately by dividing the circumference of the tommy bar by the pitch of the screw. For example, if L is the length of the tommy bar and the pitch is P, then

$$VR = 2\pi L / \text{Pitch} \tag{1.21}$$

Neglecting frictional forces, we have the ideal mechanical advantage as $2\pi L/P$.

Gears

Gears are meant to increase or decrease speed. Gears are used in cars, tractors, bicycles and cranes. In a tractor for example, gears may be made to change the speed of the rear wheels in spite of the fact that engine speed remains constant. The driving gear is called the driver while the gear being driven is called the driven (Figure 1.8 right). It is important to note that for two gears in mesh, the larger gear has a greater number of teeth than the small gear. The smaller gear rotates faster. The velocity ratio is given as:

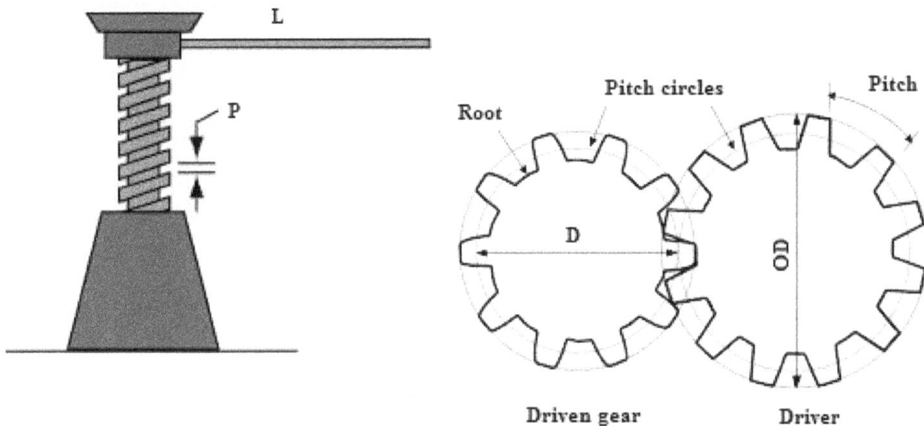

Figure 1.8. A Screw Jack (Left) and Gears (Right).

VR = Number of teeth on driven wheel/Number of teeth on driving wheel

VR can also be given in terms of angular speed. It is the ratio of the angular speed of one to that of the other.

$$W_1 N_1 = W_s N_s \tag{1.22}$$

$$W_1/W_s = N_s/N_1 \tag{1.23}$$

Here W_1 is angular speed of large gear, W_s is the angular speed of small gear, N_1 are numbers of teeth on large gear and N_s are the number of teeth on small gear.

QUESTIONS (THEORY)

1. Define force, work, power, energy, kinetic energy and potential energy.

2. Name the various sources of energy used in agriculture. Write a brief note on animate and mechanical energy.

3. What are the advantages and disadvantages of animal power in agriculture?

4. Write a short note on renewable sources of energy and their application in agriculture.

5. Write a short note on the use of solar power in agriculture.

6. Define simple machine. Name some of the simple machines and discuss the working of a lever.

7. Define the terms mechanical advantage, velocity ratio and efficiency used in simple machines.

8. Explain the working of a pulley system, wedge, gear, wheel and axle and the inclined plane giving examples of such machines from daily use.

Chapter 2

Internal Combustion Engines

HEAT ENGINE

Heat energy can be easily transferred from one body to another. A heat engine is capable of converting the energy locked in the fuels into force and motion. For example, fuels like coal, gasoline, natural gas, wood, and peat when burnt in a heat engine, release their energy to power industrial machinery and locomotives. Hence, a heat engine can be defined as *'an engine that converts the chemical energy of the fuel into thermal energy for performing some useful work'*. Heat engines have been characterized into two types on the basis of where the combustion of fuel takes place.

1. External combustion engine
2. Internal combustion engine

External Combustion (EC) Engine

Combustion of fuel takes place outside the engines in external combustion engines (Figure 2.1 left). Examples of external combustion engine are old steam locomotives, and steam turbines used in many power plants *etc.* In these engines, the heat of combustion of fuel is used to generate high pressure steam which is supplied to the engine through the valves. It expands and pushes the pistons to perform the mechanical work.

Internal Combustion (IC) Engine

In IC engines, the combustion of fuel or its mixture take place in the engine itself (Figure 2.1 right). Due to the heat of combustion, the pressure and temperature of fuel/mixture increases tremendously. As a result, high pressure mixture further expands and pushes the piston (or moves a turbine) and does the mechanical work. Examples are petrol engine, diesel engine, gas turbine, and jet engines *etc.*

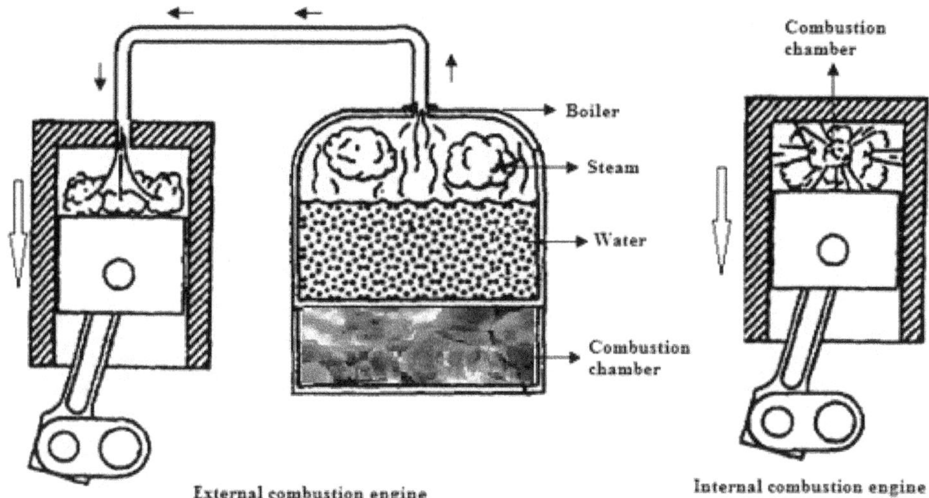

Figure 2.1. Schematic Diagram of an External and an Internal Combustion Engine.

A comparison of an external and an internal combustion engine is illustrated in the following Table.

EC Engines	IC Engines
Combustion of air-fuel occurs outside the engine cylinder	Combustion of air-fuel occurs inside the engine cylinder
The engines run smoothly and silently due to outside combustion	Very noisy operation of the engine
Ratio of weight and bulk to output is high because of the presence of auxiliary apparatus like boiler/ condenser. Hence, it is heavy and cumbersome	Ratio of weight and bulk to output is lower and as a result these are light and compact
Ordinary alloys can be used to manufacture the engine cylinder and its parts since the working pressure and temperature inside the engine cylinder is low	Special alloys are used to manufacture the engine cylinder because the working pressure and temperature inside the engine cylinder is very high
It is possible to use cheaper fuels such as the solid fuels	High grade fuels need to be used, unable to make use of solid fuels
Lower efficiency about 15-20 per cent	Higher efficiency about 35-40 per cent
Water requirement is high owing to need to dissipate the energy through cooling system	water requirement is less
High starting torque	IC engines are not self-starting
Require more space	These units require less space being compact
A lot of flexibility can be exercised in arranging various units because of the external combustion	Mechanically simpler than an EC engine
High initial cost	Low initial cost

Advantages of IC Engines

- ☆ Greater mechanical simplicity
- ☆ Because the auxiliary units like boiler, condenser and feed pump are not required, these engines have high power output per unit weight
- ☆ Low initial cost
- ☆ Have higher brake thermal efficiencies because only a small fraction of heat energy is dissipated to the cooling system
- ☆ These units require less space, being compact
- ☆ Easy starting during cold conditions

Disadvantages of IC Engines

- ☆ IC engines can efficiently use only liquid or gaseous fuels of given specification. These fuels are relatively more expensive than cheaper solid fuels
- ☆ Balancing in IC engines is a problem having many reciprocating parts. As a result these engines are susceptible to mechanical vibrations

CLASSIFICATION OF IC ENGINES

IC Engines have been classified in a number of ways. Some classification criteria are as follows:

Based on Working Cycle

According to the cycle of operations IC engines are categorized as two-stroke and four-stroke engines. The engine is a two stroke cycle engine, if the working cycle is completed in one revolution of the crankshaft. On the other hand, it is a four stroke cycle engine, if the working cycle is completed in two revolutions of the crankshaft.

Based on Cycle of Combustion

According to the cycle of combustion, IC engines are categorized as Otto cycle named after its inventor Nikolaus Otto (constant volume combustion, or spark ignition engines), Diesel cycle named after its inventor Rudolf Diesel (Combustion at constant pressure, or compression ignition engine) and dual cycle (combustion partly at constant pressure, partly at constant volume).

Based on Method of Ignition

Engines are categorized as spark ignition engines (SI engine or carburetor type engines) and compression ignition engines (CI engine or injector type engines). In the spark ignition engine, a mixture of air and fuel is drawn into the engine cylinder. Fuel is ignited by a spark plug, which as the name suggests produces a spark that ignites the air-fuel mixture. Such combustion is called constant volume combustion (CVC). In the compression ignition engine, air is compressed into the engine cylinder. As a result, the temperature of the compressed air rises to 700-900°C. At this stage diesel is sprayed into the cylinder in fine mist. The fuel gets ignited because of the very high temperature of the air. This type of combustion is called constant pressure combustion (CPC) because the pressure inside the cylinder is almost constant when combustion takes place. Besides these, several other classification criteria and types are given as follows:

Criteria	Types
On the basis of number of cylinders	Single cylinder, Multi cylinder
Based on arrangement of cylinders	Horizontal engine, Vertical engine, V-type engine, Radial engine
Based on the use	Stationary engine, Portable engine, Marine engine, Automobile engine, Aero engine *etc.*
Based on fuel used	Oil engine, Petrol engine, Gas engine (LPG, CNG), Kerosene engine, Alcohol engine (ethanol, methanol *etc.*)
Based on the speed of engine	Low speed, Medium speed, High speed engines
Based on method of cooling	Air cooled engines, Water cooled engines
Method of governing	Quality governed engines, Quantity governed engines, Hit and miss governed engines
According to the valve arrangement	Lower head type engines, Overhead valve engines. T-head type engines, F-head type engines

IC ENGINE COMPONENTS

An Internal combustion engine consists of hundreds of different parts, all of which are important for its proper working. It may not be possible to describe all these parts, being beyond the scope of this book. However, the main components, which are important from an academic point of view, are shown and discussed in the following sections (Figure 2.2).

Cylinder block: It is the solid casting body of the engine and includes the cylinder and water jackets. In air cooled engines it includes the cooling fins.

Cylinder: It is the most basic part of an engine where piston operates. Air or air-fuel mixture is sucked in the cylinder. It is compressed by the piston movement in the cylinder space. Combustion also takes place in the cylinder and the expanding gases remain confined within the cylinder space. The power generated in the cylinder is transferred to do useful work. The materials used in the construction of an engine cylinder should be able to retain sufficient strength at high pressures and temperatures. Note that pressure exceeding 50 bars and temperature exceeding 2000°C may be normally experienced in the engine cylinder. Whereas the cylinder is made of cast iron in ordinary engines, for heavy-duty engines it is made of steel or aluminium alloys. In a multi-cylinder engine, the cylinders are cast as one block known as cylinder block. Sometimes, a liner or sleeve is inserted into the cylinder, so that when worn out only the liner needs to be replaced rather than the whole cylinder.

Cylinder liner or sleeve: It is a cylindrical lining inserted in each cylinder of the cylinder block. It is either wet or dry type. In a dry liner, there is metal to metal contact with the cylinder block casing. The dry liners do not come in contact with the cooling water. On the contrary, wet liners come in contact with the cooling water.

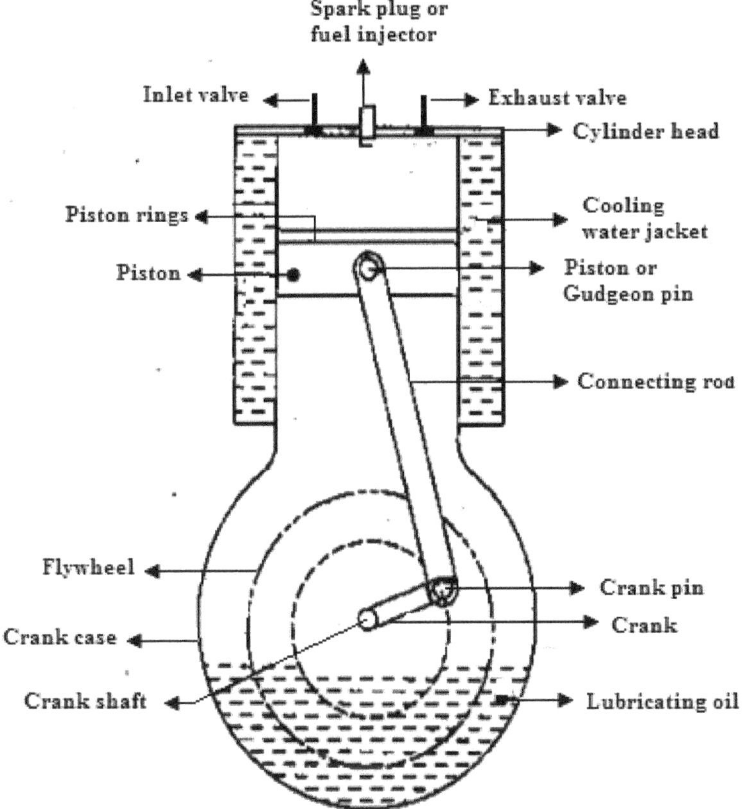

Figure 2.2. Schematic Diagram of an IC Engine Depicting its Various Parts.

Cylinder head: The cylinder head usually cast as one piece is the detachable portion of the cylinder block/engine. It is used to cover the cylinders of the cylinder block. The inlet valves for admitting fresh charge and the exit valves for exhausting the burnt gases are mounted on the cylinder head. A spark plug for igniting the fuel-air mixture in petrol engines and nozzle, the fuel valve for injecting the fuel into the cylinder in diesel engines are also provided on the cylinder head. The material used in its construction is same as the cylinder block. A copper or asbestos gasket is used to ensure an air-tight joint between the engine cylinder and the cylinder head.

Piston: Piston, the heart of the engine, is a cylindrical part closed at one end which maintains a close sliding fit in the engine cylinder. It is connected to the connecting rod by a piston pin also known as gudgeon pin. The expanding gases force the piston to move down in the cylinder. This causes the connecting rod to rotate the crankshaft (Figure 2.2). Pistons are either made of cast iron or aluminum and its alloys. Cast iron is used mainly because of its high compressive strength, while the aluminum and its alloys because of their light weight.

Head (Crown) of piston: It is the top portion of the piston (Figure 2.3).

Skirt: The cylindrical wall of the piston is the piston skirt. It is designed to absorb the side movements of the piston. The surface of the skirt is left slightly rough during manufacturing to help retain lubrication.

Piston rings: These are split expansion rings, placed in the grooves of the piston. The portion between the two ring grooves is known as ring lands. Piston rings are usually made of cast iron or pressed steel alloy (Figure 2.3), which retain elastic properties even at high temperatures. The functions of the rings are as follows:

☆ Piston rings help to form a gas tight combustion chamber

☆ These reduce contact area between cylinder and the piston walls preventing excessive friction losses and excessive wear

☆ These control the cylinder lubrication

☆ Help to transmit heat away from the piston to the cylinder walls

Piston rings are of two types: (1) Compression rings and (2) Oil rings (Figure 2.3)

Compression rings are normally single piece plain rings. These are always placed in the grooves nearest to the piston head. They help in increasing compression pressure inside the cylinder by preventing any leakage of gases. Oil rings are grooved or slotted rings. They are located either in the lowest groove above the piston pin or in a groove above the piston skirt. The oil rings control and distribute the lubrication oil in the space between the cylinder and the piston.

Piston pin: Piston pin, also called wrist pin or gudgeon pin, is used to join the connecting rod to the piston.

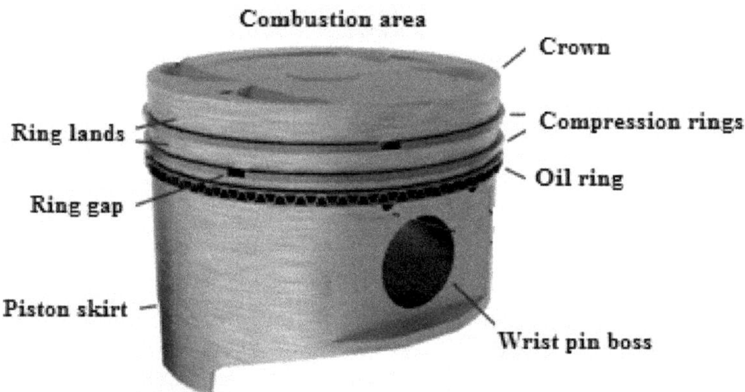

Figure 2.3. Piston and its Various Parts.

Connecting rod: Connecting rod, made of drop forged steel, is used to connect the piston and the crankshaft (Figure 2.2 and Figure 2.4). Whereas the small end of the connecting rod is connected to the piston with a piston pin, the big end is connected to the crankpin journal to provide a pivot point on the crankshaft. It functions as a lever arm and transfers motion from the piston to the crankshaft to make it rotate continuously.

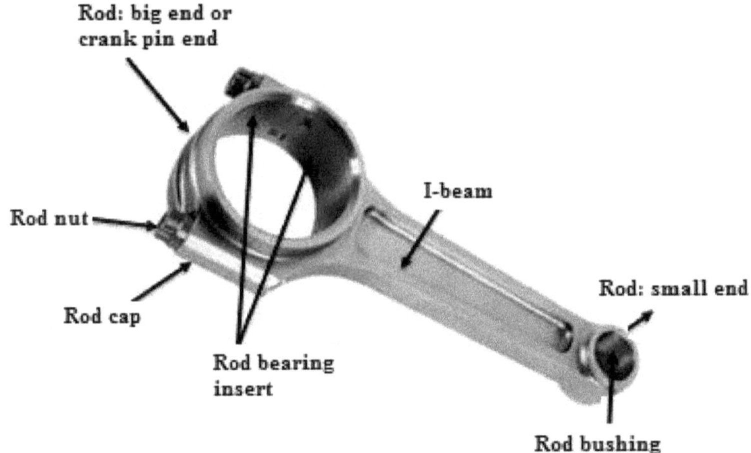

Figure 2.4. Connecting Rod and its Parts.

Crankshaft: Crankshaft, made of drop forged steel or cast steel or special steel alloys, is the backbone of an IC engine. It is the main shaft of the engine that converts the reciprocating motion of the piston into rotary motion of the flywheel (Figure 2.2). A crankshaft has two types of journals (the part of a shaft that rotates inside a bearing). The main bearing journal supports the crankshaft in the cylinder block, whereas the crank bearing journals are attached to the connecting rod. Counterweights are provided throughout the length of the crankshaft to ensure counterbalance of the unit.

Flywheel: Flywheel, made of cast iron performs the following main functions.

☆ It stores energy during power stroke and returns the energy during the idle stroke, providing a uniform rotary motion of flywheel

☆ The pressure plate is bolted to the rear surface of the engine flywheel such that it serves as one of the pressure surfaces for the clutch plate

☆ Engine timing marks are usually stamped on the flywheel, that helps in adjusting the timing of the engine

☆ Flywheel in many situations can serve as a pulley to transmit power

Crankcase: The crankcase supports and houses the crankshaft and camshaft. The bottom part of the crankcase, usually made of cast iron or cast aluminum, acts as a reservoir for the lubricating oil that holds the major part of the engine's oil. Accessories such as the oil pump, oil filter, starting motor and ignition components are mounted on it. In modern engines, crankcase is integrated into the cylinder block.

Camshaft: This shaft is housed in the crankcase parallel to the crankshaft, raises and lowers the inlet and exhaust valves at proper times. The crankshaft drives the camshaft by means of gears, chains or sprockets. In a four stroke engine, the camshaft speed is exactly half the crankshaft speed. The ignition timing mechanism, lubricating oil pump and fuel pump are all operated by the camshaft.

Timing gear: Timing gear is a combination of gears. One of the gears, the bigger one is mounted at one end of the camshaft. The other smaller one is located at the crankshaft. Since the camshaft gear has twice as many teeth as that of the crankshaft gear, it is why it is also known as half time gear. Timing gear serves the following purposes:

- ☆ Controls the timing of ignition
- ☆ Controls the timing of opening and closing of valves
- ☆ Controls the fuel injection timing

Inlet manifold: Air or air-fuel mixture enters the engine cylinder through the inlet manifold, which is fitted by the side of the cylinder head (Figure 2.5).

Exhaust manifold: The exhaust gases go out of the engine cylinder through the exhaust manifold, which is fitted by the side of the cylinder head. Since the exhaust gases have high temperature, this manifold should be able to withstanding such temperatures (Figure 2.5).

Scavenging: *The process of removal of burnt gases from the engine cylinder is known as scavenging.* Since in a two stroke cycle engine all the burnt gases do not go out of the engine cylinder in normal stroke, as may be the case in a four stroke engine, some type of blower or compressor is used to remove the exhaust gases.

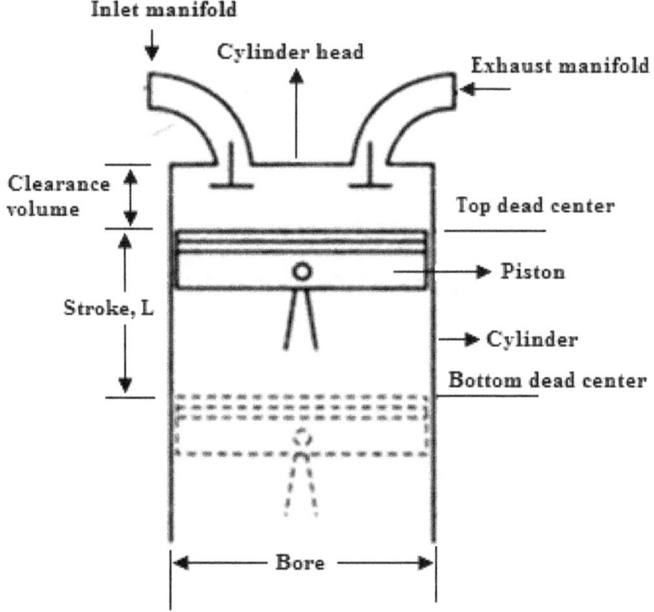

Figure 2.5. Schematic Diagram of IC Engine Showing Few Important Parts/Terms.

TERMINOLOGY OF IC ENGINES AND TRACTOR

Bore: The inside diameter of the engine cylinder is called bore (Figure 2.5). Cylinder bores vary in size, but typically range from 75 to 100 mm (3"–4").

Stroke: As the piston reciprocates inside the engine cylinder, it has limiting upper and lower position beyond which it cannot move. The reversal of motion takes places at these limiting positions. *The distance the piston moves from top dead center (TDC) to bottom dead center (BDC) is called stroke* (Figure 2.5). Crankshaft journal offset (throw) controls the piston stroke. Stroke is two times the offset.

Stroke-bore ratio: Stroke-bore ratio (L/D) is the ratio of length of stroke (L) to the diameter of bore (D) of the cylinder. Normally, this ratio varies from 1 to 1.45 but it is about 1.25 for tractor engines.

Dead centers: Dead centers are the positions of the piston and mechanical parts connected to it at the moment when the direction of the piston motion is reversed at either point of the stroke. These are the positions of the piston when it is either farthest from, or nearest to the crankshaft.

Top dead center (TDC): *The lowest position of the piston towards the cover end side of the cylinder or farthest from the crankshaft is called TDC* (Figure 2.5). At this time, the piston is at the top of its stroke. In horizontal engines it is termed as inner dead center. In some sources, it is also mentioned as Head-End-Dead-Center (HEDC).

Bottom dead center (BDC): The lowest position of the piston towards the crank side of the cylinder is called BDC. At this time, the piston is at the bottom of its stroke. In a horizontal engine, it is known as outer dead center. In some sources it is called as Crank-End-Dead-Center (CEDC) because it is not always at the bottom of the engine.

Clearance volume: The volume contained in the cylinder above top of the piston, when it is at TDC is called clearance volume (Figure 2.5).

Swept volume: The volume swept by the piston in moving from TDC to BDC is called swept volume or piston displacement. It is given as (A x L) where A is the cross-sectional area of the piston and L is the length of stroke (Figure 2.5). In terms of diameter of the cylinder, it is given as:

$$V_s = (\pi/4) \, D^2 \, L \tag{2.1}$$

Total volume of the cylinder $= V_s + V_c$, such that V_c is the clearance volume.

Compression ratio: *It is the ratio of the volume of the cylinder at the beginning of the compression stroke to that at the end of the compression stroke or in other words it is the ratio of total volume of the cylinder to the clearance volume.* Mathematically, it is written as

$$\text{Compression ratio, } r = (V_s + V_c)/V_c \tag{2.2}$$

The Compression ratio of diesel engine varies from 14:1 to 22:1 and that of carburetor type engine (spark ignition engine) from 4:1 to 8:1.

Cubic capacity or engine capacity: Swept volume of a cylinder times the number of cylinders in the engine is known as cubic capacity or engine capacity. Cubic capacity is given as:

$$\text{Cubic capacity} = V_s \times n \tag{2.3}$$

Here n is the number of cylinders.

Mean piston speed: The average speed of the piston of a reciprocating engine is mean piston speed. It is a function of stroke length and RPM. If a *piston* has a stroke length of L mm and is operating at N revolution per minute (RPM), then

Average speed of the piston, mm/s = 2 x L x N/60 $\qquad$ (2.4)

Example 2.1

An engine with a bore of 75 mm diameter has a stroke length of 90 mm. Determine its swept volume. If the clearance volume is 65 cc, determine the compression ratio.

Using eq. (2.1), we have

Swept volume = $(\pi/4)$ x 90 x 90 x 75/1000 = 477.2 cc

1000 is used to convert D and L in cm (10 x 10 x 10 = 1000).

Using eq. (2.2), compression ratio is given as:

Compression ratio = (477.2 + 65)/65 = 8.34 or we can say 8.34:1

Example 2.2

An engine has a stroke length of 90 mm and is operating at 2900 RPM. What is the average speed of the piston?

Average speed given by eq. (2.4) is

Average speed = 2 (90/1000) (2900/60) = 8.7 m/s

1000 is used to convert mm to m and 60 to convert RPM to revolutions per second (RPS).

Indicator diagram: Indicator diagram also known as pressure-volume (PV) diagram *'is a chart used to measure the thermal or cylinder performance of reciprocating steam or IC engines. It is used to calculate the work done and power produced in the engine'*. It is the graph between pressure and volume, pressure being on the y-axis and volume on x-axis. A pressure-volume diagram is divided into five parts namely compression, combustion, expansion, exhaust and induction. It is obtained by an indicator, an instrument to graphically record pressure versus piston displacement through an engine stroke cycle (Also see sections on indicator diagrams in this chapter).

Indicated horsepower (IHP): It is the power generated in the cylinders of an engine. It is calculated from the average pressure of the working fluid, the piston area, length of the stroke, and the number of working strokes per minute. It is power received by the pistons developed in a cylinder without considering the frictional losses. The general equation to determine IHP developed by an IC engine is written as

IHP = PLAN (2 n/S)/4500 $\qquad$ (2.5)

Where, P is the mean effective pressure (MEP) in kg/cm^2, L is the length of stroke, m, A is area of piston in cm^2, N is the RPM of crankshaft, n is the number of cylinders and S is the number of strokes either 2 or 4. The mean effective pressure (MEP) is calculated as the average pressure during the power stroke minus the average pressure during the other three strokes. It is the pressure that actually forces the piston down during the power stroke. If area A is defined in m^2 instead of cm^2, Equation takes the following form

IHP = 2.22 PLAN (2 n/S)   (2.6)

For a four stroke engine IHP is given as

IHP = 2.22 PLAN (n/2)   (2.7)

For a two stroke engine, IHP is given as:

IHP = 2.22 (PLAN x n)   (2.8)

In SI unit, indicated horsepower for a four stroke engine is given as below:

IHP, kW = 1.634 (PLAN x n/2)   (2.9)

For a two stroke engine, we have

IHP, kW = 1.634 (PLAN x n)   (2.10)

Brake horsepower (BHP): It is the power actually delivered by the engine without taking into account the power loss caused by the gearbox, alternator, differential, water pump, and other auxiliary components such as power steering pump, and muffled exhaust system, *etc.* In fact, it is the capacity of the engine and is the measure of an engine's horsepower. It is the power generated at the end of the crankshaft (belt pulley) and available for useful work. It is measured with the help of a dynamometer. The output delivered to the driving wheels is less than obtainable at the engine's crankshaft. Similarly, the power developed in the combustion chambers of the engine is more than the delivered power because of friction and other mechanical losses.

BHP = 2πNT/4500   (2.11)

Here T is the torque in kg-m, and N is the speed, RPM. Brake horsepower is almost about 70 to 85 per cent of the actual power developed inside the engine.

Mechanical efficiency: Mechanical efficiency is defined as the ratio of BHP to IHP

Belt horsepower: Power of the engine measured at a pulley receiving direct drive from the PTO shaft of the tractor.

Power take-off horsepower (PTO HP): It is the power delivered by a tractor through its PTO shaft. In general, the belt and PTO horsepower of a tractor are approximately the same. The PTO HP is around 80-85 per cent of tractor engine power

Drawbar horsepower (DBHP): It is the power available for the tractor to haul or pull an implement. To determine the maximum power available, a controlled load normally a second tractor with its brakes applied, in addition to a static load is used. Its efficiency is about 50-55 per cent of the engine power.

Drawbar horsepower = Pull x speed/270   (2.12)

Here pull is expressed in kg and speed in km/hr. If speed is expressed in m/min, then

Drawbar horsepower = Pull x speed/4500 (2.13)

Frictional horsepower (FHP): It is the power required to run the engine at a given speed without any load or without doing any useful work. It translates into the friction and pumping losses of an engine.

FHP = IHP - BHP or IHP = BHP + FHP (2.14)

Example 2.3

A four cylinder four stroke cycle engine having 4 cylinders has a piston diameter of 7.5 cm. It is operating at 1650 RPM with mean effective pressure of 6.5 kg/cm^2. If the stroke length is 10 cm, calculate the IHP developed by the engine.

Area = π D^2/4 = 3.142 x 7.5 x 7.5/4 = 44.2 cm^2 = 0.00442 m^2

Using eq. (2.7), we have

IHP = 2.22 x PLAN x n/2 = 2.22 x 6.5 x 0.1 x 0.00442 x 1650 x 4/2 = 21.04

Example 2.4

Calculate the IHP of the engine from the following data of a single cylinder four stroke oil engine: Cylinder diameter = 250 mm, Stroke length = 400 mm, Mean effective pressure= 6.5 bar, and Engine speed = 250 RPM.

Area A = π x 25 x 25/4 = 490.9 cm^2 = 0.0491 m^2

Stroke length, L= 400 mm = 0.4 m

Using eq. (2.7), we have

IHP = 2.22 x 6.5 x 1.02 x 0.4 x 0.0491 x 250 x 1/2 = 36.13

Note that pressure in bar needs to be converted to kg/cm^2 by multiplying it by 1.02

IHP = 36.13 x 746/1000 = 26.95 kW

Using direct formula, eq. (2.9)

IHP, kW = 1.634 x PLAN x n/2 = 1.634 x 6.5 x 1.02 x 0.40 x 0.0491 x 250/2 = 26.6 kW

Example 2.5

In a two stroke engine the diameter of the piston is 150 mm and stroke length is 180 mm. If the RPM of the engine is 300, calculate the indicated horsepower for mean effective pressure of 6.5 kg/cm^2. If the clearance volume is 1 L, determine the compression ratio.

Area, A = πD^2/4 = π (15)^2/4 = 176.73 cm^2 = 0.0177 m^2

Clearance volume V$_c$= 1000 cm^3

Using eq. (2.8), we have

IHP = 2.22 x PLAN x n = 2.22 x 6.5 x 0.0177 x 0.18 x 300 = 13.79

Swept volume, V$_s$ = A L = 176.73 x 18 = 3181.14 cm^3

Compression ratio, r = Total volume/clearance volume = (3181.14+1000)/1000 = 4.18

Example 2.6

A four stroke engine developed a BHP of 20. If the mechanical efficiency of the system is 85 per cent, what is its IHP?

IHP = BHP/Mechanical efficiency = 20/0.85 = 23.53

Example 2.7

A rope brake was used to measure the brake power of a single cylinder, four stroke cycle petrol engine. It was found that the torque due to brake load is 165 N-m and the engine makes 525 RPM. Determine the brake power developed by the engine.

Torque due to brake load, T = 165 N-m

Engine speed, N = 525

Brake power, BHP = 2πNT/4500 = 2 x 3.142 x 525 x 165/(9.81 x 4500)

Brake power, BHP = 12.33

Note: Torque in N-m is converted to torque in kg-m by dividing it by 9.81 m/s^2

Example 2.8

A 4-cylinder four stroke petrol engine develops 14.4 kW at 1000 RPM. The mean effective pressure is 5.5 bar. Calculate the stroke length of the engine, if the bore diameter is 90 mm.

IHP, kW = (1.634 x PLAN x n)/2

A = 3.142 x 0.09 x 0.09/4 = 0.00633 m^2

L = IHP x 1.634 x 2/(P x A x N x n)

L = 14.4 x 2/(5.5 x 1.02 x s1.634 x 1.634 x 0.00633 x 1000 x 4) = 0.124 m or 124 mm

Example 2.9

Calculate the IHP and BHP of an engine if its mechanical efficiency is 85 per cent and frictional horsepower is 20 kW.

The solution of this problem is the simultaneous solution of the following two equations.

Mechanical efficiency = BHP/IHP

Frictional horsepower = IHP – BHP

Thus, we have

BHP = 0.85 IHP

20 = IHP – 0.85 IHP = 0.15 IHP

IHP = 20/0.15 = 133.3 kW

BHP = 133.3 x 0.85 = 113.3 kW

Important Conversions

In most equation to calculate IHP, MEP is used in kg/cm^2. On the contrary pressure is reported in several other units. Therefore, students must be familiar with various units and conversion factors for converting the MEP in kg/cm^2.

Unit	Approximate Conversion Factor for kg/cm^2
Bar	1.02
1 standard atmosphere	1.03
1 kilo Pascal	0.01
Meter of water	0.1
1 psi	0.07

CONSTRUCTION AND WORKING OF AN IC ENGINE

Construction of an IC engine is described in Figure 2.2. It consists of cylinders bored in a cylinder block. A gasket, made of copper sheet or asbestos, is inserted between the cylinder and the cylinder head to avoid any leakage. The combustion takes place in the space at the top of the cylinder head above the TDC. The piston in the cylinder reciprocates as a result of the energy released during combustion of fuel or air-fuel mixture. Piston rings in the circumferential grooves of the piston prevent leakage of gases from the sides of the piston. A pin called gudgeon pin or wrist pin connects the piston and the connecting rod at its small end. The other end of the connecting rod called the big end is connected to the crank shaft. The connecting rod transmits the up and down motion of the piston to the crank shaft. It converts the reciprocating motion into the rotary motion. The crankshaft rotates in main bearings journals fitted in the crankcase. A flywheel fitted at one end of the crankshaft helps to smoothen the uneven torque produced by the engine. An oil sump containing lubricating oil is provided at the bottom of the engine. The oil in the sump is used to lubricate different parts of the engine. Cooling arrangement in the form of air fins in air cooled and water jackets in water cooled engines ensure that the engine does not get heated up beyond a certain temperature.

Four Stroke Cycle Engine (Petrol/Diesel Engine)

The basic principle of four-stroke petrol engine cycle, also known as Otto cycle after its inventor Nicolaus Otto, was used to build the first engine in 1876. In a four strokes cycle engine, the four events namely suction, compression, power and exhaust inside the engine cylinder are completed in four strokes of the piston or two revolutions of the crank shaft (Figure 2.6). The engine has two valves namely the suction valve that controls the inlet of charge and the exhaust valve that controls the exhaust event through which exhaust gases are released from the cylinder. The opening and closing of these valves is controlled by the cams fitted on the camshaft driven by the crankshaft with the help of suitable gears or chains. The camshaft runs at half the speed of the crankshaft. The working cycle of the engine comprises a sequence of events that takes place in the engine. Let us look at the four events.

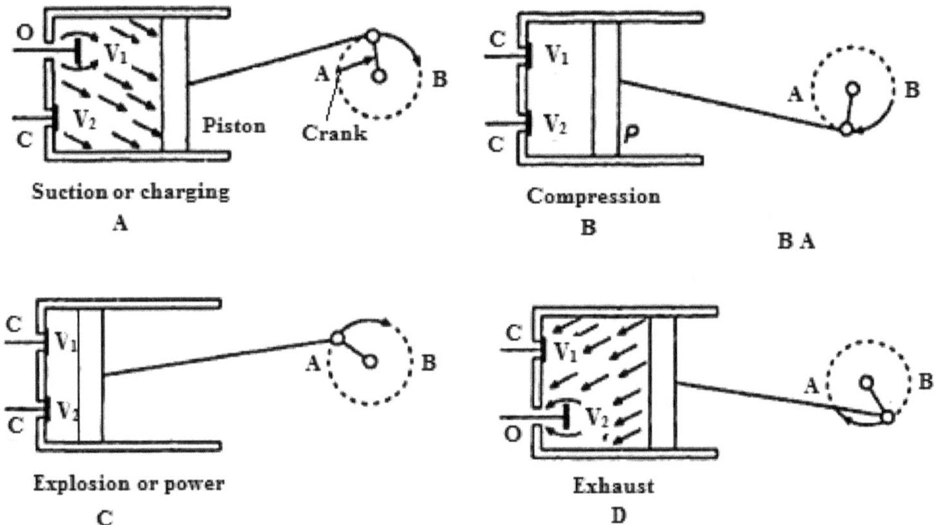

Figure 2.6. Four Stroke Cycle Operation in IC Engines.

Petrol Engine

First Stroke (Suction Stroke or Charging Stroke)

In this stroke, piston moves from TDC to BDC or crank moves from A to B during which inlet valve V_1 opens (O) and a mixture of air and gaseous fuel is drawn inside the cylinder (Figure 2.6 A). This operation is carried out at a pressure slightly lower than the atmospheric pressure because of the resistance offered by the valve. The valve V_2 remains closed (C) during this stroke.

Second Stroke (Compression Stroke)

The piston makes its return stroke, moving inwards with the crank moving from B to A (Figure 2.6 B). Piston compresses the air-fuel mixture to 1/5 to 1/10 of its initial volume. In this process, the temperature and pressure of the air-fuel mixture rises. At or around the end of the stroke, an electric sparks is produced by the spark plug that burns the fuel and air mixture. This completes one revolution of the crank shaft.

Third Stroke (Working Stroke)

Once the fuel is ignited, tremendous amount of heat is generated. In fact, the temperature may be as high as 2000°C. The very high pressure in the cylinder at this time causes the piston to move downward subjecting the gaseous mixture to an adiabatic expansion (Figure 2.6 C). The stroke is called power or the working stroke as heat energy is converted to mechanical energy in this stroke. The power from the piston is transmitted to the crank shaft by the connecting rod. As a result, the crank shaft rotates. By that time the mixture attains its initial volume and the pressure is also lowered down. Until the end, both valves remain closed but at the end of the stroke, the valve V_2 opens.

Exhaust Stroke

The piston further moves inwards and forces the spent gases out of the valve V_2. During this operation the valve V_1 remains closed (Figure 2.6 D). Once the spent gases escape out of the cylinder, the outlet valve V_2 closes. The piston begins to move outward and the initial condition is restored. This completes the cycle, and the engine cylinder is ready to suck the charge again.

Diesel Engine

We will discuss the diesel engine using the same diagram as Figure 2.6.

First Stoke (Suction Stroke or Charging Stroke)

The piston moves from top end (TDC) to bottom end (BDC) of the cylinder or crank moves from A to B. During this stroke inlet valve V_1 opens (O) and air is drawn inside the cylinder through the valve (Figure 2.6 A). This operation is carried out at a pressure slightly lower than the atmospheric pressure. The valve V_2 remains closed (C) during this stroke.

Second Stroke (Compression Stroke)

The piston makes its return stroke, moving inwards with the crank moving from B to A. Piston compresses the air to $1/15^{th}$ to $1/22^{th}$ of its initial volume. Both the valves V_1 and V_2 remain closed during this stroke (Figure 2.6 B). Both the temperature and pressure rises in this process. The temperature may be in the range of 600-900° C while the pressure in the range of 30-45 kg/cm^2. At or around the end of the stroke, the injector injects the fuel in the compressed air, which is already heated to a high temperature. It makes the injected fuel to burn. No spark plug is needed to ignite the fuel in this kind of engine. At this point, the crankshaft has completed a full 360° revolution.

Third Stroke (Working Stroke)

Tremendous amount of heat is generated following ignition of fuel. It causes very high pressure in the cylinder, which violently pushes the piston downward (Figure 2.6 C). The downward movement of the piston at this instant is called power or the working stroke. During this stroke heat energy is converted to mechanical energy. As the power is transmitted from the piston to the crank shaft, it makes the crank shaft to rotate. As the piston moves from TDC to BDC, the pressure and temperature of the mixture are lowered down. Until the end, both valves remain closed. At the end of the stroke, the valve V_2 opens.

Exhaust Stroke

It is last of the four strokes of the engine. The piston moves inwards from BDC to TDC and forces the spent gases out of the valve V_2. During this operation the valve V_1 remains closed (Figure 2.6 D). Once the spent gases escape out of the cylinder, the outlet valve V_2 closes. The piston begins to move outward and the cylinder is ready to receive the fresh air to begin the new cycle.

It may be noted that out of four strokes, there is only one power stroke and three idle strokes in a four stroke cycle engine. It is only during the power stroke necessary momentum to do useful work is produced. A flywheel is used to smoothen this intermittent supply of power. A comparison of four stroke diesel and petrol engines is reported in Table 2.1.

Table 2.1. Comparison of Diesel and Petrol Engines

Diesel Engine	*Petrol Engine*
Works on Diesel cycle	Works on Otto cycle
No carburetor, ignition coil and spark plug is needed but instead a fuel injection pump and injector is used	It has got carburetor, ignition coil and spark plug but no fuel injection pump and injector is required
Compression ratio is high in the range of 14:1 to 22:1	Compression ratio is low varying from 5:1 to 10:1
Diesel oil, a low volatility fuel, is used as fuel	Petrol (gasoline) or power kerosene, the high volatility fuels, are used as fuel
Only air is sucked in the suction stroke	A mixture of fuel and air is sucked in suction stroke
Fuel injected in combustion chamber burns due to heat of compression	Air fuel mixture is compressed in the combustion chamber when it is ignited by an electric spark
Thermal efficiency varies from 32 to 38 per cent	Thermal efficiency varies from 25 to 32 per cent
Engine weight per horse-power is more	Engine weight per horsepower is relatively low
Operational cost is low but high initial and maintenance costs	Operational cost is high but low initial and maintenance costs
Compression pressure inside the cylinder at the end of compression is around 30-45 kg/cm^2	Compression pressure at the end of compression is in the range of 6 to 10 kg/cm^2
Normally used in heavy vehicles	Normally used in light vehicles

INDICATOR DIAGRAM

Indicator diagram as already stated describes the relationship between pressure and volume in the IC engine. The indicator diagrams for Otto and Diesel engines are shown in Figure 2.7.

Otto Cycle

The actual indicator diagram for a four stroke Otto cycle or petrol engine is shown in Figure 2.7 (left). The suction stroke is shown by the bottom line which lies somewhat below the atmospheric pressure line. The fuel-air mixture flows into the engine cylinder because of this pressure difference. As a result, pressure inside the cylinder remains somewhat below the atmospheric pressure during the suction stroke. The compression stroke is shown by the line 1-2. It shows that the inlet valve closes (IVC) slightly beyond 1. At the end of this stroke, there is an increase in the pressure inside the engine cylinder. Shortly before the end of compression stroke, the charge is ignited (Ign) with the help of a spark plug as shown in Figure 2.7 left. The sparking suddenly increases pressure and temperature, but the volume, practically, remains constant as indicated by the line 2-3. The expansion stroke is indicated by the line 3-4. The exit valve opens (EVO) a little before 4. As a result, burnt gases exit into the atmosphere through the exit valve. The exhaust stroke is shown by the bottom line, which lies above the atmospheric pressure line. This pressure difference makes the burnt gases to flow out of the engine cylinder. Note that for brevity, the suction and exhaust strokes have been shown by the same line but in practice the exhaust line is slightly above the suction line and above the atmospheric pressure.

Diesel Cycle

The indicator diagram for a four-stroke cycle diesel engine is shown in Figure 2.7 (right). The suction stroke is shown by the bottom line, which lies below the atmospheric pressure line. This pressure difference allows the fresh air to flow into the engine cylinder. The compression stroke shown by the line 1-2 indicates that the inlet valve closes (IVC) slightly beyond 1. The pressure inside the engine cylinder increases until the end of this stroke. Shortly before the end of compression stroke fuel injection begins (FIO) and the fuel is injected into the engine cylinder. The fuel is ignited. While the ignition suddenly increases volume and temperature of the combustion products, the pressure practically remains constant (Line 2-3). The expansion stroke is shown by the line 3-4. The exit valve opens a little before 4 and the burnt gases are exhausted into the atmosphere. The exhaust stroke is shown by the bottom line, which lies above the atmospheric pressure line. Again note that the suction and exhaust strokes have been shown by the same line for brevity, but in practice the exhaust line is slightly above the atmospheric pressure and suction slightly below the atmospheric pressure.

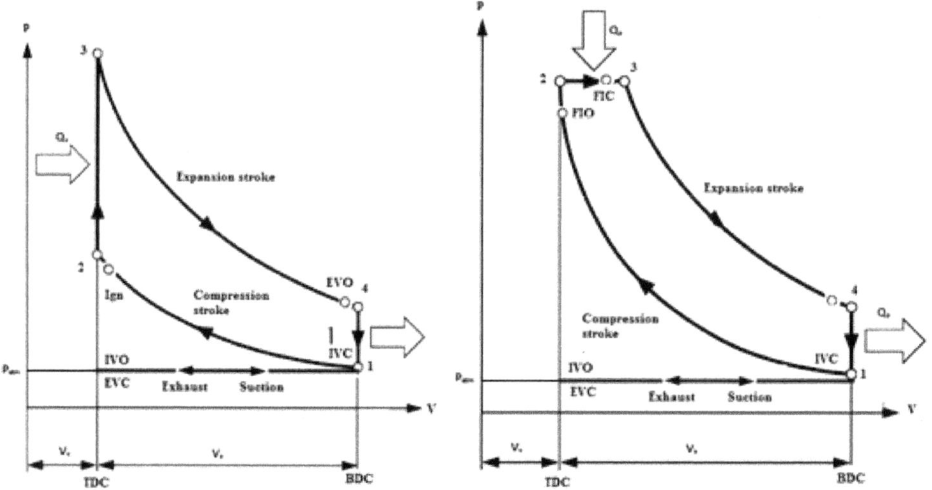

Figure 2.7. Indicator Diagrams: Otto Cycle (Left) and Diesel Cycle (Right).

Remarks

The diesel or compression-ignition engine is known for its high efficiency, lower fuel consumption and relatively low total gaseous emissions. The diesel engine has better slogging or lugging ability *i.e.* it maintains higher torque for a longer duration of time at a lower speed.

TWO STROKE PETROL AND DIESEL ENGINES

The whole sequence of events *i.e.* suction, compression, power and exhaust in a two stroke cycle engine are completed in two strokes of the piston or one revolution of the crankshaft. There are no separate piston strokes for suction and exhaust operations. Suction of the charge is accomplished by air compressed in crankcase or by a blower. Induction of compressed air removes the products of combustion through exhaust port. There are no valves in this type of engine. Holes called ports in the cylinder accomplish the gas movement. The crankcase of the engine in which the crankshaft rotates is airtight. The air-fuel mixture in SI and pure air in diesel engine enters from the inlet port into the crankcase. Salient features of a two stroke petrol engine and its parts are shown in Figure 2.8.

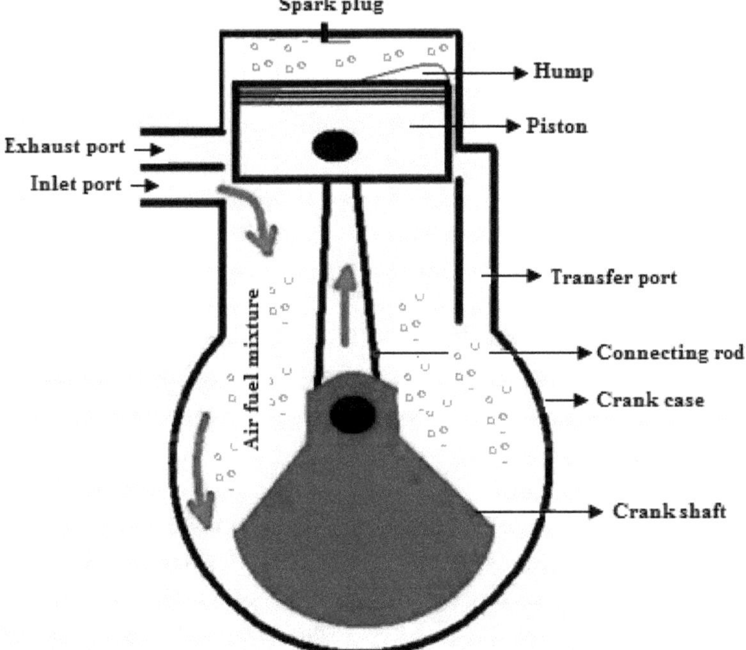

Figure 2.8. Principal Parts of a Two Stroke Engine.

Upward Stroke of the Piston (Suction + Compression)

Figure 2.9 (left) shows this stroke just before its completion. As the piston moves upward, it covers both the exhaust port and transfer port. Note that these ports are normally almost opposite to each other. This traps the charge of air- fuel mixture (or clean air alone in the case of diesel engine) drawn already into the cylinder. As the piston moves further upward, it compresses the charge besides uncovering the inlet (suction) port. Fresh mixture is drawn into the crankcase through this port. Just before the end of this stroke, the mixture in the cylinder is ignited by a spark plug (Figure 2.9 left). Thus, during this stroke both suction and compression events are completed. In a diesel engine, fuel is injected at this time as in the case of four stroke engine.

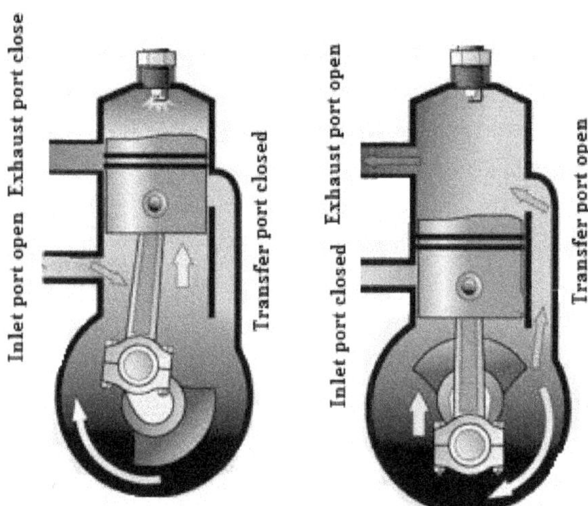

Figure 2.9. Working Cycle of a Two Stroke Engine: Going Up (Left) and Coming Down (Right).

Downward Stroke (Power + Exhaust)

At the end of the first stork, an electric spark or high air temperature in the diesel engine burns the fuel. As a consequence of burning of the fuel, the temperature and pressure of the gases rise. This forces the piston to move down the cylinder. When the piston moves down, it closes the suction port and traps the fresh charge that was drawn into the crankcase during the previous upward stroke. Further downward movement of the piston first uncovers the exhaust port followed by the transfer port (Figure 2.9 right). The fresh charge in the crankcase moves into the cylinder through the transfer port. It also drives out the burnt gases through the exhaust port. In order to facilitate the removal of exhaust gases, the piston crown is designed with a hump that helps to deflect the incoming mixture upwards around the cylinder (Figure 2.8). It helps in driving out the exhaust gases. Thus, both power and exhaust events are completed during the downward stroke of the piston. The overlapping of the injection and exhaust strokes in two-stroke cycle engines results in some fresh charge going directly out of the cylinder during the scavenging process.

A comparative statement of differences between a four and two stroke engines is reported in Table 2.2.

INDICATOR DIAGRAMS -2 STROKE ENGINES

Indicator diagrams for two stroke petrol and diesel engines are shown in Figure 2.10 and described in the following paragraphs.

Table 2.2. A Comparison of Four Stroke and Two Stroke Engines

Four Stroke Engine	*Two Stroke Engine*
Four strokes of the piston are completed in two revolution of crankshaft	Two strokes of the piston are completed in one revolution of crankshaft
One power stroke in every two revolutions of crankshaft	One power stroke in each revolution of crankshaft
Crankcase is not fully closed and airtight	Crankcase is fully closed and airtight
Full consumption of the fuel	Partial consumption of the fuel
Flywheel is heavier to take care of non-uniform turning movement	Flywheel is lighter due to more uniform turning movement
Power produced is less	Theoretically, for the same size of engine power produced is twice that of the four stroke engine
Heavy and bulky per HP	Light and compact per HP
Lesser cooling and lubrication requirements	Greater cooling and lubrication requirements
Wear and tear is less	Wear and tear is more
There are valves and valve mechanism	There are ports arrangement
High initial cost	Low initial cost
Volumetric efficiency is more due to greater time of induction	Volumetric efficiency is less due to lesser time of induction
Thermal efficiency is high and also part load efficiency better	Thermal efficiency is low, part load efficiency less
The four stroke engine can be operated in one direction only	The two stroke engine can be operated both in clockwise and counterclockwise directions
It is used where efficiency is important such as in cars, buses, trucks, tractors, industrial engines, aero planes, power generators *etc.*	This engine is used where low cost, compactness and light weight are important *e.g.* lawn mowers, scooters, motorcycles, mopeds, propulsion ship *etc.*

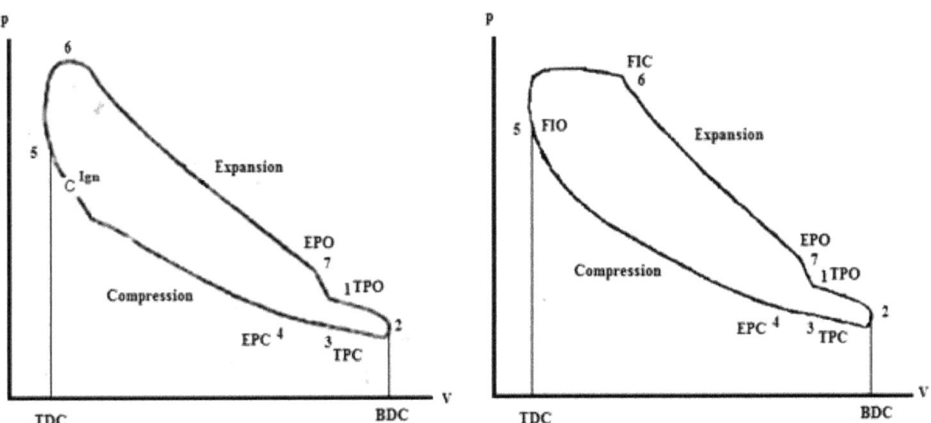

Figure 2.10. Indicator Diagrams of 2-stroke Engines: Petrol (Left) and Diesel (Right).

Petrol Engine

The line 1-2-3 shows that the exhaust port is open during the period transfer port is open (TPO) and it closes at 3 (TPC). In the first half of the suction stage as the piston moves from 1 to 2, the volume of fuel-air mixture and burnt gases increase. In the second half of this stage, the volume of charge and burnt gases decreases as the piston moves upwards from 2 to 3. The exhaust port closes (EPC) at 4, a little beyond 3. The line marked 4-5 shows that the charge inside the engine cylinder is compressed. At the end of the compression, there is an increase in the pressure inside the engine cylinder. Shortly before the end of compression the spark plug ignites the charge (Ign). It suddenly increases pressure and temperature of the combustion products. But the volume, practically, remains constant as shown by the line 5-6. The line marked 6-7 shows the expansion. Now the exhaust port opens (EPO) at 7, and the burnt gases are exhausted into the atmosphere through the exhaust port. It reduces the pressure. As the piston moves towards BDC, the volume of burnt gases increases from 7 to 1. At 1, the transfer port opens (TPO), suction begins and the cycle is repeated.

Diesel Engine

The initial process remains the same until the point 5. At the end of the compression, there is an increase in the pressure inside the engine cylinder. Shortly before the end of compression the fuel is injected and it is ignited because of the high temperature of the compressed air. This suddenly increases volume and temperature of the combustion products. The pressure as shown by the line 5-6, practically remains constant. The expansion is shown by the line 6-7. Now the exhaust port opens (EPO) at 7, and the burnt gases are exhausted into the atmosphere resulting in lowering down of the pressure. As the piston further moves towards BDC, the volume of burnt gases increases from 7 to 1. At 1, the transfer port opens (TPO) and the process is repeated.

QUESTIONS (THEORY)

1. Define bore, stroke, stroke-bore ratio, compression ratio, and cubic capacity of engine.

2. List the advantages of diesel engine over petrol engine.

3. Name the various types of engine classification and list the types of engines under each classification.

4. Draw a neat diagram of IC engine, label its parts and discuss any 5 of them.

5. Show four strokes of an IC engine with a neat sketch and explain their working for (a) Petrol engine and (b) Diesel engine.

6. Write a short note on the working of two stroke cycle engine.

7. Draw neat diagrams of indicator diagrams of four stroke Otto and Diesel engines and discuss them clearly bringing out the differences between the two.

8. Draw a neat diagram of two stroke petrol engine, label its parts and discuss any 3 of them.

9. Differentiate between four and two stroke cycle engines.

10. Write short notes on piston and piston rings, timing gears, and compression ratio.

11. Differentiate between internal and external combustion engines, compression and oil rings, spark engine and compression ignition engine, top dead center and bottom dead center, camshaft and crankshaft, clearance volume and swept volume, IHP and BHP.

Chapter 3

Tractor Systems

The word tractor is related to words like traction and tractive. Both the words have their origin from the Latin word tractus meaning drawing (pulling). A tractor is essentially a powerful engine-driven vehicle used to pull farm machinery, or other equipment such as trolleys *etc.* Basically, it is a machine that provides power to perform agricultural operations. Tractors, originally designed to replace working animals such as oxen and horses, are now much more sophisticated. These are capable of performing multifarious activities, although even now the main job is to pull things. An internal combustion engine is the heart of a farm tractor that drives the wheels or continuous tracks as in a crawler tractor. A number of different systems are integrated in the design of a tractor. These systems help in efficient functioning of the tractor and ensure the safety and comfort of the operator. The engine also supplies power to the electrical system, including the ignition system, lights, and even air conditioners in the new designs. Power to the implement is transmitted through a number of outlets such as drawbar, power take-off shaft or belt pulley. The tractor, as already stated, is an assembly of several systems and sub-systems which forms the subjects of discussion of this chapter.

AIR SUPPLY SYSTEM

Air intake system in the tractor allows the air to enter the engine. It comprises air cleaner, intake manifold, intake port or valve and a supercharger, an auxiliary unit. As agricultural operations are undertaken under very harsh conditions, unfiltered air in such environments may contain millions of particles of abrasive dust and other matter. These particles may damage the tractor engine and reduce its working life. Filtering systems to clean the air, the fuel and the oil are the first line of defense to protect engine, hydraulics and transmissions systems. In current designs of tractors, the operator's cabin is also protected with the help of air filters. Thus, air cleaning may be grouped in two categories namely air filters for combustion engine and filters for air cleaning of the cabin (driver chamber). This section deals only with air filters for the combustion engine.

An IC engine uses large quantities of air for combustion, the ratio being 30-35 kg of air for every kg of fuel burnt. In terms of volume, it may be about 2500-3000 L/L of fuel. Filtered air in an SI engine enters the engine through the carburetor, the manifold and the inlet valve (See fuel supply section). In CI engines, it enters through the inlet manifold and valve system. An air filter in essence is a device composed of fibrous materials which removes solid particulates such as dust, pollen, mold and bacteria from the air to prevent them entering the engine's cylinder. Filters having an absorbent such as charcoal (carbon) or some catalyst may also remove odour and gaseous pollutants such as volatile organic compounds or ozone. It helps to prevent the mechanical wear and tear of the engine as well as any contamination of the oil. Amongst many kinds of filter available in the market, 3 types are commonly used in tractors.

Dry Air Cleaner

Dry air cleaners are either mounted vertically in front of the tractor radiator or horizontally on the overhead engine. This type of air cleaner, also known as pleated paper air filter consists of three main parts *i.e.* Pre-cleaner, main housing and cleaning element (Figure 3.1) all sealed into one unit. Depending upon the pre-cleaning process these are grouped as the cleaner with the dust loading valve and with the dust air cup. Principally, both use centrifugal force by means of vanes and baffles, which removes about 80 per cent of the particulate particles. It greatly reduces the load on the dry air cleaner built for two-stage cleaning. The main housing contains the cleaning element. The cleaning element is normally made of multi-wire netting, but may also be made of nylon hair or pleated paper. After the pre-cleaning stage, air is allowed to pass through the holes in the metal jacket that surrounds the filter. As the air passes through the filter element, it filters out almost all the remaining small particles. The air passing through the cleaning element goes to the inlet manifold. Major advantages of a dry air cleaner are as follows:

☆ Easy to service

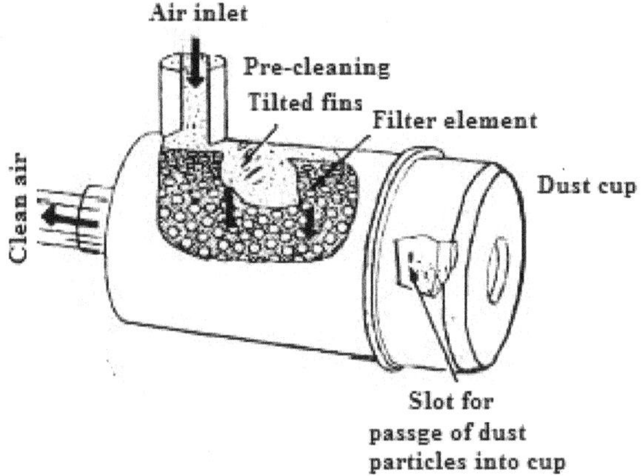

Figure 3.1. Dry Air Filter with Dust Cup.

☆ Good performance on gradient and in rough fields

☆ More efficient at high speeds

☆ Straw and chaff present may not result in much restriction to air passage

Its main disadvantages are:

☆ It is costlier to maintain than an oil bath because the filter elements need to be replaced frequently. It is recommended that the paper filter element should be cleaned after 50-100 hours of service

☆ Sometimes, dust particles enter the cylinder

Oil Wetted Mesh Cleaner

It consists of a copper mesh or nylon wire element as in the dry air filter but the filter element in this case is wetted with oil. The filter retains the dust particles as the air pass through the filter. This type of filter gets quickly clogged with dust, seriously affecting the air flow besides reducing its efficiency of removing the fine dust particles.

Oil Bath Air Cleaners

As the name suggests, oil is used to clean the air. In the past, water bath cleaners were in use, but now oil has replaced the water. Oil bath air cleaners are always mounted vertically to the engine so that the oil remains in the cup at the bottom of the cleaner. It is often mounted either in front of the radiator or by the side of the engine. An oil bath air cleaner comprises a sump containing a pool of oil. Besides, it is provided with an insert filled with fiber, mesh, foam, or any other coarse filter media (Figure 3.2). The media-containing body of the insert is located somewhat

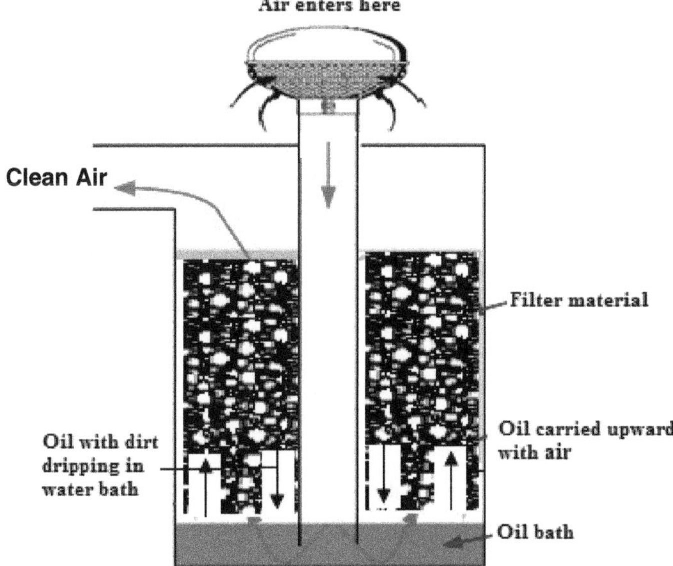

Figure 3.2. A Schematic View of an Oil Bath Air Filter.

above the surface of the oil pool such that the rim of the insert almost overlaps the rim of the sump. Principally air having dust particles enters the intake stack and pass down the inlet passage over the oil surface. Larger and heavier dust and dirt particles in the air cannot make the turn due to their inertia. As such, they fall into the oil and settle at the bottom of the bowl. Besides some oil is picked up, atomized and carried up into the filtering element. As the air passes through the screen (filter) most of the finer dirt remaining in the air is attached to the oil wetted surfaces. The excess oil drains back into the sump. A removable cup at the bottom is provided for convenient cleaning and servicing of the filter.

FUEL SUPPLY SYSTEM

An internal combustion engine inevitably uses some kind of fuel to produce power. The commonly used fuels are:

1. Petrol
2. Power kerosene
3. Light speed diesel
4. High speed diesel

Quality of Fuel

The quality of the fuel mainly depends upon the following properties:

1. Volatility of the fuel
2. Calorific value of the fuel
3. Ignition quality of the fuel

Volatility: Volatility of fuel has considerable effect on the performance of the engine as it affects: (i) Ease of starting the engine, (ii) Degree of crankcase oil dilution, (iii) Formation of vapour lock in the fuel system, (iv) Accelerating characteristics of the engine, (v) Distribution of fuel in multi-cylinder engine. In IC engine, all the liquid fuels must be converted into vapour fuel before burning. High speed diesel oil is most difficult to vapourize. Vapourizing temperature of high speed diesel oil is higher than that of the petrol, hence the petrol vapourizes quicker than diesel oil in the engine cylinder. This helps in easy starting of petrol engines.

Calorific value: *The heat liberated by combustion of a fuel is known as calorific value or heat value of the fuel.* It is expressed in kcal/kg of the fuel. The heat value of a fuel is an important measure of its worth, since this is the heat which enables the engine to do the work.

Ignition quality: Ignition quality refers to ease of burning the oil in the combustion chamber. Octane number and cetane number are the measures of ignition quality of the fuel.

Octane number: It is a measure of knock characteristics of a fuel. *The percentage of iso-octane (C_8H_{18}) in the reference fuel consisting of a mixture of iso-octane and normal heptane (C_7H_{16}), when it produces the same knocking effect as the fuel under test, is called octane number of the fuel.* Iso-octane has excellent antiknock qualities and is given a rating of 100. Normal heptane would knock excessively and hence it is assigned a value of zero.

Cetane number: *The percentage of cetane in a mixture of cetane $(C_{16}H_{34})$ and alphamethyl naphthelene $(C_{11}H_{16})$ that produces the same knocking effect as the fuel under test is called cetane number of the fuel.* Diesel fuels are rated according to cetane number which is the indication of ignition quality of the fuel. The higher the cetane numbers the better the ignition quality of the diesel fuel. The commercial diesel fuels have cetane rating varying from 30 to 60.

Detonation (Knocking): Detonation or engine knocking refers to violent noises heard in an engine giving a pinging sound during the process of combustion. It occurs during the process of combustion of the mixture within the cylinder after the ignition has taken place. It is an undesirable combustion and results in sudden rise in pressure, a loss of power and overheating of the engine. It is caused by improper combustion chamber, high compression pressure, early ignition timing, improper fuel and inadequate cooling arrangement.

Pre-ignition: Burning of air-fuel mixture in the combustion chamber before the piston has reached the top dead centre is called pre-ignition. Pre-ignition occurs when the charge is fired too far ahead of the top dead centre of the piston due to excessive spark advance or excessive heat in the cylinder.

The quality of the fuel is governed by its calorific value, the amount of heat liberated by combustion of fuel, ignition quality, the ease of burning the oil in the combustion chamber and volatility of fuel. The calorific values of petrol is highest at 11100 kcal/kg followed by power kerosene at 10850 kcal/kg followed by high and light diesel oils at 10550 and 10300 kcal/kg respectively as shown in the following table.

Name of Fuel Oil	American Petroleum Institute (API) Degree	Specific Gravity (g/cm^3)	Calorific Value (kcal/kg)
Light diesel oil	22	0.920	10300
High speed diesel oil	31	0.820	10550
Power kerosene	40	0.827	10850
Petrol	63	0.730	11100

Spark Ignition Engine

The fuel supply system of the spark ignition (SI) engine consists of fuel tank, sediment bowl, fuel filter, fuel lift pump, carburetor and related fuel pipes and Inlet manifold (Figure 3.3 top). In some SI engines the fuel tank is located above the level of the carburetor. In such a case fuel flows from the fuel tank to the carburetor under the action of gravity. A transparent sediment bowl and one or two filters between the fuel tank and the carburetor are provided. If the fuel tank level is below the carburetor, a fuel pump is used in between the tank and the carburetor. The fuel from the fuel tank moves through the sediment bowl to the lift pump and to the carburetor. A suitable pipe network is used for the fuel movement. From the carburetor the air-fuel mixture goes to the engine cylinder through the inlet manifold (Figure 3.3 bottom). Note that air from the air cleaning system and fuel from the fuel supply system enter the carburetor and mixed before they enter the intake manifold.

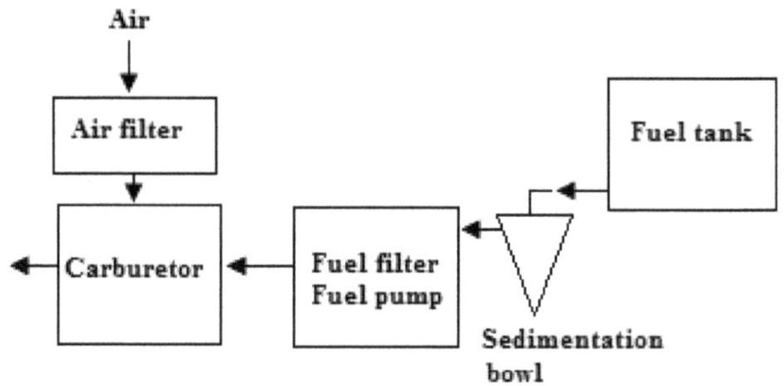

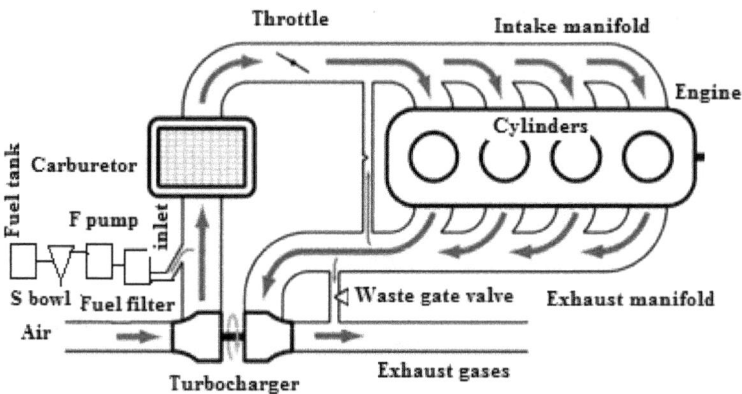

Figure 3.3. The Fuel Supply System on SI Engine: Block Diagram (Top) and Schematic Diagram (Bottom).

Fuel tank: It is a storage tank for the fuel having a wire gauge strainer under its cap. The strainer prevents foreign particles from entering the tank.

Fuel filter: Mostly two stages filter *i.e.* a sediment bowl and a preliminary filter is used. A bowl-like filter known as sediment bowl assembly is used to prevent water to move along the line with fuel. Water being denser than petrol settles at its bottom. A valve at the bottom of the bowl is used to drain off the water. The water is allowed to flow until the bowl contains only fuel. Preliminary filter consisting of a glass cap with a gasket is mostly fitted on the fuel lift pump. It prevents foreign materials from reaching inside the fuel line.

Fuel lift pump: It transfers the fuel under pressure to the inlet gallery of the fuel injection pump. As the fuel leaves the feed pump its pressure must be in the range of 1.5 to 2.5 kg/cm^2.

Carburetor: *The process of preparing a combustible fuel-air mixture outside engine cylinder in SI engine is known as carburetion.* The process of carburetion takes place in a device named as the carburetor. Its major functions are:

☆ To thoroughly mix the air and the fuel

☆ To atomize the fuel

☆ To regulate the air- fuel ratio for various speeds and loads on the engine

☆ To supply correct amount of air-fuel mixture at various speeds and loads

The basic principles of all carburetor designs are similar. A basic fixed venturi carburetor is shown in Figure 3.4. The carburetor essentially comprises a float chamber that ensures a constant level of the fuel (Figure 3.4). Fuel to the float chamber is delivered either by gravity or with the help of a pump. As the air flows over the end of a narrow tube or jet containing liquid, some liquid is drawn into the air stream. The quantity of liquid drawn is in proportion to the speed of air flow. The liquid drawn increases with increasing air flow and increasing jet size. The fuel levels in the jet and in the float chamber remains the same. The venturi helps to increase the velocity of the air flowing over the jet. A throttle butterfly valve, an adjustable obstruction in the induction pipe, is used to control the flow of air-fuel mixture to the engine. When the butterfly valve is in accelerate position, the airflow over the jet increases resulting in drawing more fuel into the air stream, mixture strength remaining constant. One of the problems encountered is with the starting of the engine in cold conditions. Rich mixture is required at this time besides at low

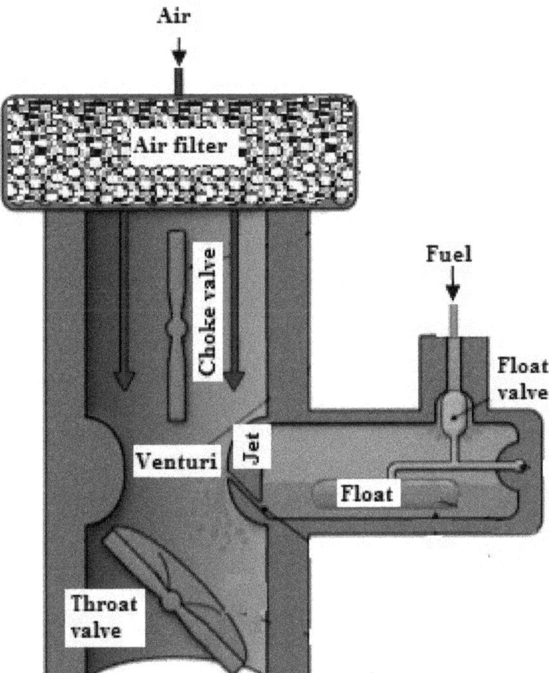

Figure 3.4. Construction of a Fixed Venturi Carburetor.

cranking speeds and before the engine warmed up conditions. For this purpose, another butterfly valve called choke is provided. It provides a richer mixture at the time of starting the engine. The choke controls the volume of air entering into the venturi. A second jet, fitted near the throttle butterfly, is used when the engine is idling or at low load operation, which requires rich mixtures of about 12:1 air: fuel ratio.

Air venting: Proper venting is required to expel the air in the system as and when air enters the fuel lines or suction chamber of the injection pump. It is accomplished by the priming pump through the bleeding holes of the injection pump.

Diesel Engine

Fuel supply system of diesel engine consists of fuel tank, fuel lift or feed pump, fuel filter, fuel injection pump, high pressure pipe, overflow valve and fuel injector also known as atomizer (Figure 3.5 top). Fuel from the fuel tank goes to the primary filter by gravity where coarse impurities are removed. From here, the fuel is drawn by a fuel transfer pump to be delivered to the fuel injection pump through the second fuel filter. An overflow valve is installed at the top of the filter that keeps the feed pressure under the specified limit. The overflow valve opens as and when the feed pressure exceeds the specified limit. The excess fuel in that case returns to the fuel tank through the overflow pipe. The injection pump supplies high pressure fuel to injection nozzles through delivery valves and high pressure pipes. The injectors atomize the fuel and inject it into the combustion chamber of the engine, where fuel is vapourized and burnt. The fuel leaking out of the injection nozzles also passes out through leakage pipes and is returned to the fuel tank through the overflow pipe (Figure 3.5 bottom).

Fuel filter: Modern tractors with tight tolerances require clean fuel. Unfiltered fuel may contain several kinds of contaminants, for example paint chips and dirt knocked into the tank while filling the fuel, or rust particles caused by moisture in the tank. These impurities must be removed before the fuel enters the system. Otherwise, the fuel pump and injectors may rapidly wear and fail to properly supply the fuel due to the abrasive action of the contaminants on the high-precision components of the injection system. The filtering of fuel is achieved through two filters in diesel engine namely the primary filter and secondary filter. Whereas the primary filter removes water and coarse particles of dirt, the secondary filter removes fine sediments from the fuel. The secondary filter consists of a hollow cylindrical element contained in a shell. An annular space is left in between the shell and the element (Figure 3.6). The filtering element essentially comprises metal gauge having appropriate media such as packed fibers, woven cloth, felt, paper *etc.* These filters need to be replaced at manufacturer's recommended intervals. Otherwise filter is liable to clog and will restrict the fuel flow. It may result in appreciable drop in engine performance.

Feed pump or transfer pump: The pump, mounted on the body of fuel injection pump, transfers fuel from the fuel tank to the fuel injection pump. The pump consists of body, piston, inlet and pressure valves. It delivers adequate quantity of fuel to the injection pump through the secondary filter.

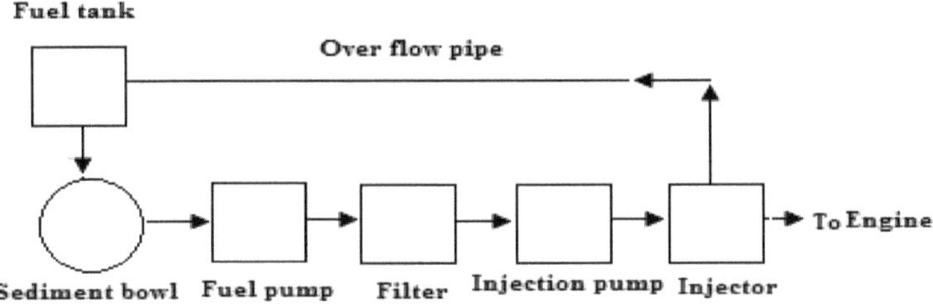

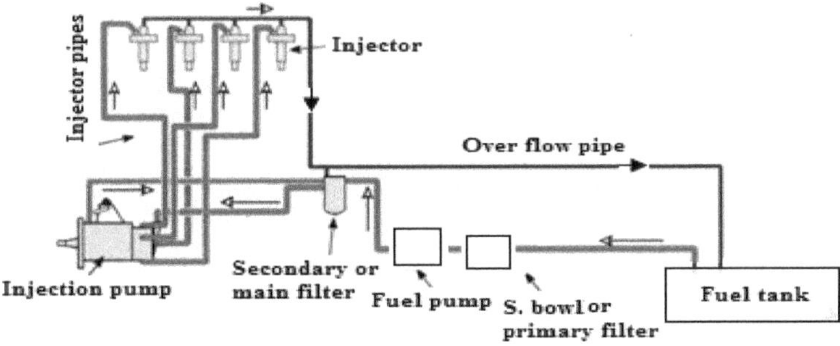

Figure 3.5. Line (Top) and Schematic (Bottom) Diagrams of Fuel Supply System in Diesel Engine.

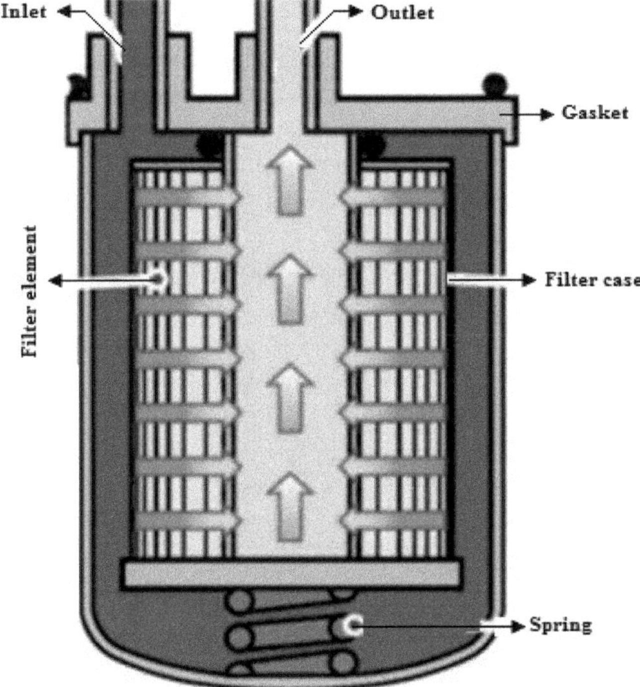

Figure 3.6. Fuel Filter for a Diesel Engine.

Fuel injection pump: It is a high pressure pump which supplies fuel to the injectors at the recommended pressure and as per the firing order of the engine. The fuel pressure varies in the range of 120 kg/cm² to 300 kg/cm².

Fuel Injector: It is used to deliver high pressure finely atomized fuel to the combustion chamber of the engine. The main functions of the fuel injection system are:

☆ To supply the correct amount of fuel required as per engine speed and load

☆ To maintain correct timing for beginning and end of injection

☆ To inject the fuel into the combustion space against high compression pressure

☆ To atomize the fuel for quick ignition

The process of fuel injection in diesel engine is also of two types namely air injection and solid injection. In air injection, compressed air is used to force the fuel into the cylinder. Being bulky, it is not suitable for tractors. In the solid injection, a high pressure pump forces the fuel into the combustion chamber. Besides, three kinds of injectors namely pintle type, throttle type and hole type are used to inject the fuel. Without going into the details, the main parts of an injector are: nozzle body and needle valve fabricated from alloy steel. A spring presses the needle valve against a conical seat in the nozzle body. Fuel injectors in modern tractor engines have multiple holes. While in operation, fuel from the injection pump enters the nozzle body through a high pressure pipe. As soon as fuel pressure becomes so high that it exceeds the set spring pressure, the needle valve lifts off its seat. As a result, the fuel is forced out of the nozzle holes and sprayed into the combustion chamber. The injection pressure may be adjusted by adjusting the screw.

Some of the essential conditions for efficient operation of the fuel system are listed as follows: These conditions if scrupulously followed will eliminate most of the troubles of the fuel supply systems.

☆ The fuel oil should be clean, free from water, suspended dirt, sand or other foreign matter

☆ The carburetor in SI engine should fully atomize the fuel and ensure vaporization, which occurs in carburetor as well as in the inlet manifold

☆ The fuel injection pump should create proper pressure so that diesel fuel may be perfectly atomized by injectors and be injected at proper time and in proper quantity in the engine cylinder

☆ Water and other deposits should be drained from the fuel tank at least every six month

Turbocharger

Turbo-charging is an important method aimed at achieving maximum mechanical efficiency and fuel-economy in automobiles. A turbocharger is a turbo-compressor to supply air under pressure to the cylinders of the engine to improve the power output of a diesel engine. If more air is delivered to the cylinders, the fuel charge increases to release more energy. The turbocharger consists of a centrifugal compressor with impellers and a gas turbine unit. The compressor impeller and the turbine wheel are rigidly fixed on a common shaft. A turbocharger uses the exhaust of the engine to run a turbine. The turbine in turn runs an air pump. The compressor impeller draws air from the atmosphere and delivers it to the intake manifold and from there it goes to the engine cylinders. It improves the volumetric efficiency of the engine. So far turbochargers are being used on diesel engines, but designs are being made to use it in SI engines as well (Figure 3.3, Gupta and Narain, 2015).

EXHAUST SYSTEM

The exhaust system collects the exhaust gases from the cylinders and expels them out. The system consists of an exhaust valve or port, exhaust manifold and muffler/silencer to reduce the noise (Figure 3.7). Note that silencer is a synonym of muffler, the former being a British term and latter an American term. The exhaust manifold collects exhaust gases from exhaust valves or ports of different cylinders and conducts them to a central exhaust passage at the end of which a muffler is connected. Mufflers mounted within the exhaust system do not serve any primary exhaust function. The muffler contains a number of chambers through which the gases flow. The gases are allowed to expand from the first passage into a much larger second one and a still longer/larger third one and finally to the passes connected at the tail known as the pipe of the muffler.

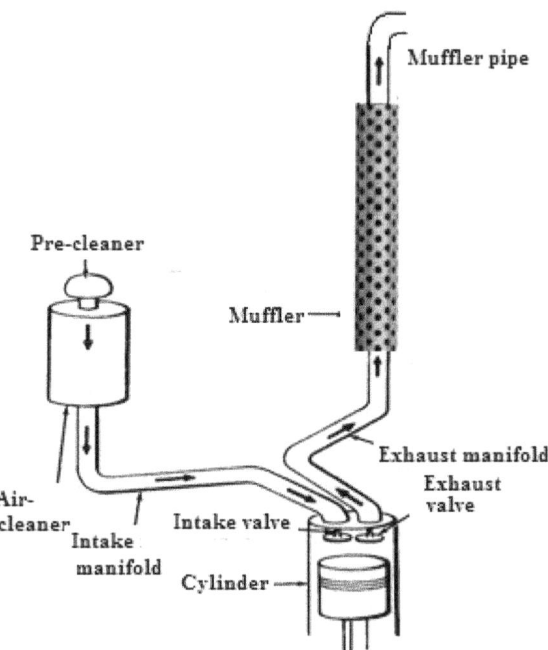

Figure 3.7. A Schematic View of an Intake and Exhaust System.

LUBRICATING SYSTEM

IC engine is made of large number of moving parts. Continuous movement of two metallic surfaces over each other results in wear and tear of these parts besides resulting in loss of power of the engine due to heat generation. Lubrication of moving parts is essential to prevent all these harmful effects. The parts of an IC engine that require lubrication are: cylinder walls and piston, piston pin, crankshaft and connecting rod bearings, camshaft bearings, valve operating mechanism, cooling fan, water pump and ignition mechanism. A lubricating system of an engine is an arrangement of mechanisms which maintains the supply of lubricating oil to the rubbing surfaces of the engine at correct pressure and temperature. Purpose of lubrication is five-fold.

Reduce friction: Two rubbing parts always produce friction and continuous friction produces heat, which causes wearing of the various parts of the engine and power loss. The primary function of the lubrication therefore is to reduce the friction and wear and tear of the two rubbing parts. The lubricating oil forms a thin oil film between moving surfaces to avoid their direct contact. This also reduces noise produced by the movement of two metal surfaces.

Cooling of the heated parts: Lubrication significantly dissipates the heat generated by piston, cylinder and bearings. Lubrication has a cooling effect on the various engine parts.

Sealing: The lubricant enters into the gap between the cylinder liner, piston and piston rings. Thus, it acts as a seal preventing the leakage of the gases.

Cleaning: Lubrication keeps the engine parts clean by removing dirt or carbon from inside the engine.

Protection from oxidization and corrosion: Lubrication protects the engine and other accessories from corrosion and rusting.

Types of Lubricants

Following three types of lubricants are commonly used.

Solid: Graphic, mica *etc.* Graphite is often mixed with oil to lubricate automobile spring. Graphite is also used as a cylinder lubricant.

Semi-solid: As grease, which is used to lubricate chassis and other parts.

Liquid: Depending upon the source, liquid lubricants are grouped as animal fat, and vegetable and mineral oils/lubricants.

Type of Lubricant	Remark
Animal fat	These do not stand much heat. They become waxy and gummy and hence are not suitable for machines
Vegetable lubricants	These are derived from seeds, fruits and plants. Cotton seed oil, olive oil, linseed oil and castor oil are used as lubricant in small machines
Mineral lubricants	These are obtained from crude petroleum found in nature. Petroleum lubricants are less expensive and suitable for internal combustion engines Nowadays synthetic oils produced from sand and coal such as poly alkaline and glycol are also used as lubricants

Properties of Lubricating Oil

Some important properties of the lubricating oil are as follows:

Viscosity: It is the property of a lubricant by virtue of which it offers resistance to flow. The lubricant should be viscous enough to maintain a thin film between the two mating surfaces. The viscosity measured by a viscometer is expressed in terms of viscosity number although its unit in SI system is Newton-second per square meter ($N\text{-}s/m^2$) or Pascal seconds (Pa-s). Viscosity is inversely proportional to temperature; it decreases with increasing temperature and vice versa.

Flash point and fire point: Vapours are released upon heating the oil. *Flash point is the lowest temperature at which oil is heated until sufficient inflammable vapours come off, which when brought to flame produces a momentary flash. The temperature at which vapours are released continuously so that the flame persists for a longer period is known as fire point.* To avoid fire hazards a good lubricant should have its flash and fire points above the temperature at which the engine operates.

Cloud point and pour point: Cloud point and pour point indicates the suitability of lubricant for use in cold conditions. *When lubricating oil is cooled, the temperature at which wax and other substance in the oil crystallize and separate out from oil or oil becomes cloudy is called cloud point.* On the other hand, *the lowest temperature at which oil is observed to flow by gravity in a specified laboratory test is known as the pour point.* Specifically speaking, the pour point is 3°C above the temperature at which the oil shows no movement when a sample container is held horizontally for 5 s. A good lubricant required to give good service at low temperatures should possess low pour point and cloud point.

Oiliness and wettability: *The ability of the lubricating oil to adhere to the surface is known as oiliness,* which depends upon its wettability and surface tension. *The ability of a liquid to maintain contact with a solid surface is known as its wettability.* It is controlled by the balance between the intermolecular interactions of adhesive (liquid to surface) and cohesive forces (liquid to liquid). A good lubricant should have enough oiliness to adhere to the surface even at very high pressure.

Volatility: A good lubricant is characterized by low volatility at normal working temperatures.

Carbon residue: A good lubricant, when used at high temperature, should not deposit carbon.

Types of Lubricating Systems

Various lubrication systems are used in IC engines.

☆ Mist lubrication system

☆ Dry sump lubrication system

☆ Wet sump lubrication system

A mist lubrication system is mostly used in 2-stroke engines where crankcase lubrication is not suitable. Most four stroke engines use either a dry sump or wet sump lubrication system. In the former, oil is stored in a separate tank rather than in the sump. The sump in this case is kept dry by a scavenging pump. Wet sump lubrication system, commonly used in tractors, has oil sump at the bottom of the crankcase. Lubricating oil from this sump is pumped to various engine components with the help of a pump. Oil, following lubrication of the desired parts, flows back to the sump by gravity. It is further grouped under the following three groups:

- ☆ Splash lubrication system
- ☆ Forced feed or pressure lubrication system
- ☆ Combination of splash and forced feed system

Besides these three methods of lubrication used in engines, another way is through the petro-oil lubrication system. Since this system is not used in tractors, only a passing reference will be made towards the end of this section.

Splash Lubrication System

In this system lubricating oil is stored in an oil sump. The oil from the oil sump is picked by scoops or dippers attached to the big end of the connecting rod. When the engine runs and crankshaft revolves, the dipper picks and splashes the oil on the cylinder walls. In some cases, oil troughs instead of dipper are provided under each big end of the connecting rods. These are filled by a pump and have an overflow to maintain a constant level of oil. As the connecting rods rotate, the oil in the troughs is splashed (Figure 3.8). The splashing action of oil creates a fog or mist of oil. This mist drenches the inner parts of the engine such as bearings,

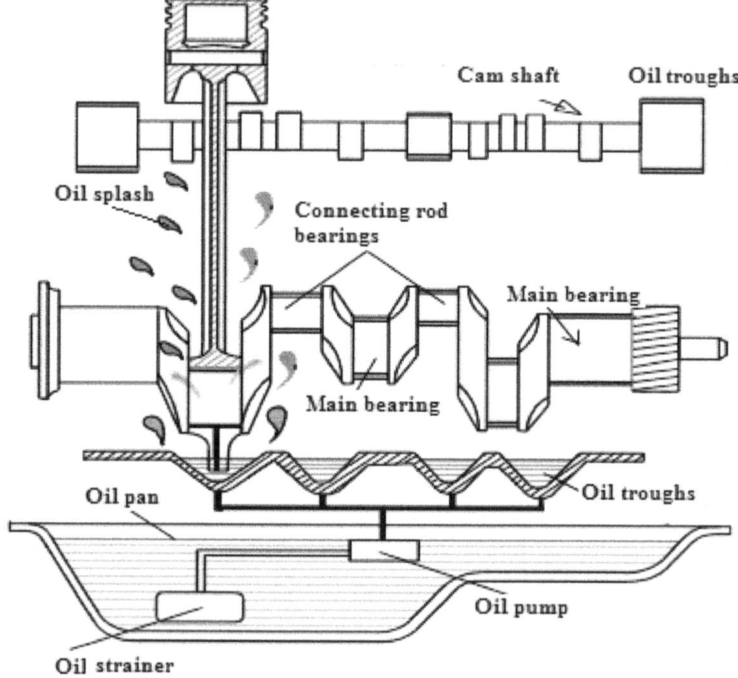

Figure 3.8.Splash Lubrication System.

cylinder walls, pistons, piston pins, timing gears *etc.* The splashed oil then drips back into the pan. This system is commonly used in a single cylinder engine with closed crankcase. For effective functioning of the system, the proper level of oil must be maintained in the oil pan. This system is very effective if the oil is clean and undiluted. Its main disadvantages are:

- ☆ Results in non-uniform lubrication
- ☆ If the rings are worn, the oil may pass the piston into combustion chamber. It causes carbon deposition, blue smoke and may even spoil the plugs
- ☆ Oil with time becomes very thin through dilution. The worn metal, dust and carbon particles also get accumulated in the oil chamber. These may be carried to different parts of the engine, causing wear and tear

Pressure or Forced Feed Lubrication System

As the splash system has certain limitations, force-feed system is used to generate additional pressure to ensure that lubricating oil reaches to all essential and desired components (Figure 3.9). In this system, lubricating oil is stored in a separate tank in case of dry sump system and in the sump itself in case of wet sump system. From here an oil pump, a positive displacement usually gear type or vane type pump, delivers the oil to the main oil gallery through an oil filter at a pressure of 2-4 kg/cm^2. The main gallery is a pipe or a channel in the crankcase casting. The oil from the main gallery goes to lubricate the main bearing. From here some of it falls back to the sump, some is splashed to lubricate the cylinder walls and the remaining goes through a hole to the crank pin. From the crank pin the lubricating oil goes to lubricate the piston pin/rings through a hole in the connecting rod.

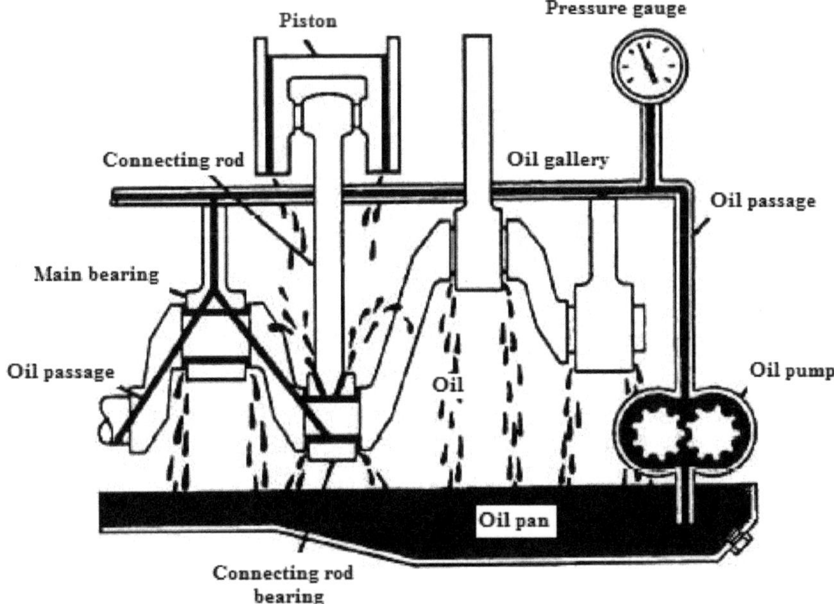

Figure 3.9. Forced Feed Lubrication System.

For the lubrication of cam shaft and gears, the oil is led through a separate oil line from the oil gallery. The oil also goes to the valve stem and rocker arm shaft under pressure through an oil gallery. The excess oil from the cylinder head comes back to the crankcase. An oil pressure gauge is used to measure the oil pressure in the system. This system is commonly used on high speed multi-cylinder engine in tractors, trucks and automobiles.

Combination of Splash and Forced Feed System

In this system, a combination of pressure and splash lubrication is used (Figure 3.10). Whereas the engine components subjected to very heavy load are lubricated under forced pressure (main bearing, connecting rod bearing and camshaft bearing), the rest of the parts like cylinder liners, cams, tappets *etc.* are lubricated by splashed oil. Note that in this figure both the dippers and pump are seen whereas in Figure (3.9), no dipper is used.

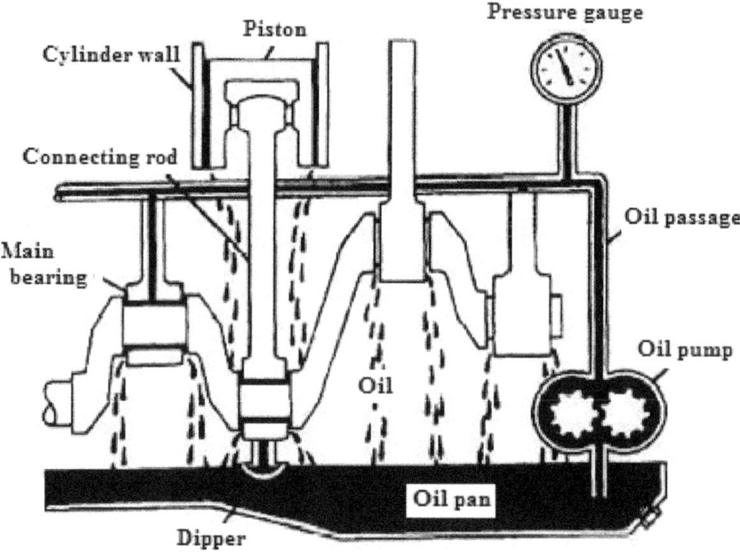

Figure 3.10. Combination of a Splash and Forced Lubrication System.

Petro-oil Lubrication System

In this method of lubrication, the lubricating oil is mixed with the petrol and fed into the engine cylinder during the suction stroke. The droplets of the oil cause the lubricating effect in the engine cylinder. This method of lubrication is used in small engines such as in motorcycles and scooters. This method is not much effective in large engines. In this method, 3 to 6 per cent of lubricating oil is added to petrol in the petrol tank. While the petrol is consumed in the engine, the lubricating oil is left behind to lubricate parts of the engine such as piston, cylinder walls and connecting rod. If the quantity of oil added to the petrol is less, it results in insufficient lubrication and may even cause seizure of the engine. On the other hand, if the added oil is more, it will result in excessive smoke in the exhaust. It will also result in excessive carbon deposits in the cylinder exhaust port and the spark plug.

Parts of Lubrication System

Major parts of the lubricating system are: oil sump, oil filter, oil pump, oil galleries, piston oil cooler and oil pressure gauge.

Oil sump: An oil pan/sump is a bowl-shaped reservoir that stores the engine oil and circulates it within the engine. It sits below the crankcase at the bottom of the engine.

Crankcase breather: The engine crankcase is fitted with some kind of breather to connect the space above the oil level and the outside atmosphere. The breather prevents build-up of pressure in the crankcase.

Oil pump: Lubricating oil pump, a positive displacement pump, is located close to the oil sump. It is driven by the camshaft of the engine. It forces the oil into the oil pipe (Figure 3.11 left). The lower end of the pump extends to the crankcase. It also has a strainer and a bypass valve to avoid damage to the pump in case filter gets clogged. The oil from the pump goes to the oil filter from where it goes to lubricate various parts of the engine.

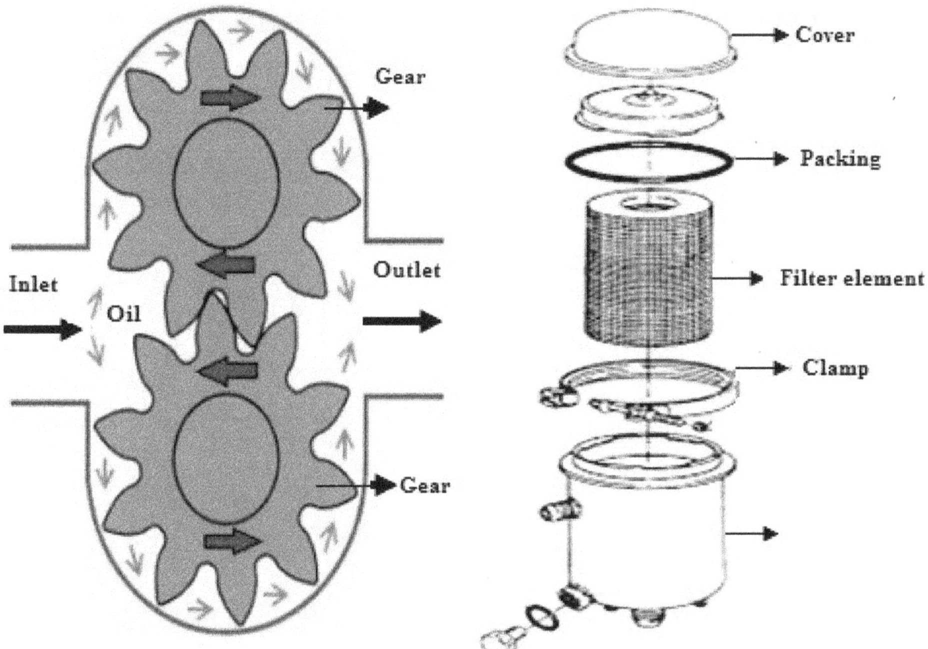

Figure 3.11. An External Gear Oil Pump (Left) and Filter Element (Right).

Oil filter: An oil filter in the system clears off the oil from dust, metal particles and other harmful particles. The inner replaceable filter cartridge is shown in Figure 3.11 right.

Oil galleries: Oil galleries are a network of large and small interconnected passages to supply the oil to various parts of the engine. These are drilled through the cylinder block.

Oil cooler: It functions like a radiator to cool down the engine oil. It transfers the heat of the oil to engine coolant through its fins. It also helps to maintain the oil viscosity so that oil quality remains under control.

Oil pressure gauge: An oil pressure gauge is used to observe the oil pressure in the lubricating system. A low oil pressure indicates some defect in the system, which should be attended to immediately.

Relief valve: Relief valve is provided to control the quantity of oil circulation and to maintain correct pressure in the lubricating system.

Troubles in Lubrication System

The list of often encountered problems in the lubrication system includes excessive oil consumption, low oil pressure; oil pressure is too high and defective indicator or gauge circuit. It is always good to make a visual inspection of the engine to identify the obvious problems.

Excessive Oil Consumption

External oil leakage out of the engine and/or internal leakage of oil into the combustion chamber are the major causes of high oil consumption. As a result, oil has to be replenished frequently. The external oil leakage is easy to detect as darkened oil wet areas on or around the engine can be easily observed. If leakage is more, oil may also be found in small patches under the vehicle. External engine oil leakage is mainly due to leaking gaskets or seals. Excessive oil consumption can be detected by the blue smoke exiting the exhaust system of the vehicle. It happens when the engine piston rings and/or cylinders are badly worn. In that case oil can enter the combustion chambers and is burned during combustion. Another plausible reason could be loss of oil in the form of vapour through ventilating system.

Low Oil Pressure

Low oil pressure is indicated when the oil indicator light glows, oil gauge reads low, or when the engine lifters or bearings rattle. The most common causes of low oil pressure are as follows:

- ☆ Low oil level in pan to cover oil pick-up. Note that it is one of the common causes of low oil pressure. It is why troubleshooting a low oil pressure problem begins with checking the oil level
- ☆ Worn connecting rod or main bearings (pump cannot provide enough oil volume)
- ☆ Thin (low viscosity) or diluted oil
- ☆ Weak or broken pressure relief valve spring so that valve opens too easily
- ☆ Cracked or loose pump pick-up tube so that air is pulled into the oil pump
- ☆ Worn oil pump

☆ Clogged oil pick-up screen

☆ Defective air pressure gauge

High Oil Pressure

High oil pressure seldom occurs but is easy to detect as the oil pressure gauge will show a high reading. Swelled condition of the oil filter is also an indicator of this problem. The most frequent causes of high oil pressure are:

☆ Pressure relief valve stuck or not opening at specified pressure

☆ High relief valve spring tension

☆ Excessively thick oil (high oil viscosity) or use of oil additive that increases viscosity

☆ Restricted oil gallery due to debris in oil passage

☆ Defective air pressure gauge

Defective Indicator

Defective indicator or gauge circuit may also show low or higher readings. It requires rectification or replacement of the defective pressure gauge.

Care and Maintenance of Lubrication System

A tractor owner is required to maintain the lubrication system. The maintenance normally requires changing the oil and filter(s). Occasionally maintenance tasks may include replacing lines and fittings, servicing or replacing the oil pump and relief valve, and flushing the system. Following care and maintenance activities will keep the system trouble free.

☆ Always use correct grade of lubricant

☆ Maintain oil level in the pan at desired level

☆ Oil should be changed regularly. Follow the manufacturer's guidelines for oil change

☆ Always clean the crankcase and flush it with a flushing oil before refilling the new oil

☆ Replace filters with the new ones at the time of oil change

☆ Connections, piping, valves and pressure gauge should be checked regularly

☆ Take due precautions to keep the oil free from dust and water

COOLING SYSTEM

Only about 25-30 per cent of heat energy generated in the engine is utilized to generate the desired output. The rest is released through the exhaust gases or is absorbed by the engine itself. A rough estimate says that 30 per cent of the heat is lost through the cylinder walls/liners, and 40 to 45 per cent through the exhaust gases and to overcome friction. This raises the temperature of the cylinder, piston and valves resulting in overheating of the engine. The adverse effects of overheating of the engine are:

- ☆ Cylinder and piston may expand to such an extent that the piston might cease in the cylinder
- ☆ Lubricating quality of the oil inside the cylinder may be reduced due to high temperature resulting in poor sucking of air in the cylinder
- ☆ Pre-ignition of fuel mixture may take place causing engine knocking as well as loss of power

A cooling system is therefore utmost necessity to dissipate the excess heat from the engine block and to maintain the temperature of the components within limit. Normally, a cooling system is designed to remove 30-35 per cent of total heat. The purposes of a cooling system are listed as follows:

- ☆ To maintain optimum temperature of the engine for efficient operation under all conditions
- ☆ To dissipate surplus heat for protecting engine components like cylinder, cylinder head, piston, piston rings and valves. Note that the temperature of the burning gases in the engine cylinder reaches up to 1500 to 2000°C, which is above the melting point of the material of the cylinder body, head of the engine and materials used in other parts
- ☆ To maintain the lubricating property of the oil inside the engine cylinder so that engine functions normally
- ☆ To maintain the volumetric efficiency of the engine as it is low at high temperatures

Besides, what has been stated, the cooling system should not over-cool the engine. Over-cooling may also reduce engine efficiency because of the following reasons.

- ☆ Loss of fuel as unburned fuel
- ☆ Loss of heat resulting in decreased thermal efficiency
- ☆ Reduced combustion efficiency because of less vapourization of the fuel
- ☆ Decrease in mechanical efficiency due to increase in piston friction as the viscosity of lubricant increases with low temperature

It is therefore recommended that the temperature of cooling system is maintained in an optimum operating range, 71° to 82°C for petrol engines and 88° to 90°C for diesel engines. On the basis of source of cooling, two kinds of cooling systems are used.

- ☆ Air cooling system
- ☆ Water cooling system

Air Cooling System

Air cooled engines are those engines in which heat is conducted from the working components of the engine directly to the atmosphere. This kind of cooling system is usually employed in light vehicles/engines. Cylinders in such an engine are not grouped in a block. An air blower or fan is used in the tractors to circulate the air for dissipating the heat from the engine surface, the cylinder in particular. In some cases the cylinder is enclosed in a sheet metal casing called cowling. The flywheel is provided with blades projecting from its face. It acts like a fan drawing air and directing it around the cylinder fins (Figure 3.12). Fins are provided to increase the surface area of the metal in contact with the air for effective cooling. Special baffles/fins are used to direct the air to reach the desired heated component so that hot spots can be avoided. Size and spacing of the fins depend upon the amount of heat to be removed, temperature of air, speed of air, thermal conductivity of metal used for fins, spacing between the fins and cylinder size. Copper and steel alloys having high heat conductivity are preferred to improve the rate of heat transfer. In general short fins in large numbers are better than large fins in smaller numbers. The length of the fins is greatest where the cylinder is the hottest say near the cylinder head. It may progressively reduce towards the crankcase. High pressure air will be required to cool the engine if the spacing between the fins is less.

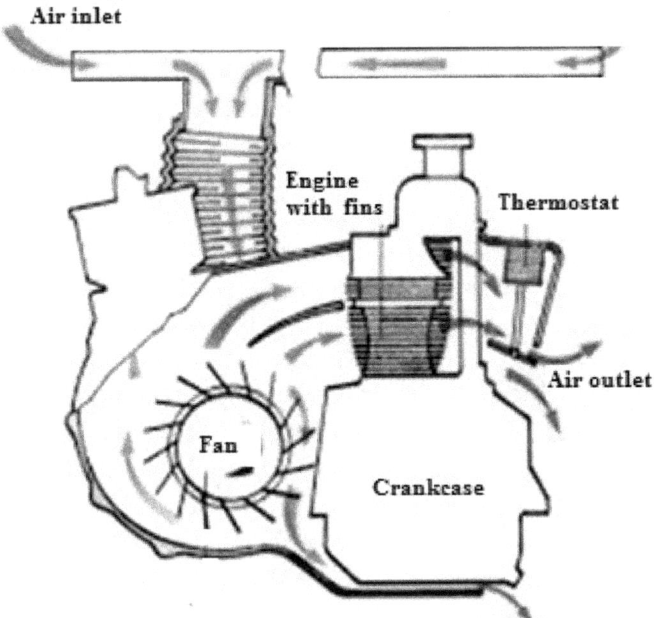

Figure 3.12. A Schematic View of an Air Cooled Engine.

Advantages of air cooled engines are:

- ☆ It is simple in design and construction because no water jackets, radiators, water pump, thermostat, pipes, hoses *etc.* are required
- ☆ It is compact and light in weight
- ☆ Air-cooled engines are able to operate well under hot and cold climates

☆ Maintenance cost is less

☆ Easy to warm up than the water cooled engines

☆ Can be used in extreme weather conditions where water is likely to freeze. Moreover, no problems related to coolant leakage are encountered

☆ Can be used in areas where water is scarce

Some disadvantages are:

☆ Uneven cooling of the engine parts

☆ Each cylinder has to be cast individually. Fin design and casting is difficult

☆ Normally suited in small engines

Water Cooling System

The engines in which water is used in the cooling process are known as water cooled engines. In such a cooling system, water jackets are provided around the engine cylinder or liners. The water circulating in these jackets absorbs the heat from the cylinder surfaces. The heated water is then cooled by the air using a radiator. The advantages of the water cooling system over the air cooled system are as follows:

☆ Uniform cooling of cylinder, cylinder head and valves

☆ Improved specific fuel consumption of engine

☆ Engine need not be provided at the front end of moving vehicle

☆ Water cooled engines are less noisy because of noise dampening by the water

Some disadvantages are:

☆ Greatly depends upon the supply of water

☆ The water pump used to circulate the water uses considerable power

☆ The water cooling system is costlier having more number of parts

☆ Requires more and regular maintenance and care

Methods of Water Cooling

Commonly used methods of water cooling are as follows:

☆ Open jacket or hopper method

☆ Non-return water cooling system

☆ Thermo-siphon method

☆ Forced circulation method

☆ Pressurized cooling

Open Jacket or Hopper Method

In this method a hopper or a jacket is provided around the engine cylinder in which cooling water or coolant is stored. The engine continues to operate satisfactorily until there is water in the hopper. As such, large hoppers are used so that the engine may run for several hours without refilling. A drain plug is provided for draining the hot water. This system is not used in present day engines.

Non-Return Water Cooling System

The water from a storage tank is directly supplied to the engine cylinder. The hot water coming out of the engine is not cooled for reuse but simply discharged. The low HP engines and diesel pumps used in agriculture are mainly cooled using this system. This kind of arrangement is suitable for large installations and/or where plenty of water is available.

Thermo-siphon Method

In this system, circulation of water takes place due to the density difference between the hot water in the engine jacket and cold water in the radiator. The system consists of a radiator, water jacket, fan, and temperature gauge and hose connections as shown in Figure 3.13. Hose pipes connect the water jacket and radiator on both sides *i.e.* at the top and the bottom. No pump is required as the water is circulated due to density difference only. Heated water surrounding the cylinder being lighter in weight rises upwards in the liquid column and goes to the radiator. Here water passes through tubes surrounded by air. A V-belt driven fan connected to the crankshaft is used to suck air through cells of the radiator unit resulting in cooling of hot water.

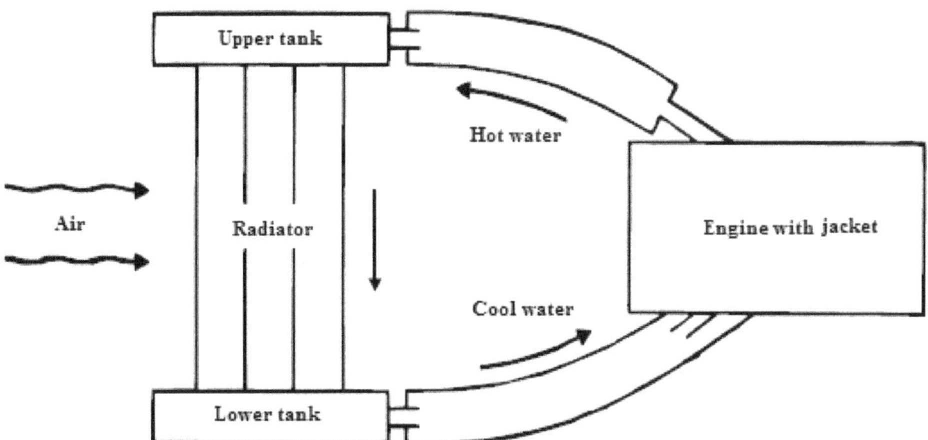

Figure 3.13. Schematics of a Thermo-siphon System.

Disadvantages of Thermo-siphon System

☆ Too slow rate of circulation

☆ Circulation begins only when there is a marked difference in temperature

☆ Circulation stops as soon as the water level falls below the top of the delivery pipe of the radiator

☆ Circulation of water is greatly reduced due to accumulation of scale or foreign matter in the passage

This system has now become obsolete and is no more used for cooling of engines.

Forced Circulation Method

In this method, water from the radiator to the water jacket of the engine is supplied under pressure using a water pump (Figure 3.14). After circulation in the jackets, water returns to the radiator where it loses its heat through the process of conduction. A thermostat valve is placed at the outer end of the cylinder head. It only opens when the cooling liquid and the engine attain the desired temperature, thus helping to maintain the engine temperature. As soon as it happens, the thermostat valve opens and the by-pass is closed such that water can now go to the radiator.

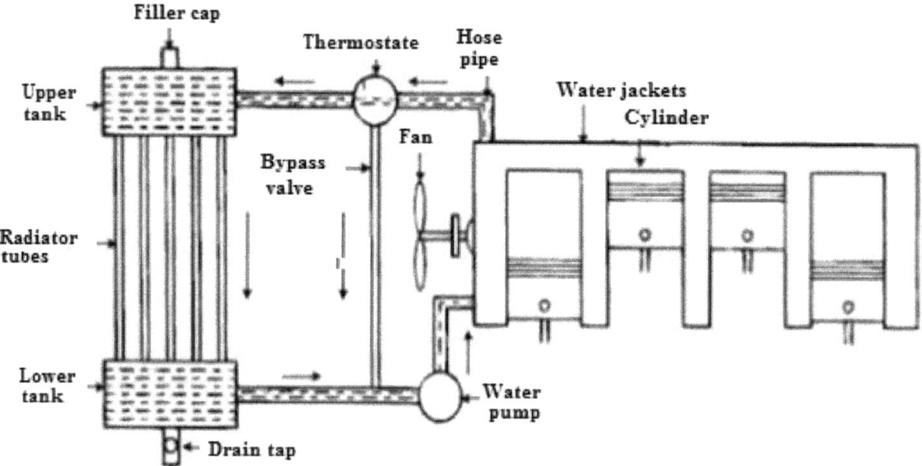

Figure 3.14. Forced Circulation System in a Water Cooling System.

Pressure Cooling System

It is an improved version of the forced circulation method. The basic principle states that in a closed radiator under high pressure, water boils at a higher temperature than in ordinary water cooling systems where the cooling water is subjected to atmospheric pressure (boiling temperature at atmospheric pressure is 100°C). The high water temperature gives more efficient engine performance and affords additional protection under high altitude, tropical conditions and for long hard driving periods. In this system, besides all other components of the forced circulation system, a special spring-loaded cap is fitted on the neck of the radiator to withstand the pressure in the forced circulation cooling system (Figure 3.15). The pressure cap with an airtight seal has an overflow pipe fitted above the spring-loaded part of the cap. The pressure cap increases the pressure within the cooling system besides preventing water evaporation. The pressure-release valve is set to open at a pressure between 0.3 and 1 kg/cm². Any increase in pressure beyond the set value is released to the atmosphere. Besides, there is a vacuum relief disc to prevent the radiator tubes from collapsing. Such a situation may arise when the engine cools. The water level in the radiator goes down and creates a vacuum. In that event, the vacuum so created operates and pushes down the disc that prevents the radiator tubes from collapsing.

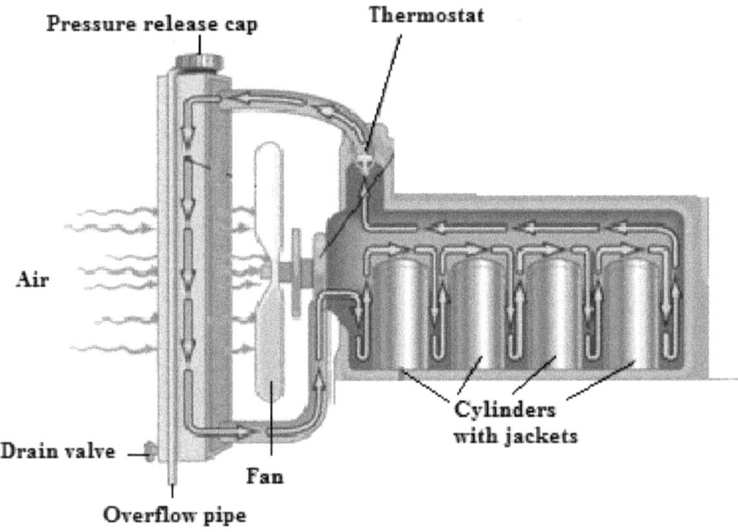

Figure 3.15. Pressurized Cooling System.

Components of a Forced Water Cooling System

Main parts of a water cooling system are: a radiator, a thermostat valve, a water pump, a fan, water jackets and some antifreeze mixtures.

Radiator

The radiator is located in the front of the tractor/vehicle. It is generally made of copper and brass with appropriately soldered joints. The main components are the upper tank and lower tank with a core in between them (Figure 3.14). The upper and lower tanks are connected respectively to the water outlets of the engines jackets and to the jacket inlet by means of hose pipes. The hot water coming from the engine cylinder fills the upper tank. The water from the upper tank comes to the lower tank through the radiator tubes with attached fins. The fins improve the effectiveness of air convection to dissipate heat. When the water is flowing down through the radiator tubes, it is cooled partially by the fan fitted at the backside of the radiator, which blows air and partially by the air flow developed by the forward motion of the vehicle.

Water Pump

A water pump is used to pump the circulating water from the lower tank to the engine jackets. It is mounted at the front end in between the outlet of the lower tank and the inlet of the engine jacket. The water pump is normally an impeller type centrifugal pump. It consists of an impeller mounted on a shaft and enclosed in the pump casing (Gupta, 2021). The pump casing has inlet and outlet openings. The belt driven pump gets its drive from the engine output shaft.

Fan

It is driven by the engine output shaft through the same belt that drives the pump. Located behind the radiator, it blows air over the radiator that helps to cool the liquid in the radiator.

Water Jackets

Cooling water jackets are provided around the cylinder, cylinder head, valve seats and other hot parts which require cooling. Heat generated in the engine cylinder is conducted through the cylinder walls to the jackets. The water flowing through the jackets absorbs this heat and gets hot. This hot water is then cooled in the radiator.

Thermostat Valve

Efficient operation of the engine is achieved in the temperature range of 80°C to 90°C. Therefore, the engine temperature should reach to this temperature as early as possible in the cool weather. On the other hand, the engine should remain in this temperature range under excessive hot weather conditions. The thermostat is designed to maintain this temperature range by regulating the temperature of water/coolant circulating in the water jackets. Generally, water below 70°C is not allowed to flow through it. It automatically begins to function once the water attains the desired operating temperature. A bellow type thermostat valve is commonly used. It contains a bronze bellow containing liquid alcohol. Bellow is connected to the butterfly valve disc through the link. When the temperature of water increases, the liquid alcohol evaporates and the bellow expands and in turn opens the butterfly valve that allows the hot water to move to the radiator. Another type of thermostat is bimetallic type consisting of bimetallic strips. Unequal expansion of the two strips causes the valve to open and allows the water to flow to the radiator.

Coolants and Antifreeze Mixture

Although, water is the most commonly used cooling agent, special coolants having better properties like corrosion free, high boiling point *etc.* are also available and used. These coolants are recommended for obtaining and maintaining higher engine efficiency.

In countries having extremely low temperature, water used in the radiator may freeze forming ice. The volume of ice being more, it may result in cracking of the cylinder block, pipes, and radiator. To prevent the freezing, antifreeze mixtures or solutions are added in the cooling water. The ideal antifreeze solutions should have the following properties:

- ☆ It should easily dissolve in water
- ☆ It should not evaporate
- ☆ No foreign material should be deposited in the cooling system
- ☆ Should not harm any part of cooling system in any manner
- ☆ It should be cheap and easily available
- ☆ It should not corrode the system

No single antifreeze satisfies all the requirements. Normally methyl, ethyl and isopropyl alcohols, a solution of alcohol and water, ethylene glycol and a solution of water and ethylene glycol and glycerin along with water, *etc.* are used as antifreeze materials. Sometimes chromates are added to prevent deposits.

Operation

In the cooling process, water is forced to circulate in the water jackets continuously with desired pressure and speed with the help of the water pump. The pump inlet is connected with the radiator at its bottom to draw the coolant/water from the radiator. When the engine is cool, the thermostat valve remains closed and the water/coolant is circulated through the water jackets. As soon as water/coolant gets heated; the thermostat valve opens and allows the water to pass to the radiator. Here heat is dissipated as the water comes in contact with the cool air passing over the radiator. Cool air is generated by the fan as well as by the forward movement of the tractor. Since the temperature difference between the air outside and water inside the radiator is high, the heat is dissipated quite quickly from water to the air.

Cooling System Troubles

Four common problems with cooling systems are:

- ✰ Water pump failure
- ✰ Leaky radiator hoses
- ✰ Radiator leaks
- ✰ Thermostat failures

Two main adverse effects of a defective cooling system are overheating of the engine and low warm-up of the engine. Over heating is mostly due to:

- ✰ Rusting and deposits of scales in the radiator and water jacket
- ✰ Defective hose pipe
- ✰ Defective thermostat
- ✰ Defective water pump
- ✰ Loose fan belt

Low warm-up or overcooling mostly occurs when the coolant bypasses a defective thermostat and flows directly to the radiator. It prevents the engine from reaching normal operating temperature.

Care and Maintenance

Proper maintenance schedules should be followed for proper functioning of the cooling system to ensure high engine efficiency and fuel economy.

- ✰ Always fill the radiator with clean, fresh and salt free water
- ✰ In cold climates use an appropriate antifreeze in the radiator
- ✰ The tension of the fan belt should be frequently checked. The maximum permissible V-belt sag when applying average finger pressure is 15 mm

☆ Never use rotten or soft hose pipes

☆ Never use oil or grease on the belt. Greasy belts should be wiped clean

☆ Lubricate the water pump bearing at regular intervals

☆ Never fill cold water when engine is very hot. It may fracture the cylinder wall and the cylinder head

☆ Flush out the radiator and water jackets using special air pressure guns. To remove scale from the system, use 1 kg of washing soda and 0.5 kg of kerosene oil. Pour it along with fresh water in the radiator and allow it to remain for 8 to 10 hr. After this, start the engine and run on medium speed. When the engine has run for 15 to 20 min, the solution should be drained out and the radiator should be flushed with clean water

☆ The radiator cap should not be opened when the engine is hot

POWER TRANSMISSION SYSTEM

Power transmission system in a tractor transfers the power of the engine to the rear wheels. It is accomplished through a speed reducing mechanism, equipped with several gears and shafts. The system uses various devices that cause forward and backward movement of a tractor. The complete path of power from the engine to the wheels is called power train comprising engine – crankshaft – flywheel – clutch – transmission box – differential – final drives – axle – drive wheels (Figure 3.16). The main functions of the power transmission system are:

☆ To disconnect the engine from the road wheels when desired

☆ To transmit power from the engine to the rear wheels of the tractor smoothly without shocks and jerks

☆ To make reduced speed available to rear wheels of the tractor based on tyre size and forward speed required

☆ To alter the ratio of wheel speed and engine speed in order to suit the field conditions

☆ To transmit power through right angle drive because the crankshaft and rear axle are normally at right angles to each other

☆ To provide for auxiliary power outlet in the form of PTO for powering the implements and also for stationary machinery

A power transmission system usually consists of the following parts:

Clutch: The device that connects or disconnects two torque transmitting devices.

Transmission: A device for transmitting power at different speed and torque.

Differential: The device, usually in the axle housing, that allows the two wheels on an axle to rotate at different speeds.

Brake: The device, usually in the axle housing, that stops the motion of the tractor.

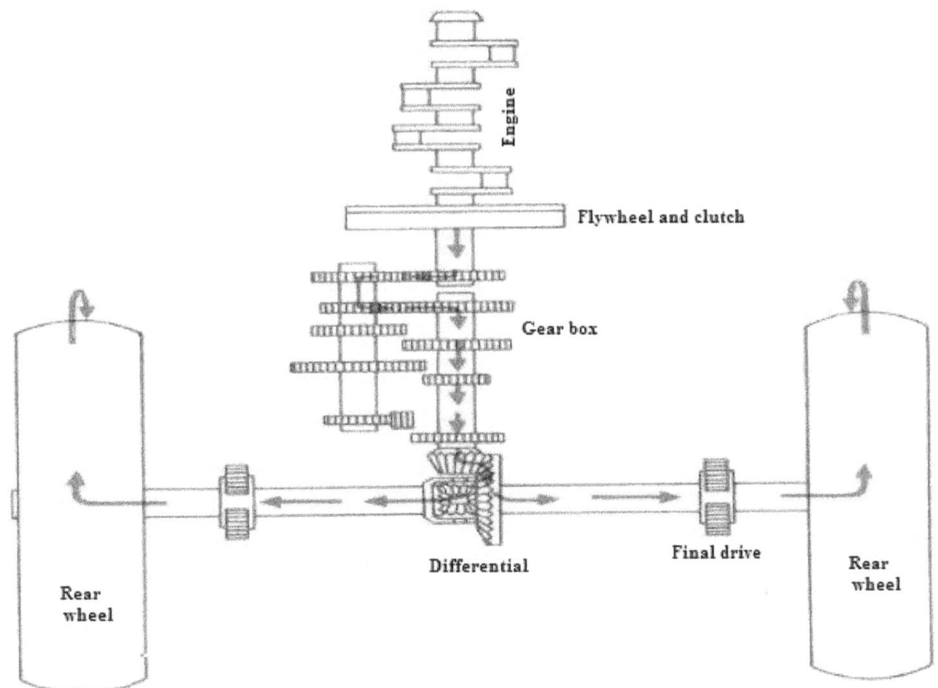

Figure 3.16. The Main Power Train in a 4 Wheeled Two Wheel Drive Tractor.

Axle: The shaft and other connecting parts that transmits torque from the differential or final gear reduction to the wheels.

Besides PTO drive sub-system transmit torque from the engine to the PTO on the rear of the tractor to drive working machines. It has a separate PTO clutch.

Clutch

Clutch is a part of the power transmission system located between the engine and the gearbox. It enables the rotary motion of one shaft to be transmitted to the other shaft, whose axis is coincident with the first one. It connects and disconnects the tractor engine from the transmission gears and drive wheels. When the clutch is engaged, the power flows from the engine to the rear wheels through the transmission system and makes the vehicle move. A clutch helps to accomplish the following functions.

- ☆ To disconnect the engine from the rest of the transmission unit for easy cranking. Once the engine is on, the clutch is engaged to transmit power from the engine to the gearbox

- ☆ To keep the gearbox free from the engine power when changing the gears. It helps in perfect engagement of the gears without any damage to gear teeth

☆ It helps to transmit the engine power to the wheels smoothly without shock to the transmission system while setting the wheel in motion

☆ To stop belt pulley of the tractor without stopping the engine

Important essential features of a good clutch are:

☆ It should be able to take the load

☆ High capacity to transmit maximum power without slipping

☆ Friction surface should be highly resistant to heat effects

☆ It should be easy to control by hand or pedal lever

Three types of clutches namely friction clutch, dog clutch and fluid coupling are used on tractors. Whereas friction clutch is mostly used in four wheel tractors, dog clutch is used in power tillers. Fluid clutch is now being used in some types of tractors.

Friction Clutch

The friction clutch as the name suggests operates on the principle of friction. When two friction surfaces are brought in contact with each other and pressed they get united due to friction between them. If one revolves, the other will also revolve. The friction between the two surfaces depends upon area of the surfaces, pressure applied upon them, and coefficient of friction of the surface materials. In a tractor, the driving member is the flywheel mounted on the crankshaft, and the driven member is the pressure plate mounted on the transmission shaft. When the clutch is engaged, the engine is connected to the gearbox. Thus, power flows from the engine to the rear wheels. While starting the engine, the clutch pedal is depressed. After the start of the engine, the clutch pedal is slowly released to gradually increase the pressure on frictional surfaces until there is no slip. At this time, the driven and driving plates are firmly gripped. Transmission of power depends upon the kind of material used for the friction members and intensity of the force pressing them together. Friction clutch is further sub-grouped into 3 classes namely single plate clutch or single disc clutch, multiple plate clutch or multiple disc clutch and cone clutch.

Single Plate Clutch

It consists of pressure plate, clutch plate, springs and release fingers. It has only one clutch plate, which is pressed against the flywheel of the engine by means of spring-loaded pressure plate. When the clutch pedal is depressed, the release fingers push back the pressure plate. This releases the pressure from the clutch plate. Then the clutch plate stops rotating, but the flywheel continues to rotate. When the clutch pedal is released, the pressure plate forces the clutch plate against the flywheel to cause the clutch plate and the flywheel to run together as one unit. Thus, the power of the engine is transmitted to the gearbox for onward transmission to rear wheels.

Multiple Plate Clutch

It has a number of thin metal plates arranged alternately to work as driving and driven members. One set is attached to the flywheel and the other set to the clutch shaft. If the plates are pressed together, the clutch is said to be engaged and the power is transmitted from the engine to the gearbox for onward transmission to the rear wheels. The pressure is obtained by a set of heavy springs fitted together in housing. This kind of clutch can be quite smoothly engaged or disengaged because of larger surface area of friction members.

Cone Clutch

A cone clutch serves the same purpose as a disc or plate clutch. However, instead of mating two spinning discs, the cone clutch uses two conical surfaces to transmit torque by friction. Small cone clutches are used in synchronizer mechanisms in manual transmissions and some limited-slip differentials.

Dog Clutch

It is a simple clutch having square jaws used to drive a shaft in either direction. A dog clutch couples two rotating shafts or other rotating components not by friction but by interference or clearance fit. Dog clutches are used inside manual automotive transmissions to lock different gears to the rotating input and output shafts.

Fluid Coupling

Fluid coupling, also known as hydraulic coupling, is also a device for transferring power from a driving shaft to another driven shaft by means of acceleration or deceleration of a hydraulic fluid. An impeller with radial-vanes constitutes the driving member and runner with radial vanes constitutes the driven member. The entire unit is housed in a suitable casing. A coupler mounted on the engine crankshaft is 3/4th filled with suitable oil. A spring loaded sealing ring is used to make the driven shaft oil tight. At the rotation of the crankshaft, the oil is thrown out by centrifugal force from the center to the outer edge of the impeller increasing the velocity and the energy of the oil. It then enters the runner vanes at the outer portion and flows towards the center causing rotation of the runner unit. As long as impeller and runner rotate at different speeds, the oil continues to circulate uniformly but when the impeller and runner start running at the same speed, the oil circulation stops. Note that a coupling does not increase the applied torque but only transmits the torque in a uniform manner. A fluid coupling is capable to absorbs shocks and vibrations, has smooth start and easy in operation.

Construction and Operation of a Friction Clutch

A friction clutch essentially comprises a release bearing, diaphragm spring, pressure plate, clutch disc all enclosed in a clutch cover (Figure 3.17). It is operated by the operator using a foot or hand pedal. The clutch transmits power by means of friction between the driving and the driven members. Clutch remains engaged when the vehicle is moving. When the clutch is disengaged, the power is not transmitted to the rear wheels although engine is still running. It makes the vehicle to stop. The clutch is disengaged at the time of starting the engine, shifting the gears and idling the engine.

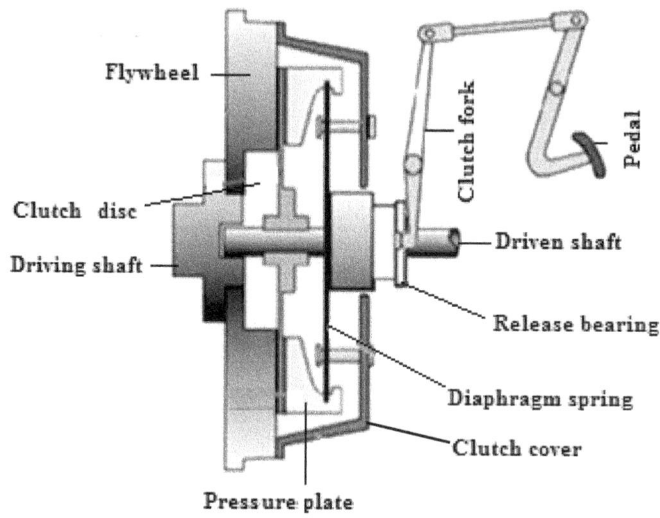

Figure 3.17. Parts of a Friction Clutch (Engaged Position).

GEARBOX

The speed of a tractor engine is quite high. On the other hand, rear wheels of a two wheel drive tractor require high torque. In order to achieve this, gears are used to reduce the speed to increase the torque available at the rear wheels as per the following relations.

$$HP = 2\,\pi\,NT/4500$$

$$HP = 2\,\pi\,NT/45000$$

$$kW = 2\,\pi\,NT/60000$$

Such that T is the torque (kg-m) and N is the RPM. Note that in the last 2 equations T is in N-m such that g is nearly 10 m/s². Obviously, to get higher torque at the wheels for the same HP, speed has to be reduced. As an illustration, when a tractor is started, the torque provided by the engine output shaft is not enough to overcome the weight of the tractor. It makes it difficult to move the tractor initially. The gearbox helps to provide initial high torque to move the vehicle. On the other hand, when the tractor is moving at high speed on the road, torque is not at all effective so a gearbox is needed that can also provide a high-speed low torque ratio so that vehicle can maintain and move at the high speed. Thus, a gearbox helps to provide variable torque and speed. A gearbox mounted between engine's output shaft and the final drive can be defined as *'a transmission device to transfer required torque and power to the wheels of the vehicle'*. Since the combination of torque and speed depends upon the field requirements, a number of gear ratios are provided to suit these requirements.

Components of Gearbox

A manual transmission gearbox is used in most tractors due to its low cost.

Basically, the main components of the gearbox are: gear lever, gearbox housing, gear shafts, gears and bearings. It has limited gear or speed ratios and the shifting of gears is done manually by the driver by pushing or pulling the gear lever in a predefined fashion. Manual transmission of course always requires the use of clutch.

Gear lever is the lever used for shifting or sliding the dog clutches/synchromesh devices *etc.* over the main shaft. It is operated by the driver from his seat. The outer casing of the gearbox is known as the gearbox housing, which is usually made of cast iron. It houses various shafts and gears and contains the gearbox oil (SAE 90) for lubrication of the gears. A gearbox contains a series of gears of different diameters. The gears may be spur, helical, bevel, worm and epicycle type depending upon the type of gearbox used. The gears are arranged in a special fashion to provide required gear or speed ratios to the final drive of the vehicle. While the gears have different number of teeth, the gear ratio determines the final speed. The ratio of the speeds of the two shafts always remains constant when connected by the gears. This constant ratio called the gear ratio is written as:

Speed of the driver/speed of the driven = No. of teeth in the driven/No. of teeth in driver

Speed is reduced in proportion to the number of teeth on the gears. The approximate gear reduction in tractors from engine drive to final drive is 1:175 (at lower gear) to 1:12 (at higher gear).

All the gearboxes have three gear shafts namely

☆ Input shaft also called primary shaft or clutch shaft

☆ Counter shaft also called lay shaft or auxiliary shaft

☆ Main shaft also called secondary shaft or out shaft

The gears on the respective shafts are called by the name of the shaft such as input pinion, counter shaft gear, main shaft gear. Note that construction and working of primary and counter shafts in case of all the three types of gearboxes is similar. Only the construction and working of the main shaft gears differ. The main shaft in all cases is splined, but the main shaft gears sit on the shaft in different ways depending on the type of gearbox. Other components will be discussed in the respective gearbox.

Kinds of Gearboxes

Three types of gearboxes namely sliding mesh gearbox, constant mesh gearbox and synchromesh gearbox are commonly used on tractors.

Sliding Mesh Gearbox

It is the oldest and the simplest of automotive gearboxes (Figure 3.18). As the name suggests, the selected gear on the main shaft is slid over the main shaft to mesh with the corresponding gear on the counter shaft. Clutch shaft is the input shaft, which takes the rotational power from the clutch mounted on one end of this shaft. Thus, the clutch shaft carries the engine output to the gearbox. Only one helical type gear (A) is provided on the clutch shaft and is located inside the gearbox housing. It is meshed to another helical gear on the lay shaft that transmits rotational motion to lay shaft (B, Figure 3.18), being always in contact. Lay shaft, also known as counter shaft as it rotates in a direction counter to the engine rotation, is an intermediate shaft between the clutch shaft and main shaft. The rigidly fixed gears on the lay shaft rotate with the lay shaft and mesh with different gears on main shaft to give different gear ratios. Lay shaft also has a reverse gear with an idler gear attached to it. The gears on the main shaft, which is the output shaft, are not rigidly fixed. While the outer surface of this shaft is splined, the gears are also internally splined. As such the gears on this shaft can easily slide over the shaft.

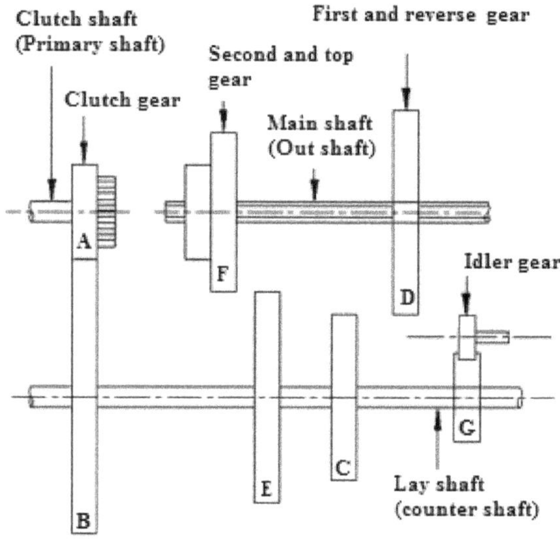

Figure 3.18. Line Diagram of a Sliding Mesh Gearbox.

Figure 3.18 shows 3- speed 1-reverse transmission system. In the neutral position, none of the main shaft gears are engaged to the counter shaft gears. The gears of the main shaft are slid to mesh with the respective gear on the counter shaft as per the speed-torque requirement. At any given time, only one set (pair) of main shaft and countershaft gear are in mesh with each other. If two pairs get meshed, they will tend to rotate the main shaft at different speeds, leading to breakage of either the main shaft or the meshed gears. To mesh a particular gear, it is slid over the main shaft by its collar and is made to mesh with the corresponding counter shaft gear. First gear provides maximum torque at low speed which is obtained when the smallest gear on the lay shaft, C meshes with the biggest gear, D on the main shaft. Second gear provides less torque and higher speed than first gear and is obtained when the middle size gear, F of the main shaft meshes with the second smallest gear on the lay shaft, E. In this case high speed and second high torque is transmitted to the final drive. Third gear provides maximum speed and minimum torque to the final drive and is also known as high-speed gear or top gear in sliding mesh gearbox. It is obtained by moving gear F towards left to mesh directly with gear A. The main shaft turn with clutch shaft and a gear ratio of 1:1 is obtained or we can say that the drive obtained a maximum speed of the clutch shaft. When the reverse gear is selected, the rotation of the output shaft is reversed, which is made possible by using an idler gear between the main shaft and lay shaft. It changes the rotation of the output shaft and the vehicle starts moving in the reverse direction. To achieve this, move the gear D on the main shaft toward right axially to mesh with idler gear mounted on another shaft. The drive is obtained from gear G while the idler gear reverses the direction of rotation of main shaft.

Merits

- ☆ Simple design
- ☆ Economical
- ☆ Highly efficient

Demerits

- ☆ Shifting of gears is difficult
- ☆ Need to slow down the speed while changing gear
- ☆ Noisy operation
- ☆ Require highly skilled driver to change the gear

Constant Mesh Gearbox

It is the modified gearbox introduced to overcome the limitations of the sliding mesh gearbox. As the name suggests, a constant mesh transmission is the type of manual transmission in which gears on the clutch shaft and main shaft are constantly meshed to the corresponding gears on the counter shaft (Figure 3.19). A new shifting device named dog clutch that transmits appropriate gear ratio to the final output is also introduced. The dog clutch helps to avoid the sliding of gears over the shaft for meshing or shifting. In the system, the pair of gears with suitable gear ratio comes in contact with the sliding dog clutches which in turn transmit the gear ratio of the pair of meshed gears to the final output shaft. Similar to the sliding mesh gearbox, the main shaft is splined in this case too. Since the gear of the main shaft are in constant mesh with the appropriate gear of the lay shaft, the selection of 1, 2, 3, 4 and reverse gear is obtained with the sliding and meshing of the dog clutches with the appropriate pair of gears. As an example, first gear is selected by pushing or pulling the gear lever, such that the dog clutch D2 is shifted to the right (D2-A). The smallest gear of lay shaft and largest gear of main shaft A then transmit power to the output shaft. If dog clutch D2 is shifted to left, it selects the second smallest lay shaft gear and second largest main shaft gear to drive the output shaft (D2-B). Similarly, third gear is selected with the help of dog clutch D1 moving towards the right (D1-C). The highest speed gear of 4-speed manual transmission is obtained by shifting D1 to the right such that it directly meshes with the clutch gear (D1-D). The reverse gear is obtained with the special gear known as idler gear mounted between the lay shaft and the main shaft. As the reverse gear is selected, the dog clutch makes contact with the idler gear and reverse gear becomes operational.

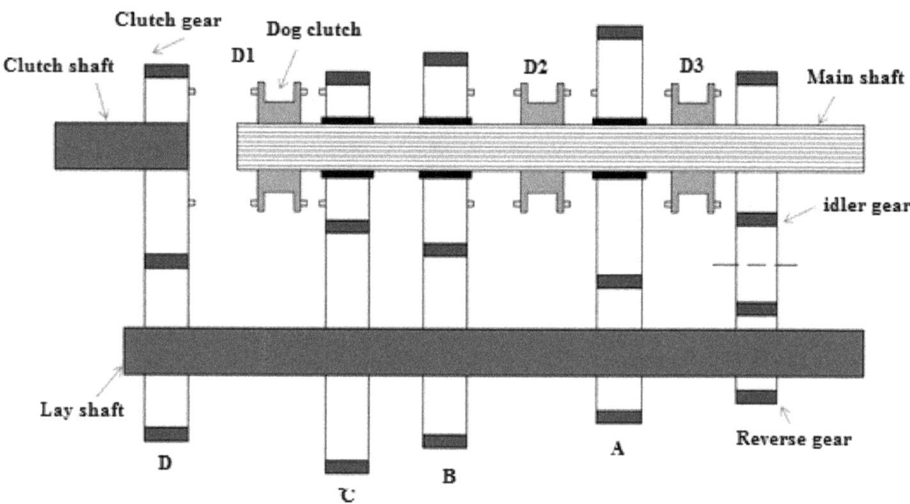

Figure 3.19. A View of the Constant Mesh Gearbox.

Merits

- ☆ Less shifting effort
- ☆ Less noise in operation

Demerits

- ☆ Less efficient than sliding mesh gearbox
- ☆ Double declutching is required resulting in fatigue to the operator

Synchromesh Gearbox

Synchromesh gearbox is similar but an advanced version of the constant mesh gearbox in which dog clutches are replaced by synchromesh devices. A synchronizer is placed between two gears. Thus, one unit can be used for two gears. The synchronizers are the special shifting devices which have conical grooves cut over its surface that provide frictional contact with the gears which is to mesh in order to equalize the speed of the main shaft, lay shaft and clutch shaft. Principally, before engaging the gears they are brought in frictional contact with each other and engagement is done only after equalizing the speed. As a result, gears are engaged smoothly without any slippage. The side of the gear to be engaged has two features. One is hollow-cone, and the other the cone surrounded by the ring of dog teeth. The gear is made the cone and teeth that the synchromesh mechanism contacts. In the figure, G1 and G2 are the ring-shaped members which are having the internal teeth that fits onto the external teeth. F1 and F2 are the sliding members of the main shaft. H1, H2, N1, N2, P1, P2, R1, R2 are the friction surfaces. Gear B is on the clutch shaft and meshes with U1 on the lay shaft. Gears C, D and E are on the splined main shaft and are free to rotate with respect to the main shaft. The gears U1,U2,U3, and U4 are fixed on the lay shaft (Figure 3.20). As long as shaft A is rotating, all the gears on the main shaft and lay shaft rotate continuously.

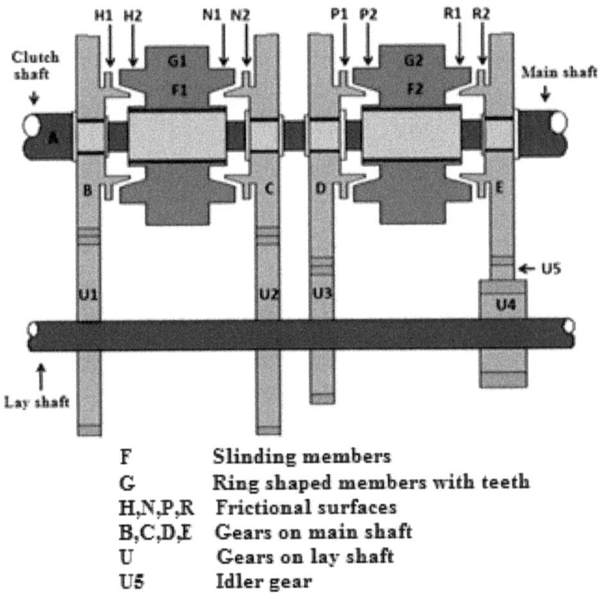

Figure 3.20. A View of the Synchromesh Gearbox.

To engage the first gear, using the shifting fork and the ring shaft, the sliding members G2 and F2 are moved towards left until the cones P1 and P2 rub each other. Friction between the two makes their speed equal. Once their speeds are equal G2 is further pushed towards left to engages with the gear. A motion is carried from clutch gear B to the lay shaft gear U1. Then it goes to U3 on the lay shaft, which further moves to the main shaft gear D. From there the motion is transferred to F2, which is the sliding member and then to the main shaft for the final drive. For second gear the ring shaft and the sliding members G1 and F1 are moved towards the right till the cones N1 and N2 rub each other. Again the friction makes their speed equal. G1 is further pushed towards the right so that it meshes with the gear. The motion is transferred from clutch gear B to the lay shaft gear U1. From U1 the motion is transferred to U2. From U2 it is shifted to the main shaft gear C. Then the motion is transferred to the sliding member F1 from where it goes to the main shaft for the final drive. For top or direct gear, the motion is shifted directly from clutch gear B to the sliding member F1 and then to the main shaft. It is done by moving G1 and F1 to the left. For reverse gear, the motion is transferred from clutch gear B to the lay shaft gear U1. From there it is transferred to lay shaft gear U4 and then to the intermediate gear U5. From there to the main shaft gear E and then to the sliding member F2 and then to the main shaft for the final drive. This is done by moving G2 towards the right. Intermediate gear helps to achieve the reverse motion.

Merits

- ☆ Smooth and noise free shifting of gears
- ☆ No loss of torque transmission from the engine to the driving wheels during gear shifts
- ☆ Double clutching is not required
- ☆ Fewer vibrations
- ☆ Quick shifting of gears without any risk of damage to gears

Demerits

- ☆ High manufacturing cost
- ☆ More number of moving parts
- ☆ While teeth make contact with the gear, the teeth fail to engage as they are spinning at different speeds resulting in a loud grinding sound
- ☆ Improper handling may easily damage the gears
- ☆ Unable to handle high loads

DIFFERENTIAL

When a tractor travels along a curved path, the outside wheels needs to travel greater distance than the inside wheels. If the wheels had been mounted on dead axles they would have turned independently of each other at different speeds to compensate for the difference in travel. Since the wheels are driven positively by the engine, a device is necessary which will permit them to revolve at different speeds without interfering with the propulsion system. *A differential unit is a special arrangement of gears in the driving axle that makes one of the rear wheels of the tractor to rotate slower or faster than the other.* The driving axle consists of a housing, a differential, two half axles (axle shafts), and final drives, only if required. The differential unit consists of (Figure 3.21):

- ☆ Input pinion gear
- ☆ Crown wheel or bevel gear
- ☆ Differential cage
- ☆ Differential pinion or star, a combination of two differential pinion and gears each
- ☆ Differential axle (sun) gear

The main functions of the differential system are:

- ☆ To reduce the rotations coming from the gearbox before the same are passed on to the rear axles
- ☆ Change the direction of axis of rotation of the power by 90° *i.e.* from being longitudinal to transverse direction

☆ To distribute equal power to both the rear driving axles when the tractor is moving straight

☆ To distribute the power to the driving axles during turning so that more rotations are given to the outer wheel as compared to the inner wheel

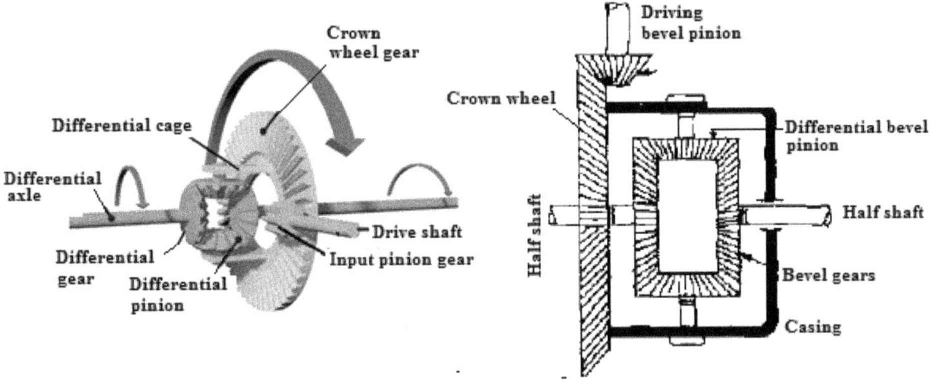

Figure 3.21. A Schematic View of a Differential.

The output drive shaft of the gearbox has a bevel pinion also known as input pinion gear at the end. The bevel pinion is in mesh with a large bevel wheel known as the crown wheel. The crown wheel assembly performs two main functions. One it transmits power through right angle drive to suit the tractor wheels and second reduces the speed of rotation. The differential casing is rigidly attached with the crown wheel and moves like one unit. Two pinions are provided inside the differential casing such that they are carried round by the crown wheel, but they are also free to rotate on their own shaft or stud. There are two or more bevel gears in mesh with differential pinions. One bevel pinion is at the end of each half shaft, which goes to the tractor rear wheel. Thus, instead of the crown wheel being keyed directly to a solid shaft between the tractor wheels, the drive is taken back from the indirect route through differential casing, differential pinion and half shaft of the tractor. When the tractor is moving in a straight line, the differential pinion do not rotate on the stub shaft but are solid with the differential casing. They drive the two bevel gears at the same speed and in the same direction as the casing and the crown wheel. When it is carried round by the casing, it drives the half shaft in the same direction but when it is rotated on its own shaft, it drives them in the opposite direction *i.e.* rotation of the differential pinion adds motion to one shaft and subtracts motion from the other shaft. Some actions and their results and motion of various components of the differential are described as follows:

Action	Results	Crown Wheel	Sun Gears (Differential Gears)	Star Gear (Differential Pinion)
One wheel jacked and rotated, gear engaged	Engine rotates	Rotates	Only jacked side rotates	All rotate
Two wheels jacked and one rotated	The other wheel will rotate in the opposite direction	Does not rotate	Both rotate	All rotate
Tractor moving in straight ahead position	Both wheels rotate with the same speed	Rotates with the cage	Both rotate with the cage	Do not rotate independently but with the cage
Tractor turning left or right	The turning side rotates with slow speed. The other wheel rotates faster	Rotates	Turning side rotates slower, other side rotates faster	All rotate
Differential locked (side dog coupling engaged with crown-wheel)	Both wheels rotate with the same speed	Rotates with the cage	Both gears rotate with the same speed along with the cage	Do not rotate independently but with the cage

Differential Lock

In normal circumstances, a differential operates like an open differential *i.e.* the wheels can rotate at different speeds. However, there is a provision in many tractors to lock the differential, which locks both rear wheels together and drives them both concurrently. It is usually done when traction is needed forcing the wheels to rotate at the same speed. Differential locking is especially helpful in off-roading situations when one wheel is off the ground or on an otherwise very low traction surface. It is necessary on a slippery or snowy surface or when the tractor is stuck in the mud. When locked, the wheel in the air doesn't receive any torque because there is no traction and the wheel on the ground receives all the torque, allowing the vehicle to move. Thus, a differential lock can lock the axles together to provide 100 per cent of the available torque to the wheel with traction. It is advisable that one should lock the differential only when necessary and only for a short period of time. Moreover, the differential should be locked only when traveling in a straight line. Attempting a turn with the differential lock engaged will put strain on the differential lock, and may damage the differential.

Final Drive

Final drive mechanism is a device for final speed reduction of tractor rear wheels. It is located in the power train between the differential and the drive wheels (Figure 3.16). The tractor rear wheels are not directly attached to the half shafts, but the drive is taken through a pair of spur gears. Each half shaft of the transmission system terminates in a small pinion gear which meshes with a large gear called bull gear. The bull gear is mounted on the shaft carrying the tractor rear wheel. The gear ratio of the final drive is given by the ratio of the number of teeth on the bull gear to the number of teeth on the pinion gear. It gives a fixed reduction in speed of the drive shaft and the axle driving the wheels. Note that the system governing the angular movement of front wheels of a tractor is known as steering system, which is discussed in a later section of this chapter.

PTO SHAFT

A tractor is not only used for pulling the implements but also used to power stationary and moving equipment. Some such operations are: threshing of crops, pumping water from tube wells, spraying insecticides and pesticides, operation of rotavators and similar others operations/equipment. A power outlet, known as Power Take Off or in short PTO, is generally provided at the rear of the tractors. It is in fact the extended countershaft of the gearbox, which acts as the PTO outlet. Thus, PTO transfers the power directly from the engine to the implement regardless of whether the tractor is moving or at rest. PTO, a splined rotating shaft receives power from the gearbox by engaging the PTO clutch. Note that the PTO's job is to power an implement, and not to pull it. It means, even if one is using a stationary implement with a PTO, it needs to be hooked up to either the drawbar or the hitch. Power transfer is accomplished by connecting a drive shaft from the machinery to the tractor's PTO stub shaft. The flexible universal joints or U joints connect the tractor to the implement, which are further connected by a square rigid shaft which turns inside another shaft (Figure 3.22). The PTO and drive shaft rotate at 540 rpm or 1,000 rpm when operating at full recommended speed. This standardization of the speed helps in designing the implements and selecting them for a particular kind of tractor. The PTOs, which can run on both 540 rpm and 1000 are known as dual PTOs. A Reverse PTO can turn in the reverse direction as well using a lever or button. It is used in operating implements such as post hole digger or when thresher get stuck.

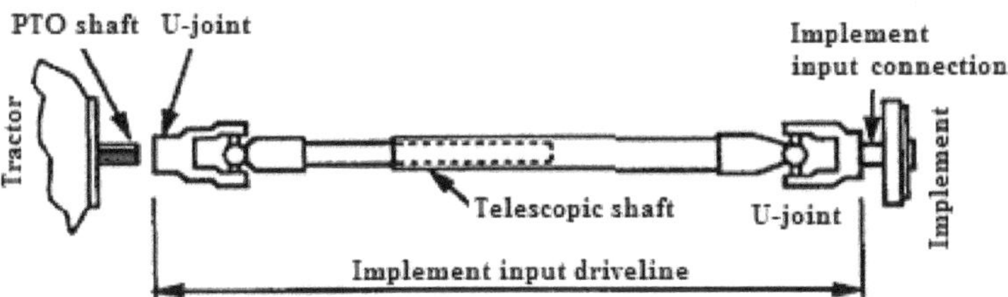

Figure 3.22. Implement Drive Line with U Joints for Connecting Tractor PTO to the Implement.

HYDRAULIC CONTROL SYSTEM

Hydraulics is the generation of forces and motion using hydraulic fluids. In a hydraulic control system, hydraulic fluid is used as a medium for power transmission. The hydraulic systems are commonly used in the automotive vehicles including tractors. The hydraulic systems in tractors are used in the braking and steering systems besides being an integral part to operate the three point hitch of the tractor for raising or lowering the implements. This section only deals with the system that helps to raise, hold or lower the mounted or semi-mounted equipment.

Other systems are discussed under the respective headings. Basic components of a hydraulic system are:

- ☆ Hydraulic tank or reservoir
- ☆ Filter
- ☆ Hydraulic pump
- ☆ Control valve
- ☆ Safety valve
- ☆ Hydraulic cylinder and piston
- ☆ Fluid, hose pipe and fittings
- ☆ Lifting arms or linkages
- ☆ Accumulator (Energy storage device not shown in Figure 3.23)

The flow diagram of the whole system is shown in Figure 3.23. High density incompressible oil used as the transmission media is stored in a storage/fluid tank. Amongst the two kinds of arrangements to store hydraulic oil, one employs a common oil reservoir for the hydraulic system and the transmission system. In the second arrangement, a special tank separate from the transmission chamber is provided for hydraulic oil. The filter is used to filter the oil to remove dust or any other unwanted particles. A hydraulic pump, operated by suitable gears connected with the engine, is used to pump the fluid. Several types of hydraulic pumps, such as gear pump, plunger pump, vane pump, and screw pump are used, although gear pump is most widely used in tractors. Its capacity depends on the hydraulic system design. The pumps deliver constant volume in each revolution of the pump shaft. The pump forces the fluid from the reservoir to the cylinder, the oil pressure being in the range of 150 to 200 kg/cm^2. Therefore, the fluid pressure can increase indefinitely at the dead end of the piston so much so that the system can fail. Relief valves or pressure regulators are used to control the system pressure to a specific set level. As and when this set level is reached, the pressure relief valve allows the excess flow from the system back to the tank (Figure 3.23).

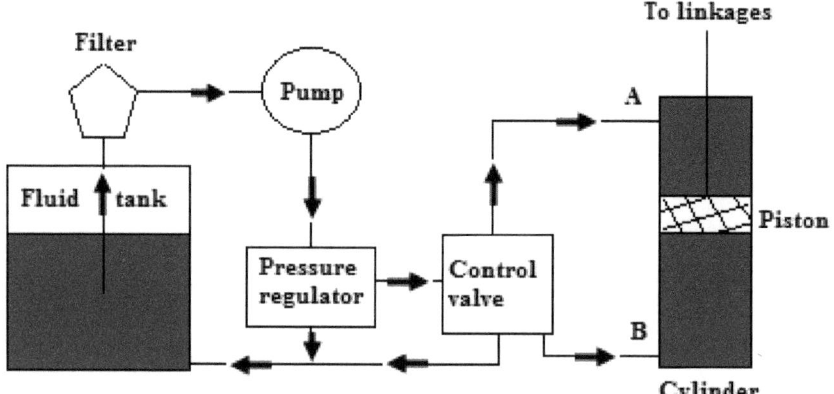

Figure 3.23. Flow Diagram of a Hydraulic Control System of a Tractor.

The leak proof piping is important to avoid environmental hazards and economical aspects besides safety. The control valve controls the movement of hydraulic oil in the desired direction, magnitude and speed of lifting. This determines the speed of movement of the actuators. The movement of the piston is controlled by changing liquid flow from port A and port B (Figure 3.22). The fluid pressure line is connected to the port B to raise the piston, and to port A to lower down the piston. The valve can also stop the fluid flow in any of the port. The control valve is operated by a hand lever, which can be one of the following two types:

1. Manual type
2. Automatic type

The manual type valve is used mostly for small tractors. It has only three positions *i.e.* up, down and neutral. In automatic type, there are more than three positions and any of these positions inside a quadrant can be chosen for operation.

Implement Control

The tractor with a built-in lift system is provided with a hitch which supports an implement through a specific type of mechanical linkage termed as three point linkages or commonly known as three-point hitch. It securely attaches an implement at three points, a central link at the top and two hydraulically powered lift arms, one on either side. Both the bottom links are connected to two lift arms through lift links. The lift arms are directly mounted on a rockshaft connected to the piston rod such that any movement of the piston is transferred to the bottom links (Figure 3.24). The hydraulic arms are used to lower or raise the implement. It can be done with the help of a switch or even automatically so that the tractor keeps pulling the implement effectively as the ground conditions change and give low/more

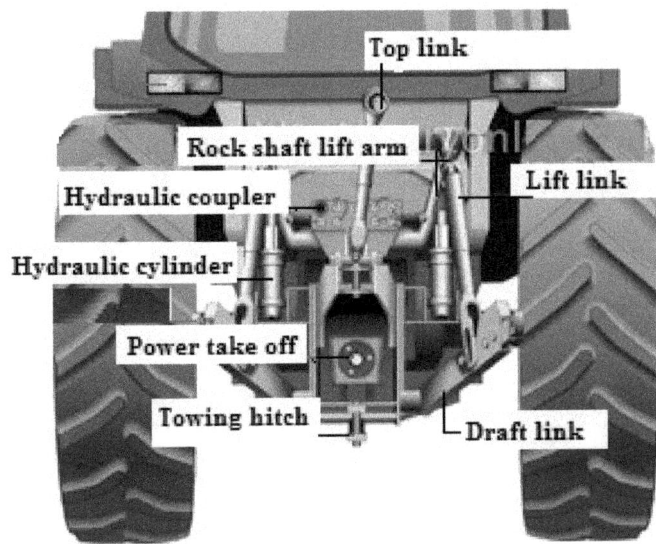

Figure 3.24. A Real View of the Tractor Showing Various Parts of 3-Point Linkage.

resistance. The top link is used to connect the third hitch point of the implement. It is adjustable to maintain the implement level and suction angle. The spring-loaded top link also senses the load for draft control. In some tractors the lower links are spring-loaded for draft sensing. Attachment at three places secures the implements in level position and prevents them from wobbling from side to side. It also transfers the implement's weight (and the load of pulling it) to the back wheels of the tractor, giving it more grip against the ground. The mechanical design of the hitch keeps the whole tractor safe and stable. It also stops it from flipping backward when the implement being pulled suddenly snags in the ground. Almost all tractors use similar hitches, so virtually any implement will work with any make of tractor. Depending upon the soil conditions and type of operation, the mounted implement is controlled either by position control or draft control.

Position Control

In this system, depth of ploughing is maintained to a constant depth by automatic adjustment of draft of the tractor. The control valve can be operated directly by the driver to raise, lower or hold the implement at any chosen height.

Draft Control

Most indigenous tractors besides the position control have a draft control system. With this system, the implement can be set for a particular draft (drawbar pull) rather than depth. The working depth of any implement is controlled continuously without the need for a depth wheel on the implement. If the implement goes too deep its draft will increase. The hydraulic control valve will sense and react to this change in the draft through the top or lower links. As a result, the implement will be raised until the draft is back to the desired level and the implement is at the original depth again using the draft control system.

Mixed Position and Draft Control

In some tractors a suitable blend of the two controls is achieved through an interlink mechanism. Thus, a desired depth of ploughing is maintained within close limits and draft control too is allowed to function for better traction.

The major benefits of a hydraulic control system are:

- ☆ It is simpler and safer than mechanical system
- ☆ Very compact and powerful
- ☆ Easy prevention of overload
- ☆ Continuous speed change
- ☆ It is possible to remote control the system
- ☆ Power can be engaged or disengaged using simple valve
- ☆ System is flexible enough

Some disadvantages are:

- ☆ Difficult pipe connection
- ☆ Possibility of oil leakages

☆ High power consumption

☆ Affected by high temperature

Maintenance and Repair of Hydraulic System

Cleaning and periodic maintenance is quite essential for the efficient operation of a hydraulic system. Some major activities are:

☆ Periodic checking of fluid and refilling with clean oil, if necessary

☆ Regular cleaning of the filter

☆ Keeping the tubes and rubber hose tight

☆ Protecting the system from dust or any other contaminants

Although small, yet timely actions make the system to work efficiently and provide trouble free service.

IGNITION SYSTEMS

In a SI engine, the charge of air and petrol mixture is ignited by means of the spark produced by the ignition system with the help of a spark plug. Conventional ignition systems are basically of 2 types.

1. Battery or coil ignition system
2. Magneto ignition system

Both these systems work on the mutual electromagnetic induction principle. Earlier battery ignition system was generally used in 4-wheelers, but now it is commonly used in 2-wheelers as well. Magneto ignition system is mainly used in 2-wheelers hving kick-start engines.

Components of Battery Ignition System

A diagram of battery ignition system for a 4-cylinder petrol engine is shown in Figure 3.25 (left). It mainly consists of a battery, ammeter, ignition switch, auto-transformer (step up transformer), contact breaker, capacitor, distributor rotor, distributor contact points, spark plugs, *etc.* A 4-cylinder petrol engine has 4-spark plugs and contact breaker cam has 4-corners. The ignition system is divided into 2-circuits.

Storage Battery

Storage battery is a device for converting chemical energy into electrical energy. Although several types of batteries are available in the market, but lead-acid battery is most commonly used in IC engines of the tractors and automobiles. It consists of: (i) Plates (ii) Separators (iii) Electrolyte (iv) Container and (v) Terminal wire. A 6 V or 12 V battery is used to supply the necessary current to the primary winding.

Distributor

It is a rotating device used to open and close the electric circuit between the battery and the ignition coil's primary circuit. A magnetic field is produced with

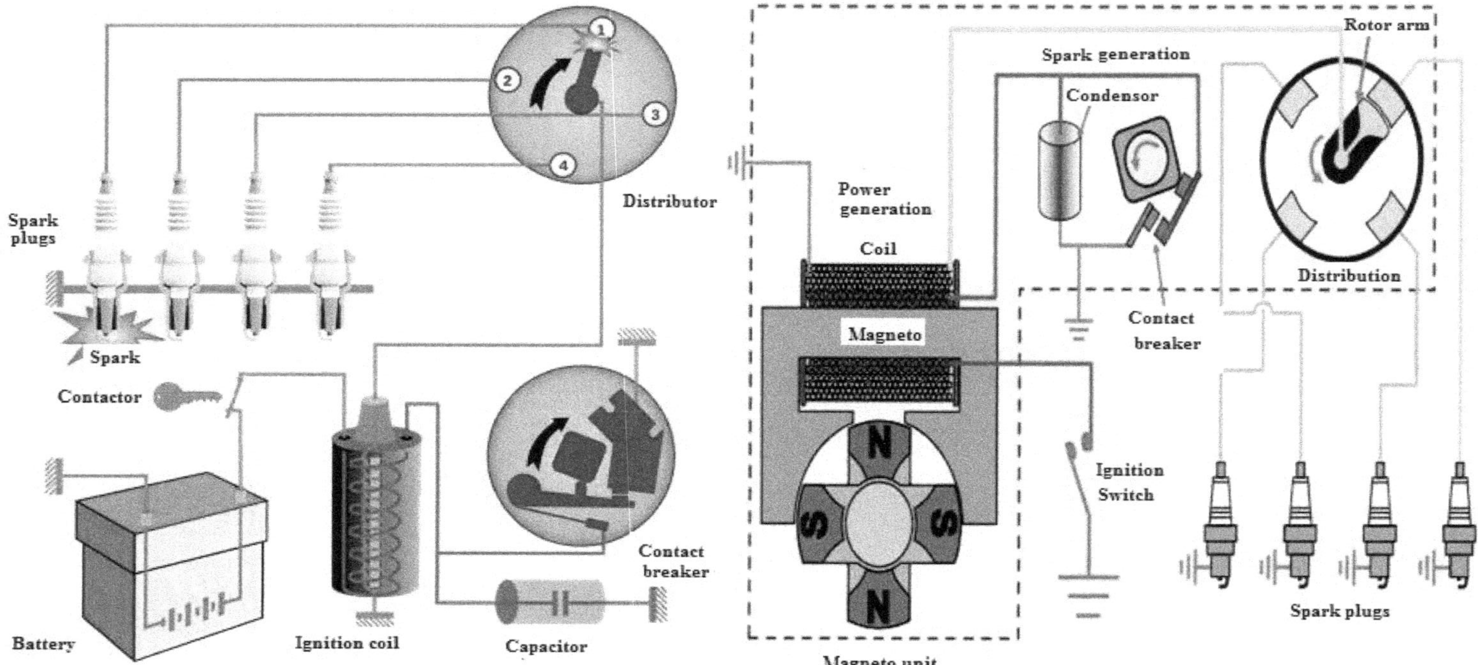

Figure 3.25. Schematic Diagram of Battery Ignition System (Left) and Magneto Ignition System (Right).

this surging of current in the primary coil and then the distributor opens the circuit to collapse the magnetic field. It generates a high voltage current in the secondary winding of the coil. The distributor rotor then guides this high voltage to the spark plug.

Ignition Coil

It is used as a voltage transfer device, which transfers the low voltage (6 -12 volts) current drawn from the battery to high voltage (around 20,000 volts) current necessary to jump the gap of the spark plug. The ignition coil is made up of laminated core having several hundred turns of heavy wire as primary winding. The primary circuit consists of 6 or 12 V battery, ammeter, ignition switch, primary winding. It has 200-300 turns of 20 SWG (Sharps Wire Gauge) gauge wire, contact breaker and capacitor. The secondary circuit consists of secondary winding. Secondary winding consists of about 21000 turns of 40 (SWG) gauge wire. Its bottom end is connected to bottom end of primary and top end of secondary winding is connected to the center of distributor rotor. Primary and secondary windings are insulated from each other and are fully sealed.

Condenser

To have sudden collapse of the magnetic field (which is mandatory to generate high voltage current in the secondary coil), a condenser is provided which is further connected with the breaker point. This condenser also helps to avoid spark being produced at the breaker points which might damage these points. As these contact points are used continuously, the gaps (0.3 to 0.5 mm) between these contact breaker points should be checked on a regular basis.

Generator

A generator is provided in the electric system of an engine to charge the battery. The generator is provided with a cutout that disconnects the generator from the battery when the engine is running at very low speed. The generator is driven by a belt. The belt should be maintained at proper tension for the generator to function properly. For the same reason, bearings of the generator should be well lubricated.

Spark Plug

This device is used to generate high voltage spark, which is transferred from the tip of the spark plug to the engine combustion chamber. Available in different sizes as per the requirement of the engine, it consists of an outer shell having threads and an electrode. The other parts of the spark plug are: terminal, insulator, ribs, insulator tip, seals, metal case/shell, central electrode and side (ground, earth) electrode (Figure 3.26). The central electrode and ground electrode (metal tongue) are its major functional parts. The central electrode is covered by means of porcelain insulating material. The spark plug is fitted in the cylinder head through the metal screw. It should be properly tightened while fitting in the head. When the high tension voltage of the order of 20000 to 30000 volts is applied across the spark electrodes (gap), current jumps from one electrode to another producing a spark. As the point of the spark plug is always exposed to the intense heat in the cylinder/

combustion chamber, it is prone to damage. Besides, carbon may get deposited around the plug. Thus, the carbon deposits should be regularly cleaned and the gap between the electrodes must be adjusted between 0.5 to 0.85 mm.

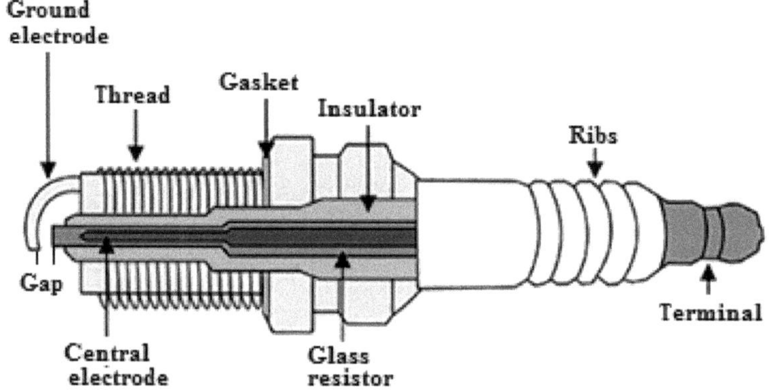

Figure 3.26. Diagram showing Major Parts of a Spark Plug.

Working of the System

When the ignition switch is closed and the engine is cranked, a low voltage current begins to flow through the primary winding as soon as the contact breaker closes. When the contact breaker opens the contact, the magnetic field begins to collapse. Because of this collapsing magnetic field, current is induced in the secondary winding. Because of more turns (21000 turns of secondary) voltage goes up to 20000-30000 volts. This high voltage current is brought to the center of the distributor rotor. The distributor rotor rotates and supplies this high voltage current to proper stark plug depending upon the engine firing order. The contact beaker cam opens and closes the circuit 4-times (for 4 cylinders) in one revolution. As soon as the high voltage current jumps the spark plug gap, it produces a spark that ignites the charge.

Notes

☆ The function of the capacitor is to reduce arcing at the contact breaker (CB) points. Also, when the CB opens, the magnetic field in the primary winding begins to collapse. When the magnetic field is collapsing, the capacitor gets fully charged. Later it starts discharging and helps in building-up of voltage in secondary winding

☆ Contact breaker cam and distributor rotor are mounted on the same shaft

☆ In 2-stroke cycle engines these are motored at the same engine speed and in 4-stroke cycle engines they are motored at half the engine speed

Magneto Ignition System

In this system, a magneto produces and supplies the required current to the primary winding. The magneto can either have rotating magneto with fixed coil or rotating coil with a fixed magneto to produce the current (Figure 3.25 right). Main components of a magneto ignition system are: a) Frame b) Permanent magnet c) Armature d) Soft iron field e) rotor f) Primary and secondary windings g) Breaker points and h) Condenser. The armature is an iron core having primary and secondary windings. As the armature rotates (it is driven by the engine), primary windings cut the lines of force of magnetic field and an induced current begins flowing in the primary circuit. As the primary current reaches its maximum value in each direction, the primary circuit is suddenly opened by a contact breaker and the current collapses. This action induces a very high voltage in the secondary winding, which causes a momentary spark to jump at the spark plug gap. A distributor carries the current to the spark plug through high tension wires. The condenser is used to eliminate the arching at the breaker points and intensifying the current in the secondary circuit. Other arrangements are the same as in a battery ignition system (Figure 3.25 left). The salient differences between the two systems are listed in Table 3.2.

Table 3.2. A Comparison of Battery and Magneto Ignition Systems

Battery System	Magneto System
A battery is required	No battery is needed
Battery maintenance problems are encountered	No battery maintenance problems are encountered
Battery supplies current to the primary circuit	Magneto produces the required current for primary circuit
A good spark is available even at low speeds	The quality of spark is poor at the time of starting due to slow speed
Occupies more space	Quite compact
Battery needs to be recharged as and when it is discharged	No such arrangement is required
Mostly employed in four wheelers, which require cranking of the engine	Used on motorcycles, scooters, *etc.*

Drawbacks of Conventional Ignition Systems

☆ Because of arcing problem, pitting of contact breaker point occurs requiring regular maintenance

☆ After few thousands of kilometers of running, the distortion in timing may result in starting troubles

☆ The system performs poorly at high engine speeds because of inertia effects of the moving parts

☆ Sometimes spark produced may not be proper particularly when the spark plugs are fouled

Electronic Ignition System

Electronic ignition systems are categorized as Capacitance Discharge Ignition System, Transistorized System, Piezo-electric Ignition System and The Texaco Ignition System. These systems help to overcome the problems enumerated in the conventional ignition systems. We end the section by listing the advantages of electronic ignition systems as more discussion is beyond the scope of this book.

Advantages of Electronic Ignition System

☆ No moving parts and therefore almost no maintenance problems

☆ No arcing as there are no contact breaker points

☆ Spark plug life increases by about 50 per cent

☆ Better combustion in combustion chamber, about 90-95 per cent of air fuel mixture is burnt compared with 70-75 per cent with conventional ignition system

☆ Higher power output

☆ Higher fuel efficiency

Compression Ignition System

Diesel engines are compression ignition engines in which no external component is required to ignite the fuel. Nonetheless an electrical system is required even in these engines to initiate the ignition in the first power stroke. It comprises a battery and starting motor. An electric current of 500 amperes at 12 to 24 volts from the battery is provided to the starting motor which further actuates the cranking of the piston and generates the first power stroke. If the engine does not start in the first attempt, subsequent attempts are made to make the connection between battery and starting motor. Continuous and frequent attempts should be avoided to prevent damage to the battery and the motor. Once the engine starts, the battery gets disconnected from the motor and power strokes come in continuation as per the firing orders in multiple cylinders. However, in a single cylinder engine, the flywheel restores sufficient energy to complete the engine cycle. Sometimes glow plugs are also provided in the diesel engines to heat the air entering into the cylinder through inlet manifolds. These are used during winters or in the excessively cold areas. The glow plugs are also connected with the battery.

Starting Motor

Motors of 12 or 24 volts are used in tractors (diesel engines) having a drive to actuate automatic engagement of the pinion with the engine flywheel ring. The engagement of the pinion with the ring gear initiates the engine cranking. Once the engine is started, the pinion is disengaged from the ring gear. These two requirements are essential for the efficient cranking of the tractor. The motor should be capable of generating enough power to crank the engine by moving the piston against heavy loads. There should be a positive meshing of pinion and ring gear so as to crank the engine. Otherwise, multiple attempts made continuously or in a short interval of time may lead to discharge of the battery.

Firing Order

The order or sequence in which the firing in different cylinders of a multi-cylinder engine takes place is called the firing order. It differs from engine-to-engine. In SI engines, the distributor connects the spark plugs of different cylinders according to engine firing order. Probable firing orders for 3 cylinder engine is 1-3-2 and 4 cylinder engine (inline) is 1-3-4-2 or 1-2-4-3.

Advantages

☆ A proper firing order reduces engine vibrations

☆ Maintains engine balancing

☆ Secures an even flow of power

STEERING MECHANISM

The system governing the angular movement of front wheels of a tractor is called the steering system. The function of a steering system is to convert the rotary movement of the steering wheel in driver's hand into the angular turn of the front wheels on road. Another function of the steering system is to provide directional stability. The motion of the vehicle being steered needs to become straight ahead when the force on the steering wheel is removed. Further, the steering system should provide mechanical advantage over front wheel steering knuckles offering driver easy turning of front wheels in any desired direction with minimum efforts. Besides, the design of the steering system should be such that it should cause minimum wear of the wheel tyres. Steering system is categorized as:

1. Mechanical steering
2. Hydraulic steering

Mechanical Steering

When the operator turns the steering wheel, motion is transmitted down through the steering tube to the steering gear. While the steering tube revolves inside the steering column, the steering gears inside the steering gearbox change the direction of motion and increases the turning force applied by driver at the steering wheel in accordance with the gear ratio. The gear rotates the drop arm (pitman arm) which transfers the motion to the steering knuckles through the steering gear connecting rod, tie-rod and knuckle arms. The linkage as a whole is called the relay steering linkage and includes various components such as steering wheel, steering shaft, steering gear, drag link, steering arm, tie rod and king pin (Figure 3.27). Steering arms are keyed to the respective kingpins which are integral part of the stub axle on which wheels are mounted. The movement of the steering arm affects the movement of the front wheel. It is also known as conventional steering system.

The other kind of the system is rack and pinion steering system. In this case, a pinion is attached at the end of the steering shaft. When the steering wheel is turned, the pinion gear spins, moving the rack- left or right, depending on which way the steering is turned. The rack forms the part of the tie rod with steering spindle at its ends which pushes or pulls the steering links to steer the wheels.

Hydraulic/Electronic Power Steering (EPS)

As vehicles become heavier and tyre width and diameter increase, the efforts required to turn the wheels about their steering axis increase. To alleviate this

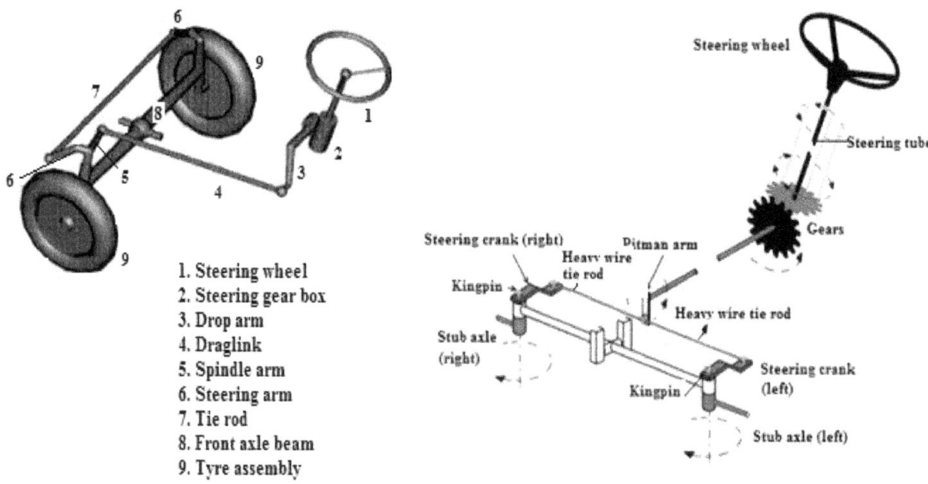

Figure 3.27. Conventional Gear Type Steering System.

problem, automakers have developed power steering systems: or more correctly power-assisted steering for off-road vehicles likes of automobiles, trucks and tractors. Power assisted steering helps the driver of a vehicle to steer by directing some of its power to assist in swiveling the steered road wheels about their steering axes. There are two types of power steering systems:

1. Hydraulic power steering (HPS)
2. Electric or electronic power steering (EPS)

The hydraulic power steering system consists of pump, oil reservoir, rack and pinion gear (Figure 3.28 left). The power steering pump is a vane type pump that delivers hydraulic pressure to operate the power steering system. The pressure relief valve in the pump controls the discharging pressure. The rotary valve in rack and pinion gear directs the oil from the power steering pump to one side of rack piston. The integrated rack piston converts the hydraulic pressure to linear movement. The operating force of the rack moves the wheels through tie rod, tie rod end and steering knuckle.

Many manufacturers are looking to cut back or eliminate the use of hydraulics, so it is becoming much harder to find spare capacity on a hydraulic pump for the steering system. If spare capacity is not available then it becomes necessary to add a hydraulic system dedicated to steering which substantially raises the cost of this approach. Electronic steer-by-wire systems, on the other hand, are completely self-contained and do not require external pumps or hoses. This means that they are usually considerably less expensive than hydraulic steering. Automotive electronic power steering system is a very popular steering system. It directly relies on the motor to provide auxiliary torque (Figure 3.28 right). Currently, most of the vehicles are installed with the electronic power steering system. It is capable of reducing fuel consumption by about 5 per cent.

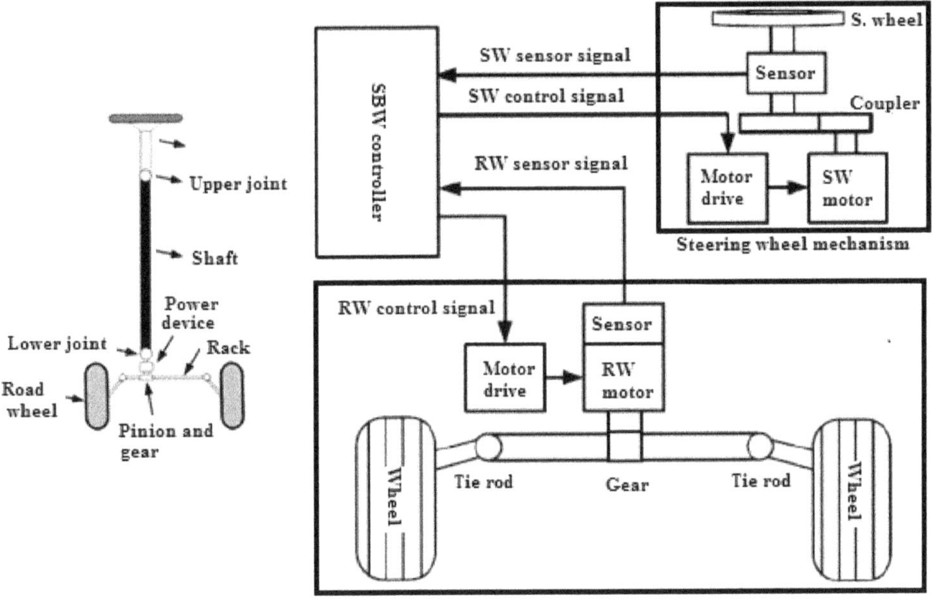

Figure 3.28. Rack and Pinion type Power Steering System (Left) and Steer by Wire System (Right).

GOVERNORS

All IC engines like any other engine are always designed to run at a particular speed. But in actual practice, load on the engine keeps on fluctuating from time to time. Any change in load changes the speed of the engine. In order to have high efficiency of the IC engine at different load conditions, as far as possible its speed must be kept constant. It is why; a governor is essentially required on a tractor engine, since the load on the tractor engine is subjected to rapid variations in the field. It is almost impossible for the operator to control the rapid changes in the engine speed without any automatic device. A governor automatically regulates the engine speed on varying load conditions and thus the operator is relieved of this duty to constantly regulate the throttle lever to suit different load conditions. Thus, *a governor is a mechanical device designed to control the speed of the engine within specified limits*. The objectives of speed control are:

☆ Maintain a nearly constant speed of engine under different load conditions. Idle and maximum speeds must be controlled to prevent the engine from stalling during low-speed idle and to keep the speed from exceeding the maximum limit fixed by the manufacturer

☆ Protect the engine and attached equipment against high speeds when the load is removed or reduced

Methods of Governing

Though there are several methods of governing IC engines, following four are the most important ones.

Hit-and-Miss Governing

The hit-and-miss system controls the frequency of cycles of fuel supply in the fuel system. It is the most widely used method in smaller capacity or gas engines. The method is quite suitable for engines frequently subjected to reduced loads making the engine to run at high speeds. In this system, as soon as an engine starts running at higher speed, some explosions are omitted or missed. It is accomplished with the help of a centrifugal governor which closes the inlet valve of fuel supply resulting in missing of some explosions until the engine speed reaches its normal value. The main disadvantage of this method is uneven turning moments due to missed explosions. As a result these engines require quite heavy flywheels.

Qualitative Governing

In this system of governing, fuel delivery pipe has a control valve that controls the quantity of fuel to be mixed in the charge. The centrifugal governor connected with the accelerator through a rack and pinion arrangement regulates the movement of the valve. While the amount of air in each cycle remains the same, the quality of charge is altered by altering the quantity of fuel. At higher speeds, the quantity of fuel is reduced while at lower speeds, the quantity of fuel is increased until the engine speed reaches its normal value.

Quantitative Governing

In this system quality of charge (air-fuel ratio of the mixture) remains constant. The quantity of mixture supplied to the engine cylinder is varied by means of a throttle valve regulated by the centrifugal governor through rack and pinion arrangement. Whenever the engine starts running at higher speed, the quantity of charge is reduced until the engine speed reaches its normal value. On the other hand, if the engine starts running at lower speed, the quantity of charge is increased. This method is commonly applied to govern large higher capacity engines.

Combination System of Governing

In this system of governing, a combination of qualitative and quantitative methods of governing is used to control the speed. It means both the quality and the quantity of the charge are varied according to the changing conditions. This system has not proved its worth so far mainly because of its complexity.

Kinds of Governors

The following 4 kinds of governors are commonly used to govern the speed in automobiles. The last one in the list is used only in high speed heavy-duty trucks.

☆ Mechanical or centrifugal governor

☆ Pneumatic governor

☆ Hydraulic governor

☆ Electronic governor

Mechanical Governor

The most commonly used governor in tractors is mechanical/centrifugal governor. It has two spring-loaded centrifugal weights mounted on the governor shaft having a sliding collar. The basic principle of working of the governor is that the governor spring and flyweights are so selected that at any designed engine speed centrifugal force and spring force are in equilibrium. They actuate the throttle and the fuel supply (Figure 3.29). As the engine speed increases, the weights fly apart with the centrifugal force against the spring tension. It actuates the fuel injection pump to reduce the amount of fuel delivered to the engine which in turn decreases the engine speed. Similarly, the fuel supply is increased by the governor when the engine speed tends to decrease.

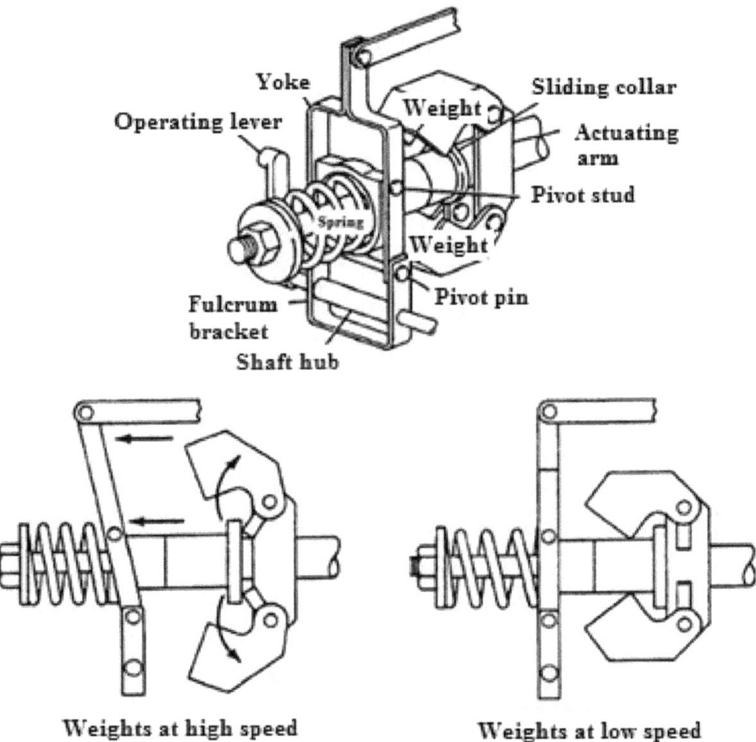

Figure 3.29. A Mechanical Governor and its Operation at High and Low Speeds.

Advantages

☆ It is inexpensive

☆ It is quite good when it is not necessary to maintain exactly the same speed regardless of load

☆ Extremely simple with few parts

Disadvantages

- ☆ They have large dead bands *i.e.* the change in speed required is large before the governor will make a corrective movement of the throttle
- ☆ Their power is relatively small unless they are excessively large
- ☆ They have an unavoidable speed droop (difference in rotational speed of the governed engine between no and full load speeds), and therefore cannot truly provide constant speed when required

Pneumatic Governor

Pneumatic governors are all speed type governors that control the whole speed range of the engine. A pneumatic governor comprises a venturi unit and a diaphragm unit which are connected by a vacuum pump. The venturi unit is connected to the engine inlet manifold and the diaphragm unit is connected with the fuel injection pump (Figure 3.30). The position of the butterfly valve in the venturi unit is controlled by the accelerator pedal that controls the amount of vacuum from the inlet manifold. The adjusting screws provided in the venturi unit are used to adjust the maximum and the idling speed of the engine. The diaphragm within the governor housing is connected to the fuel control linkage that changes its setting with increase or decrease in the vacuum produced by the venturi unit.

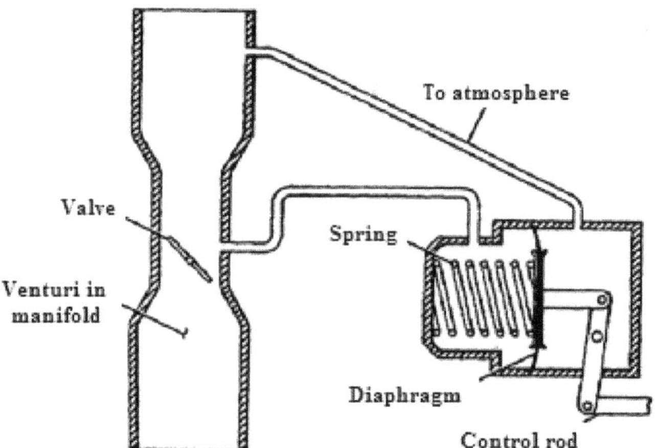

Figure 3.30. Construction of a Pneumatic Governor.

Hydraulic Governor

This kind of governor is highly sensitive, have greater power to move the fuel control mechanism of the engine, and can be timed for identical speed for varying loads. It can virtually be used in any type of engine. A hydraulic governor works on the simple principle of pressure change. When the governor is operating at control speed, the pilot valve closes the port and there is no oil flow. When the speed falls due to an increase in engine load, the flyweights move in and the pilot valve moves down. This opens the port to the power piston to control the oil supply of oil under pressure. This forces the power piston to move upward to increase the fuel supply. On the other hand, when the speed increases due to decrease in engine load, the flyweight moves out and pitot valve moves up. This opens the port from the power piston to the drain into the sump. The spring above the power piston forces the power piston down, thus decreasing the speed. The process is illustrated in Figure 3.31. A simple hydraulic governor is inherently unstable, which prevents its practical use. It keeps moving continually making unnecessary corrective actions, which in general terms is called hunting. These kinds of governors have more moving parts and are expensive. In spite of this, they are used in many applications because of the benefits listed in the opening lines of this section.

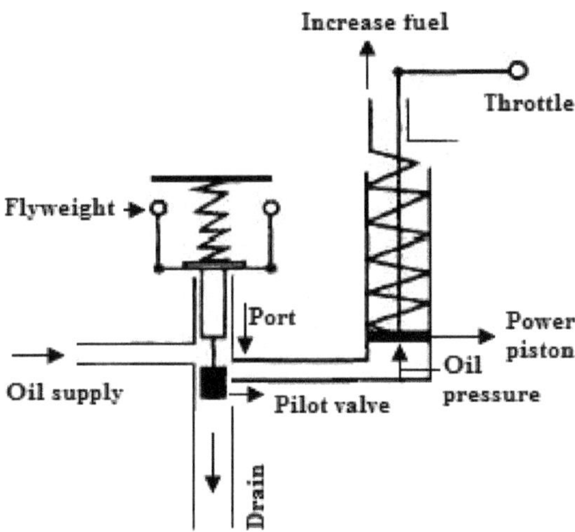

Figure 3.31. Working Principle of a Hydraulic Governor.

Electronic Governors

Although used only on heavy-duty high-speed truck engines, it allows the speed of the diesel engine to be controlled electronically, rather than mechanically. The balancing conditions are similar to a mechanical governor. The major difference is that in the electronic governor, electric currents (amperes) and voltages (pressure) are used together instead of mechanical weight and spring forces. Electronic speed governing systems are set up to provide six basic governing modes:

- ☆ Idle speed control
- ☆ Maximum speed control
- ☆ Power-take-off speed control
- ☆ Vehicle speed cruise control
- ☆ Engine speed cruise control
- ☆ Road speed control

Governor Hunting

A governor is said to be hunting, if the speed of the engine fluctuates continuously above and below the mean speed. In other words, governor hunting is the erratic variation of the speed of the governor when it over compensates for speed changes. It is caused by a too sensitive governor, which changes the fuel supply by a large amount when a small change in the speed of rotation takes place. Hunting may also be due to governor being too stiff or due to some obstruction in free movement of governor components. The other reasons may be due to incorrect adjustment of fuel pump or carburetor, improper adjustment of the idling screw and excessive friction.

Governor Regulation

The function of a governor is to maintain a constant engine speed, but still variations occur in engine speed at no load and maximum load. This variation is expressed by the term governor regulation and is expressed in per cent.

Governor regulation = $100 (S_0 - S_1) / [(S_0 + S_1)/2]$

Here, S_0 is the speed at no load and S_1 the speed at maximum load.

A governor with a high degree of precision or stability is known as a "dead deal" governor. The speed variation from no load to full load is known as "steady state regulation" or speed drop or RPM (revolution per minute) drop.

Speed drop = $[(S_0 - S_1)/S_1] 100$

Example 3.1

Find the percentage regulation of the governor, if speed at no load is 1580 RPM and at full load is 1475 RPM.

Governor regulation = $100 (1580 - 1475)/[(1580 + 1475)/2] = 6.9$ per cent

BRAKE SYSTEM

The brake system is one of the important systems in a tractor. It is used to slow down or completely stop the tractor motion. It is also used to prevent a parked tractor from moving. During field operations, it helps in taking sharp turns using differential brakes on the two rear wheels. In this case one of the two independent pedals is used to assist in turning the vehicle. In normal driving, the two pedals are locked together by means of a lock. A brake works on the principle of friction. When a moving element is brought into contact with a stationary element, the motion of the moving element is affected. The frictional force, which acts in the opposite direction of the motion, converts the kinetic energy into heat energy. The stationary components are mechanically or hydraulically moved so that they come in contact with the rotating component. The stationary components that come in contact with the rotating components are lined with a hard wearing friction material. Brakes must be strong enough to stop the vehicle within a minimum distance in an emergency. Brakes must have good anti-fade characteristics *i.e.* their effectiveness should not decrease with prolonged application. This requirement demands cooling of brakes should be very efficient.

Categorization of Brakes

Brakes have been classified according to Purpose, Location, Construction and Method of actuation

Purpose: From this point of view brakes are classified as service or primary and parking or secondary brakes.

Location: From this point of view brakes are located at wheels or at transmission.

Construction: From this point of view brakes are drum brakes and disc brakes. Drum brakes are further categorized as internal expanding shoe type or external contracting shoe type.

Method of actuation: According to this criterion brake type are mechanical brakes, hydraulic brakes, electric brakes, vacuum brakes, air brakes and by-wire brakes.

In tractors mostly drum and disc brakes actuated by mechanical or hydraulic means are used.

Mechanical Brakes

Mechanical brakes are assemblies consisting of mechanical elements using levers or linkages to transmit force from one point to another. There are several types of mechanical brakes. In olden days band brakes, the simplest brake configuration, had a metal band lined with heat and wear resistant friction material. These have now become obsolete. The other two types are the drum brake and the disc brake.

Drum Brake

A drum brake is a traditional brake in which friction is caused by a set of shoes or pads that press against a rotating drum-shaped part called a brake drum. Drum brakes of internal expanding shoe type are commonly used on wheeled tractors on the rear wheels. The parts include (Figure 3.32 left):

- ☆ A cylindrical drum made of cast iron
- ☆ A brake shoe actuating mechanism – either a cam or a hydraulic wheel cylinder
- ☆ A pair of brake shoes
- ☆ Shoe adjuster
- ☆ Retracting springs
- ☆ Anchor plate/pins *etc.*

In the mechanical braking system, the brake force applied on the brake pedal is carried to the final brake drum or disc rotor by various mechanical linkages like cylindrical rods, fulcrums, springs *etc.* When the brake pedal is released, the retracting springs bring the brake shoes back to their original position. This results in a gap between them and the drum so that drum again spins freely. External contracting shoe type drum brakes are normally used on crawler tractors. The brake band directly surrounds the drum mounted on the drive axle. When the pedal is depressed, the band tightens the drum.

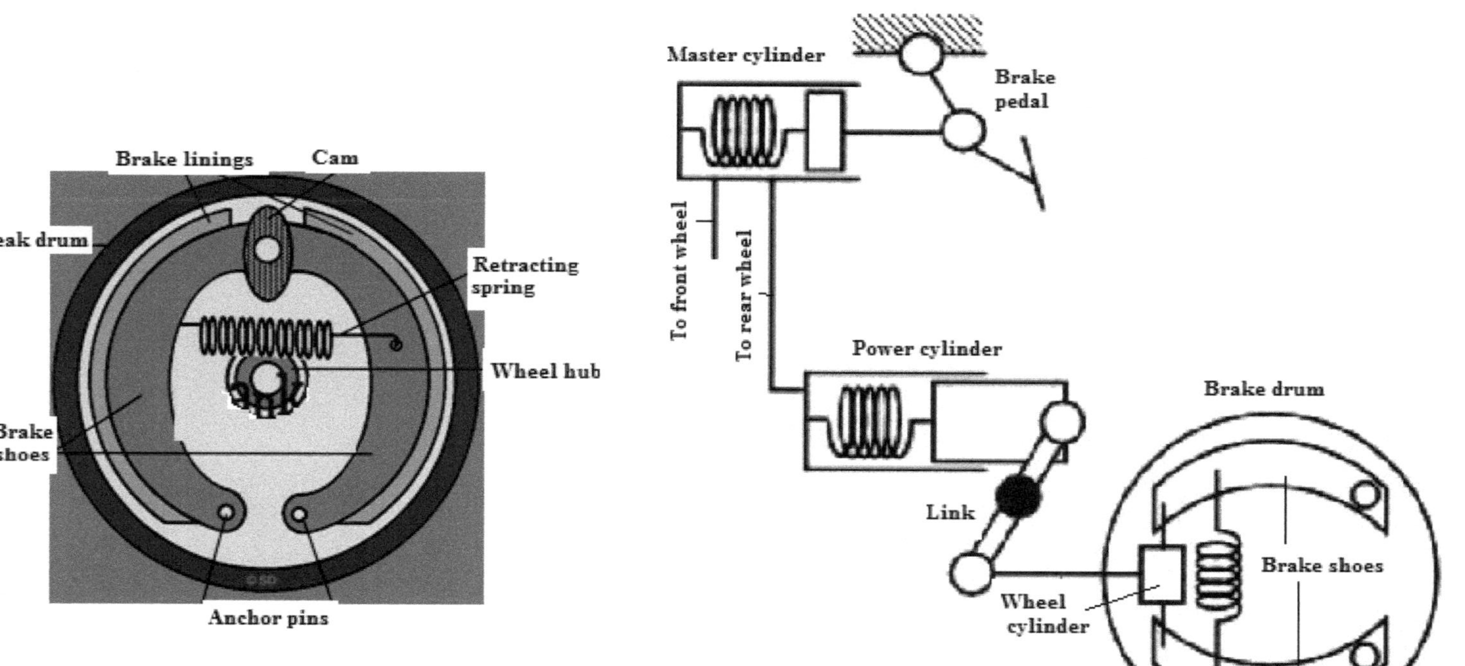

Figure 3.32. Drum Brake: Mechanical (Left) and Hydraulic (Right).

Hydraulic brake system is based on the principle of Pascal's law. The foot pedal of the brake is connected to the brake master cylinder through the linkage. The master cylinder is filled with brake fluid, glycol ethers or diethylene glycol, which is operated by foot pedal via brake control valve. Again the master cylinder is connected to the wheel cylinder through the fluid line inside the brake drum mechanism. A brake drum mechanism consists of brake shoes, spring and rotating drum on which wheels are mounted (3.32 right). When the driver presses the brake pedal, it pushes the piston in a forward direction inside the brake master cylinder and forces the fluid to the wheel cylinder inside the brake drum. Liquid coming from the master cylinder, forces the piston in outward direction. The wheel cylinder piston end is connected to the brake shoes. When the piston goes outwards, the shoes also slide in an outward direction towards the rotation of the drum and stop it with the help of a friction pad mounted inside it. The advantages of the hydraulic system are:

- ☆ The hydraulic brake system is considered as an efficient braking system for modern vehicles
- ☆ The force generated in the hydraulic brake system is more compared to the mechanical brake system
- ☆ There are very less chance of brake failure because of the direct connection between the actuator and the brake disc or drum

Advantages and disadvantages of drum brake system over the disc brake systems are:

- ☆ Simple design
- ☆ Fewer parts
- ☆ Easy and cheaper to manufacture
- ☆ Low maintenance cost
- ☆ Comparatively longer life

Some disadvantages are:

- ☆ Low braking force compared to disc type
- ☆ Brakes 'fade' when the driver applies them for a prolonged time
- ☆ The brake shoe lining made of asbestos is harmful to humans
- ☆ The braking grip is considerably reduced when the system is wet
- ☆ Non-asbestos linings catch moisture causing the brakes to grab suddenly

Disc Brake

Major parts of a disc brake are shown in Figure 3.33 left. These brakes like drum brakes can also be operated mechanically or hydraulically. A disc brake consists of a cast iron disc but in some cases, it is also made of composites such as carbon-carbon or ceramic-matrix composites. The disc is bolted to wheel hub and stationary housing called caliper. Caliper is connected to some stationary part of vehicle like axle. During operation, the brake pads are squeezed against the rotor.

To stop the wheel, friction material is forced against both sides of the disc. Friction caused on the disc wheel slows down or stops the vehicle. On releasing brakes, the rubber sealing rings act as return springs and retract the piston and friction pads away from the disc.

The main parts of the hydraulically actuated disc brake are brake pedal, push rod, master cylinder assembly and brake caliper assembly. Caliper is connected to some stationary part of vehicle like axle. As the brake pedal is pressed, a push rod exerts force on the piston(s) in the master cylinder. This forces the fluid through the hydraulic lines toward the calipers. The brake caliper piston(s) then apply force to the brake pads. This causes them to be pushed against the spinning rotor, and the friction between the pads and the rotor generates a braking torque that slows down the vehicle (Figure 3.33 right). When the pedal is released, the return spring of the master cylinder moves the piston back to its original position.

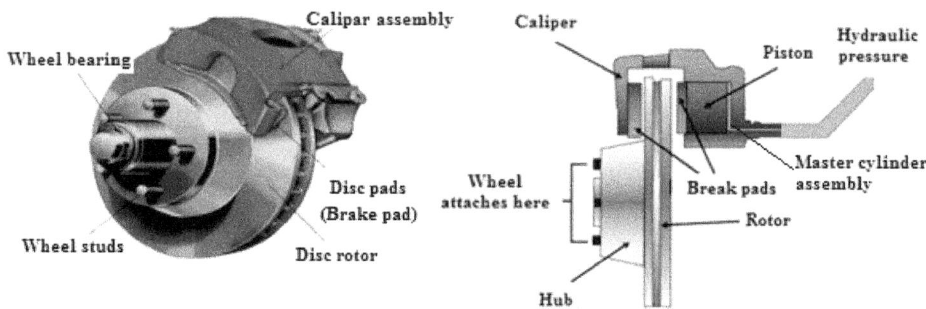

Figure 3.33. Disc Brake: Major Parts (Left) and Its Hydraulic Functioning (Right).

Disc brakes are more efficient, provide better stopping power, dissipate heat easier and work better in wet conditions, all while being less complex. A comparative statement of characteristics of a drum and a disc type brakes is given in Table 3.3.

Table 3.3. A Comparative Statement of Drum and Disc type Brakes

Drum Brake	*Disc Brake*
Low ventilation	Better ventilation than drum type
Non-uniform wear due to friction forces	Uniform wear due to friction forces
Complex design	Easy design
Higher weight than disc type, about 20 per cent higher	Lower weight than drum type
Poor anti-fade characteristics	Better anti-fade characteristics
Replacement of pads is difficult	Pads can be replaced easily
Frictional area of pads in the drum is more	Frictional area of pads over disc is less

Emergency Brake

A vehicle's emergency brake is usually applied to the rear axle. This feature is easier to install on a drum brake than to a caliper or inside the hub of a disc brake rotor.

Braking Efficiency

Braking distance is the distance a vehicle travels from the point when the brakes are fully applied to a place where it comes to a complete halt. The original speed of the vehicle and the coefficient of friction between the tyres and the road surface govern the braking distance. High braking efficiency helps to reduce the braking distance especially during emergencies. On the other hand, it may result in injuries to the driver due to high decelerating forces and dislodging of loads in the trolley. Higher braking efficiency also causes rapid wear and tear of the brakes. There is greater risk of losing the control over the vehicle. It has been observed that braking efficiencies in the range of 50-80 per cent enables the operator to stop the vehicle within reasonable distance.

TRACTOR HITCHING SYSTEMS

Implements are needed to be hitched properly for efficient and safe operation of the tractor. The tractor operated implements are categorized as trailed, semi-mounted and mounted types. Implements can be hitched in two ways:

1. Drawbar hitch
2. Three point linkage

A drawbar is a device by which the pulling power of the tractor is transmitted to the trailing implements. Normally there are two kinds of drawbars fitted at the rear part of the tractor. One drawbar consists of a crossbar with suitable holes, attached to the lower 3-point hitch links. A trailing type of machine can be attached to the drawbar with a thick metal rod called a drawbar pin, held in place by a metal security clip. The other is the sliding drawbar located below the PTO shaft (Figure 3.24). It is used to pull things like trailers or wagons that don't need any kind of lifting. It makes a secure but flexible link between the tractor and whatever is following it. This drawbar can be pulled out or pushed in, fixed in place (stationary) or allowed to pivot, so it swings from side to side. The swinging drawbar produces a slight lateral movement, so that towing devices can turn easier with the motion of the tractor. It allows a more fluid turn instead of staying rigid, which could cause possible tipping issues. A drawbar can be slid out of the way or removed altogether if the farmer wants to use a more elaborate and secure form of attachment. The tractor drawbar is the only safe place to connect a load. Do not hitch higher than the drawbar so that all pulling forces stay below the tractor's center of gravity. For most operations, the drawbar should be placed midpoint between the rear tyres to maximize pulling power. Hillside operations may require a drawbar adjustment to one side to balance the pulling forces.

Three point link hitch comprises a combination of three links, one upper and two lower links, the links articulated to the tractor and the implements at their ends in order to connect the implement to the tractor (Figure 3.24). Advantages of three point linkage are:

☆ Easy control of working implements

☆ Quick setting of implements

✰ Automatic hydraulic control of implements such as position control, draft control *etc.*

✰ Good balancing of attached implements

Besides, we have stationary implements that need to be coupled with the tractor for operation. The kinds of outlets available for this purpose are discussed in Chapter 4 of this book.

QUESTIONS (THEORY)

1. Discuss the construction and working of a dry air filter. List its advantages and disadvantages.

2. Discuss the construction and working of oil wetted mesh type and oil bath air cleaners.

3. Discuss the various components of fuel supply system of spark ignition engines.

4. Discuss the various components of fuel supply system of compression ignition engines.

5. Briefly describe the splash and force feed lubrication systems.

6. Discuss various troubles encountered in a lubrication system along with their causes. List the care and maintenance activities for proper functioning of the lubrication system.

7. What is the need of a cooling system in an engine? What are the adverse effects of overheating or overcooling of engines?

8. Discuss the air cooling system listing its advantages and disadvantages.

9. Discuss the water cooling system for engines. Name and describe the various methods used for this purpose.

10. Name and describe the various components of forced water cooling system of engines.

11. What do you understand by the power transmission system? List its main functions. Draw a neat line diagram showing the power train.

12. Name and describe the various kinds of clutches.

13. Write descriptive notes on sliding mesh, constant mesh and synchromesh gearboxes.

14. What are the main functions of a differential? Discuss its construction and working. Use a neat diagram to illustrate your answer.

15. Discuss the construction and working of a battery or coil ignition system.

16. List the differences between a battery and a magneto ignition system. List the drawbacks of conventional ignition systems.

17. What is a steering system? Briefly describe the mechanical and hydraulic steering systems.

18. What do you understand by governing of an engine? List and describe the various methods of governing.

19. Name and describe various kinds of governors used to govern the speed.

20. Describe the functions of a brake system in a tractor. How will you characterize the brakes?

21. Write a descriptive note on drum brakes. List its advantages and disadvantages over disc brakes.

22. Write a descriptive note on disc brakes. How this system differs from drum braking system?

23. Write short notes on: carburetor, radiator, anti-freeze mixtures, hydraulic system and spark plug.

24. Differentiate between plate and cone clutches, position control and draft control in mounted implements and flash point and fire points of lubricant oils.

Chapter 4

Farm Tractors and Power Tillers

A tractor is traditionally used on farms to mechanize agricultural operations. It plays a vital role in the life of rural communities where leading occupation is farming. It is in fact the best friend of the farmers, as it makes their tasks much easier. A tractor is a self-propelled power unit available in a wide range to suit specific tasks and requirements. Tractors in a horsepower range of 15 to 40 HP are ideal for performing landscaping jobs and tasks such as ploughing, digging, or hauling on average sized farms, fields and pastures. Utility tractors in a horsepower range of 45 to 110 HP are recommended for mechanizing complex farming tasks. The engine power of tractors is also used to operate stationary farm machinery through power-take off (PTO) or belt pulley. Modern tractors besides being used for ploughing, tilling, planting, harvesting and threshing are also widely used for varying activities such as moving or spreading fertilizer, clearing bushes, routine lawn care, and landscape maintenance in urban settings. To meet such requirements an equally wide range of farming implements is available, which forms the subject matter of discussion in forthcoming chapters of this book. Some of the most important benefits of farm mechanization in general and tractors in particular are listed below:

☆ Farm mechanization reduces time to accomplish a farming operation by about 15-20 per cent

☆ Reduces resource use such as seeds approximately 15-20 per cent and fertilizers approximately 15-20 per cent

☆ Cuts down the expenses while increasing profits

☆ Increases cropping intensity (approximately 5-20 per cent)

☆ Provides multiple farming uses and even used as multipurpose transport carrier, even like a family car

☆ Custom hiring helps to earn rental income

☆ Results in several social benefits

☆ Helps in conversion of uncultivable land to agricultural land through advanced tilling techniques

☆ Decrease in workload of women as a direct consequence of the improved efficiency of labour

☆ Improves safety of farm practices

☆ Helps in encouraging the youth to join farming and attract more people to work and live in rural areas

☆ Adds to the social status of the tractor owner

HISTORY OF TRACTOR DEVELOPMENT

The tractor, as we see today, has been a gradual development process passing through several stages of design and development. Historically speaking, tractor development began in 1890, when the word tractor first appeared on record in a patent issued to a traction engine (tractor) invented by George H. Harris of Chicago. By 1920-1924, an all-purpose tractor was already in operation followed by introduction of diesel engine and pneumatic tyres during 1936-1937. Tractors were imported and introduced in Indian agriculture by the British Government in 1914 with the main objective of bringing more land under agriculture. It was only during 1960-1961 that M/s Eicher Good Earth began tractor manufacturing in India. It soon expanded to 5 manufacturers producing just 880 units per year. Later on many firms began manufacturing tractors. The country emerged as one of the global leaders in tractor production in 1997 with a yearly production of more than 255,000 units. The production of 619,000 tractors in 2013 accounted for 29 per cent of the total output worldwide. A total of 64730 tractors were sold in India in August 2020, a whopping increase over 37050 tractors in August 2019 (https:// www. tractorjunction.com/tractor-news/domestic-tractor-sales). The tractor market is amongst the biggest segments in the category of farm equipment in the country with annual sales of 600,000-700,000 units over the last several years (Table 4.1). It is estimated that 6.85 million tractors were in operation in 2017-18 (MoAFW, 2019). The number is projected to increase to 11 million by 2023-24. Accordingly, tractors are likely to contribute about 62.5 per cent of the farm power used in the year 2023-24 (Figure 4.1, Data source: The working Group Report, 2018). Relatively speaking, the contribution of farm power by power tillers is quite less and their number is only around 0.57 million in 2017-18. Since the usage of tractors in construction and other industries is increasing, the demand for tractors in the coming years will multiply several folds. It will substantially increase the manufacturing and sales of tractors in India. A list of BIS standards on various parts of tractors, power tillers and agricultural machines is given in Annexure I. Readers interested in details especially related to design aspects may consult these standards.

CLASSIFICATION OF TRACTORS

As per structural design, tractors have been categorized into 3 categories.

Wheel tractor: Tractors with pneumatic wheels are called wheel tractors. They may have three, (now obsolete) or four wheels. Four wheel tractors have now fully replaced the three wheel tractors.

Table 4.1. Year-wise Sales of Tractors and Power Tillers in India

Year	Tractors	Power Tillers
2004-05	247531	17481
2005-06	296080	22303
2006-07	352835	24791
2007-08	346501	26135
2008-09	342836	35294
2009-10	393836	38794
2010-11	545109	55100
2011-12	535210	60000
2012-13	590672	47000
2013-14	696828	56000
2014-15	551463	46000
2015-16	571249	48882
2016-17	582662	45200
2017-18	797000	52000
Total	**6849812**	**574980**

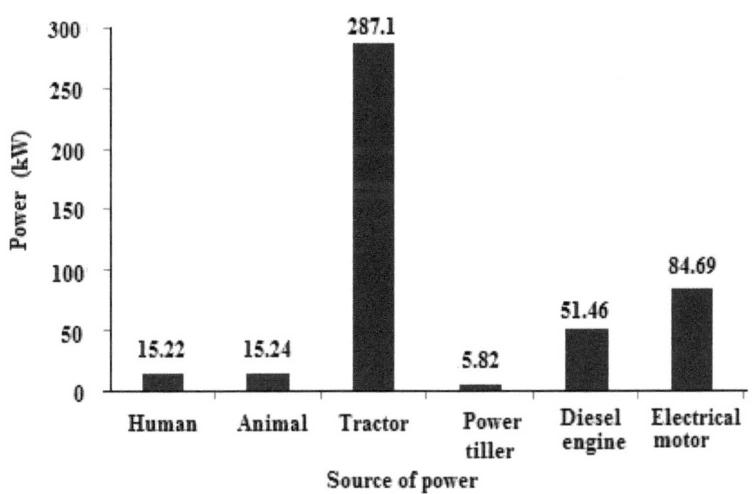

Figure 4.1. Projected Farm Power Scenario on Indian Farms during 2023-24.

Crawler tractor: It is also called track type or chain type tractor. In such a tractor, an endless chain or track is used in place of pneumatic wheels.

Walking tractor (Power tiller): A walking type tractor, also called power tiller is usually fitted with only two wheels. The direction of travel and controls required for field operations are performed by the operator, while walking behind the tractor. It is a perfect machine for small paddy fields, dry land and hilly areas and field operations in orchards and gardens.

On the basis of purpose, wheeled tractors are further classified into 5 groups:

General or all purpose tractors: It is the most common kind of tractor used in major farm operations such as ploughing, harrowing, sowing, harvesting and transporting work. Such tractors have (i) low ground clearance (ii) relatively high engine power (iii) good adhesion and (iv) wide tyres.

Row crop tractor: These tractors used for crop cultivation are provided with replaceable driving wheels of different tread widths. Wheel tracks can be adjusted to suit inter row crop spacing. These have high ground clearance to avoid damage to the crops.

Orchard tractors: These are special types of tractors used to carry out field operations in orchards. They are quite high so that the driver is able to perform the operations on the trees while sitting on his seat. Moreover, no part of the tractor is protruded outside so that the tractor can easily move in between the tree rows.

Special purpose tractor: These are used for well defined special tasks like cotton fields, marshy land, hillsides, garden *etc.* Special purpose tractors have special designs to perform a specific task under specific conditions.

Earth moving tractors: These tractors are used for earth moving work on dams, quarries and other constructional works. These tractors are heavy (in weight) and have strong built. These are available both in track and tyre type.

On the basis of available power, tractors are classified as:

1. Small tractors – 15 to 25 HP
2. Medium tractors – 25 to 45 HP
3. Large tractors – more than 45 HP

Characteristics of a Good Tractor

A good tractor should have the following characteristics.

- ☆ Greater vertical and horizontal clearances
- ☆ Adaptable to the usual row-widths of crops
- ☆ Quick – short turning ability
- ☆ Convenient and easy handling
- ☆ Quick and easy attachment of implements
- ☆ It must have essential accessories such as hydraulic controls, three point linkage and PTO
- ☆ Should be equipped with the latest features
- ☆ Environment friendly
- ☆ High speed, if work entails mainly transportation

SELECTION OF TRACTOR

Following factors aid in the selection of tractors and their size.

Land holding/cropping pattern: For single crops (cropping intensity of 100 per cent), 1 HP may suffice for an area of 2 ha. In other words, a tractor of 20-25 HP is able to serve the farming needs of a 40 ha farm. On the other hand, if adequate irrigation facilities are available and more than one crop is taken then 1 HP is recommended for an area of 1.5 ha. In such cases, a 30-35 HP tractor may be required for a 40 ha farm land.

Soil condition: A tractor with less wheel base, higher ground clearance and low overall weight suits areas with light soils but may not be able to give sufficient depth in black cotton soils.

Climatic conditions: For very hot zone and desert area, air cooled engines are preferred over water-cooled engines. Similarly, air cooled engines are also preferred for high altitude areas because water is liable to freeze at higher altitudes.

Repairing facilities: Before purchasing a given type of tractor, ensure that its dealer is available at a nearby place. Moreover, the dealer should have adequate manpower with technical skills for repair and maintenance of tractor.

Running cost: Tractors with less specific fuel consumption should be preferred over others so that running cost is less.

Initial cost and resale value: The initial cost should not be very high as it will have a great bearing on the fixed cost per hour. Some compromise on this issue may be made keeping the resale value in mind.

Test report: One should preferably consult and take guidance from test reports of tractors released by farm machinery testing stations.

Tractor Components

Every tractor is fitted with an IC engine. It may be SI or CI type but currently almost all the tractors are diesel tractors or CI types. Major components of a tractor are listed in Figure 4.2 left and Figure 4.2 right. Important amongst them are: Chassis, IC engine, clutch, transmission gears, differential unit, final drive, rear wheels, front

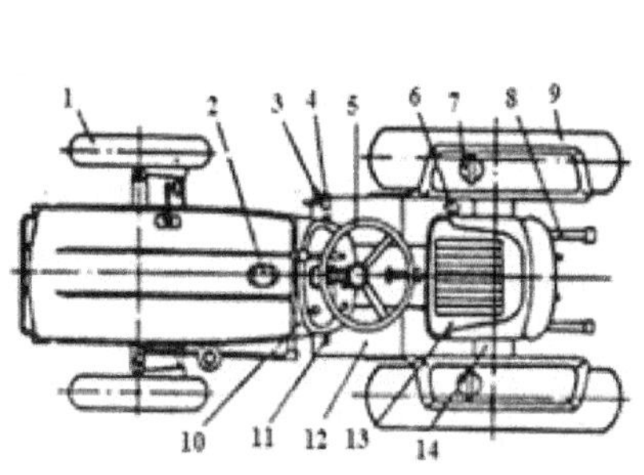

1. Front wheel, 2. Fuel tank cap, 3. Accelelrator pedal, 4. Brake pedle
5. Steering wheel, 6. Hydraulic control lever, 7. Turn signal lamp
8. Lift arm, 9. Rear tyre10. Instrument panel, 11. Clutch pedal,
12. Step, 13. Seat, 14. Rear axle housing

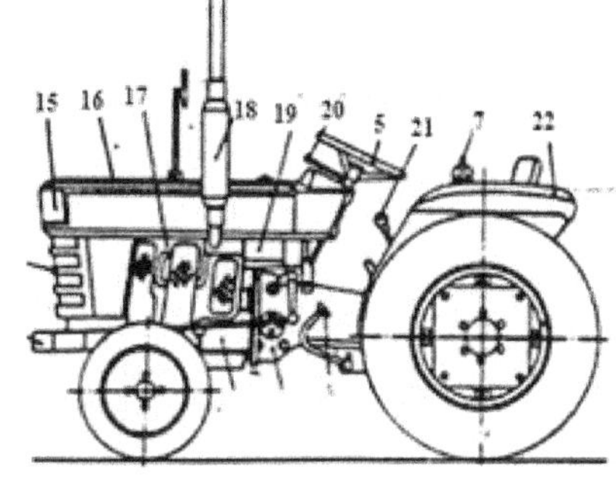

15. Side clearnace, 16. Engine hood, 17. Fan cover, 18. Muffler
19. Fuel tank, 20. Throttle lever, 21. Main speed change lever
12. Fender

Figure 4.2. Major Parts of a 4-wheel General Purpose Tractor.

wheels, steering mechanism, hydraulic control and hitch system, brakes, power take-off unit, tractor pulley and control panel. Only those parts are included in this chapter, which have not been discussed in chapter 3.

Dash Board or Control Board

The dashboard also known as instrument panel or control board of a tractor generally consists of: main switch (starting the engine), throttle lever (controlling the speed of the engine), decompression lever (releases compression pressure when stopping the engine), hour meter (shows engine hours run and RPM), light switch (light control) and horn button, battery charging indicator (shows the battery charging status), oil pressure indicator (lubricating oil pressure) and water temperature gauge (temperature of cooling water). A view of a modern dashboard is shown in Figure 4.3. Note that not all the facilities shown are included in all the dashboards.

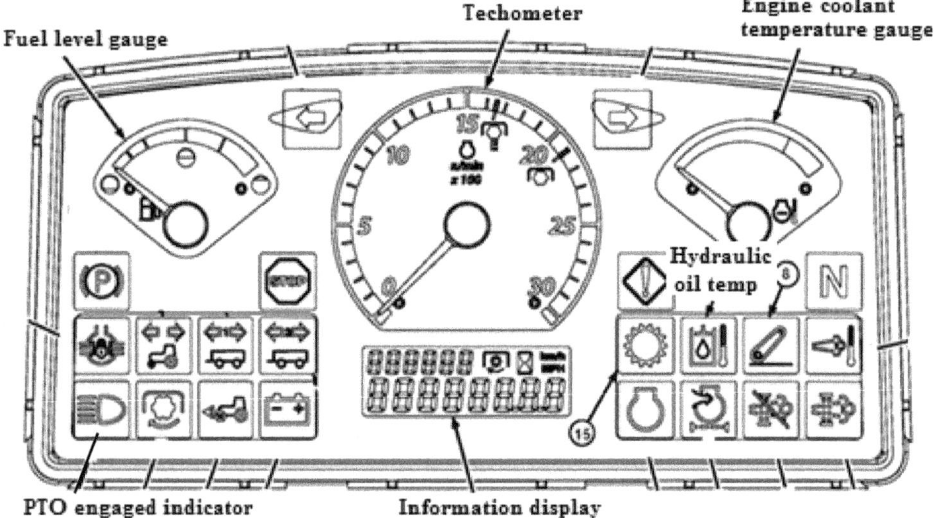

Figure 4.3. A View of Modern Dashboard.

Main switch: Electrical current flows in the electrical circuit when it is in on position.

Throttle lever: It is used to increase or decrease the speed of the engine.

Decompression lever: This lever is used to decompress or release the compression pressure from the combustion chamber of the engine to help start the engine.

Information display: Displays information on speedometer, hour meter, transmission speed indicator (Hi, Lo, or R) (if equipped).

Light switch: It helps to switch on the lights.

Horn button: To use the horn of the tractor.

Battery charging indicator: This indicates the charge and discharge of the battery.

Oil pressure indicator: This gauge indicates the lubricating oil pressure in the system.

Water temperature gauge: This indicates the temperature of water in the cooling system.

PTO engaged indicator: Illuminates when rear PTO is engaged.

Tractor Tyres

The tyres are available in many sizes with the ply ratings of 4, 6 or 8. The ply rating indicates the relative strength of tyres, higher the rating stronger the tyres. The tyre size is designated as A- B. It means that the sectional diameter of tyre is A″ and it can be mounted on a rim of B″ diameter. Thus, tyre size 12- 38 means that the sectional diameter of tyre is 12″ and it can be mounted on a rim of 38″ diameter. Whereas the inflation pressure in the rear wheels of the tractor varies between 0.8 to 1.5 kg/cm^2, in the front wheels it varies from 1.5 to 2.5 kg/cm^2. Useful life of the pneumatic tyres is around 6000 working hours under normal operating conditions for doing drawbar work.

Front Axle

Front axle is a unit on which front wheels are mounted. These wheels are idler wheels (Unless the tractor is 4 wheel drive commonly mentioned as 4 WD or 4 x 4) by which the tractor is steered in various directions. The axle is a rigid tubular or I-section made from steel. It is pivoted at the center. Various adjustments can be made in the front wheel.

IMPORTANT DIMENSIONS AND TERMS OF TRACTORS

Length and width: Length of a tractor is the distance between two vertical planes at right angles to the median plane of the power unit and touching its front and rear extremities (Figure 4.4). Width of the tractor is the distance between two vertical planes parallel to the median plane of the power unit, each plane touching the outermost point of the power unit on its respective side (Figure 4.4). Length and width are measured when the power unit is stationary. Moreover removable attachments such as cage wheels are not included in the width.

Height: Height of a tractor is the distance between a firm horizontal supporting surface and the horizontal plane touching the uppermost part of the tractor (Figure 4.4).

Ground clearance: It is the height of the lowest point of the tractor chassis from a firm horizontal supporting surface, the tractor being ballasted as used for drawbar test *i.e.* the tractor is loaded to its maximum permissible weight (Figure 4.4).

Track: Track of a wheeled tractor is the distance at the ground level between the median planes of the wheels on the same axle of the tractor when stationary and with the wheels in position for travelling in a straight line. The track is defined both for front and rear wheels (Figure 4.4). In case of twin wheels, the track shall be taken as the distance between two planes, each being the median plane of the pairs of wheels (Figure 4.4).

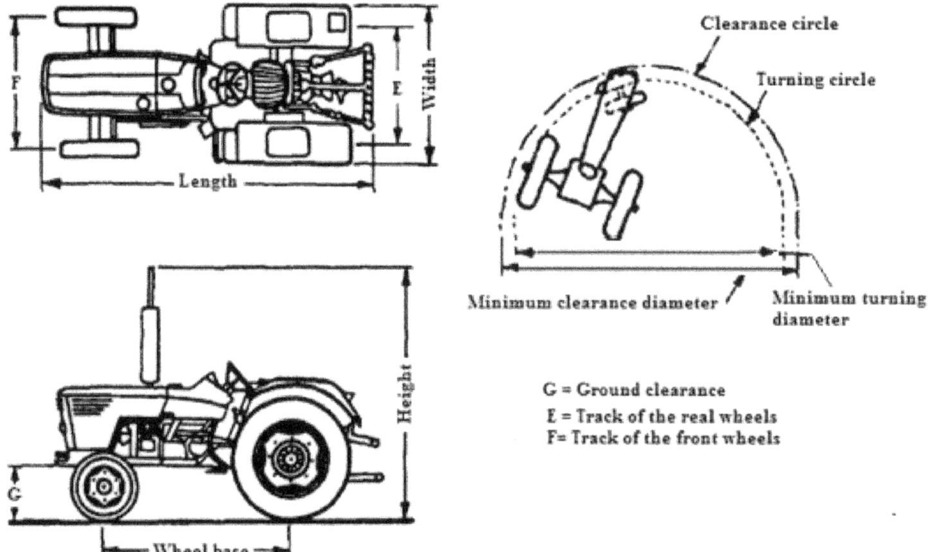

Figure 4.4. Depiction of Important Dimensions of a Tractor.

Wheel base: It is the horizontal distance between the front and the rear wheels in the same plane measured at the centre of their ground contact (Figure 4.4).

Minimum clearance diameter: The minimum clearance diameter of a tractor is the diameter of the smallest circle described by the outermost point of the projections of the tractor while executing its sharpest practical turn (Figure 4.4).

Minimum turning diameter: It is the diameter of the circular path described by the center of tyre of the outermost wheel of the tractor while executing its sharpest practical turn (Figure 4.4). In other words, it is the diameter of the smallest circle, described by the outermost point of the tractor while moving at a speed not exceeding 2 km/hr with the steering wheels in full lock.

Mass of the tractor: Mass of the tractor is given as dry mass and operational mass. Dry mass is the mass of the tractor fitted with all components necessary for its operation, but without water, fuel, oil and operator. On the other hand, the operational mass of tractor is given by its mass in normal working condition with fuel tank and radiator full, lubricants *etc.* filled to the specified levels including the mass of the driver (normally assumed 75 kg). Note that name and mass of any accessory fitted should be clearly stated while reporting any of the mass of the tractor.

Cage wheel: A cage wheel is a wheel or an attachment to a wheel with spaced crossbars for improving the traction of the tractor in a wet field. These are commonly used in paddy fields for puddling operation.

Hitching Systems of Implements

Tractor drawn implements are of three kinds namely a) Trailed type b) Semi-mounted type and c) Mounted type. These implements are of higher working capacity and can be operated at higher speeds.

Trailed type implements: These are pulled and guided from a single hitch point, but their weight is not supported by the tractor.

Semi-mounted type implements: These implements are attached to the tractor along a hinge axis and not at a single hitch point. These are controlled directly by the tractor steering unit, but their weight is only partly supported by the tractor.

Mounted type implements: These implements attached to the tractor are directly controlled by the tractor steering unit. When out of operation in the field, its load is fully carried by the tractor.

POWER OUTLETS

Any outlet provided in the tractor which transmits the engine power to the tractor in order to make it functional is known as power outlet. There are four power outlets of tractor.

Power Take-off Unit (PTO)

A part of tractor transmission system discussed in chapter 3 consists of a shaft, a shield and a cover. The shaft is externally splined to transmit torsion power to another machine (Figure 4.5). A rigid guard called power take-off shield covers the power take-off shaft as a safety measure. Agricultural machines are coupled to this shaft at the rear end of the tractor. As per ASABE standard, PTO speed is 540±10 rpm when operating under load. In order to operate a 1000 RPM drive machine, modern tractors are provided with PTO speed of 1000±10 rpm or both.

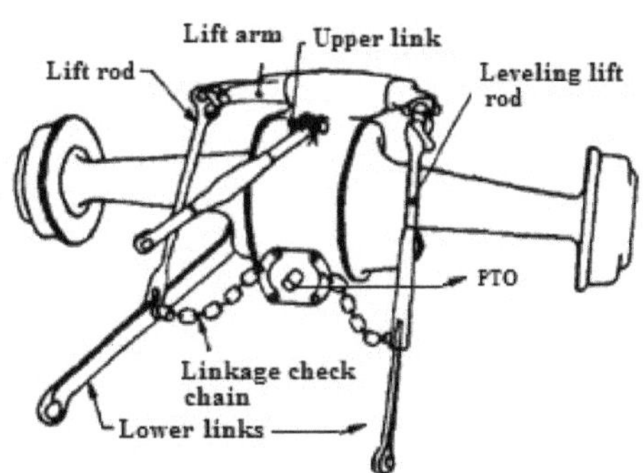

Figure 4.5. A View of Various Outlets in a 4-wheeled Tractor.

A Hook and a Drawbar

The hook is provided at the back of the tractor for hitching trailers, and other stationery machines operated by tractor like thresher, winnower *etc.* A drawbar is a bar having holes along its length and is supported by the two lower links of the 3 point linkage (Figure 4.5 right).

Three Point Linkage

It is a combination of three links, one upper and two lower links. The links are articulated to the tractor. All the mounted type tractor drawn implements are connected at their ends which are supported by the tractor. The hitch points and hence the implement is controlled by the hydraulic system (refer Chapter 3, and Figure 4.5).

Belt Pulley

All tractors are provided with a belt pulley generally made of cast iron. It is used to transmit power from the tractor to stationary machines by means of a belt. It is used to operate thresher, centrifugal pump, silage cutter, and several other machines. The pulley can be located on the left, right or rear side of the tractor. Many times it is provided with PTO shaft to tap the rotary power output for operating stationary machines. The pulley drive can be engaged or disengaged from the engine by means of a clutch.

SAFETY PRECAUTIONS FOR HANDLING AND DRIVING TRACTORS

Tractors require proper handling and operation, periodic servicing, and maintenance and repairs for efficient and economical performance (Wadhwa *et al.*, 2003, Sharma and Mukesh, 2004). Although, most tractor manufacturers have appointed dealers to provide sufficient operational know how, the services rendered by the dealers are awfully inadequate. One of causes of abnormal break downs and wear and tear is improper operation resulting in reduced effective life of the tractor. It also results in injuries to operator and other workers and is a major cause of road accidents particularly tipping of tractors/trailers. Carelessness in checking the oil and water results in seizures and overheating of engine (https://www.escortsgroup. com/agri-machinery/international/templates/agriintl_home/images/pdf/FT-6045.pdf). It has been observed that many tractors have been rendered unserviceable within a short period of 5000 hours or even less due to their improper operation/ handling. Following sections lists some of the general operational and driving guidelines related to tractor at the time of start, at the farm and on the road.

Precautions while Starting the Tractor

- ☆ Run and maintain the tractor according to the operator's manual provided by the tractor manufacturer
- ☆ Before alighting the tractor, check the oil level in the sump and water in the radiator
- ☆ Immediately after occupying the driver's seat, check the working of all controls of the tractor

☆ Release the parking brakes before starting the tractor

☆ Bring the gear-shift lever to neutral position whenever tractor is stopped even if it is only for a short period

☆ Always park the tractor with gear shift lever in the neutral position and with parking brake on

☆ Operate the tractor smoothly avoiding jerky starts, turns and stops

☆ Drive slowly in difficult conditions

☆ Look at the rear while reversing the tractor

☆ Attend immediately to oil and fuel leakages

☆ Listen to the noise or sound in the engine, power transmission, etc. If any abnormal noise is noticed stop the tractor and investigate the causes

☆ Always keep a watch ahead of the tractor

☆ Refuel the tractor only when the engine is cool

☆ Don't spill fuel and never smoke while refueling

☆ Hitch implements only to drawbar or specified hitch points of the tractor

☆ Remove the air intake assembly before raising the bonnet

☆ Never drive after consuming alcoholic drink

☆ Never run the tractor engine in a closed shed or garage

☆ Don't permit unauthorized persons to ride the tractor

☆ Never operate the hand accelerator of tractor from the ground

☆ Do not allow the tractor wheels to run over sharp objects

☆ Do not keep foot (ride) on the clutch and brake pedals while the tractor is running

☆ Do not sit or stand on the implement while the tractor is in motion

☆ Do not attempt the dual selector lever when the tractor is in motion

☆ Avoid overloading of the tractor during operations

☆ Do not get off or on the tractor when the tractor is in motion

☆ Do not remove the radiator cap when the engine is hot

☆ Never leave the key in the starting switch

Precaution during Operation at the Farm

☆ Set the wheels as wide as required for the job. Use wider wheel track on slopes for stability

☆ For proper traction, add weights on rear or front, as the case may be

☆ Keep PTO and belt pulley shields in proper place

☆ Do not hook load at a point above the drawbar

☆ Reverse the tractor in low gear

☆ Drive tractor in low gears while overcoming obstacles like small dykes or ditches

☆ Draft control should not be used for raising or lowering the implements at the end of trip/row

☆ Do not ride on the drawbar of tractor during operation

Precautions during Operation on the Road

☆ Obey the traffic rules while driving on road

☆ Drive slowly while making turns

☆ Use lower gear during up and down-hill driving

☆ Be careful at road crossings

☆ Stop the tractor on the left side of the road

☆ Keep brake pedals interlocked when driving on the road

☆ Give way to automobile vehicles

☆ Make extensive provision of lights at the rear and on the sides while driving at night with trolley

☆ Never overload the trolley over and above the recommended load specified by the manufacturer

☆ Never run the tractor down- hill in neutral gear

☆ Never depress clutch pedal while driving down-hill

☆ Do not tend to take sharp turn using independent brakes when travelling at high speeds

☆ Do not drive without rear-view mirrors

CRAWLER TRACTOR

A crawler tractor also named as track-type tractor, tractor crawler, or track-laying vehicle runs on continuous tracks instead of wheels (Figure 4.6). The principal advantage of crawler tractors over wheeled tractors is that they are in contact with a larger surface area than the wheeled tractor. As such, these tractors exert much lower force per unit area on the ground than conventional wheeled tractors of the same weight. This makes them suitable for use on soft, low friction and uneven ground such as mud, ice and snow. Moreover, these are more efficient than wheeled tractors. For farm operations, these tractors are usually fitted with a 3-point linkage to fit mounted implements. But most earth machines just use trailed implements attached to the drawbar. These tractors are most often employed for land clearing and land leveling operations. These are widely used in fields other than agriculture. Depending upon the attachments and designs, the crawler tractors are named differently such as dozers, scrapers, graders, backhoe loaders, front-end loaders, excavators, fork lifters *etc*. A comparative statement of wheeled and crawler tractors highlights the differences between the two kinds of tractors.

Wheeled Tractor	Crawler Tractor
Relatively high speed	Low speed
Suitable for light duty works	Suitable for heavy duty works
Low initial cost	High initial cost
Operation and maintenance cost is low	Operation and maintenance cost is high
Self-driven for long distances	Trailer required for transporting long distances
Low stability during working	High stability during working
Can be used on roads and pavements	Cannot be used on roads and pavements

Figure 4.6. A View of the Crawler Tractor (Left) and Field Operation using a Crawler Tractor (Right).

Crawler type tractors run on tracks comprising track frame assembly, track chain, steering clutch and steering (Figure 4.6). Track frame assembly consists of a continuous chain surrounding the track frame and drive sprockets. The rollers support the whole unit. The links of the chain offer a flat surface to the track rollers to pass over. In summary, major characteristics of a crawler tractor are as follows:

- ✰ It is designed to secure good adhesion and transmit high drawbar pull in difficult field conditions, where wheel tractors may fail to secure adequate grip. Therefore, it is useful at places where adhesion is difficult and rolling resistance is high
- ✰ It provides large area of contact with the ground
- ✰ It is most suited for heavy work especially earth moving and reclamation works
- ✰ It is capable of pulling heavy implements

The principal disadvantage is due to the complexity of the track mechanism as compared to wheels. It is relatively more prone to failure such as snapped or derailed tracks.

POWER TILLER

A power tiller, also known as hand tractor or walking type tractor, is a prime mover in which direction of travel and its control for field operations is performed by the operator while walking behind it. The concept of power tiller came from the European garden tractors, which was adapted by Japanese farmers in the 1920s. The first successful model of power tiller was designed in the year 1947 and then imported to South Korea in the early 1960s. Currently, power tillers are extensively used for paddy cultivation in Japan, where the average farm size is smaller than India. Production of power tiller rapidly increased during 1950 to 1965. Manufacturing of several makes of power tillers like Iseki, Sato, *Krishi*, Kubota, Yanmar and Mitsubishi began in India following its introduction in 1963. But still the power tiller market is most serviced by two Indian companies *i.e.* VST Tillers Tractors Limited, Bangalore and Kerala Agro Machinery Corporation (KAMCO) Ltd. These two companies collectively cater to more than 68 per cent of the market and have the capacity to produce 90000 power tillers per annum (MoAFW, 2017). The remaining market is catered by power tillers imported mainly from China. Although the power tiller is a single axle walking type tractor, yet a riding seat for the operator is provided in many models. In spite of the fact that average size of holding in India is about 2.5 ha, even small and marginal farmers prefer to own or hire a tractor rather than own an easily affordable power tiller. In spite of this the power tillers may prove to be the most ideal multi-utility devices for small and marginal farms.

Components of Power Tiller

All the power tillers are fitted with an IC engine, mostly the diesel engine. Few main parts of a power tiller are shown in Figure 4.7. A list of other components, some of which are not shown in Figure 4.7, is provided in Table 4.2. Few of the important components are discussed in the following sections.

Table 4.2. A Select List of Components Used in a Power Tiller

No.	Component	No.	Component
1	Handle grip	17	Front lights
2	Throttle lever	18	V – belt
3	Auxiliary handle	19	Tension pulley
4	Main gear shift lever	20	Main clutch pulley
5	Front stand operating lever	21	Tine speed change lever
6	Auxiliary chain case	22	Main gear shift lever
7	Steering clutch lever (right)	23	Hand light
8	Rear wheel height adjusting lever	24	Steering clutch lever (left)
9	Rear wheel pipe attaching handle	25	Steering clutch wire
10	Ridger set screw	26	Rubber guard
11	Mud -guards	27	Protector
12	Side covers	28	Fender
13	Tilling tines	29	Rubber tyre

No.	Component	No.	Component
14	Magic bar	30	Hexagon wheel
15	Front frame	31	Side Frame
16	Protector	32	Reverse switch

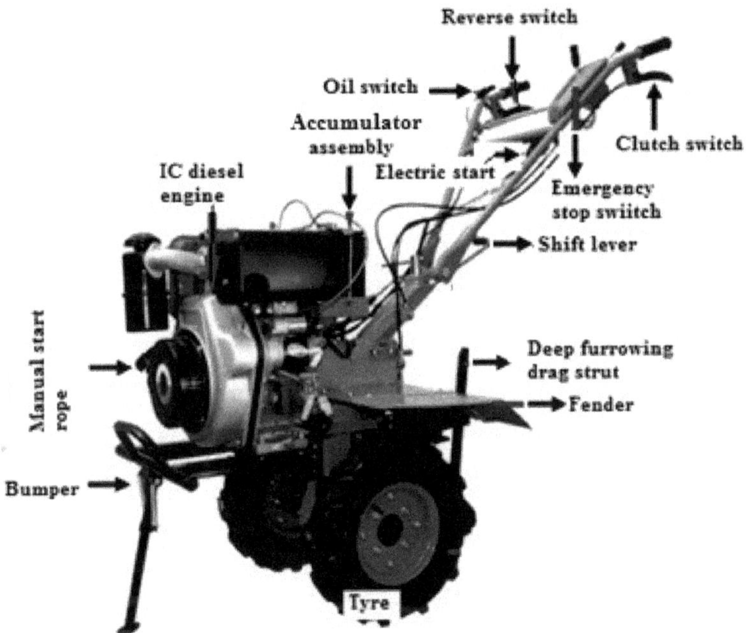

Figure 4.7. A view of the power tiller showing its select parts.

Main clutch: Power goes from the engine to the main clutch, which may either be friction clutch or V-belt tension clutch. Whereas the former is generally used for a bigger power tiller, V-belt tension clutch is used for small power tillers. The main functions of the clutch are:

☆ To transmit engine power to transmission gears

☆ To make the power transmission gradual and smooth

A major problem with the clutch arises when it slips continuously. It could be due to oil inside the clutch or worn lining or insufficient spring tension. Poor adjustment of the clutch may also result in this problem.

Steering clutch lever: Steering clutch is provided on the grip of the right and left handles. When the left side is gripped, power is cut-off on the left side of the wheel and the power tiller turns to the left. Similarly, when the right side is gripped, the power tiller turns to the right.

Transmission gears: Transmission box consists of gears, shafts and bearings. The speed change device may be gear type or belt type.

V-belt: It is normally used to transmit power from the engine to the main clutch. It has high efficiency and works as a shock absorber.

Brakes: All power tillers have some braking arrangement to stop the movement of the power tiller. Most of them use inner side expansion type brake.

Wheels: Two to four ply rating pneumatic tyres are used on power tillers. The pressure of the tyres range from 1.1 to 1.4 kg/cm^2.

Rotary unit: Power tiller has a rotary unit for field operations. It may be center drive type or side drive type. The center drive type has transmission at the center while the side drive type has transmission on one side. The center drive type has the following characteristics:

- ☆ Tilling width can be widened
- ☆ Rotary unit is light (in weight)
- ☆ Fixing of the attachment is easy
- ☆ The tine shaft can be detached easily
- ☆ Mounting and dismounting of rotary unit is easy
- ☆ Has one point support on the ground

One of the limitations being that it may leave some portion of the field untilled. The characteristics of the side drive type are:

- ☆ Deep tilling is possible
- ☆ The arrangement is useful for hard soil
- ☆ It has two points support on the ground

Power Transmission

The power from the IC engine goes to the main clutch with the help of belt or chain. From here the power takes two routes, one goes to transmission gears, steering clutch and then to the wheel as shown in Figure 4.8 (Cherian *et al.*, 2016). The other part goes to the tilling clutch and then to the implement (Figure 4.8).

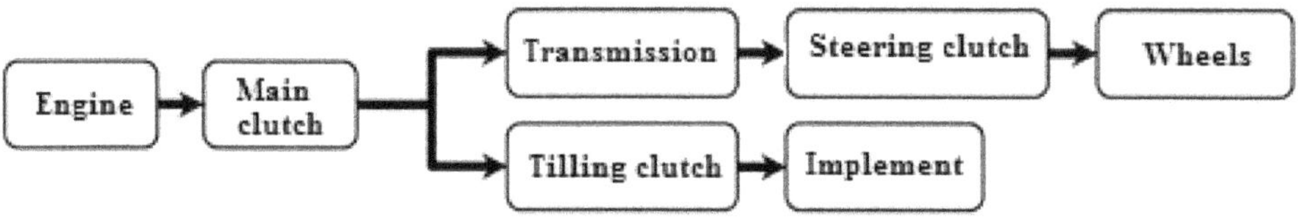

Figure 4.8. Schematics of Power Transmission in Power Tillers.

Similar to tractor, important dimensions of the power tiller are illustrated in Figure 4.9. Other dimension and mass can be defined in a similar manner as in tractors.

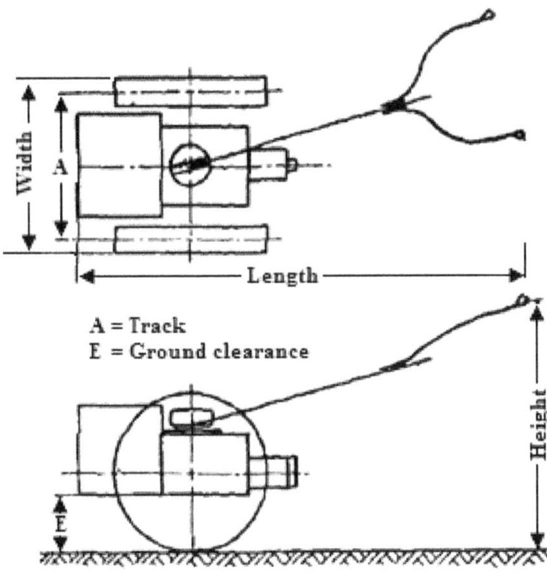

Figure 4.9. Important Dimensions of a Power Tiller.

Operation

The power tiller is mostly operated by walking behind the machine. Realizing the strenuous operation of walking long distances during a day's operation, now some power tillers provide seats for the driver. When the main clutch, a lever on the handle is in on position, the power from the engine is transmitted through the main clutch to the various parts of the power tiller. The power from the engine is cut-off from the rest of the system when it is in off position. The operator guides/controls the forward movement of the machine by actuating hand clutches provided on each handle or sometimes by pushing/pulling the handles towards the sides. Sometimes, he is required to lift the rear portion of the machine while taking sharp turns.

The power tiller, a multi-purpose machine, is capable of performing primary and secondary tillage and most other field operations. External attachments can be attached on the tiller depending upon the nature of work. A host of implements and attachments such as ploughs, extension wheels, water pumps, diggers, tine cultivators, seed-cum-fertilizer drills, cage wheels, sprayer units, ridger and slasher-cum-*in-situ* shredders are now available for use with power tillers. It is most suited for field operations in hilly regions, wet conditions and for small holdings. The light weight of power tiller favours its use in wet and dry land conditions. The owner must maintain the machine as per manufacturer's instructions so that the machine performs at its optimum efficiency besides giving trouble free service throughout its life. Some maintenance activities include:

- ☆ Clean the machine thoroughly including the rotor and the inner panels after each day's work
- ☆ Drain and refill gearboxes with new oil as per operator's manual

☆ Remove and service the PTO shaft. Follow the instructions in the PTO shaft's operator's manual. Check all tools and their fasteners

☆ Store the machine in a covered dry place

☆ Before using the machine in the new season, check the condition of the friction slip clutch, check all nuts and bolts and make sure that all the guards are in place.

POWER TRANSMISSION FROM POWER SOURCE TO MACHINES

Mechanical energy possessed by rotating elements can be utilized at the desired place by transmitting the same say from prime mover to machine or from one shaft to another. It is achieved through a number of power transmitting systems. Following three are the most commonly used methods of power transmission.

☆ Belts (flat and Vee)

☆ Chains

☆ Gears (Also see chapter 1)

Major factors that determine the kind of system to be used are the distance between the driver and driven pulley, operational speed and power to be transmitted.

Belts

Belt pulleys systems are used to transmit power between the parallel shafts located at relatively larger distances. The system consists of two pulleys one the driver pulley and the other the driven pulley and a belt. The belts are normally made of rubber, leather, canvas, cotton and steel. An endless belt connects and encircles both the pulleys. Two kinds of drives namely the open drive and crossed drive systems are used to drive the driven pulley (Figure 4.10). While selecting a particular system, keep the following characteristics of the two drives in view.

☆ In the open drive system, the driven and the driver pulleys revolve in the same direction. On the other hand, in the crossed drive system the driver and the driven pulleys rotate in the opposite directions

☆ The grip in a crossed drive is greater than in a open drive

☆ The length of the belt is less in the open than the crossed system

☆ The belts used in the open system have longer life than the cross drive system. It is because the life of the belt at the rubbing point in the cross belt system is less

☆ If the drive and driven pulleys are of different sizes, then the angle of contact at the pulleys is different in the open but same in the crossed system

The power transmitted by the belt system is a function of the strength of the belt and the friction between the belt and pulleys. It is obvious because the grip depends upon the friction offered by the two surfaces, and the total angle by which the belt laps the pulley.

Speed Ratio

In a belt driven system, the speed ratio of the two pulleys on the shaft of motor and machine to be driven is calculated by the following relation:

Speed of driver/Speed of driven

$$= \text{Diameter of driven/Diameter of driver} \tag{4.1}$$

Speed of driver/Speed of driven

$$= \text{Radius of driven/Radius of driver} \tag{4.2}$$

Obviously, it means if the driven pulley is smaller than the driver pulley, speed of the driven pulley will be high. On the other hand, speed will be less if the driven pulley is larger than the driver pulley. The statements hold good only if there is neither slip nor creep between the belt and pulleys.

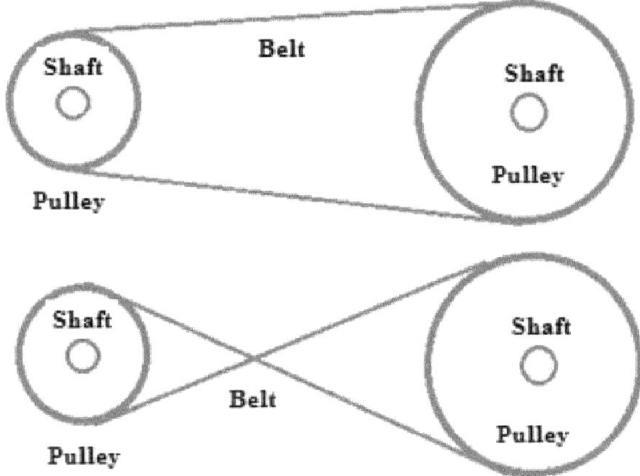

Figure 4.10. Open Belt Drive System (Top) and Crossed Belt Drive System (Bottom).

Example 4.1

The driver pulley of 240 mm diameter is running at 250 revolutions per minute. What sized pulley will be required to run the drive shaft at 1500 revolutions per minute in an open belt drive system? Assume the ideal state with no friction.

Using the speed ratio eq. (4.1)

$$N_1/N_2 = D_2/D_1$$

Such that N_1 is the speed of driver and N_2 the speed of driven. Similarly, D_1 is the diameter of driver and D_2 the diameter of driven.

$$D_2 = N_1 \, D_1/N_2 = 240 \times 250/1500 = 40 \text{ mm}$$

Vee Belt

Vee belt, or simply written as V-belt, is a wedge shaped belt that runs on split pulleys. The V-belt drives are quite useful when the driver and driven shafts are quite close (normally less than 1 m). V-belt runs in grooves turned in the pulleys

(Figure 4.11). Moreover, a provision is made to adjust the pulleys to take care of the slack in the belt. The V-belts are made of rubber impregnated with fabric cords and are covered with a protective layer. A V-belt is able to carry more power without slip because of its shape that increases the area of contact resulting in higher friction than a plain belt drive system. V-belt is always used in open configuration. The V-belt drives are preferred over flat belts drives because of the following reasons.

☆ Power transmission is high due to wedging action in the grooved pulley

☆ They permit a wide range of driven speeds allowing high velocity ratios (normally up to 4:1 in flat and up to 10:1 in V-belts)

☆ V-belt drive is more compact, quiet (with little noise) and shock absorbing

☆ The drives have smooth start and run with negligible slip

☆ These are rugged capable of giving trouble-free performance for years even under adverse conditions

☆ They are clean and require no lubrication

☆ They are efficient with an average efficiency range of 94-98 per cent

☆ They cover extremely wide horsepower range

☆ They dampen vibrations between driving and driven machines

☆ V-belts and sheaves wear gradually to allow timely preventive corrective maintenance

Figure 4.11. V-Belt Pulley (Left) and Pulley System (Right).

Chains

Chain drives, sometimes called chain gears, are employed for a wide range of power transmission applications, like in bicycles, motorcycles, rolling mills, agricultural machinery, machine tools, conveyors, coal cutters, *etc.* Accordingly chains are classified as hoisting chains, conveyor chains and power transmission chains. A power transmission chain drive consists of a chain and two wheels, called sprockets. The sprockets are toothed wheels over which an endless chain is fitted with the teeth of the gear meshing with the holes in the links of the chain (Figure

4.12). Thus, sprocket wheels are designed to fit each type of chain. Chain drives are suitable for power transmission at small distances normally up to 3 m, which in special cases may go even up to 8 m. The velocity ratio can be as high as 8:1. There is no slip in these drives being positive drives. The velocity ratio therefore, remains constant. Types of chains used in power transmission are categorized as roller chain, block chain and silent chain or inverted tooth chain. The roller chain is made of special high-grade steel and is extensively used on agricultural machines, manure spreader, cotton picker, combine harvester and similar other machines. Block chain is used mostly in low-speed systems such as planters because it produces noise due to the sudden contact between sprocket and chain. The following precautions should be exercised while fixing chains on sprockets:

☆ Sprockets chains must be properly tensioned. These should not be too tight to avoid excessive wear and tear of chains. The chains should not be too slack as slackened chains repeatedly jump off the sprockets

☆ Never use a new chain on a worn out sprocket because the chain will wear out sooner than expected

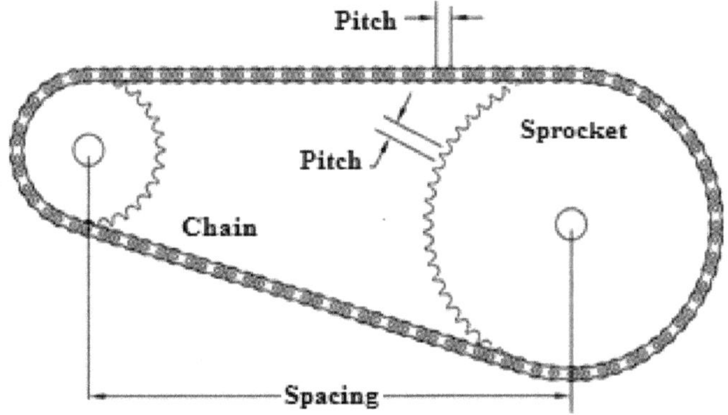

Figure 4.12. Chain Drive with Sprockets.

Power transmission chains are usually pre-lubricated with rust prevention oil. Nonetheless, lubrication is essential for increased life of the chain. Properly lubricated chain will have a longer life and is less likely to rust. Major advantages of chain drives over other power transmission methods are:

☆ Unlike belts they transmit large amounts of torque

☆ Unlike meshing gears they don't need constant lubrication

☆ Unlike belts and mesh-gears the spacing between centers can be easily adjusted by shortening/lengthening the chain

Some disadvantages of chain drives are:

☆ They cannot be used where systems require some slip to occur

☆ They require precise alignment compared to belt drives

☆ They require frequent lubrication

☆ They have less load capacity compared with gear drives

☆ Their operation sometimes may be noisy and can cause vibrations

Following expression can be used to calculate the speed of the driven sprocket.

$$T_2/T_1 = S_1/S_2 \tag{4.3}$$

Here T_1 is the numbers of teeth on the driving sprocket, and T_2 the numbers of teeth on the driven sprocket. S_1 is the speed of the driving sprocket and S_2 is the speed of the driven sprocket.

Gears

The mechanism of power transmission by gear drives is quite simple. The teeth cut on the blanks of the gear wheel mesh with each other to transmit power. The projections on one disc perfectly mesh with the recesses on other disc of the gear drive. Thus, the use of gears resolves the problems of slip and creep encountered in belt drive system. A gear drive has three main functions: to increase or decrease the torque of shafts, to increase or decrease the speed, and/or to change the direction of the rotation. It may be noted that direction of drive in two gears meshed together is reversed. Different types of gears are used in power transmission although the most common are the spur gear, the bevel gear and the worm gear. Some gear types are shown in Figure 4.13 and characteristics of some of them listed in Table 4.3.

Figure 4.13. Types of Gears used in various Machines.

The speed ratio of a gear drive depends on the number of teeth in the gears, the one with the less number of teeth revolving faster. The speed ratio for gears is given as

$$\text{Revolutions of Q/Revolutions of P = Teeth in P/Teeth in Q} \tag{4.4}$$

Here Q and P are driven and drive shafts gears respectively.

Table 4.3. Selected Types of Gears and their Characteristics

Type of Gear	Characteristics
Spur gear	Spur gears transmit power through shafts that are parallel. The teeth of the spur gears are parallel to the shaft axis
Helical gear	Helical gears have teeth that are oriented at an angle to the shaft. Thus, more than one tooth is in contact during operation. As such, helical gears are capable of carrying more load than spur gears
Double helical gear	Double helical gears are a variation of helical gears in which two helical faces are placed next to each other with a gap separating them. Each face has identical, but opposite helix angles
Herringbone gear	Herringbone gears are very similar to the double helical gear, but they do not have a gap separating the two helical faces. Herringbone gears are typically smaller than the comparable double helical. These are ideally suited in drives with high shock loads and vibrating applications
Bevel gear	Bevel gears are most commonly used to transmit power between shafts that intersect at a 90° angle
Worm gear	Worm gears transmit power through right angles on non-intersecting shafts. Worm gears produce thrust load and are good for high shock load applications but have low efficiency
Hypoid gear	Hypoid gears look very much like a spiral bevel gear but they operate on shafts which do not intersect, which is the case with a spiral bevel gear

COST OF TRACTOR POWER

Tractor and machinery costs are major items of costs in any farm business. The objectives of any cost assessment in farm business are 3 folds.

- ☆ To have an idea whether to purchase or hire the tractor/implement on custom hiring
- ☆ To have an idea of the payback period of the investment
- ☆ To understand the cost for fixing the custom hiring charges of the tractor/ implement

The cost of a tractor or for that matter any machine can be divided into two categories:

- ☆ Annual ownership costs or commonly known as fixed cost, which is borne by the owner irrespective of whether he uses the tractor or not. This cost comprises depreciation, interest (opportunity cost), taxes, insurance, and housing facilities.
- ☆ Operating costs, which are directly related with the time a tractor or machine is used. Operating costs include the costs of fuel, lubricants, replacement of tyres, equipment maintenance and repairs. Labour costs are associated with direct wages of the labour employed, food contribution, transport and social costs, including payments for health and retirement of the staff. The supervision cost may also be included in the labour costs.

Although it may not be possible to determine the true cost of a tractor yet a fair value of the cost can be assessed using certain assumptions namely the likely machine life, its annual use, and fuel and labour prices. The cost per hour operation of the tractor is then calculated by summing up the fixed costs and the operating costs including the labour costs. Cost of tractor is usually calculated as cost per hour or cost per hour per HP and cost of tractor + machinery is calculated as cost per hour or cost per ha.

Ownership Cost

Depreciation

Depreciation cost results from wear, obsolescence, and age of a machine. The degree of mechanical wear may cause the value of a particular machine to be somewhat above or below the average value for similar machines when it is traded or sold. The introduction of new technology or a major design change with new features may make an older machine obsolete. It may result in sharp decline in its salvage value. But the most important factors in determining the remaining value of a machine are its age and/or accumulated hours of use. Before an estimate of annual depreciation can be made, an economic life for the machine and its salvage value at the end of the economic life must be specified. The economic life of a machine is the number of years for which costs are to be estimated. It is the period over which the equipment can operate at an acceptable operating cost and productivity. For farm tractors or machines it is generally measured in terms of years or hours. More often, it is less than the machine's service life. Most farmers wish to replace the machine for a new one before the older one is completely worn out. A good rule of thumb is to use an economic life (Age) of 10 to 12 years for most new farm machines and tractors (Ambast *et al.*, 2015). Norms given in Table 4.4 may be applied to determine the life of various machines in terms of years and hours (BIS, 1979). Salvage value (S), defined as the estimated value of a tractor/implement at the end of its useful life is taken as 10 per cent of the purchase price (A) of the machine.

Depreciation per annum = $(A - S)/Age = (A - 0.1A)/Age$ (4.5)

Depreciation per hour, $D = 0.9\,A/(Age \times H)$ (4.6)

Here H is the number of hours the tractor runs per annum.

Interest

If the operator borrows money to buy a machine, the lender determines the interest rate to be charged. But if the farmer uses his or her own capital, the rate will depend on the opportunity cost of the capital use elsewhere in the farm business. If only part of the money is borrowed and part is used from own sources, it is reasonable to use an average value of the two rates. In the current scenario an average interest rate of 8-10 per cent is reasonable.

Table 4.4 Useful life of some commonly used farm machines

Machine	Useful Life	
	Hours	years
Stationary engine	10000	10
Electric motor	15000	15
Power tiller	8000	10
Tractor (wheeled and crawler)	10000	10
Combine (self-propelled)	3000	6
Combine (mounted and drawn)	2000	7
Seed drill	2500	10
Seed-cum-fertilizer drill	2000	8
Planter	2000	10
Plough	3000	10
Disc harrow	3000	10
Cultivator	4000	10
Front-mounted dozer attachment for wheeled tractor	3000	10
Towed scraper for wheeled tractor	2000	10
Power sprayer (knapsack and tractor mounted)	2000	8
Seed cleaner	2500	5
Agricultural trailer	3600	12
Power thresher	2500	8
Centrifugal pump	10000	10
Power chaff cutter	5000	8
Rotavator	2400	8
Ridger	1500	12
Blade terracer	2000	10
Puddler	2500	10
Cane crusher	10000	10

Interest per annum $= [(A + S)/2]\,(I/100)$

Here I is the interest rate per annum in percent and $(A+S)/2$ is the average cost of the tractor over its life span.

Interest component per hour, $I_h = [(A + S)/2]\,(I/100)/H$ $\qquad$ (4.7)

Taxes, Insurance and Housing

Although this cost is much smaller than depreciation and interest, but must be considered in the calculations. Road tax can be distributed over the life of the machine. Tractor is insured against loss by theft or damage and so is the case with farm machinery. Actual amount paid or likely to be paid annually for insurance and annual taxes should be charged. If such information is not available, it may be calculated on the basis of 2 per cent of the average cost price per annum. The charge for housing is taken as 1 per cent of the average cost price of the machine. So the charges for taxes, insurance and housing can be taken as 3 per cent per year of the average cost of the machine.

$$TIH = 3 \times (A+S)/(2 \times 100 \times H) \tag{4.8}$$

Here TIH is taxes, insurance and housing charges, Rs/hr. Sometimes, the percentage charges of these items are given in terms of initial purchase price of the tractor/machine. In such cases calculations may be made accordingly. The total fixed cost is the sum of depreciation (D), interest on investment (I_h) and Taxes, Insurance and housing (TIH) charges.

$$\text{The total fixed cost, Rs/hr} = D + I_h + TIH \tag{4.9}$$

Operational Cost

Fuel Cost

Fuel consumption mainly depends on three factors namely the size of the power unit, load factor and operating conditions. It is always useful to monitor the actual average consumption of the tractor while operating under different loads. It can also be taken from the results published by official testing stations. Average fuel consumption can also be estimated by the following formulae (BIS, 1979).

$$A = 0.15 \times B \tag{4.10}$$

Here A is the average diesel consumption in L/hr, B is the rated power in kW and

$$C = 0.25 \times B \tag{4.11}$$

Here C is average petrol consumption in L/hr and B has the same meaning as in the previous equation.

$$\text{Fuel cost, F} = \text{Quantity of fuel consumed per hour (L/hr)} \times$$
$$\text{Cost of fuel (Rs./L)} \tag{4.12}$$

The market value of the fuel should be used in calculations using eq. (4.12).

Repair and Maintenance Cost

Repair and maintenance costs are incurred to keep the machine in perfect working condition due to wear, part failure, renewal of tyres and tubes and accidents. The repair and maintenance costs shall be calculated at 10 per cent of the initial cost of the machine per year.

$$RM = 10 \times A/(100 \times H) \tag{4.13}$$

Here RM is the repair and maintenance costs, Rs/hr.

Lubricating Oil Cost

The actual oil consumption should be recorded during various operations in the field. If oil consumption data are not available, oil consumption may be taken as 2.5 to 3 per cent of the fuel consumption on volume basis (BIS, 1979). Other guideline state that it may be taken as 30 per cent of fuel cost.

Oil cost, O = 30 F/100 $\qquad$ (4.14)

Operator Wages

The actual number of operators engaged for carrying out the operation should be used for calculation of operator charges. The prevailing rate of wages should be adopted for calculation. While employing the labour on daily wages an 8 hour day is commonly used for calculation purposes.

Operator cost (Rs/hr), L = (Number of persons engaged x wages per day)/8 $\qquad$ (4.15)

The total variable cost is the sum of repair and maintenance cost, fuel cost, oil cost and operator charges.

The total variable cost, Rs./hr = F + RM + O +L $\qquad$ (4.16)

Total cost of operation of the tractor, Rs./hr = Total fixed cost + Total variable per unit time

Cost per HP-hr = Total cost (Rs./hr)/HP of the tractor $\qquad$ (4.17)

Custom hiring cost per hour, CHC is determined by adding 20 per cent profit margin. Thus

CHC = Total Cost + (20/100) Total cost = 1.2 x Total cost $\qquad$ (4.18)

Summarizing the above steps fixed cost and variable cost components are listed in Table 4.5. The aforementioned procedure is illustrated with the help of the two examples.

Table 4.5. Components of Fixed and Variable Costs for Calculating the Power Cost

Fixed/Ownership Cost	*Variable/Operational Cost*
Initial cost	Fuel consumption
Salvage value	Fuel cost
Life of the machine (Years/hours)	Lubrication cost
Interest rate	Repair and maintenance
Taxes, insurance and housing	Labour wages

Example 4.2

A farmer purchases a 40 HP tractor at a cost of 530000. Calculate the cost of tractor power per hour of its use, if the tractor life is assessed at 12000 hr of use. Also calculate the cost per HP-hr and custom hiring cost. Skilled labour wages are Rs. 400/day. Assume other values wherever required.

Before proceeding to calculate the hourly cost of tractor, let us make few assumptions first.

Average life 10 years (Table 4.3), Salvage value 10 per cent of the initial cost

Fuel consumption on full load 6.5 L/hr, fuel cost Rs. 80/L

Interest rate on agriculture 6 per cent, Taxes insurance and housing 1 per cent of average price each

Repair and maintenance cost as 10 per cent of purchase value and lubricating oil cost as 30 per cent of the fuel cost

Fixed Cost

Since the average life of tractor is 10 years, average annual use is 1200 hr/annum

Depreciation per hour (eq. 4.6), D = (530000 – 53000)/(10 x 1200) = 39.75

Interest per hr (eq. 4.7), I_h = [(530000 + 53000)/2] x 6/(100 x 1200) = 14.58

TIH (Eq. 4.8) = 3 x [(530000+53000)/2]/(1200 x 100) = 7.29 (Note that taxes, insurance and housing costs are 1 per cent each making a total of 3 per cent)

Total fixed cost per hr (eq. 4.9) = 39.75+14.58+7.29 = 61.62 ≈ 61.6

Variable Cost

Fuel cost (eq. 4.12), F = 6.5 x 80 = 520

Repair and maintenance cost (eq. 4.13), RM = 10 x 530000/(100 x 1200) = 44.2

Oil cost (eq. 4.14), O = 30 x 520/100 = 156

Operator cost (eq. 4.15), L = 1 x 400/8 = 50

Total variable cost per hour (eq. 4.16) = 520+156+44.2+50 = 770.2

Tractor power cost per hour = 770.2 +61.6 = 831.8

Cost per HP-hr (eq. 4.17) = 831.8/40 = 20.95 ≈ Rs. 20.8

CHC (eq. 4.18) = 1.2 x 831.8 = Rs. 997 ≈ 1000/hr

Note that this cost per hour is valid if tractor runs for 1200 hours in a year. If the tractor runs less or more number of hours, the cost needs to be recalculated.

Example 4.3

A farmer purchases a 15 horsepower diesel power tiller at a list price of Rs 250,000. An economic life of 10 years is assumed, and the tiller is expected to be used 800 hours per year. Other values given are: Salvage value (SV): 10 per cent of the initial cost, interest rate: 10 per cent per annum, Insurance and taxes: 1 per cent of purchase price, Housing: 1 per cent of purchase price, fuel consumption: 2 L/hour, Fuel cost = Rs. 75/L, Lubrication oil cost: Rs 150/L, Lubrication consumption: 6 per cent of fuel, Repair and maintenance: 5 per cent of purchase price, Labour: Rs 50 per hour. Assume any other value, if required.

Fixed Cost

Assuming salvage value at 10 per cent of the purchase price

Depreciation per hour (eq. 4.6), D = (250000 – 25000)/(10 x 800) = 28.1

Interest per hr (eq. 4.7), I_h = [(250000 + 25000)/2] (10/(100 x 800) = 17.2

TIH = 2 x 250000/(1200 x 100) = 6.3 (Note TIH is given as 2 per cent of the purchase price)

Total fixed cost per hr (eq. 4.5) = 28.1+17.2+6.3 = 51.6

Variable Cost

Fuel cost (eq. 4.12), F = 2 x 75 = 150

Repair and maintenance cost (eq. 4.13), RM = 5 x 250000/(100 x 800) = 15.6

Oil required = 6 x 2/100 = 12/100 = 0.12 L

Lubrication oil cost, O = 0.12 x 150 = 18.0

Operator cost/hr = 50

Total variable cost per hour (eq. 4.16) = 150+15.6+18+50 = 233.6

Power tiller cost per hour = 51.6 + 233.6 = 285.2

Cost per HP-hr (eq. 4.13) = 285.2/15 = 19

CHC (eq. 4.18) = 1.2 x 285.2 = Rs. 342.24 ≈ Rs. 345/hr

Note that this cost per hour is valid if power tiller runs for 800 hours in a year. If the power tiller runs less or more number of hours, the cost needs to be recalculated.

QUESTIONS (THEORY)

1. Name and discuss various types of tractor on the basis of structure design. List the characteristics of a good tractor.

2. Give a brief history of tractor development and its current status in India.

3. List various factors that should be considered while selecting a tractor.

4. Name and describe the various outlets of tractor.

5. Write a descriptive note on the crawler tractor.

6. Make a comparative statement of the differences between the wheeled and crawler tractors.

7. Name and briefly discuss various components of a power tiller.

8. Explain the various steps and assumptions to be made in calculating the tractor cost in Rs./hr.

9. Write descriptive notes on power transmission devices namely belts, chain and gears.

10. Differentiate between the row crop and orchard tractors, dry mass and operational mass of tractors, minimum clearance and turning diameters of tractors, fixed and operational cost.

Chapter 5

Introduction to Electric Motors

Electrical motors have brought about a perceptible change in the life of mankind ever since the invention of electricity. *It is an electromechanical machine that converts electrical energy to mechanical energy.* In agriculture, electrical motors are used to pump water, perform stationary works such as threshing, winnowing *etc.* The efficiencies of electrical motors are much higher in the range of 50 to 90 per cent compared to IC engines having thermal efficiencies between 28 to 38 per cent. Some important features of the electrical motors are:

- ☆ Low initial and low operating costs
- ☆ Clean and highly efficient source of power
- ☆ Easy to start even at loads, and designed to take up temporary over loads
- ☆ Long life
- ☆ Compact construction having low noise levels
- ☆ No exhaust smoke
- ☆ Operational simplicity and minimum safety issues

CLASSIFICATION/TYPES OF MOTORS

The primary classification of motors shown in the Figure 5.1 categorizes the motors as direct current (DC), alternating current (AC) and special motors. As the names suggest, DC motors run on direct current, all other motors including special motors run on alternating current (AC). Special motors are designed to meet specific demand of the industry or for some specific purposes. DC and AC motors are further sub-grouped and some of these groups form the subject matter of discussion in the forthcoming sections. Electric motors are further classified on the basis of armature designs, rotor design, speed of operation (constant, variable and adjustable), structural features (open, enclosed, ventilated, pipe-ventilated and riveted frame-eye) and kilowatt power output *etc.* Although both DC and AC motors are used in agriculture but in irrigation AC motors find wide applications. More than 1.34 million electrified tube wells were in operation in Punjab as on 31.03.2019 indicating the extensive use of electrical motors to supply irrigation water.

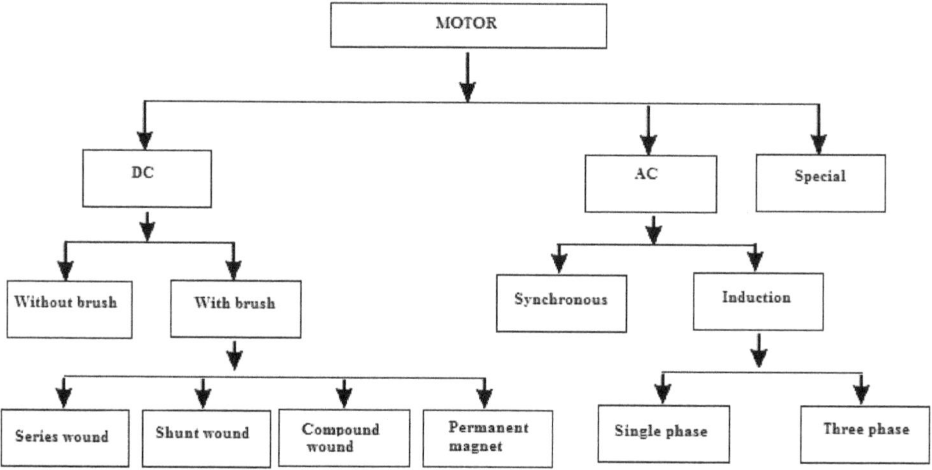

Figure 5.1. Classification of Electric Motors.

Basic Principle of Motors

An electric motor consists of a permanent stator (external field magnet) and a rotor (coiled conducting ammeter) which is free to rotate within the field magnet. Brushes and a commutator (designed differently for AC or DC current supplied to the armature) connect the armature to an external voltage source. When armature coils are supplied with electric current from the supply, the field magnets get excited experiencing a force that tends to rotate the armature either in a clockwise or anticlockwise direction. The forces produce a torque, a twisting effect that causes the armature to rotate in one of the directions. The motor speed is a function of many factors namely the amount of current flowing, the number of coils on the armature, field magnet strength, the permeability of the armature, and the load connected to the shaft.

DC MOTORS

Direct current motors, as the name implies, operate on direct-unidirectional current. These motors are used in applications requiring high starting torque or where smooth acceleration over a broad range of speed is required. Such situations are most often encountered in the automotive and agricultural industries. A DC motor comprises a wound armature, commutator, graphite or carbon brushes and magnets- all within a totally-enclosed housing (Figure 5.2). Current is conducted to the armature winding by connecting leads from the armature winding to the commutator that feeds the commutator with current by means of brushes. Rotation occurs when the magnetic field of the motor created by the permanent magnets in the housing interacts with the commutator through brushes. The main features of DC motors are listed as under:

- ☆ Runs on DC power or on AC with a rectifier
- ☆ Operating speeds of 1,000 to 5,000 rpm
- ☆ Efficiency in the range of 60-75 per cent
- ☆ High starting torque
- ☆ Low no-load speeds
- ☆ Sizes range from 1/100 HP to thousands of HP

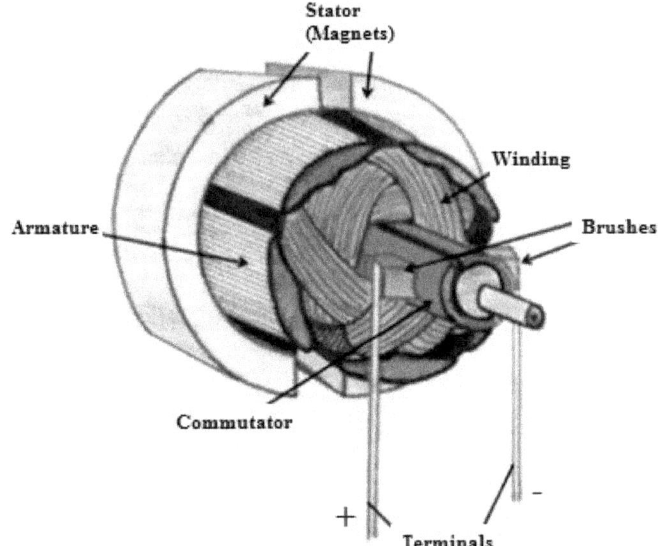

Figure 5.2. Construction of a DC Motor.

DC motors are classified as brushless or brush motors, although the principle behind the internal working of both these kinds of motors is essentially the same. Whereas brushes deliver current to the motor windings through commutator contacts, brushless motors do not have the current-carrying commutators. In the brushless motor, the field is switched via an amplifier triggered by a commutating encoder. Moreover, windings are on the rotor (rotating part of motor) for brush motors and on the stator (stationary part of motor) for brushless motors. Further classification of the brush DC motor depends on the way of connecting the armature and field winding of the motor. Accordingly, DC motors are categorized in 4 main groups namely the series motors, the shunt motors, the compound motors and permanent magnet (Figure 5.1). Even further sub-groupings have been made. DC series wound motor has its field winding connected in series with the armature. In other words, rotor windings in this kind of motor are connected in series. The principle of operation of this electric motor is governed by the simple electromagnetic law. This law states that whenever a current carrying conductor is placed in a magnetic field, it experiences a mechanical force that generates rotational motion. The change in magnetic environment can be achieved by changing the magnetic field strength. It can also be achieved by moving the magnet toward or away from the coil, moving the coil into or out of the magnetic field, or rotating the coil in the field. This motor has very high starting torque, but has a tendency to 'run-away' when lightly loaded or unloaded, and has poor speed regulation. These motors are mainly used in starter motors used in elevators and cars. In a DC shunt motor the windings like the armature windings and field windings are linked in parallel across the supply voltage, like a shunt. It is why the winding of this kind is known as shunt winding and the motors as shunt wound DC motor. This motor has very good speed regulation. The DC compound motor has a series field winding connected in series with the armature and a shunt field in parallel with the armature. Thus, it is a combination of the series and the shunt motors. This combination allows the motor to have the torque characteristics of the series motor and the regulated speed characteristics of the shunt motor. The field poles and the armature poles of a permanent magnet DC (PMDC) motor are made of permanent magnets. The magnets, mounted on the inner periphery of the cylindrical steel stator, are radially magnetized. While the stator serves as a return path for the magnetic flux, the rotor has a DC armature with commutator segments and brushes. These motors have high starting torque, good speed regulation and a definite maximum speed.

Motor Starter

Small DC motors of less than 1/2 kW size consume very little current and therefore, can be started by applying full voltage across the motor terminals. But in large DC motors, excessive current may be as high as 15 to 20 times higher than the full load current, will flow when full voltage is applied. It might damage the motor. The high starting current can also be harmful in the following manner:

- ☆ The supply voltage will fluctuate as a result of high starting current

- ☆ The insulation of the armature winding may burn due to high current

- ☆ The fuses may blow or circuit breakers might trip

- ☆ Excessive large starting torque put a heavy mechanical stress on the winding and shaft of the motor, which may result in the mechanical damage to the motor

In order to avoid all these problems, the starting current of the motor is kept below the safe limit. It is achieved with the help of a starter. In principle, a starter connects a resistance unit in series with the armature, so that the starting current is reduced to a safe value. The starter of a DC motor is usually operated manually. As already stated, starter is basically a resistance, which remains in the circuit only at the time of starting. As soon as the motor speeds up to a desired speed, it becomes ineffective. Let us look at its working. While starting, the starter is in the start position (Figure 5.3 left). As such, the full starter resistance appears in series with the armature. It limits the initial current to a safe value. At this time starter is pushed to the run position as shown in Figure 5.3 (right) under the normal operating condition. In this position, the starter resistance is zero. The counter emf produced in the armature at high speeds prevents the excessive current flow. When the resistance is finally removed, the operating handle rests against a holding magnet mounted alongside the last point. The holding magnet is connected in series with the main-shunt field of the motor. The operating handle is also provided with a heavy spring. In the event, the holding magnet is demagnetized due to failure of the power supply or break in the motor field circuit, the spring instantly returns the operating handle to the off position. It stops the motor and remains so until started again manually.

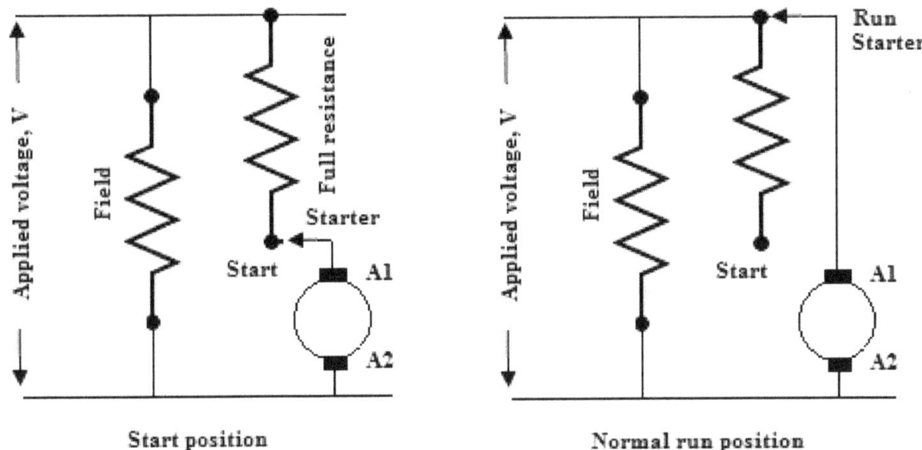

Figure 5.3. Operation of a DC Motor Starter.

AC MOTORS

AC motors have been classified as synchronous, induction and special kinds of motors. Induction motors are generally used for most farm works. Induction motors are further classified as single phase and three-phase motors (Figure 5.1). Further sub-grouping of single and three-phase motors has also been made. For example, single phase motors are available in following types:

Shaded pole: Low starting torque. Usually used in direct-drive fans and blowers.

Permanent split capacitor (PSC): Performance and applications similar to shaded pole but more efficient, with lower line current and higher horsepower capabilities.

Induction run (Split-phase motors): Moderate starting torque, high breakdown torque. Used on easy starting equipment such as belt-driven fans and blowers, grinders, centrifugal pumps, gear motors, *etc.*

Split-phase start, capacitor run: Its performance is similar to the induction run motor, but for higher efficiency.

Capacitor start, induction run (or Capacitor start): High starting and breakdown torque, medium starting current. Used on hard-starting applications: compressors, positive displacement pumps, farm equipment, *etc.*

Capacitor start, capacitor run: It is similar to capacitor start, induction run, but has higher efficiency. It is generally used in higher HP single-phase ratings.

Without going into these details, we will discuss split phase and capacitor start type motors in single phase and three-phase induction motors, because of their widespread use.

Split-phase Motor

Split-phase motor, also known as resistance-start motor, is operated from a single phase power circuit. It has four main parts, namely rotor, stator, end plates, and centrifugal switch. The rotor consists of laminated core, the shaft and the squirrel cage winding. The squirrel cage winding consists of heavy copper bars placed in slots of the iron core, which are connected to each other by means of heavy copper rings placed on both ends of the core. The stator consists of a laminated iron core with semi-closed slots, a heavy cast iron frame, and two windings of insulated copper wires wound into the slots. These windings are known as main winding and starting winding. The main winding has low resistance and high reactance while starting winding has high resistance and low reactance. The resistance of the starting winding can be increased or decreased by connecting a resistance in series with the starting winding (Figure 5.4 left). Both the windings are displaced 90° in space. When the motor is started, both the winding are connected to the power line. As the current starts flowing through both the windings, a magnetic field is formed inside the motor. This magnetic field rotates and induces a current in the rotor winding which in turn results in another magnetic field. These two fields combine in such a manner to cause the rotation of the rotor. No sooner does the motor reach approximately 70-80 per cent of the full speed, the starting windings is cut out of the circuit by means of centrifugal switch or a relay. The end plates are fastened to the stator frame mainly to keep the rotor in position. End plates are fitted with either ball bearings or sleeve bearings in which the rotor shaft rotates.

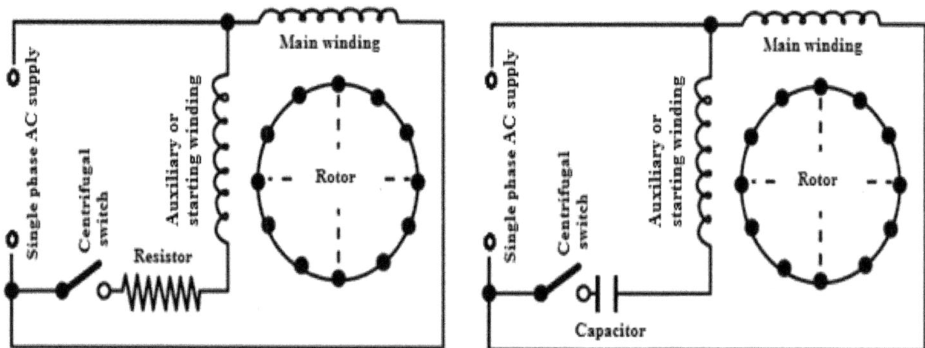

Figure 5.4. Single Phase AC Motors: Resistance-start (Left) and Capacitor-start (Right).

Capacitor Motor

As the name suggests, a capacitor is used for starting the motor. Let us first look at the capacitor. A capacitor has capacity to store electricity. It is formed when two conductors are separated by an insulator, such as a waxed paper. Capacitors are rolled and placed in a metal container. They are rated in microfarads. The capacitor motor has similar construction as that of the split phase motor. Like in resistor split phase motor, there are main and auxiliary winding but the basic difference between the two is that in capacitor split phase motors a capacitor of suitable value is connected in series with the auxiliary winding (Figure 5.4 right). Depending upon the size of the motor, the capacitor may vary from 10 to 150 microfarads (μF). The capacitor is usually mounted on the top of the motor. Capacitor in series with the auxiliary winding results in desired time phase displacement between the auxiliary winding current and main winding current. A centrifugal switch is provided that cut outs auxiliary winding when the speed of the motor reaches 70 to 80 per cent of synchronous speed. As such, the motor develops high starting torque with low starting current. In order to produce a starting torque, a revolving magnetic field must be established inside the capacitor motor. It is accomplished by placing the starting winding 90° out of phase with the running winding. The capacitor helps the current in the starting winding to reach the maximum value before the current in the running winding reaches the maximum value. As a result, a revolving magnetic field is produced in the stator, which in turn induces current in the rotor winding. Thus, the two windings act like a two-phase stator and produce the rotating field required to start the motor. The capacitor-start motor develops about 3 to 4.5 times higher starting torque than the full load torque. In order to obtain a high starting torque, the starting capacitor value must be large say 250 μF and the starting winding resistance must be small.

Three-phase Induction Motor

AC motors which operate on three-phase AC supply are known as 3-phase motors. These are categorized as synchronous and induction types (BIS, 1996). *A synchronous motor runs at synchronous speed i.e. the rotation of the shaft at steady state is synchronized with the frequency of the supply current, or the rotation period is exactly equal to an integral number of AC cycles.* Induction motors are asynchronous motors. *In asynchronous motors speed is always less than the synchronous speed, and the speed decreases with increasing load.* The parts of an induction motor shown in Figure 5.5 reveal that it has a stator and a rotor very similar to that of the split phase motor. The stator of the motor consists of a cast iron frame having slots on its internal side to carry the winding circuitry called stator winding. This circuit is supplied with AC power. The coils in the slots of the iron core are connected in such a way that these form three separate windings called phases. The rotor of the motor has also slots to carry conductors. These conductors make up the rotor winding. Based on the rotor construction, a 3 phase induction motor is further classified into two types namely squirrel cage and wound rotor induction motors. The former is commonly used because of its less cost and low maintenance requirement. The rotor of a squirrel cage type motor is made of copper bars, which extend out from the rotor length and are fixed in slots of the rotor core. The extending bars are short- circuited together by means of copper rings on each side. These windings are connected so that a magnetic field is formed inside the stator that causes the rotor to turn at a constant speed. In a wound rotor induction motor, a 3 phase winding is placed uniformly on slots of the rotor, which is quite similar to the stator's winding. The ends of these windings are connected to 3 slip rings on the shaft, each phase out of the three phases being connected to one slip ring. The slip rings of each phase rotate with the rotor and are further associated with stationary brushes. The end plates in each case are meant for supporting the rotor shaft. These motors do not have centrifugal switch.

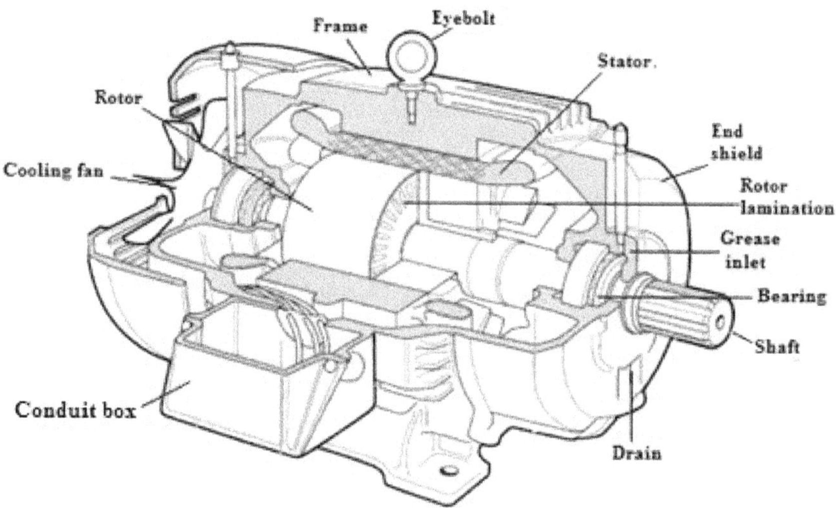

Figure 5.5. Construction of a 3-phase Induction Motor.

When three-phase AC voltage is supplied to the stator windings, 3 phase current starts flowing in it. The current flowing in windings produces a magnetic field that rotates due to AC oscillations. The flux of this rotating magnetic field crosses the rotor conductors and a voltage is induced in them which depend on the relative speed of conductor and magnetic field, magnetic flux density and length of the conductor. This induced voltage generates induced current in the rotor, which lags behind the rotor's induced voltage. This leads to the production of the rotor's magnetic field. Finally, the torque induced is the scaled product of stator magnetic field and rotor magnetic field and the rotor of 3 phase induction motor starts rotating.

SUBMERSIBLE MOTORS

Submersible pumps are now commonly used to develop groundwater at deeper depths beyond the suction reach of centrifugal pumps. For this purpose a submersible motor is used to drive the pump. The motor is normally a two-pole, three-phase, squirrel-cage induction electric motor. Submersible motors are rugged and corrosion resistant with cast iron frames and end plates. Frames are precision machined for tight fits between the parts. Most motors use stainless steel for its hardware and shafts. Motor size is dictated by the amount of power required to drive the pump to lift the estimated volume of fluid to be pumped out of the well. Submersible motors differ from the normal motors in the sense that these are hermetically sealed (airtight) motor close-coupled to the pump body (Gupta, 2020). Moreover, unlike the normal motors that are air cooled, submersible motors are cooled by the pumping fluid, water in the case of groundwater. These are designed to run continuously while submerged, but may only be able to run for about 15 minutes in the absence of fluid flow. Since the motors are designed to run continuously, they include features to keep water out. The oil in an oil filled chamber acts as a barrier to trap moisture and provide sufficient time for shut down if for any reason water enters the motor. It also lubricates the upper part of the mechanical seal. Grooves with O-rings are used to prevent water ingress to the motor. A seal section, which also houses the thrust bearing between the pump intake and the motor, isolates and protects the motor from well fluids and equalizes the pressure in the well bore with the pressure inside the motor. A non-wicking cable cap assembly filled with epoxy, a cured end products of *epoxy* resins, protects the motor from moisture entering through the opposite drive end of the motor. Power is provided from the surface to the motor via a specially made three-phase electric cable. To minimize cable movement in the well and to support its weight, the cable is banded or clamped to the tubing. This cable should never be used to lift the motor.

AC MOTOR STARTERS

Like a DC motor, if an AC motor is started on full line voltage, it draws more current, may be about 5-8 times its normal rated current. On large motors, it is always desirable to reduce the starting current to avoid damage to the motor and to minimize line disturbances that may adversely affect operation of other motors or equipments on the same line. Single phase motors below 1 kW are controlled by a hand operated switch. Medium size motors of 2 HP or more require some device in the line which may reduce the starting current. The starter is a specially designed electromechanical switch having an overload protection for the motor. The purpose of the starter is twofold.

1. It switches the power either manually or automatically
2. Protect the motor from overload or faults

Types of Motor Starters

In a three-phase induction motor, when the motor is standstill, the speed of the motor is zero and hence slip is at maximum. This induces very high electromotive force (emf) in the rotor at starting time and thereby a very high current flows through the rotor. To meet the high current requirement of the rotor, the stator winding draws a very high current from the supply. This huge current can damage the motor windings. It also causes a large voltage drop in the line. A starter reduces the initial high current of the motor by reducing the supply voltage applied to the motor. Such reduction is applied for a very short duration. Once the motor accelerates, a normal voltage can then be applied as slip is reduced. Several kinds of starters are available in the market. Some of these are:

☆ Stator resistance starter
☆ Auto transformer starter

☆ Push button or direct on-line starter

☆ Star-delta starter

☆ Soft starter

Push button type or direct online starters and star-delta-starters are widely used for 3-phase motors up to 7.5 kW but above 7.5 kW only star-delta starter is used. It is the simplest, cheapest and most reliable and hence widely used. Motors directly connected using direct on-line (DOL) starters are provided with overload, low voltage and single phasing protection. It is because such motors can withstand high starting current for short duration.

Push Button Type or Direct On-line Starters

At the starting time, normally open contact (NO) is pushed for a fraction of a second and the magnetizing coil becomes energized (Figure 5.6). This magnetic flux produced by the coil attracts the contactor (C1, C2 and C3) so that the motor is now connected to the supply. The contactor maintains this position while the coil gets supply from the additional switch, SW. When a normally closed (NC) switch is pressed, the coil becomes de-energized and the contactors get separated by spring arrangement thereby the supply to the motor is removed (Figure 5.6).

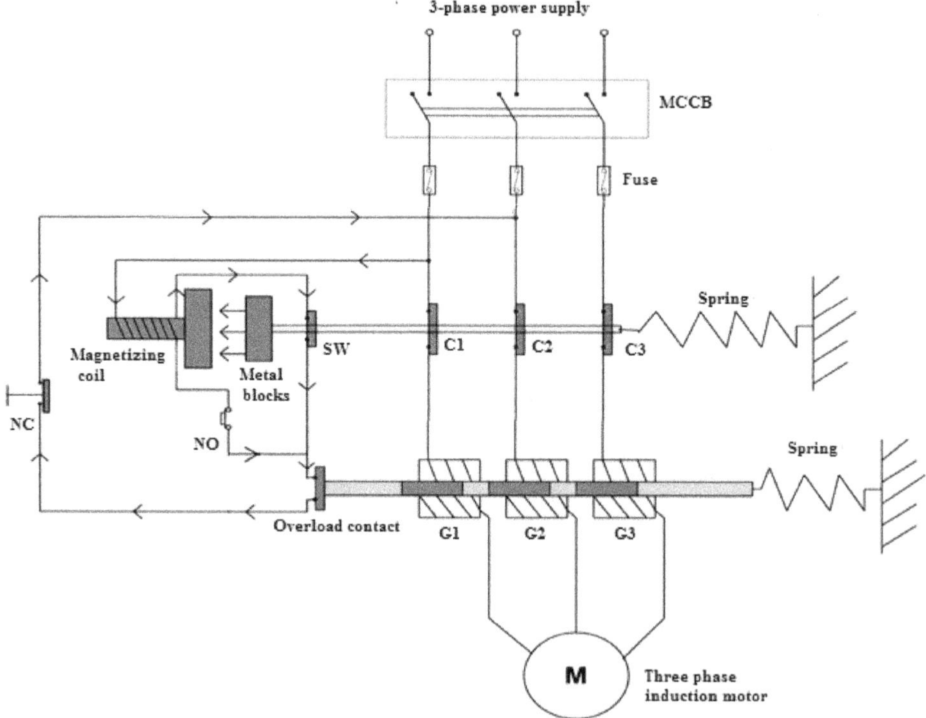

Figure 5.6. Working of Direct On-line Starter.

DOL is an isolated motor switching setup that provides the following protections.

- ☆ Over-current protection or short circuit protection (MCCB and fuses)
- ☆ Overload protection (G1, G2, G3 relays)

The motor draws heavy current during overloading resulting in overheating. Overload sensors sense this excessive heating and operate the thermal relays that disengage the current supply to the motor.

Advantages of DOL Starter

- ☆ Most economical, compact, reliable and cheapest starter
- ☆ Easy to operate through ON and OFF and a knob for setting the overload safety making it easy to operate
- ☆ Easy maintenance
- ☆ Since there is no start-up protection, the motor operating with DOL starter provides 100 per cent starting torque

Disadvantages of DOL Starter

- ☆ Only suitable for low and medium power motors
- ☆ Since there is no start-up protection, the DOL starter does not limit the starting current and unnecessary high starting torque during motor start-up
- ☆ The power line on which motor is connected will experience voltage dips at the time of starting the motor. This may harm other electrical equipment feeding on the same supply
- ☆ The motor is subjected to thermal stress which affects the life of the motor
- ☆ The motor experiences high mechanical stress because of unnecessary high starting torque at the time of starting the motor

Star-delta Starter

The star-delta starter is used on three-phase delta connected motors. When a large induction motor directly starts in delta, the starting current in the coils can be over 5 times higher than the full load current before the motor stabilizes and runs normally. In order to reduce this current a star-delta starter is used, which is available in manual, semi-automatic and automatic operational modes. We always start the motor in star before switching over to the delta configuration. The starter is connected to the electrical terminal box on the top or sometimes on the side of the motor. The terminal box contains 6 terminals having corresponding letter and number U1, V1, W1 and V2, W2 and U2 (Figure 5.7). Note different arrangement of terminals on top and the bottom of the terminal box. Inside each three phase induction motor, 3 separated coils are used that produce a rotating magnetic field. Phase 1 coil is connected to the two U terminals, the Phase 2 coil to the two V terminals and the phase 3 coil to the two W terminals. The supply side is connected to terminal U1, V1 and W1 (Figure 5.7). For the delta connection, the terminals are

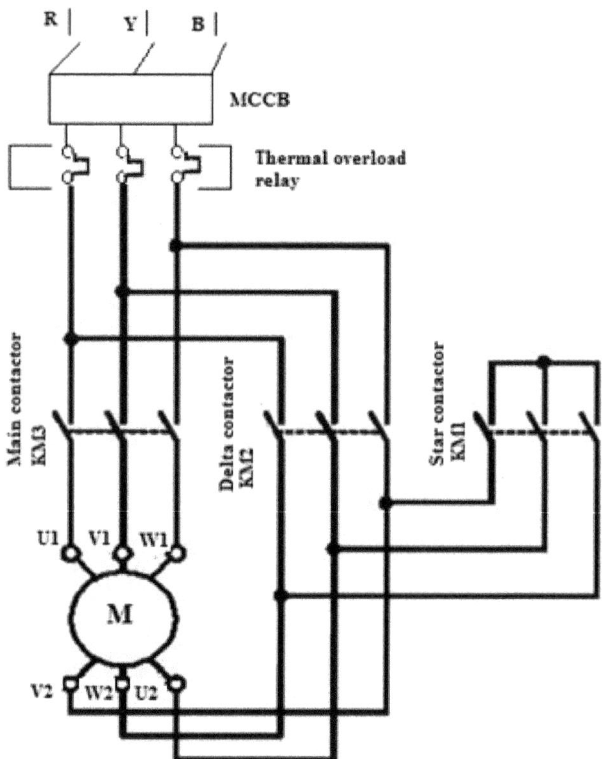

Figure 5.7. Schematic Connections in a Start Delta Starter.

connected as per order: U1 to W2, V1 to U2 and W1 to V2. When current is passed through the phases, electricity flows from one phase into another as the direction of AC power in each phase reverses. That's why the terminals have different arrangements so that it is possible to connect across and allow electricity to flow between the phases. For star equivalent design connect between W2, U2 and V2 on only one side of the motor terminals (Figure 5.7). Main, delta and star contactors are used to automate this, which activates to make or break a circuit to simultaneously control the flow of electricity in all three phases. As already stated, motor is always started in star connection by activating the main and star contactor terminals so that they close to complete the circuit. The coil voltage in star configuration is around $1/\sqrt{3}$ or about 58 per cent on each phase of the motor (400 line voltage reduces to 230 volts) compared to the delta configuration. A lower voltage results in a lower current. The current in the coil in star configuration will be around ~33 per cent of the delta configuration. It results in reduced torque, being about 33 per cent compared to delta connection. This set-up runs for a few seconds before switching over to delta. For the delta connection, the star contactor is disconnected and delta connection is on. The control of the contactors or the changeover is automatic through a timer relay built into the starter (Not shown in the diagram). To operate the starter, the lever is pressed down such that KM3 and KM1 are on. The motor starts rotating as a star connected motor (Figure 5.7). After the motor accelerates and attains about 90 per cent of rated speed, the starting lever is raised up so that motor is brought from star to delta state. Now KM3 and KM2 are on and KM1 is disconnected (Figure 5.7). At this time full line voltage is applied to the motor, and thereafter it rotates at full speed.

Advantages of Star Delta Starter

☆ Being cheap, star-delta starters are quite popular amongst the users

☆ They can be operated any number of times without any limit

☆ Compared to DOL starter, the starting current is around 1/3 of the inrush current

☆ Produces high torque per ampere of line current

Disadvantages of Star Delta Starter

☆ Star-delta starter can only be useful in motors where the six motor terminals can be accessed

☆ The supply voltage and the rated motor voltage must be the same for delta connection

☆ Starting torque is also about 1/3 of the full load torque

SELECTION OF ELECTRIC MOTORS

Wide varieties and sizes of motors are available in the market. Each motor is designed to operate under a standard set of conditions. Two most important requirements for selecting an appropriate motor are: the starting requirements of equipment and the maximum current that may be drawn from the single phase power service (Desai and Sivakumar, 2017). Besides, the motor should be able to meet the load requirements without exceeding temperature and torque limits (https://beeindia.gov.in/sites/default/files/3Ch2.pdf). Therefore, assessment of the load characteristic such as power or torque requirements, speed and duty cycle of the motor is the first step in motor selection. To select the right type of motor for any particular job, the following factors should be considered.

1. The motor should conform to the kind of power supply whether AC or DC, single or three-phase, 50 or 60 Hz supply

2. Understand the operational requirement for which motor is being purchased. The motor rating (HP or kW) should be such that the motor is not overloaded. Slightly higher but not too big size should be preferred

3. Duty requirement whether continuous or intermittent use should be kept in view

4. Speed of motor should be selected with due consideration of the machine to be driven

5. Starting current requirement should be known before selecting the motor for a particular use

6. Atmospheric and environmental conditions-humidity, splashing of liquids, excess dust *etc.* should be kept in view

7. Motor should be equipped with an over load protection device

8. Logically the lowest priced motor that will do the job satisfactorily should be selected, but it should not be the main consideration. An ISI marked motor should be preferred

9. Understand and read the label/nameplate on the motor critically as most of the information is included on the plate as illustrated below: For detailed information readers may refer to BIS (2004).

Rated Output	kW	Rated Voltage	V	Current	Amp
Efficiency class	IE 3 (Premium)	Frequency	Hz	Speed	RPM
Class of insulation		No. of phases		No. of poles	
Connection		Type of duty		Type of enclosures	
Type of motor	Induction motor	Power factor		Degree of protection	
Ref standard number	IS 12615-2011	Frame size		Method of cooling	
Motor serial No.		BIS certification mark		Mounting	
				Year of manufacturing	

CARE OF ELECTRIC MOTORS

Electricity is a clean, economical and dependable source of power. Electric motors can provide years of service, if properly selected, operated and maintained (Beard and Hill, 2000). The following list summarizes major points for maintenance of electric motors.

☆ Appropriately sized and type of motor should be selected to prevent overloading. Never overload the motor to avoid damage due to burning of the windings. A motor may be purchased with an overload protective device or one may be installed in the circuit during its installation

☆ The motor should always be lifted using the lifting hook at the eye bolt of the motor

☆ The machine when not in use for a long period should be stored in a clean and dry place. A well ventilated and screened shed should be used to house the pump and motor as it protects the equipment from undesirable natural elements

☆ Repaint or re-varnish the protective paint and varnish as and when necessary to maintain them in good condition

☆ Avoid overheating of the motor as it is the main cause that reduces the motor life. Some points discussed in this section protect the motor from overheating

☆ Operate motor at proper voltage. Both low and high voltages are harmful

- ✫ Proper electrical motor protection devices must be installed in service panel and/or motor enclosure
- ✫ Excessive motor or pump noise may be due to wear and tear of motor bearings. Remember to follow the manufacturer's recommended maintenance schedule. In general, motor bearings should be lubricated once every year. Open the drain plug in the bearing housing. Never force the grease into the bearing housing without opening the plug. It can damage the bearing shields. It may also force old grease, moisture and dirt into the motor housing. Pump the grease into the bearing housing until the new grease begins to come out of the drain (refer Figure 5.5). Replace the drain plug. Replace the bearings before they begin to fail. It can save the cost of shaft repair, replacement of bearing housings and repair of damaged pumps
- ✫ Dip and bake windings at regular intervals as it adds insulation to the motor windings. The cost of the operation is just a fraction of the cost of rewinding the motor
- ✫ Keep electrical connections tight. Loose connections can build-up heat at electrical terminals, which may deteriorate insulation of motor winding. It may also cause electrical short circuits and/or motor failures
- ✫ Keep motor and area around the motor clean. Keep water away from the motor so that area around the motor remains dry. Don't allow plants and weeds around the pumping plant
- ✫ Shade the motor so that it remains cool
- ✫ Protect the motor from extreme weather conditions
- ✫ The belt between motor and driven machine should be kept moderately tight
- ✫ Ensure that the motor is well grounded
- ✫ Provide adequate support to the electrical panel to prevent loosening of electrical connections due to motor vibrations
- ✫ Maintain and service instead of repair

How to Clean a Motor?

Regular cleaning and proper lubrication schedule of the motor is necessary not only for the motor to run efficiently but also for its extended life. The following procedure should be observed to keep a motor clean (BIS, 1992a).

1. After having disconnected the power, disassemble the motor by removing the rotor, shaft and fan assembly. All possible care should be taken not to break any wire connection or centrifugal switch mounted on the rotor shaft or elsewhere

2. Blow out the loose dirt from the motor windings and from inside the end plates using pressurized air from a cycle pump or a blower. A dry paint brush or cloth should be used to wipe off any dirt that might remain after cleaning with the pressurized air

3. Air passages in the frame and in the rotor should be made dirt free. Metal parts may be washed with non- inflammable fluid

4. Before reassembling, accessories like the starting switch, the commutator and brushes should be thoroughly checked and repaired, if need be

5. Lastly the motor bearings should be lubricated with the recommended type of lubricant before the connection is provided to check its operation

SAFETY OF THE MOTORS

Every motor needs to be protected from all possible faults to ensure its prolonged life and safe operation. It also saves time that may be lost in repairs following breakdowns. Some commonly used devices to protect the circuits and motor are fuse, MCB, MCCB and over load relays.

Fuse

A fuse is a fuse wire placed in a fuse holder. It is the most crucial device employed in various electrical circuits to protect electrical/electronic circuits against overloads, short circuits and earth faults. When the current taken by the circuit exceeds the rated current of the fuse wire, the fuse wire melts and disconnects the supply from the circuit. Thus, it protects the circuit and the components connected in the circuit from damage. The rating of the fuse wire is expressed in amperes, which is the maximum current that a fuse can carry without being burnt. For example 5 A fuse, 10 A fuse, 16 A fuse, 60 A fuse, 100 A fuse, *etc.* Current rating of the fuse should be equal to the maximum current rating of the machinery, appliance or component connected in the circuit.

MCB

The MCB is the short form of miniature circuit breaker. *It is an electromechanical automatically operated electrical switch.* It is designed to protect an electrical circuit from damage caused by excess current resulting from an overload/short-circuiting. Its main function is to break the current flow. Fuse may also sense these conditions, but it has to be replaced but an MCB can be easily reset.

MCCB

MCCB stands for Molded Case Circuit Breaker. It is another type of electrical protection device used when load current exceeds the limit of a miniature circuit breaker. The **MCCB** provides protection against overload, and short circuit faults. It is also used for switching the circuits. MCCB are quite useful in industrial applications because of their wide current ratings and high breaking capacity.

Relay

As already stated, an overloaded motor draws excess current that causes overheating and may damage the windings of the motor. An overload relay protects the electric motor against overloads, phase loss/failures and phase imbalance. It interrupts the power flow to the motor upon sensing any overloading of the motor. Overload relays protect the motor, motor branch circuit, and its components from excessive heat caused by overloading. Overload relays being part of the motor starter are wired in series with the motor. It ensures that the current flowing through the overload relay is the same as is flowing through the motor. If the current rises above a certain limit over a certain period of time, the overload relay trips. It then operates an auxiliary contact which interrupts the motor control circuit that de-energizes the contactor that disconnects the power to the motor. Some relays need to be reset manually, the others reset themselves automatically after a certain period of time that automatically restarts the motor. Kinds and working principles of automatic relays are beyond the scope of this book and therefore, are not included in this section.

QUESTIONS (THEORY)

1. Discuss the classification of electrical motors. Explain the basic principle of motor operation.

2. Write a descriptive note on DC motors.

3. Write short notes on single phase AC split phase and capacitor motors clearly bringing out the differences between the two. Illustrate your answer with neat circuit diagrams.

4. Discuss the construction and working of a 3-phase induction motor.

5. Write a descriptive note on star-delta starter. List its advantages and disadvantages.

6. List various activities required to take care of the motor at the farm.

7. Name and describe the various protection devices used on motors.

8. Write short notes on motor starter (DC motors) and direct on-line starter.

Chapter 6

Farm Mechanization

MECHANIZATION DEFINED

Several people have defined farm mechanization in their own way but the crux of all the *definitions is replacement of bullocks, horses and other draught animals or human labour by machine power for performing farm operations.* Thus, mechanized agriculture may be defined *as the process of using agricultural machinery to mechanize the agricultural operations to increase the productivity of farm workers.* Depending upon the nature of the works performed, farm mechanization has been grouped as mobile mechanization and the stationary types of mechanization. Similarly, mechanization is said to be partial when only a part of the farm work is done by machines. On the other side, mechanization is said to be complete when animal and human labour is completely dispensed by machines.

NEED OF FARM MECHANIZATION

India's population is likely to stabilize at or slightly higher than 1.6 billion by the end of 2050. It will roughly make about 17 per cent of the world's population at that time. India is also home to 15 per cent of the global livestock. Even though food grains production in India is increasing significantly over the years, yet land and water being limited, the stress on their availability and rising food demand are subjects of hot discussion. Rising population, urbanization, industrialization, improved living standards, changing eating habits, increasing land use changes, and increasing pollution of land and water resources are likely to change the demand supply scenario in a big way. The only way to ensure food and nutritional security of the nation is through higher productivity per unit land and water, being quite low in comparison to other developed nations. Although multi-disciplinary approaches are being advocated, success of all these approaches hinges on timely completion of farming activities and their quality. Enhanced farm mechanization is the only key to meet these two requirements.

The agriculture sector in India, for a long time, has been dependent upon cheap and surplus labour in the rural areas. It is not going to continue for long. Many states have already started feeling the pinch of labour shortages or high labour wages or both. Urbanization, industrialization and increasing opportunities in other sectors are weaning away the farm labour from the agricultural sector. Clearly, shortage of labour and high labour wages will strongly propel farm mechanization. Consequently, the more labour-intensive operations, such as pumping of irrigation water, land preparation and threshing are being mechanized first followed by other operations. An added advantage of farm mechanization is that it will improve the efficiency of limited farm manpower.

Drudgery and physical exertion are typical to Indian agriculture farm operations. Often a farmer is associated with a man with a *kassi* in his hand. This picture scares away educated persons to opt for agriculture. There is an urgent need to minimize human labour especially when there is a great potential of doing that. For example, development of improved riding type animal drawn machines can lessen the drudgery of walking behind, although it may not be a proper substitute for the tractor drawn machines. With mechanical power a farmer is able to control a large area and perform all activities in lesser time permitting him to have some free quality time. Farmers of India like their counterparts elsewhere want to improve their income, lifestyle and general well-being. They have a ray of hope in farm mechanization to achieve this goal.

As already discussed in the first chapter, the level of farm power has seen strong growth through the last two decades, although much remains to be achieved. India is way behind the USA, Europe, Brazil and even China in farm mechanization with a low penetration of less than 50 per cent. Moreover, the level of mechanization varies greatly by region. Northern states such as Punjab, Haryana and Uttar Pradesh have high level of mechanization (70-80 per cent overall; and 80-90 per cent for rice and wheat) mainly attributed to highly productive lands and declining labour force. It is because the development in farm mechanization is very closely related to the shortage of labour and industrial development of a region. The state governments in these states have also provided timely support in promoting mechanization. The eastern and southern states have lower level of mechanization (35-45 per cent). It is because of smaller and more scattered land holdings. For example; power availability in Odisha had been only 0.60 kW/ha compared to 3.7 kW/ha in Punjab during 2012. In the north-eastern states, the level of farm mechanization is extremely low mainly due to hilly topography, high transportation cost of farm equipment, and socio-economic condition of the farmers. These factors together make a good cause of strengthening farm mechanization in already mechanized states and encouraging the mechanization in states with low penetration. It is projected that if the states having reasonable water resources can have the same level of mechanization as in Punjab, India's agricultural production can possibly touch 450 million tonnes. In short agricultural mechanization is aimed at one or more of the following objectives.

☆ To increase the efficiency of production processes

☆ To minimize farm drudgery

☆ To help the farmers to practice large scale farming

☆ To increase production and productivity

☆ To minimize losses due to natural hazards

☆ To bridge the gap between the demand and supply to control prices of food items

☆ To obtain better quality of farm produce

☆ To ensure timeliness of farm operations

☆ To maximize the profits from farm enterprise

☆ To increase farmer's income and improve their standard of living

☆ To save labour for release to other sector of economy such as industries

☆ To make farm work easy, interesting and attractive to youths

☆ To supply agro-based industries with adequate and good quality raw materials

☆ To encourage private investments in agriculture

☆ To facilitate the establishment of small scale industries and processing units in the rural areas thus taking care of employment of youth in the village

BRIEF HISTORY AND CURRENT STATUS

Historically India faced a severe food crisis during the sixties. Expansion of irrigation, introduction of high yielding varieties and assured procurement of main crops at minimum support price brought a sea change in agricultural practices, farmer's income and farmers' attitude to farm mechanization. As high yielding varieties needed assured irrigation, farmers not owning tube wells began purchasing water from their neighbours. Probably, this was the beginning of custom hiring of farm equipment (tube well, engine and electric motor driven irrigation pumps) and an important milestone in farm mechanization in India. Irrigation pumps in India have increased from 0.4 million to more than 25 million during 1960 to 2010. Then there came the era of tractorization. Although, tractors alone are not the farm mechanization but is the basic mechanical input that largely determines the extent of use of allied machinery and equipment. There were only about 37000 tractors (1 tractor/3600 ha) in India in 1967, which went up to 4 million tractors that translates into 36 ha/tractor in 2010 (Table 6.1). It is likely to increase to 20.36 million by 2032-33. Today Indian tractor industry has emerged as the largest in the world and account for about 1/3rd of the total global tractor production.

Introduction of new wheat varieties posed a new challenge as most of the varieties were not amenable to the traditional way of threshing. Necessity is the mother of invention. Farm industries came with the idea of thresher. In fact, initially many threshers were fabricated by local black-smiths in small towns using locally available materials. The threshers became popular and a new era of custom hiring began in the farm sector. Threshing was normally done on share basis during the early periods until more and more farmers procured their own threshers. Without going into the details of each and every machine, many new innovations have been introduced. Combine harvesters, laser land leveling, straw management systems,

Table 6.1. Over the Years Changes in some Important Aspects of Indian Agriculture

Item	1960	Year (as stated)
Agricultural land (M ha)	133	140.1 (2014-15)
Irrigation pumps (million)	0.4	25 (2010)
Irrigated area (per cent)	19	68.4 (2014-15)
Cropping intensity	1.15	1.42 (2017-18)
Fertilizer use (kg/ha)	2	128 (2017-18)
Grain yield (kg/ha)	700	2223 (2017-18)
Tractors (thousands)	37 (3600 ha/tractor)	4000 (36 ha/tractor in 2010)
Power tillers (thousands)	0.9	155 (2010)
Draft animals (million)	80.4	50 (2010)

sub-soilers, maize threshers, auger boring machine for plantations are some such introductions that may go a long way in ensuring farm mechanization of the country. It looks that India is on the right path and going steadily in the direction of farm mechanization. In fact, farm mechanization in India has followed the same trend as observed elsewhere around the globe (Table 6.2).

☆ Farm operations like tillage, transport, water pumping, milling, threshing, *etc.* were mechanized first, being high power input and low control operations

☆ Farm operations like seeding, spraying, intercultural operations, *etc.* were mechanized next, requiring medium levels of power and control

☆ Farm operations that require high degree of control and low power inputs (such as transplanting, planting of vegetables, harvesting of fruits and vegetables, *etc.*) are the ones that were mechanized last

Table 6.2. Relative Requirement of Power Inputs or Control in different Farm Operations

Operation	Highly Power Intensive	Intermediate Level	High Degree of Control but less Power Intensive
Water pumping	√	-	-
Tillage	√	-	-
Direct seeding	-	√	-
Transplanting	-	-	√
Weeding	-	-	√
Plant protection	-	√	-
Harvesting	-	√	-
Threshing	√	-	-
Milling	√	-	-
Transport	√	-	-

Similarly, mechanization technologies were first adopted by the large farmers followed by medium scale farmers. Punjab, Haryana and western Uttar Pradesh witnessed greater mechanization because large numbers of such farmers in these states proved catalytic and instrumental in bringing change in the mindset of farming community. They also helped establish a viable agricultural machinery and implements distribution services sector.

Agricultural policy of the Punjab outlines that due to intensive agriculture and short window of time available between harvesting of one crop and sowing of the next, and shortage of labour, farm machinery is crucial for timeliness and precision in farm operations. Increase in cropping intensity, change in cropping pattern, use of high yielding varieties, good cultivation practices, *etc.* demand mechanization of field operations. Water saving, cost-effective and yield promoting farm machinery has already been introduced and made available to the farmers on custom hiring basis through cooperatives. Singh *et al.* (2013) compared the growth of farm machines in Punjab during 1960-61 and 2012-13 (Table 6.3). The department has provided specifications of various machines commonly used in Punjab for quality control and subsidy purposes (Table 6.4). Singh *et al.* (2015) reported the annual market size of agricultural machines in India (Table 6.5). These data sets prove that India is on right track of farm mechanization.

Table 6.3. Growth of Farm Machines and Equipments (000 numbers) in Punjab

Machines	1960-61	2012-13
Tractors	7.9	476.8
Disc harrows	4.0	201.1
Rotavators	0.0	8.7
Cultivators	0.0	460.0
Laser levelers	0.0	7.2
Seed-cum-fertilizer drills	0.0	175.3
Spray pumps	0.0	625.0
Combine harvesters	0.0	13.8
Reapers	0.0	5.5
Straw reapers	0.0	38.7
Threshers	0.0	660.3
Maize shellers	0.0	1.9
Potato planters	0.0	5.6

The crop-wise and operation-wise farm mechanization in the country reported in Table 6.6 shows wide variations (Mehta *et al.*, 2019). For example mechanization in harvesting and threshing of rice and wheat is around 60-70 per cent while it is in the range of 0 to 30 per cent for other crops (Table 6.6). The figures may also vary widely from state to state, region to region and even within the sources of information. It is because of the way the values are calculated by the various authors. For example PWC and FICCI (2019) report that seed-bed preparation activity for most crops is mechanized to the extent of 80-90 per cent.

Table 6.4. List of Commonly Used Machines Approved by Punjab Government for Subsidy

Type of Farm Operation	Name of the Machine
Land leveling	Laser guided land leveler
Tillage and seedbed preparation	Rotavator, Rotary plough (Power harrow), Pulverizing roller an attachment for cultivator, Reversible mould-board plough
Seeding and planting equipment	Seed-cum-fertilizer drill, Zero till drill, Spatial zero till drill, DSR drill/planter, Pneumatic planter, Happy seeder, Raised bed planter, Multi-crop planter, Ridge planter, Self-propelled paddy transplanter, Walk behind paddy transplanter
Harvesting and threshing	Self-propelled reaper binder, Self-propelled fodder harvester, Multi-crop thresher, Vertical conveyer reaper, Forage reaper
Weeding and spraying equipment	Power weeder, Knapsack sprayer, Battery operated knapsack sprayer, Power operated spray pump, Tractor operated hydraulic sprayer, Aero blast sprayer, Orchard aero blast sprayer
Crop specific machines	Sugarcane trencher, Sugarcane *tirphali*, Sugarcane ridger, Potato digger, Semi-automatic potato planter, Automatic potato planter, Combine harvester with maize header, Maize sheller, Maize thresher, Portable maize dryer, Fodder chopper- cum-loader, Cotton see drill/planter
Others	Sub-soiler, Post hole digger, Tractor operated single row harvester cum shredder, Rotary mulcher, Straw reaper, Paddy straw chopper shredder, Bailer, Disc mower, Gyro rakes, Nursery trays, Nursery seeding machine

Table 6.5. Annual Markets of Major Farm Machines used in India

Items	Numbers	Items	Numbers
Tractors	450000 - 500000	Power tillers	50000 - 60000
MB plough	45000 - 50000	Rotavator	100000 - 120000
Cultivators	150000 - 200000	Harrows	120000 - 150000
Seed-ferti drills	60000 - 75000	Planters	15000 - 25000
Rice transplanters	2000 - 3000	Threshers	60000 - 75000
Combine harvesters	3500 - 4000	Trailers	150000 - 175000
Sprayers	10000 - 15000	Laser land levelers	2500 - 3500
Potato diggers	25000 - 30000	Rotary hoes/Power weeders	20000 - 25000

Benefits of Farm Mechanization

Farm mechanization provides a number of economic and social benefits to farmers. Some of these benefits are discussed in the following sections.

Economic Benefits

The most visible and important direct benefit is in the form of higher yields. The crop yields are known to increase with increasing farm mechanization. A direct relationship between farm mechanization (farm power availability) and farm yield has been established through various studies. Food grain productivity in India has

Table 6.6. Crop-wise and Operation-wise Farm Mechanization in India

Major Crops	Mechanization Levels for Field Operations (Per cent)			
	Seed-bed Preparation	Sowing/Planting/ Transplanting	Weeding, Intercultural and Plant Protection	Harvesting and Threshing
Rice	70	20	30	60
Wheat	70	60	50	70
Maize	60	40	30	30
Sorghum/millets	50	30	15	10
Pulses	50	40	20	25
Oilseed	50	40	20	25
Cotton	50	30	25	0
Sugarcane	55	10	20	10

increased from 0.636 t/ha in 1965-66 to 2.20 t/ha in 2013-14, as the farm power availability increased from 0.32 kW/ha to 2.11 kW/ha during the same period. It is estimated that food grain productivity will increase to 3.96 t/ha with farm power availability of 4.81 kW/ha by 2032-33. The increase in farm productivity is attributed firstly to the timeliness of operation and secondly the good quality of work. In the absence of mechanical power, operations like seedbed preparation and cultivation are either partially done or sometimes completely neglected, resulting in low yields due to poor growth or untimely harvesting or both. Improved implements can help to increase productivity up to 30 per cent. Timely operations also result in increased cropping intensity to the extent of 5-20 per cent. All these add to the income of the farmer. From an economic point of view, reduced costs and higher profits make farm mechanization a win-win situation.

Timeliness of Operations

There is very little scope of horizontal expansion of cultivated area, which has remained around 140 M ha over the last few decades. Increasing competition for best quality land resource from other sectors may reduce the cultivated area. The only way to increase the area under cultivation is through vertical expansion. It is projected that cropping intensity needs to increase from an average of 142 per cent in 2017-18 to 200 per cent or more. As the cropping intensity increases, the turnaround time will be drastically reduced. Thus, it may not be possible to harvest and thresh the standing crop on one hand, and prepare seedbed and do timely sowing operations of the subsequent crop without a fair degree of farm mechanization. It will need adequate farm power and matching implement to complete all activities in time.

Input Savings

Farm mechanization is known to result in input savings. Savings of seeds (approximately 15-20 per cent) and fertilizers (approximately 15-20 per cent) can be expected from use of seed-cum-fertilizer drills mainly due to higher nutrient use efficiencies.

New Avenues of Employment

Farm mechanization provides different streams of employment related to handling of farm machines thus resulting in increased rural employment. Increased farm mechanization is a key step towards doubling farmer's income and better rural prosperity.

Increased Efficiency

Farm machinery helps in increasing the efficiency of farm labour and reducing drudgery and workloads. Mechanization enhances the farm production per worker. By its nature it reduces the quantum of labour needed to produce a unit output. It is estimated that farm mechanization can help reduce time by approximately 15-20 per cent. Various factors help to reduce the cost of cultivation up to 20 per cent.

Reduced Post-harvest Losses

Farm mechanization helps in timely harvests saving the farmers from vagaries of nature. Timely harvest helps in maintaining the quality of produce as well as reduces the post-harvest losses. These benefits help the farmers to earn more income.

Improved Agricultural Operations

Some activities considered to be unmanageable can be accomplished through farm mechanization. The tractor implements can help to smoothen hillocks, fill depressions and gullies and eradicate deeps-rooted weeds. Sub-soiling can be easily performed to break hard pans. Mechanical contour *bunding* and terracing using self-propelled graders and terracers can help prevent soil erosion. Thus, besides improved productivity, hitherto uncultivated lands can be brought under cultivation.

Social Benefits

Some social benefits of farm mechanization are listed as under.

- ☆ Ownership of tractor and implements raises the social status of the farmer in rural setting
- ☆ Encourages farmers to do custom hiring and other allied jobs for additional income
- ☆ Rural youth is encouraged to join farming besides attracting more people to work and live in rural areas
- ☆ Decrease in workload of women as a direct consequence of the improved efficiency of labour freeing them for household or other leisurely activities
- ☆ Improved safety in performing farm operations

Besides what is stated in the foregoing points, farm mechanization leads to the following.

- ☆ Commercialization of agriculture
- ☆ Solves the problem of labour shortage at critical junctures and makes them available for other sectors

A summary of some important benefits discussed in the foregoing sections is reported in Table 6.7.

Table 6.7. Benefits of Agricultural Mechanization

Benefits	Value (per cent)
Saving in seed	15-20
Saving in fertilizer	15-20
Saving in time	20-30
Reduced labour requirement	20-30
Increase in cropping intensity	5-20
Productivity increase	10-15

Ill-effects of Farm Mechanization

Any technology comes with its benefits and ill-effects. In the case of farm mechanization, ill-effects are only due to improper use of the machines or poor implementation strategies. Some ill-effects are listed so that we can draw lessons for making appropriate corrections in future plans.

Accidents: When threshers were introduced there were large numbers of accidents. Many labourers lost their limbs. It was mainly because the illiterate labours were engaged on the machines without giving them proper training and guidance. Later on attachments were introduced to minimize or altogether eliminate the possibilities of such accidents.

Over-exploitation of groundwater: Due to increasing number of electric motor operated pumps (estimated to be about 25 million in the year 2015) and number of diesel operated irrigation pumps (about 7 million in 2015), groundwater table across states is declining. At few places in Punjab and Haryana water table is declining at the rate of 1 m or more per annum. In both the states groundwater development exceeds 100 per cent of replenishable recharge. It can mainly be attributed to farmers switching over from traditional to water guzzling cropping patterns, poor delivery of canal water, government policy of free electricity for agriculture and similar other causes rather than the mechanization of irrigation.

Environmental pollution: Environmental pollution due to burning of crop residues is a recent phenomenon often attributed to combine harvesting of crops. The combine harvesters have become popular mainly because of potential cost savings, reduced labour management problems and timeliness of farm operations. On the other hand, residue recovery from combine harvested fields is difficult because combines leave more residues that too unevenly spread over the harvested fields. Therefore, farmers resort to straw burning especially after rice (rice residues are burnt on a much larger scale than wheat residues) and to some extent in wheat. It is a major phenomenon in the Indo-Gangetic plains and is an important source of atmospheric pollution in this region. Since combine harvesters have come to stay, numbers of alternatives are emerging to manage residues. Straw management systems on combine harvesters, direct sowing of wheat crop in rice residue using happy seeder, use of bio-decomposers to decompose the residues and mixing in soils, use of bailers for transport to industries and feed preparation are some of few alternatives that are being tried to resolve the issue of residue burning.

Farm suicides: There is a significant change in the psych of the farming community. Ownership of the tractor has become a prestige symbol in the village. A farmer invests on the tractor without taking into account whether he needs it or not. Small farmers with lesser holdings can do without a tractor unless they have plans to do custom hiring work. Extension organizations can help the farmers in taking knowledgeable decisions on this issue.

CONSTRAINTS IN FARM MECHANIZATION

India is a very diverse country having widely varying agro-climatic conditions, land holdings, and socio-economic conditions of the farming communities. A common factor to most farmers is their low income per capita because of low yield per hectare. The only way to increase their income is through farm mechanization. Though a lot of efforts are made by the center and the state governments, NGOs and other agencies to implement farm mechanization policies, but there are a number of constraints which hinder the effective implementation of farm mechanization schemes. So much so that few people have formed an unfounded opinion that Indian agriculture cannot be fully mechanized because of the following constraints:

1. Few people think that there is a surplus of agricultural labour and enough draft animals in the country to do the farm work. A large labour and animal force will be displaced, if farm mechanization proceeds as envisaged. This fear is completely unfounded. Contrarily, India's increased food requirements must be met through increased productivity of the land from higher yields and multiple cropping. It will generate additional labour demand to carry out different farm operations. Additional avenues of employment for the rural youth may come from establishing farm machinery repair and maintenance infrastructure

2. The size of farm holdings of the majority of the Indian farmers is too small to justify the use of a tractor on their farms. Moreover, abundance of dry land agriculture with low productivity may also affect the pace of farm mechanization

3. The socio-economic conditions of the farmers may not permit them to buy a tractor and tractor-drawn implements

4. The farmers may not be able to adapt to technicalities involved in farm mechanization

5. Mechanization may not prove to be as effective as in advance countries because of inadequate and ill-equipped repair and maintenance shops, poor after sales service, inadequate infrastructure for agro-processing and marketing of agro produce

6. Mechanization may not result in lowering the cost of production

7. There is little liaison between government departments and industries

8. Absence of long-term agricultural mechanization policy

9. Lack of proper mandatory safety standards and legislative measures

10. Proliferation of improper and sub-standard machinery designs

11. High operating cost due to low annual use of machinery

12. Absence of farm machinery management data regarding use-patterns, annual-use, breakdown frequency, repair and maintenance cost, reliability *etc.* to take informed decisions for guiding the farmers

Government of India flagged some key challenges for the agricultural sector before initiating the 12th Five-Year Plan. The two key challenges included shortage of farm labour and inadequate mechanization. Most of the constraints in farm mechanization can easily be solved without much difficulty. But there is no dispute regarding the need to strengthen mechanization to boost agricultural production and productivity to stabilize the economy. Fortunately, evidences generated so far have also belied many of the listed constraints. Mechanization adopters have shown that mechanization has positively impacted their income, and improved their lifestyle and general well-being.

Small and Marginal Farmers and Farm Mechanization

One of the most favourable features of the success of any farm mechanization program is the existence of large farms that allows proper utilization of agricultural machines. As against the farm sizes in the USA, Europe and Russia where mechanization is to the tune of 95 per cent, India has extremely small size of holdings being less than 5 ha. This segment constitutes over 80 per cent of the land holdings and the bulk of them having land holdings of less than 2 ha. The penetration of farm mechanization is lowest in this segment. Since this segment has an important role to play in improving agricultural production, there is a great potential for increasing the penetration by resolving specific problems of this segment. The major problems faced by this group and suggested solutions are listed in Table 6.8.

Table 6.8. Constraints and Proposed Solutions for Farm Mechanization on Small and Marginal Land Holdings

Constraints	Solutions
Small farm size	Land clustering, consolidation and cooperative and contract farming
Lack of capital	One window easy and adequate financing credit facilities, subsidy
Fear and lack of risk taking capacity	Confidence building through tours and training
Lack of appropriate sized machines	Investments in R and D for developing machines for various sized fields, establishment of custom hiring centers, machine banks and self-help groups
Lack of knowledge	Strengthening the extension wing of department of agriculture
Lack of training facilities	Strengthening of extension wings and increasing association of research organizations and NGOs
Other problems	Innovative solutions by field officers

In summary, small farms usually do not present an economically lucrative proposition to permit extensive use of agricultural machinery given the constraint of limited use of machinery. In such situations, operational and capital costs may be optimized for them by making the machinery available through custom hiring centers. In that case even small farmers may be able to get the benefit of agricultural mechanization.

GOVERNMENT INITIATIVES ON FARM MECHANIZATION

The Government of India initiatives on promoting farm mechanization have come through various national programs. Important amongst them are: Macro Management of Agriculture (MMA), *Rashtriya Krishi Vikas Yojana* (RKVY), National Food Security Mission (NFSM), National Horticulture Mission (NHM), *Gramin Bhandaran Yojana* (GBY), Integrated national mission on agricultural mechanization aiming at catalyzing an accelerated but inclusive growth of agricultural mechanization in India, Scheme on Promotion of Agricultural Mechanization and Machinery for *in-situ* Management of Crop Residues in the States of Punjab, Haryana, Uttar Pradesh and NCT of Delhi, *etc.* Just to illustrate, MMA launched in 2001 focused on provision and promotion of hybrid seeds, farm mechanization and integrated cereal-development programs. National Food Security Mission (NFSM) launched in 2007 aimed to increase production of rice, wheat and pulses through technologically superior seeds, soil management and farm mechanization techniques. The interventions covered under NFSM-Rice include demonstrations of hybrid rice and HYVs, cono-weeder, manual sprayer, power sprayer, drum seeder, pump set (up to 10 HP), seed drill, multi-crop planter, zero till multi-crop planter, power weeder, rotavator, paddy thresher/multi crop thresher, laser land leveler, self-propelled paddy transplanter and assistance for custom hiring. A sub-mission on Agricultural Mechanization (SMAM) under the National Mission on Agricultural Extension and Technology envisages to demonstrate the agricultural machinery on the farmers' fields to increase farm mechanization and productivity, test and evaluate machines through identified testing centers to ensure quality and performance, support the custom hiring centers (CHCs) of agricultural machinery, create hi-tech hubs to ensure the availability of agricultural machinery, and increase trained and skilled personnel, *etc.* A list of equipment available on subsidy under SMAM is reported in Annexure II. It gives a fair view of wide range of equipment being encouraged by the Government of India under this program. The SMAM, a seven-year plan effective from 2016-17 to 2021-22 has identified new targets such as (FASAR, 2016):

☆ Farm power availability of 2.5 kW/ha to be achieved by 2022 from the level of 2.02 kW/ha in FY18

☆ Average level of mechanization for different farm operations to be increased to 50 per cent

☆ 148,000 trainees to be trained to develop skilled manpower in farm mechanization sector

☆ 10,270 agricultural machines to be tested

☆ To establish 280,000 CHCs at the village level

☆ 19,000 demonstrations to be organized on farmer fields

☆ Farm machinery to the tune of 1900,000 to be distributed under SMAM

☆ 8 new Farm Machinery Training and Testing Institutes to be established

☆ 200,000 beneficiaries to get farm machinery for individual ownership in north-eastern and Himalayan region

☆ Establish hi-tech hubs and farm machinery banks to manage high value crops at village level

Business and enterprise friendly policies, laws and regulations, physical and institutional infrastructures to encourage commercial activities and entrepreneurship in farming, input supply and produce handling, and processing and marketing are some key factors through which government is encouraging agricultural mechanization. Government, NGOs, research organizations and private entrepreneurs should join hands to work on the following policy framework to improve the state of farm mechanization in India.

POLICY FRAMEWORK

Mechanization of agriculture in India has advanced considerably. The level in several regions has gone far ahead of the average national level. The predominance of human and animal power sources on Indian farms is no more visible. The annual addition of various equipment is quite impressive (Table 6.5). Nonetheless, we have not reached the desired penetration levels in several regions. Following policy framework may be considered for promoting farm mechanization in the country.

☆ Government subsidies may be reduced or eliminated altogether. The funds so saved should be used to provide soft loans and to reduce taxes on equipment and machinery for agricultural operations and food processing

☆ To take care of small and marginal farmers custom hiring centers and machine banks should be established in cooperative mode or by involving private entrepreneurs

☆ Manufacturing facilities should be developed in areas with low level of farm mechanization by providing incentives to manufacturers to establish such facilities in these areas

☆ Cooperative and contract farming should be promoted to provide economical mechanization solutions to the farmers

☆ Research and development organizations should be mandated to develop machines for mechanization of horticulture, hill agriculture and gender friendly equipment for women farmers

☆ Promotion of crop-and state-specific mechanization priority plan, introduction of specialized equipment for small and marginal farmers through in-house manufacturing, focus on scale- and gender-neutral machines, farm waste management machinery and climate smart mechanization should be the new initiatives for future ready India

☆ Special incentives may be provided to the small and marginal farmers for purchase of power tillers and related equipment

✩ Skill development in the area of operation, repair and maintenance of farm machines should be taken up to strengthen local farm mechanic networks

✩ In order to reduce the input costs through higher input use efficiency of farm inputs, equipment for high-tech interventions such as site-specific nutrient management should be developed and made available to the farmers

✩ Post harvest technologies of on-farm storage, processing and marketing facilities deserves special attention on Indian farms

✩ Bureau of Indian Standards (BIS) must frame stringent quality standards of farm machines to bring confidence amongst farmers for adoption of farm mechanization

✩ Integrate and strengthen support services for research and development; testing and standardization; and human resource development to support agricultural mechanization

QUESTIONS (THEORY)

1. Discuss the need of farm mechanization especially in the Indian context.

2. Discuss the brief history of farm mechanization and its current status in India.

3. List and briefly describe the benefits of farm mechanization. List its ill-effects, if any.

4. List various constraints that act as impediments in farm mechanization.

5. List various issues that should form a part of policy framework to promote farm mechanization in India.

Chapter 7

Primary Tillage and Related Equipment

TILLAGE DEFINED

Tillage operations in various forms have been practiced since the beginning of agriculture. Pre-historically, it is believed that man used crude tools of wood or other material for loosening the soil. Later, he made use of stone tools to disturb the soils to place the seeds. The word tillage is derived from 'Anglo-Saxon' words Tilian and Teolian, meaning to plough and prepare soil for seed to sow, to cultivate and to raise crops. Thorough ploughing is necessary to make the soil into fine particles, suggested Jethrotull considered to be the father of tillage. *Tillage has been defined as the mechanical manipulation of the soil for the purpose of crop production.* Soil tillage consists of breaking the compact surface of earth to a certain depth to loosen the soil mass to enable the roots of the crops to penetrate and spread into the soil. Tillage significantly affects the soil characteristics such as soil water absorption and conservation besides altering soil temperature distribution in the profile and infiltration and evapotranspiration processes. A good seedbed usually implies finer particles and greater firmness in the vicinity of the seed. Tilth, on the other hand is the physical condition of the soil resulting from tillage. The tilth has been categorized as coarse tilth, moderate tilth or fine tilth, the cost of operation increasing as the tilth goes from coarse to fine. Until very recently, Indian farmers believed that finer the tilth, better is the productivity. But now this viewpoint is changing giving rise to several new tillage technologies and terminologies.

Objectives of Tillage

- ☆ To remove the hard pan to prepare a deep seedbed or a root bed suitable for different types of crops
- ☆ To improve soil structure

☆ To increase water absorbing capacity of the soil resulting in rapid infiltration and good retention of rainfall

☆ To provide adequate air capacity and exchange within the soil and to minimize resistance to root penetration

☆ To destroy weeds or to remove unwanted crop plants (thinning)

☆ To mix and manage plant residues from decomposition standpoint as well as from the standpoint of reducing soil erosion in such a way that it adds more humus and fertility to soil by covering the vegetation

☆ To minimize soil erosion by following contour tillage, listing and proper placement of trash

☆ To establish specific surface configurations for planting, irrigating, drainage and harvesting operations etc.

☆ To incorporate and mix fertilizers, pesticides, and soil amendments etc. into the soil

☆ To destroy the insects, pests and their breeding places. Tillage exposes the lower soil to external environment of the atmosphere. As a result, insect-pests come out to the upper layer of soil and are eaten by birds and predatory animals

All these objectives are achieved by a suitable combination of disturbing, opening up and turning over the soils.

Tools/Implement/Machine

The terms tool/implement/machine are commonly used to describe various equipment that are used in the process of agricultural production. Strictly speaking there is little difference between the terms tools, implement and machine. Most of the times these terms are used synonymously, but a difference is sometimes made on the basis of size or on the basic of working element. For example *a tool is small equipment having an individual working element such as a khurpa or shovel*. It is mostly used manually. *Implement on the other hand, is larger equipment generally having no driven/moving parts or have only simple mechanism such as harrow and plough*. Implements are pulled or pushed by the machinery to be able to perform their designated tasks. *The term machine is used to describe equipment that consists of a combination of rigid or resistant bodies having definite motions and capable of performing useful work such as a tractor or combine harvester*. In other words, it is a mechanized equipment which can run on electric or diesel or petrol or hydraulic to apply force and control movement for an intended action.

TILLAGE CLASSIFICATION

Tillage operations have been characterized in various groups based upon the time, purpose, and kinds. On the basis of time, tillage has been characterized as seasonal tillage and off-season tillage. Tillage operations performed at the onset of the crop season for raising crops are known as seasonal tillage. These operations include preparatory tillage, planking, layout of the seedbed including intercultural operations. Tillage operations for conditioning the soil before the forthcoming main season crop are called off-season tillage. Off-season tillage may include post harvest tillage, summer tillage, winter tillage, sub-soiling and fallow tillage. Tillage normally referred as preparatory tillage is classified as primary or secondary tillage discussed in the forthcoming sections. Currently, tillage is used in a wider context and covers the whole sequences of operations that manipulate the soil in order to produce a crop. Operations include tilling, planting, fertilization, pesticide application, harvesting, and residue chopping or shredding. Following sections discuss various kinds of tillage operations.

Preparatory Tillage

Preparatory tillage refers to tillage operations that are performed to prepare the field for raising crops. It comprises deep opening and loosening of the soil in order to achieve desirable tilth. It also helps uproot weeds and incorporate these and other crop stubble into the soil. Preparatory tillage is further categorized as primary tillage and secondary tillage. The essential prerequisite for both these operations is to bring the soil to a workable condition.

Primary Tillage

The operations performed to open up any cultivable land with a view to prepare a seedbed for growing crops is known as primary tillage. More often it is also referred as ploughing since it involves opening of the compact soil with different kinds of ploughs. It is the first major soil working operation after the last harvest designed to deep plough the soil. It is normally performed when the soil is wet enough to allow ploughing and strong enough to give reasonable levels of traction. The objectives of primary tillage are

☆ To reduce soil strength

☆ To rearrange aggregates

☆ To cover plant materials and bury weeds

☆ To kill insects and pests

Primary tillage is accomplished through primary animal or tractor drawn tillage implements. Amongst the animal drawn implements are the indigenous plough and mould-board plough. Tractor drawn implements include mould-board plough, disc plough, sub-soil plough, chisel plough and other similar implements. Rotavators and reversible mould-board ploughs are new additions to this group.

Secondary Tillage

Tillage operations performed to create proper soil tilth for seeding and planting following primary tillage are known as secondary tillage. These operations are generally restricted to the surface soil and are lighter but finer. These operations help to condition the soil, and improve the seed bed through increased soil pulverization to meet different tillage objectives.

Secondary tillage is usually shallower and less aggressive than primary tillage. These operations do not cause much soil inversion and shifting of soil from one place to other. Compared to primary tillage operations, the implements used in secondary tillage, often called as secondary tillage implements, consume less power per unit area. They include different types of harrows, pulverizers, cultivators, levelers, weeders, sweeps, clod crushers, *bund* formers, ridge ploughs and special tools for surface tillage to conserve moisture and destroy weeds *etc.* Planking is done to break/crush the hard clods, level the soil surface and to lightly compact the soil. The main objectives of secondary tillage are:

☆ To break big clods and prepare uniform and leveled soil surface as required for the preparation of a seedbed

☆ To provide adequate seed-soil contact to permit water flow to seed and seedling roots

☆ Surface configuration for seeding and planting

☆ Manage water and air in the soil

☆ Incorporation of fertilizers

☆ Puddling

☆ Control weeds and soil-borne insects pests and diseases

☆ To destroy grasses and weeds in the field

☆ Manipulate plant residues and farm wastes, to cut and mix them with topsoil

☆ Establish a surface layer which prevents wind and soil erosion

Kinds of Tillage Operations

Layout of Seedbed

This operation is also a component of the preparatory tillage. Leveling board, buck scrapers *etc.* are used for leveling and markers are used for layout of seedbed.

Inter Tillage (after cultivation)

The tillage operations carried out in the standing crop after the sowing or planting operations but prior to the harvesting of the crop plants are called after tillage. A common term for these operations is inter-cultivation or post seeding/planting cultivation. It includes hoeing, weeding, earthing-up, drilling or side dressing of fertilizers *etc.* Spade, hoe, weeders *etc.* are used for inter cultivation.

Off-season or Special Purpose Tillage

Tillage operations intended to serve some special purposes that helps to rejuvenate the soils are known as special purpose tillage.

Clean Tillage

It refers to working of the soil over the entire field in such a way that no living plant is left undisturbed. It is practiced mainly to control weeds, soil borne pathogen and pests.

Blind Tillage

It refers to a tillage operation after seeding or planting the crop (in a sterile soil) either at the pre-emergence stage of the crop plants or at early growth stages so that crop plants (say sugarcane, potato *etc.*) are not damaged, but extra plants and broad leaved weeds are uprooted.

Dry Tillage

Dry tillage is practiced to make the soil more porous and soft for crops that are sown or planted in dry land condition having sufficient moisture for germination of seeds. It is suitable for crops like broadcasted rice, jute, wheat, oilseed crops, pulses, potato and vegetable crops. Dry tillage is done in a soil having sufficient moisture in the range of 21-23 per cent. It also helps to improve the water holding capacity of the soil and in improving soil aeration. The soil condition becomes quite favourable for soil micro-organisms.

Wet Tillage

Wet tillage is performed mainly as a land preparation operation for transplanting semi-aquatic crops such as rice. This operation, also known as puddling, is done on a land having standing water. The operation consists of ploughing repeatedly in standing water until the soil becomes soft and muddy. Planking helps to make the soil level and compact. Puddling prepares a soft seedbed for planting rice seedlings while creating an impervious layer below the surface to reduce deep percolation losses of water. Puddling hastens transplanting operation as well as helps in the establishment of seedlings. The operation significantly reduces the weed problem. Puddling is also performed to incorporate green manures and weeds. On the minus side, wet tillage is known to destroy soil structure. The soil particles separated during puddling settle later to form a hard layer affecting the free water movement through the soil.

Sub-soiling

Hard pans may occur naturally or may be formed beneath the plough layer as is commonly observed in the rice-wheat systems where conventional rice transplanting is practiced. Special tillage operation (chiseling) is performed to break these hard pans and reduce soil compaction so that roots proliferate to greater depths. Where heavy machines are used for field operations, such as seeding, harvesting and transporting, sub-soiling operation should be performed once in four to five years. The main advantages of sub-soiling are:

- ✰ Greater volume of soil is available for the crops
- ✰ Roots are able to penetrate deeper to extract moisture and nutrients from the lower layers
- ✰ Infiltration rate increases allowing excess water to percolate downward to recharge the water table

☆ Reduced temporary water logging conditions in the cropped fields

☆ Reduced runoff and soil erosion

Conservation Tillage

Recent classification characterizes the tillage operation into two groups.

☆ Conventional tillage

☆ Conservation tillage

Conventional tillage also known as clean tillage comprises ploughing the entire field several times to prepare a seedbed. Conventional-till system leaves less than 15 per cent or less than 560 kg/ha residues cover after planting. The system generally involves ploughing or some other form of intensive tillage. This crop-focused tillage is a major cause of the soil degradation and loss of organic carbon in soils. The old approach is now being replaced by soil-focused tillage approach that improves the damaged soils and preserves soil quality on a long-term basis. The practice known as conservation tillage involves ploughing the field with lesser number of passes over the entire land or ploughing only in the required space of the land needed for sowing of the crop. *Any tillage and planting system covering 30 per cent or more of the soil surface with crop residues after the crop is planted is known as conservation tillage.* It makes about 1,120 kg/ha of flat, small grain residue equivalent on the surface. Thus, conservation tillage practices focus upon managing crop residues. The purpose is to reduce soil erosion by water or wind and improve soil organic carbon besides hosts of other benefits. Different types of conservation tillage are in vogue and are named accordingly. Few categories include no-till or zero-till, ridge till, mulch till and reduced tillage systems.

No-till

No-till or zero tillage is an extreme form in which primary tillage is completely avoided. Secondary tillage is restricted to seedbed preparation in the row zone only. Planting or drilling is accomplished in a narrow seedbed or slot created by coulters, row cleaners, disc openers, in-row chisels, or roto-tillers. Thus, no-till is defined as a system in which the soil is left undisturbed from harvest to planting except for nutrient application. Large populations of perennial weeds appear in zero tilled plots. Although, weed control is accomplished primarily with herbicides but cultivation may be practiced for emergency weed control. Major advantages of zero tillage are: erosion control, moisture conservation, seedling protection, reduced labour requirement, time saving, reduced cost, and oil fuel economy resulting in the environmental protection. Yields may be around the same or even higher in dry years. One of the problems associated with zero till is reduced seedling establishment around 20 per cent less than the conventional methods. Secondly, high dose of nitrogen needs to be applied as mineralization of the organic matter is slow in zero tilled fields. Higher number of volunteer plants and build-up of pests are the other problems.

Ridge-till

In ridge-till also, the soil is left undisturbed from harvest to planting except for nutrient application. Planting is completed in a seedbed prepared on ridges with sweeps, disc openers, coulters, or row cleaners. Residue is left on the soil surface between ridges. Weed control is accomplished with herbicides and/or cultivation. Ridges are rebuilt during cultivation.

Mulch-till

Stubble mulch tillage or stubble mulch farming is a new approach wherein soil is kept protected at all times whether by growing a crop or by crop residues left on the surface during fallow periods. In this system tillage tools such as chisels, field cultivators, discs, sweeps or blades are used. A disc type implement is used for the first operation, if large amount of residues are present. It helps to incorporate some residues into the soil keeping enough residues on the surface. Incorporated residues decompose early and easily. Zone-till and strip-till practices also fall into this category. Strip till is a tillage system in which only isolated bands of soil are tilled. Less than 25 per cent disturbance of row width is considered no-till. More than 25 per cent row width disturbance is considered mulch-till or "other tillage type," depending on the amount of residue left after planting. The traditional tillage and sowing equipment cannot be used for this kind of tillage operation.

Minimum or Reduced-till

Minimum tillage causes considerably less soil disturbance than the conventional tillage. It is aimed at reducing tillage to bare minimum necessary to ensure a good seedbed, rapid germination, a satisfactory crop stand and favourable growing conditions. Reduced-till system leaves 15-30 per cent residue cover after planting or 560 to 1120 kg/ha of small grain residue equivalent throughout the critical wind erosion period. Tillage can be reduced in two ways; firstly, by omitting the operations which do not give much benefit when compared to the cost and secondly by combining agricultural operations. Minimum tillage can be practiced by using improved tilling tools such as rotavators or by row zone tillage where after the primary tillage secondary tillage is done in the row zone only. Plough plant tillage is another method wherein a special planter is used for one run over the field, such that the row zone is pulverized where seeds can be sown.

Other Tillage Systems

Tillage systems that leave less than 30 per cent crop residue after planting are not classified as conservation tillage. However, these systems may meet erosion control goals with or without other supporting conservation practices, such as strip cropping, contouring, terracing, *etc.* Rotary tillage employs rotary action to cut, break and mix the soil. Combined tillage simultaneously utilizes two or more different types of tillage tools or implements to simplify, control or reduce the number of operations over a field.

PLOUGHING

Ploughing is the process of cutting, breaking and inverting the soil either partially or completely. Ploughing essentially means opening of the upper crust of the soil, breaking the clods and making soil suitable for sowing seeds. Normally ploughing is of two kinds. One is normal ploughing wherein ploughing up to a depth of about 15 cm or more is performed. Depending upon the depth of ploughing, it is further categorized as shallow (5-6 cm), medium deep (15-20 cm) and deep ploughing (25-30 cm). The second kind is contour ploughing, where fields are ploughed along the contours. Few important terms frequently used in connection with ploughing of land are as follows:

Furrow slice: The mass of soil cut, lifted and thrown to one side is called furrow slice (Figure 7.1A).

Furrow wall: It is an undisturbed soil surface by the side of a furrow (Figure 7.1A).

Crown: The top portion of the turned furrow slice is called crown (Figure 7.1A).

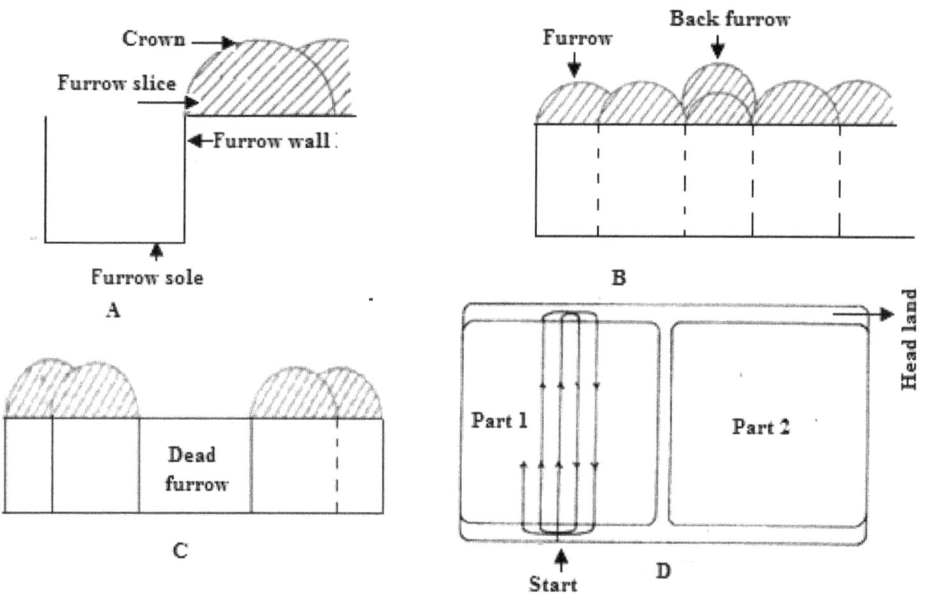

Figure 7.1. The Line Diagrams Showing Few Terminologies used in Ploughing.

Furrow: A trench cut by an implement in the soil during the field operation (Figure 7.1B).

Back furrow: When ploughing is started from center to side, a raised ridge left at the center of the strip of land is called back furrow (Figure 7.1B).

Dead furrow: An open trench (a double furrow) left in between two adjacent strips of land after finishing the ploughing is called dead furrow (Figure 7.1C).

Head land: When a tractor is used to plough the land, a strip of unploughed land is left on each end of the field for the tractor to turn. It is called head land (Figure 7.1D). At the end of each trip, the plough is lifted and turned until in a position to start the return trip. The head land is about 6 m for two or three bottom tractor plough and 1 m more for each additional bottom.

Methods of Ploughing a Land

Soil tillage is an expensive and critical operation in the whole process of crop cultivation. Therefore, attempts are made to use optimal tillage patterns that minimize the time spent in unproductive work. One of the objectives of tillage patterns is to minimize the number of turns and maximize the length of tillage runs. Ploughing pattern to be adopted is governed by the following factors:

- ☆ Kind of implement used
- ☆ Desired field levelness
- ☆ Shape of the fields
- ☆ Efficiency in completing the work

Two main patterns/methods of ploughing the fields are circuitous pattern and headland pattern.

Circuitous Pattern

The plough in this case moves round and round in a field. It is why, circuitous ploughing is also known as round and round ploughing. This system is effective in areas where ridges and furrows are likely to interfere with cultivation. It is commonly used with mould-board, disc and offset disc ploughs. The ploughing can be started in two ways.

Starting at the Center

A small plot of land is marked in the middle of the field, and it is ploughed first. After that, the ploughing is carried out around this small plot until the entire plot is completed. This method is quite uneconomical.

Starting at the Outer End

Conventional ploughing is normally performed by this method. Tractor in this case begins ploughing at one end of the field and then moves on all the sides of the field. Gradually it comes from the sides to the center of the field. To avoid turning with the plough, wide diagonals remain unploughed. There are no back furrows in this method.

Headland Pattern

Three different headlands pattern namely one way pattern, gathering, casting and continuous ploughing, a combination of gathering and casting, are used.

One Way Pattern

The ploughing in this case is done in runs parallel to each other. It starts at one boundary of the field and ends at the opposite with turns being made on the headlands. This system is practiced on big pieces of land and is mainly used for tined implements, rotavators, harrows and reversible ploughs. It is considered to be the most field efficient system. It does not leave any furrows in the field for a correctly set-up and operated equipment.

Gathering

A plough working round a strip of ploughed land is said to be gathering. Every time the tractor and the plough reach the head land, they turn right (Figure 7.2 Top). A raised ridge (double width ridge) is formed at the center of the field with this method of ploughing. This is an uneconomical way of working as time is wasted at the start in making awkward turns, while later, total idle running is increased along the head land. It can prove to be a good practice, if the field has lower elevation in center so that it helps to level the fields.

Casting

Whenever a plough works round a strip of un-ploughed land, it is said to be casting. The tractor and plough turns to the left each time the head land is reached (Figure 7.2 Bottom). This kind of ploughing leaves a wide furrow (double width furrow) at the center, which is termed as finish or open furrow or dead furrow. It is good practice for fields having higher elevation at the center.

A general recommendation is that a long field should be ploughed by gathering in one season and casting in another season or a combination of gathering and casting. It avoids building up of a ridge in the center and an open furrow at each side or vice versa. It is normally uneconomical to plough the land either by casting or by gathering alone. The continuous ploughing method as described in the following section has proved to be quite convenient and economical.

Continuous Ploughing Method

In this method the tractor and plough never run idle for more than three-quarter land width along the headland and never turn in a space narrower than a quarter land width (Dogra and Singh, 2020, Figure 7.3). First the headland is marked and the first ridge is set-up at three-quarter of a land width from the side. The other ridges are set at full width over the field. Ploughing begins between the first ridge and the side land. The operator continues to turn left and cast in the three-quarter land until a quarter land width of ploughing is complete on each side. At this stage, operator lifts the plough to half depth for last trip down the side land of the field. It leaves a shallow furrow at the finish. After this, the driver turns right and gathers round the land already ploughed on the first ridge. Gathering is continued till the unploughed strip in first three-quarter land has been ploughed and completed. This gathering reduces the first full land by a quarter. The remaining three-quarter land is then treated in the same manner as the original three-quarter land.

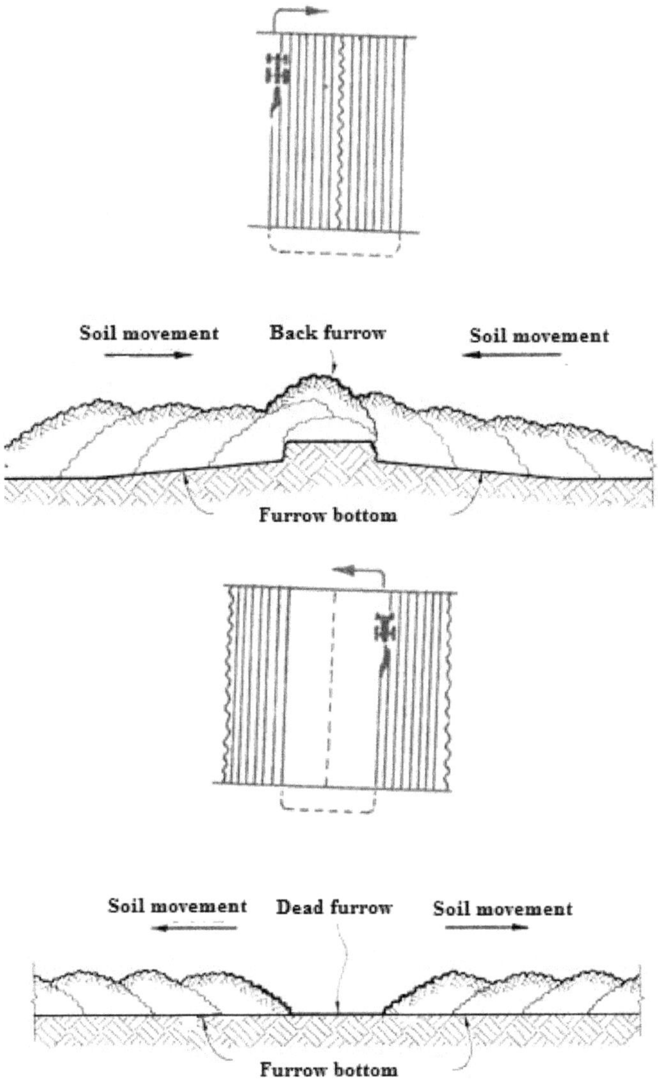

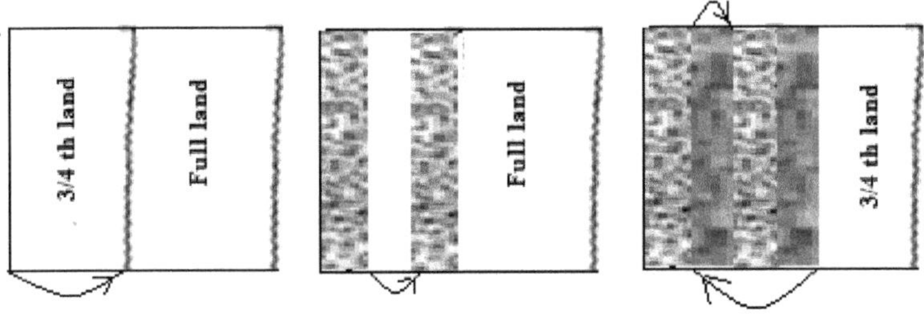

Figure 7.2. Ploughing Patterns: Gathering (Top) and Casting (Bottom).

Figure 7.3. Schematics of Continuous Ploughing Method.

PLOUGHS

Farming tools vary widely from one region to another. Plough is not an exception. Some commonly used ploughing or primary tillage tools are discussed in the following sections.

Country or Indigenous Plough

An indigenous plough has been used for centuries, and millions of them are still being used. It is quite versatile being capable of working under the dry land as well as wetland conditions. A wooden indigenous plough has five functional components namely share, body, shoe, handle and beam (Figure 7.4). Share is the working part of the plough with which it penetrates and breaks open the soil. It is attached to the shoe that supports and stabilizes the plough at the required depth. Body of the plough is main part to which the shoe, beam and handle are attached. Beam, a long wooden piece, connects the plough to the yoke. The handle is made of a wooden piece vertically attached to the body that enables the operator to control the plough. The size of the plough is given by the width of the body. It is commonly drawn with bullocks or other animal power sources. It cuts a V shaped furrow but there is no inversion. Ploughing operation is somewhat imperfect because some unploughed strip is always left between furrows. Although some improvement can be achieved through cross ploughing, even then small squares will remain unploughed. It is good only for cultivation of small land holdings. The shape and size of these ploughs vary from one region to another although basic parts and functions remain similar. Operational adjustments are achieved by raising or lowering the beam with respect to the plough body. It changes the angle of the share with the horizontal plane. It in turn increases or decreases the depth of ploughing. Changing the length of the beam (body to yoke on the beam) may also increase or decrease the depth of operation.

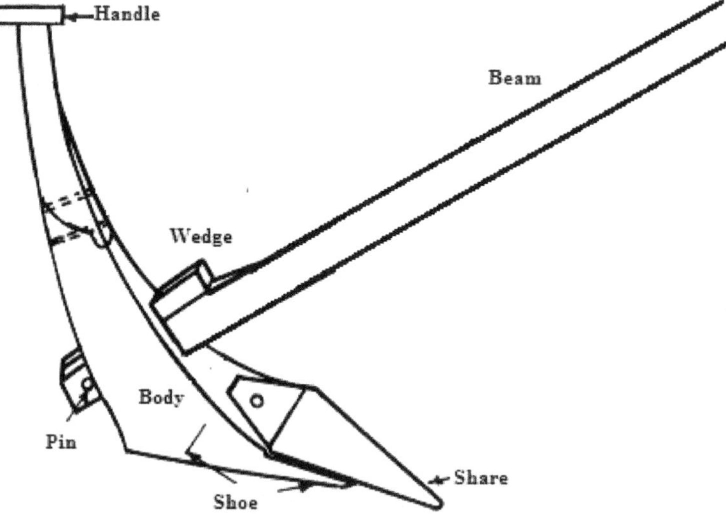

Figure 7.4. An Indigenous Plough Commonly used in West Bengal.

Animal Drawn Mould-board Plough

Ploughs that actually invert and mix soil layers weren't invented until the 17th century. Mould-board (MB) plough, curved mould-board plough or soil turning plough made their appearance in 1760. By the 18th century, US farmers had also switched over to oxen or horses to pull crude wooden ploughs. Real breakthrough happened only after 1900 upon introduction of the farm tractor. MB plough is a primary tillage implement especially used in high rainfall areas and in heavy soils. It cuts the soil, inverts the furrow slices and pulverizes the soil to some extent. Both animal and tractor drawn versions are available, the former being normally drawn by a pair of bullocks. In both cases, the constructional features are almost similar except that the animal drawn plough is smaller, the size of plough being 100, 125, 150, 175, 200, 225 and 250 mm. It can plough to a depth of about 15 cm. The parts include frog or base of the plough bottom (not shown), mould-board or wing, share, landside, connecting rod, beam and handle (Figure 7.5, BIS, 1998a). This type of plough does not leave any portion of the field unploughed. Furrow slices are cut clean and inverted to one side resulting in some pulverization. Complete details of mould-board plough are included in a subsequent section on mould-board plough.

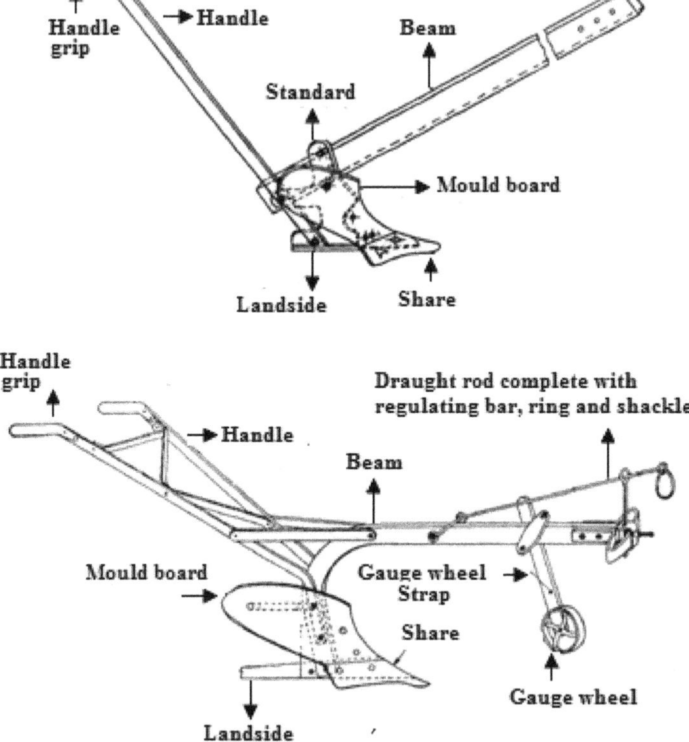

Figure 7.5. Animal Drawn Mould-Board Ploughs: Long Beam (Top) and Short Beam (Bottom).

Tractor Operated MB Plough

Like the animal drawn MB plough, it also consists of share point, share, mould-board, landside, frog, shank, and frame besides the hitch system to hitch the plough to 3 point linkage (Figure 7.6). It cuts the furrow slice, lifts the soil, turns the furrow slices and pulverizes the soil.

The most commonly used specifications are as follows:

No. of bottom	: 2 - 4
Length (mm)	: 1778-2392
Width (mm)	: 889-1194
Height (mm)	: 1092-1092
Weight (kg)	: 253-386
Capacity (ha/day)	: 1.5-2.0
Power requirement (HP/kW)	: 30-40/22.5-30

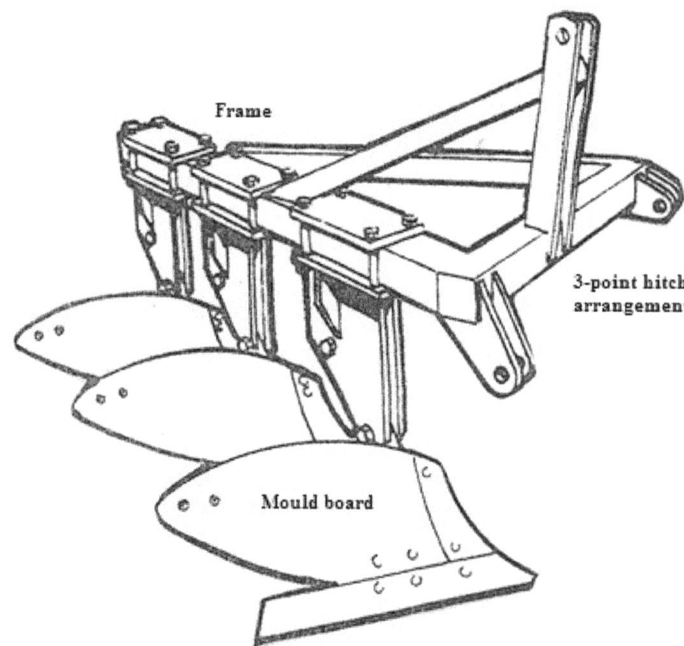

Figure 7.6. A Line Drawing of a Tractor Operated Mould-Board Plough.

Kinds of MB Plough Bottoms

Different soil and field conditions require mould-boards of varying shapes and sizes to give proper tilth. Based on the length and curvature of the plate, the mould-board bottoms are made of different types *viz.* general purpose, stubble bottom, sod or breaker, high speed and slat type (Davidson, and Chase, 1908).

General purpose: It has medium curvature, which is in between the curvatures of stubble and sod types. The sloping of the surface is gradual (Figure 7.7). It turns the well-defined furrow slice and pulverizes the soil thoroughly. Mould-board is fairly long having a gradual twist with the surface being slightly convex. It performs fairly well in heavy soils and stubble lands.

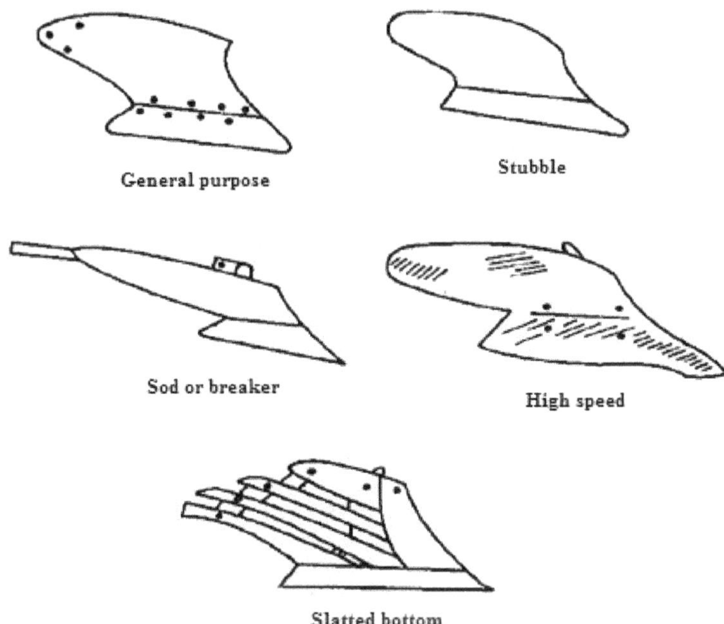

Figure 7.7. Kinds of MB Plough Bottoms.

Stubble type: As the name suggests, it is best suited to work in stubble soils, which are under cultivation for years together and where good pulverization is required. The stubble soil is a soil in which stubble of the plants from the previous crop are still left on the land at the time of ploughing. The plough is short with broader mould-board having abrupt curvature along the top edge (Figure 7.7). It lifts, breaks and turns the furrow slice, the abrupt curvature causing the furrow slice to be thrown off quickly; pulverizing it much better than other types of MB ploughs.

Sod or breaker type: This plough has a long mould-board with gentle curvature. It lifts and inverts the unbroken furrow slice (Figure 7.7). Furthermore, it is used in tough soil with much grass. It turns over thickly covered soil. It is very useful when complete inversion of soil is needed.

High-speed bottom: It has slightly less curvature than the general-purpose type and is suitable for general farm use.

Slat bottom: The surface of this plough is made of slats placed along the length of the mould- board, so that there are gaps between the slats (Figure 7.7). This type of mould-board is used in sticky soils, where solid mould-board fails to scour well.

Components

Parts of a mould-board plough can be categorized into two major groups *viz.*, plough bottom and plough accessories. The plough bottom, the main working body of the plough, comprises different parts like share, mould-board, landside and frog as shown in Figure 7.8.

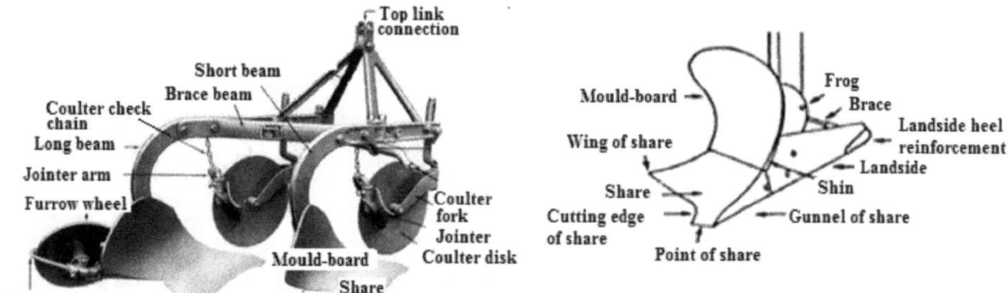

Figure 7.8. Components of Mould-board Plough.

Share

Share, the cutting part of the plough bottom, is a sharp and pointed component. The shares are made of chilled cast iron or steel containing 0.70 to 0.80 per cent carbon and 0.50 to 0.80 per cent manganese besides other minor elements. It consists of the point of share, cutting edge, wing and gunnel (Figure 7.8 right). The shares are classified as slip share, slip nose share, shin share and bar point share. The most commonly used is the slip share, which is a one single metal piece. As a result, whole piece has to be replaced once the slip share is worn out. Slip nose share has a replaceable share point. As the share point wears out, only share point needs to be replaced instead of the complete share. Shin share is almost similar to the slip share, but it extends over the side of the mould-board. This protects the mould-board inner edge from wear and tear. Bar point share consists of a long bar in addition to the main share and is used as the share point. The bar share can be pushed forward as and when the share point is worn-out. Thus, the share can be used for a much longer period of time than other kinds of shares.

Mould-board

The mould-board, a curved iron plate, is a part of the plough that receives the furrow slice from the share (Figure 7.8). It lifts, turns and breaks the furrow slice. As previously discussed, mould-boards are categorized in 5 types and used as per soil conditions and crop requirements.

Landside

Landside is the flat metal plate which slides along the furrow wall and absorbs or transmits lateral thrust of the plough bottom to the furrow wall (Figure 7.8 right). It provides stability to the plough bottom during the ploughing operations.

Frog

It is an irregular piece of metal made of cast iron in case of cast iron ploughs or welded steel in steel ploughs. It is the part of the plough bottom to which other components of the plough bottom are bolted together (Figure 7.8 right).

Tailpiece

It is an important extension of mould-board and helps in turning the furrow slice.

Plough Accessories

Plough accessories help the plough to efficiently perform various operations. While handle, beam and standard are found in animal drawn plough, accessories such as coulter, jointer, furrow wheel and gauge wheel are part of the tractor drawn mould-board plough (Figure 7.8).

Vertical Clevis

It is a vertical plate with a number of holes at the end of the beam to control the depth of operation and to adjust the line of pull.

Horizontal Clevis

It helps the plough to make lateral adjustment relative to the line of pull.

Jointer

It is a small irregular piece of metal having a shape similar to an ordinary plough bottom (Figure 7.8 left). It looks like a miniature plough. Its purpose is to turn over a small ribbon like furrow slice directly in front of the main plough bottom. This small furrow slice is cut from the left and upper side of the main furrow slice and is inverted so that all trashes on the top of the soil are completely turned down and buried under the right-hand corner of the furrow.

Coulter

Coulter is a small disc provided in front of the share (Figure 7.8 left). It is used to cut the furrow slice vertically from the land ahead of the plough bottom. It leaves a clear wall. It also cuts the trash that gets covered under the soil by the plough. The commonly used coulters are rolling type disc coulter and sliding type knife coulter. Ploughing can be made smoother and neater by correctly setting the disc coulters. It also helps to pull the plough in an easy manner.

Gauge Wheel

Gauge wheel, also known as a depth control wheel, is an auxiliary wheel that maintains uniform depth of working. It is mostly used when working conditions show extreme variability during ploughing. It is usually placed in a hanging position.

Land Wheel

It is the wheel of the plough, which runs on the ploughed land.

Front Furrow Wheel

Furrow wheel, mostly used in tractor drawn ploughs is the front wheel of the plough, which runs in the furrow. Its main function is to absorb the side thrust of the plough against the furrow wall.

Rear Furrow Wheel

It is the rear wheel of the plough, which also runs in the furrow. It also supports some weight of the rear end of the plough in case of large mounted type ploughs.

Materials of Construction

The materials used in manufacturing of various parts of mould-board plough are given in Table 7.1.

Table 7.1. Material of Construction of Various Parts of MB Ploughs

Name of the Part	Material	
	Animal Drawn	Tractor Drawn
Beam	Wood/mild steel	Forged steel
Handle	Wood/mild steel	-
Standard	Cast iron/mild steel	-
Share	Chilled cast iron	Chilled cast iron/high carbon steel
Mould-board	Chilled cast iron/mild steel	Soft center steel/chilled cast iron/ crucible steel
Landside	Cast iron/mild steel	Cast iron/mild steel
Frog	Cast iron/mild steel	Cast iron/mild steel
Braces	Mild steel	Mild steel
Gauge wheel	Cast iron	Cast iron
Furrow wheel	Cast iron/mild steel	Cast iron/mild steel

Adjustments of Mould-Board Plough

Vertical Suction

Vertical suction, also known as vertical clearance or down suction, is measured as the maximum distance between the landside (at the joining point of share and landside) and ground surface when the plough is kept in normal working position over a leveled ground. This clearance varies according to the size of the plough but is usually kept in the range of 3 to 5 mm. In practical terms, it is the amount by which the share point of the plough is bent downward to help the plough penetrate into the soil when the plough is pulled forward (Figure 7.9 top)

Horizontal Suction

Also known as horizontal clearance or side suction, it is the amount by which the point of the share is bent off line with the landside or the maximum clearance between the landside and a horizontal point touching the point of share at its gunnel side and heel of landside (Figure 7.9 bottom). It is measured by placing a scale on the side of the plough extending from the heel of the landside to the point of share. It is usually kept at 5 mm but may vary with size of the plough. This suction helps the plough to cut proper width of furrow slice.

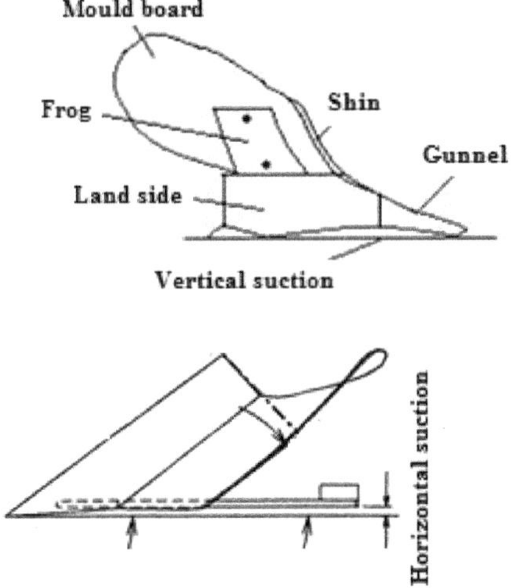

Figure 7.9. Vertical and Horizontal Suctions in a MB Plough.

Throat Clearance

It is the perpendicular distance between point of share and lower position of the beam of the plough (Figure 7.10 left).

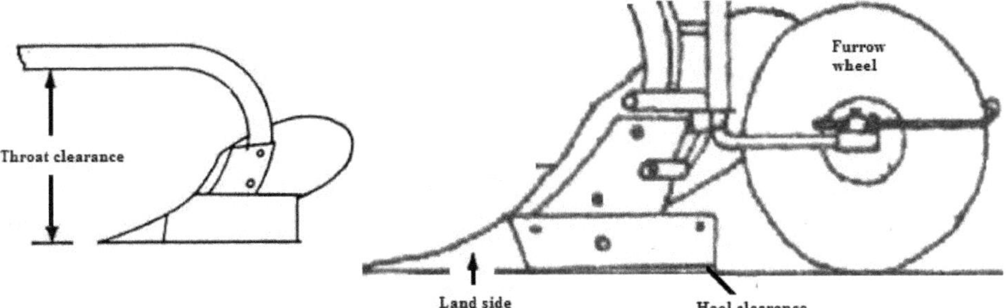

Figure 7.10. Throat and Heel Clearances in a MB Plough.

Heel and Landside Clearance

In plough with rear furrow wheel, the heel of landside must be 8-12 mm above the bottom of the furrow so that the rear furrow wheel carries one third of the plough weight. This clearance is known as heel clearance (Figure 7.10 right). The landside is adjusted to have a clearance of 12-20 mm between the furrow wall and the rear of the landside.

Coulter and Jointer Setting

To obtain a neat furrow wall, the coulter for average conditions is set between 1.2 mm and 1.6 mm to the left of the landside. Coulter is set directly over the share point and cuts half the depth of ploughing. For loose ground and tough scouring conditions, the coulter is set a little wider say 2 mm. The jointer should be set to cut 4-5 cm deep furrow. It should be placed as near the coulter as possible. Coulter mould should be set directly above the share point for average ploughing. Coulter mould is set 20 mm away from the landside. Jointer mould is set 12-16 mm outside the landside.

Plough Leveling

The fore (front) and rear (behind) leveling of the plough can be done by adjusting the length of the top link. Length of the top link is increased, if the front plough bottom is deeper than the rear one. On the other hand, the top link is shortened, if the rear bottom is deeper than the front one. The lateral leveling can be done by adjusting the right lower link length to make the plough frame parallel to ground.

Furrow Wheel Setting

The width of the cut of the front bottom is adjusted by changing the position of the furrow wheel. It is also used to adjust the landside.

Plough Size

The size of the tractor drawn mould-board plough is expressed by the width of cut or the width of furrow that it is designed to cut (Figure 7.11). It is measured as the horizontal distance between the wing of the share and the landside. In tractor drawn ploughs, 30, 35 and 40 cm plough bottoms are commonly used. In multi-bottom ploughs, the rated width or the working width of the plough is expressed as the product of the number of bottoms times the size of one bottom (Figure 7.11).

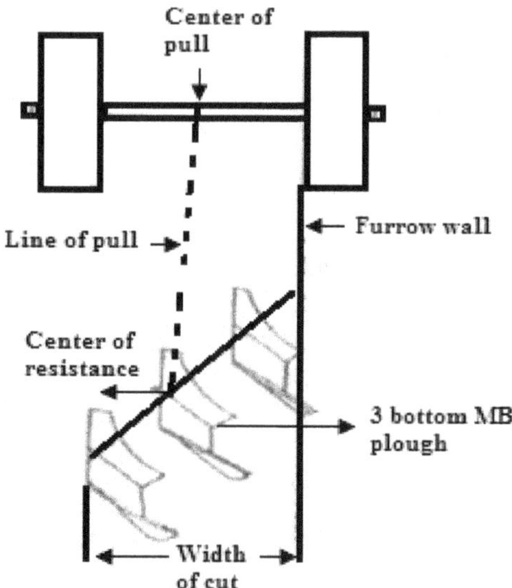

Figure 7.11. Commonly Used Terms in Ploughing.

TERMS USED IN PLOUGHING

Center of power/pull: It is the true point of hitch of a tractor (Figure 7.11).

Center of resistance: It is the point at which the resultant of all the horizontal and vertical forces acts. The center lies at a distance equal to 3/4th size of the plough from the share wing (Figure 7.11).

Line of pull: It is an imaginary straight line passing from the center of resistance through the clevis to the center of pull (Figure 7.11).

Pull: It is the total force required to pull an implement.

Draft: Draft is the horizontal component of the pull parallel to the line of motion. It is governed by various factors such as sharpness of cutting edge, working speed, working width, working depth, type of implement, soil condition and attachments.

$$D = P \cos \theta$$

Here D is draft (kgf), P is the pull in (kgf) and θ is the angle between line of pull and horizontal.

$$HP = Draft \ (kgf) \times Speed \ (m/s)/75$$

Side draft: Horizontal component of the pull perpendicular to the direction of motion is called side draft. This draft is caused when the center of resistance is not directly behind the center of pull.

Unit draft: It is the draft per unit cross-sectional area of the furrow.

Soil inversion: It is defined as per the following relation:

$$Soil \ inversion = 100 \ (W_{bp} - W_{ap})/W_{bp}$$

Here W_{bp} is the number of weeds before ploughing and W_{ap} is the number of weeds after ploughing

Soil pulverization: It is the quality of work in terms of soil aggregates and clod size. Soil strength measured using a cone penetrometer is an indicator of this quality.

DISC PLOUGH

The disc plough bears little resemblance to the common mould-board plough. It is designed to reduce friction by making a rolling rather than a sliding plough bottom as in the case of MB plough. A large, revolving concave steel disc replaces the share and the mould-board. The disc cuts, turns the slice to one side with a scooping action and in some cases breaks the furrow slice. The disc plough is highly suitable for lands where there is much fibrous growth of weeds as the disc is capable of cutting and incorporating the weeds into the soil. Although a disc plough works very well in soils free from stones but it works equally well in rough and stony soils. In fact, a disc plough works well where mould-board plough does not work satisfactorily such as in heavy clay, hard pan and loose sandy soils. No harrowing is required to break the clods of the upturned soil as in a mould-board plough. The usual diameter of the disc is 60 cm and it turns a 35 to 30 cm furrow slice.

Advantages of Disc Plough

☆ A disc plough can penetrate in soils that are too hard or too dry to work with a mould-board plough

☆ It works equally well in sticky soil where mould-board plough fails to scour

☆ It can be used for deep ploughing

☆ It can be used safely in stony and stumpy soil without fear of plough breakage

☆ A disc plough works well even after a considerable part of the disc is worn off

☆ It effectively works in loose soil also (such as peat) without much clogging

☆ With scrapers fitted, it can invert the soil. Without scraper, disc plough has more mixing action rather than inversion

☆ The maintenance cost of disc plough is low. There are no shares to replace or sharpen as in the case of mould-board ploughs

Disadvantages

☆ It does not cover surface trash and weeds as effectively as a mould-board plough does

☆ Comparatively, the disc plough leaves the soil in rough and cloddy condition than the mould-board plough

☆ For equal capacities, a disc plough is much heavier than the mould-board plough because penetration of the disc plough is largely due to its weight rather than suction. To emphasize this point, note that a mould-board plough is forced into the ground by the suction of the plough, while the disc plough is forced into the ground by its own weight

☆ Disc plough cannot be used at high speeds as slow speed is necessary for its cutting action

Commonly Used Terms

Disc: It is a circular, concave revolving steel plate used for cutting and inverting the soil. Disc is made of 5 mm to 10 mm thick heat treated steel. The concavity varies with the diameter of the disc, being approximately 8 cm for 60 cm diameter and 16 cm for 95 cm diameter discs. The number of discs and diameter determine the plough capacity. Depth of cut depends on diameter of discs, the limiting value being about $1/3^{rd}$ of blade diameter. Width of cut also depends on diameter of disc, normally being 0.4 times the diameter of disc blade.

Tilt angle: *It is the angle at which the plane of the cutting edge of the disc is inclined to a vertical line* (Figure 7.12 left). The tilt angle of a good plough varies from 15° to 25°.

Disc angle: *It is the angle at which the plane of the cutting edge of the disc is inclined to the direction of travel* (Figure 7.12 right). Disc angle of good plough is in the range of 42° to 45°.

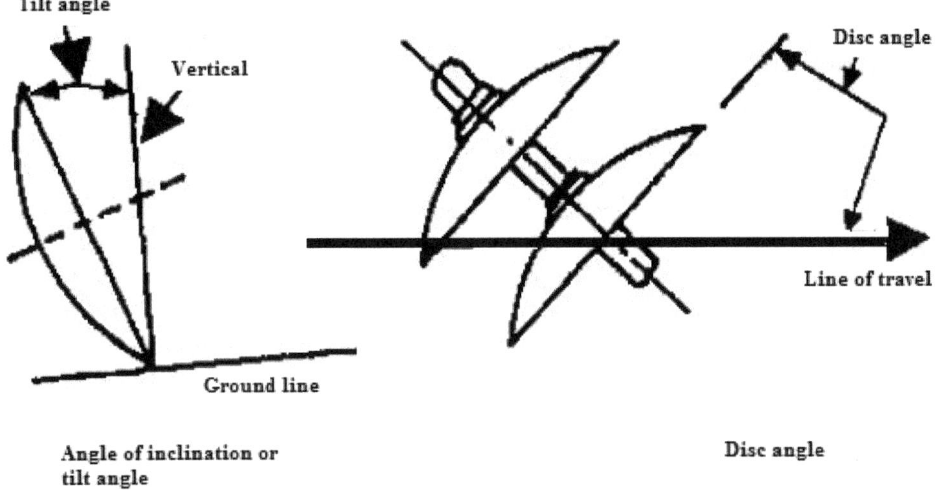

Figure 7.12. Description of Tilt and Disc Angles in a Disc Plough.

Standard: It connects disc bearing and plough frame. Sometimes the beam is bent for disc attachment mainly to reduce the cost.

Plough frame: Standards are attached to the plough frame. It has provision for disc angle adjustment, adding or removing standards and discs.

Scraper: The device to remove soil that tends to stick to the working surface of a disc (Figure 7.13). It also helps in pulverization of furrow slice, invert the furrow slice and cover the trash better. Amongst the 3 kinds of scrapers, disc scraper is used in non-scouring soils, mould-board scraper is used to turn over the soil and trash and hoe scraper is used in sticky soils.

Concavity: It is the depth measured at the center of the disc when the concave side of the disc is placed on a flat surface. Concavity affects the disc angle and soil turning.

Rear furrow wheel: The wheel fixed at the end of the plough to make it stable and bear the side thrust.

Bearings: Since disc blades are at an angle to the direction of travel, both radial and thrust forces act on the axle. Radial forces push the axle at right angle while thrust forces push along the axis. It is why taper roller bearings are used to take care of both these forces.

Draft of disc plough: The draft of a disc plough is lighter than the mould-board plough. It turns the same volume of soil under similar conditions with lesser draft. In very hard soils, some extra weight may be added to the wheel to increase the draft. Draft is also affected by the kind of bearings and use of scrapers.

Types of Disc Ploughs

Disc ploughs are divided into two main categories *i.e.* bullock drawn and tractor drawn ploughs. Bullock drawn ploughs are further grouped as sulky and gang types. The sulky disc plough like a sulky mould-board plough (single bottom riding type instead of walk behind type) is a plough with only one bottom. The gang disc plough on the other hand has two or more bottoms. Many sulky ploughs are so constructed that they can be converted into a gang plough by adding additional bottoms. Thus, it is possible to make them either two or three-disc ploughs. Tractor drawn ploughs are categorized as direct mounted, semi-mounted and trailed type. According to mounting of discs these are categorized as standard disc plough and vertical disc plough also termed as harrow plough or wheat land plough.

Standard Disc Plough

Animal drawn standard disc plough is attached to a universal frame, which is mounted on two wheels. These ploughs have only one disc blade that can be tilted backward from 15° to 25° (tilt angle) in the vertical plane. It makes an angle of about 45° (disc angle) with the direction of travel. The diameter of the disc is 45 cm. A rear furrow wheel is provided to take care of the side thrust of the plough. The frame is pulled by a pair of bullocks, the weight of the plough being about 30 kg per disc.

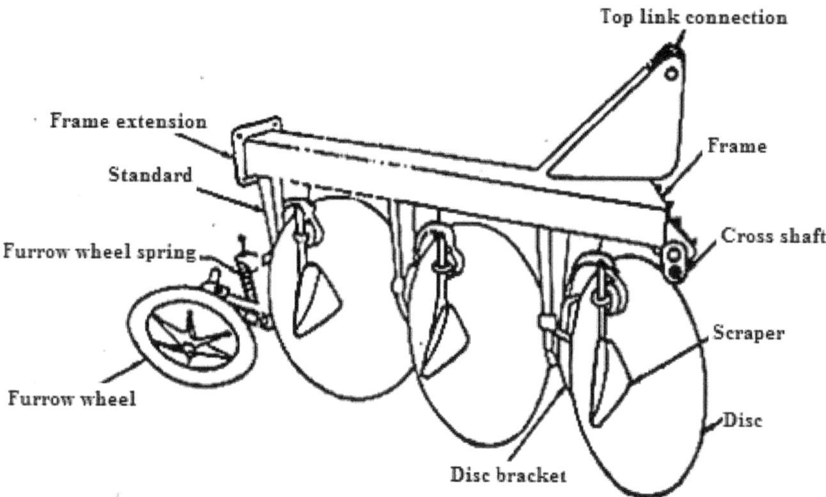

Figure 7.13. Major Components of a Disc Plough.

The tractor drawn standard disc plough is a series of inclined disc blades individually mounted on a frame. The whole frame is supported by a wheel (Figure 7.13). Each disc revolves on a stub axle in a thrust bearing, carried at the lower end of a strong stand which is further bolted to the plough beam. These are normally multi-disc ploughs having 3 to 6 discs, disc diameter ranging from 60 to 90 cm. The discs are so spaced that each disc cut about 18 to 30 cm slice. The discs are tilted backward at an angle of 15 to 25° from the vertical. Discs make an angle of 42 to 45° (disc angle) with the direction of travel. The angle of the disc to the vertical and to the furrow wall can be adjusted. Scrapers are used to cover trash and prevent any soil build-up on the discs under sticky soil conditions. Mounted or semi-mounted reversible disc ploughs have an arrangement to reverse the disc angle at each end of the field. Such an arrangement permits one way ploughing. Tractor drawn disc ploughs weigh from 200 to 600 kg per disc. Sometimes a rear wheel is fitted to take the side thrust of the plough to some extent. On the other hand, in mounted type ploughs wheels of the tractor take the side thrust. Furrow wheel bears the side thrust in trailed type ploughs. In action, the disc cuts the soil, breaks it and pushes it sideways.

Vertical Disc Plough

It is also known as harrow plough or one way disc plough and wheat land plough as these are preferred in wheat growing areas, where moisture conservation for winter crops is the main objective. It is generally used in plain areas where shallow ploughing and mixing of stubble with soil is required. The one-way disc plough combines the principles of the standard disc plough and the disc harrow. Whereas it has a frame, wheel arrangement, and depth adjusting devices of the disc plough, the discs and their attachment to the frame is like the disc harrow. However, all the discs are set to throw the soil one way and turn together as a unit similar to a gang plough. Thus, its action is intermediate between regular disc plough and disc harrow. The number of discs in this kind of plough may vary from 2 to 32 that are spaced about 18 to 25 cm apart on a gang (Figure 7.14). The diameter and curvature of the individual disc of the plough is slightly smaller. Diameter of the disc varies from 40 to 65 cm. Although the disc angle may range from 30 - 65° but normal range is 40 to 45° as in the case of a standard disc plough. For a given width of cut, disc angle in this range has minimum draft. Width of cut per disc depends upon the angle between gang axis and direction of travel. Width of cut obtained from various sizes of vertical disc plough ranges from 2 to 6 m. A comparison of standard and vertical disc plough is illustrated in Table 7.2.

Table 7.2. Comparison between Standard and Vertical Disc Ploughs

Sl.No.	Item	Standard Disc Plough	Vertical Disc Plough
1.	Mounting of disc	Individual axis	Common axis
2.	Number of discs	1-6	2-32
3.	Size of discs	60-90 cm	40-65 cm
4.	Spacing	18-30 cm	18-25 cm
5.	Concavity	More	Less
6.	Depth of cut	30-40 cm	8-10 cm
7.	Angle to direction of travel	42-45 degree	30-65 degree
8.	Tilt angle	15-25 degree	0 degree
9.	Weight/disc	200-600 kg	50-100 kg
10.	Draft requirement	High	Low

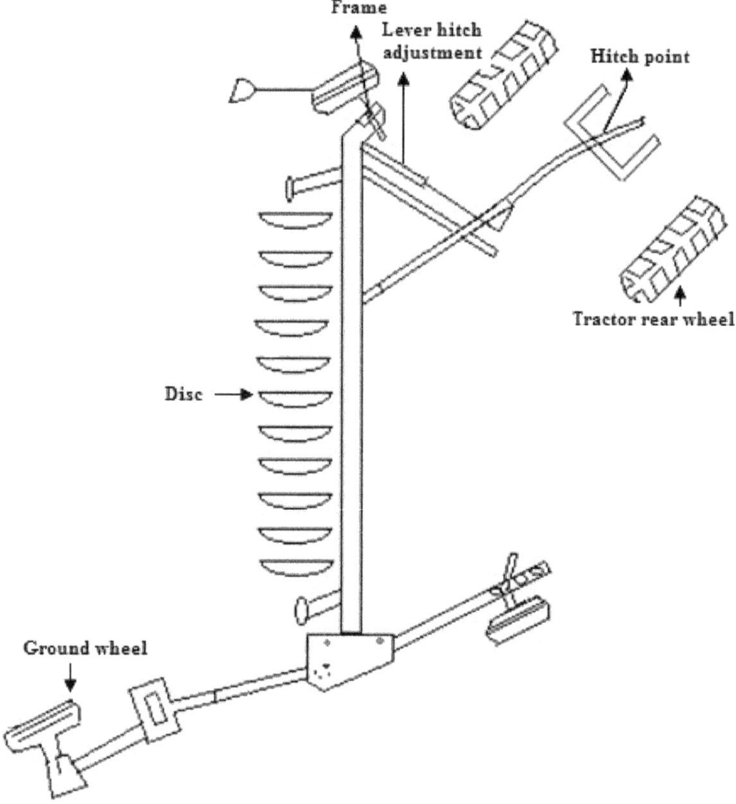

Figure 7.14. Line Diagram of a Vertical Disc Plough.

Plough Adjustments

Leveling the plough: The tractor top link is used to control the level of the plough. Lengthening the top link levels the plough if the rear end of the plough beam is higher than the front end of the beam. On the other hand, shortening the top link will help to level the plough if rear end is lower than the front end. Lateral leveling is adjusted by the tractor's right lower link. These adjustments need to be made before the plough is put in operation.

Setting the scrappers: Scrapper is adjusted low enough to turn the furrow slice before the slice falls away from the disc. A little higher setting is required for deeper ploughing. In sticky soils scrapers are set closer to the disc.

Adjustment for deep ploughing: Position and draft control levers of the tractor hydraulic system are used to adjust the depth of ploughing. Since a correctly tilted disc plough tends to penetrate better, following adjustments of the tilt angle besides other recommendations can improve penetration.

- ☆ Increasing the tilt angle improves disc penetration in heavy, sticky soils
- ☆ Decreasing the tilt angle improves disc penetration in loose and brittle soils

☆ If the ground is covered with trash, set the disc in almost vertical position. In such soils notched discs give better results

☆ Add weight to the plough

Adjusting the width of the cut: The width of the cut for the front disc is adjusted with the help of the cross shaft. The cross shaft has an index line which can be lined up with different (1, 2, 3) markings on the cross shaft carrier. Increasing the disc angle will reduce the width of cut.

Trouble Shooting

While working with a disc plough, few commonly encountered problems, their causes and remedies are listed in Table 7.3.

Table 7.3. Commonly Encountered Problems and their Remedies in Disc Ploughs

Problem	Causes	Remedies
Improper plough penetration (low)	It could be due to blunt discs, or higher tilt angle or plough being too light	Sharpen the edges of the discs, or set the tilt angle or add weight to the plough respectively
Draft too heavy	It could be due to blunt discs, or furrows may be too wide	Sharpen the edges of the discs, or reduce the tilt angle respectively
Excessive side draft	Improper hitching	Properly hitch the plough
Less scouring	Defective scrapper setting	Adjust the scrapper
Uneven furrows	Disc angles are not uniform or loose bearings or defective hitching	Properly set the disc angles, set bearings and make proper hitching

SPECIAL PLOUGHS

Special ploughs are also part of the primary tillage equipment. These ploughs are modified ploughs to suit some special conditions or to perform some special operations. As such, these ploughs are called special ploughs. For example even a slat mould-board plough discussed previously may be called a special plough. With this modification, friction between the mould- board and the furrow slice reduces considerably because of the reduced surface area. Thus, it makes the plough useful to handle very stiff soils.

Rotavator

Rotavator comprises a steel frame, a rotary shaft on which blades are mounted, power transmission system, and a gear box (Figure 7.15). A 3-point hitch system is used to mount the implement on the tractor. The L-shaped blades are made from medium carbon steel or alloy steel, hardened and tempered to suitable hardness. It works on power derived from tractor PTO. Rotavator can be used as primary and secondary tillage implement. It ensures a good seedbed and pulverization of the soil in a single pass. Thus, it is a means of minimum tillage practiced by the farmers. It is used in both dry land and wetland conditions. It is also suitable for incorporating straw and manure in the field. The power requirement varies with the width of the rotavator. Detailed information on this equipment is included in Chapter 8.

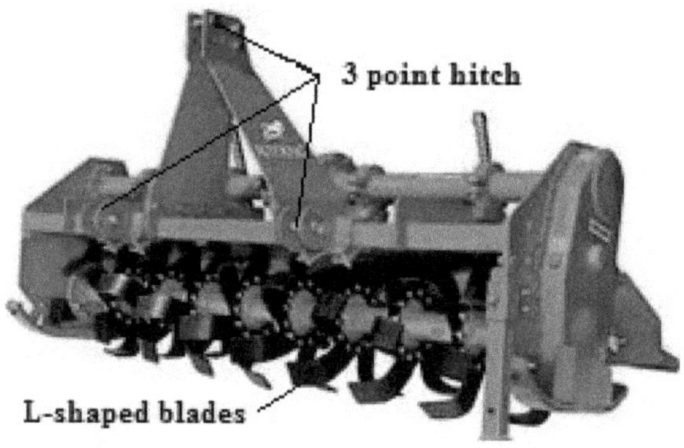

Figure 7.15. A Pictorial View of a Rotavator.

Reversible MB Plough

In one of the previous sections, we discussed one way MB plough that throws the furrow slice to one side of the direction of motion. On the other hand, a two-way or reversible plough turns the furrow slice to the right or left side of direction of travel as per requirement. Such ploughs have two sets of opposing bottoms. In such a plough, the entire furrow can be turned towards the same side of the field by using one bottom for one direction of travel and the other bottom on the return trip. Two sets of bottom are so mounted that they can be raised or lowered independently or rotated along an axis. The main advantage of a two-way ploughs is that they neither upset the slope of the land nor leave dead furrows or back furrows in the middle of the field. It has a central body to which the mould-board is hinged. The body is fixed along with the share on either the right or the left side with a hook. A lever provided on the distributor is used to operate the bottom reversing mechanism. When the implement is hitched, the plough bottom is free to rotate 180° along the axis of the hollow shaft. Ploughing is commenced on one side of the field. The plough is turned at the end of the field, the mould-board is shifted to the opposite side and the ploughing is continued till the field is completed, shifting the mould-board to the left and right alternately at the end of each furrow. A two bottom reversible plough is shown in Figure 7.16.

Figure 7.16. Schematic and Pictorial Views of a Reversible Plough.

Ridge Plough

The name given to this plough is a misnomer, as it forms furrows and not ridges. It is a double mould-board plough, the two mould-boards meeting along a central line. Because of the common share for both the mould-boards, it is sometimes called double winged plough (Figure 7.17). While the single mould-board plough lays the furrow slice on one side, the ridge plough lays the earth equally on both the sides, leaving an open furrow in the middle. The inter-furrow space in close furrows takes the form of ridges. When the furrows are formed wide apart for planting the sugarcane or cotton crops, inter furrow spaces take the shape of raised flat beds. The ridge plough is used to split the field into ridges and furrows and even for earthing-up of crops. These are also used to make broad bed and furrows by attaching two ridge ploughs on a frame with 150 cm spacing between them.

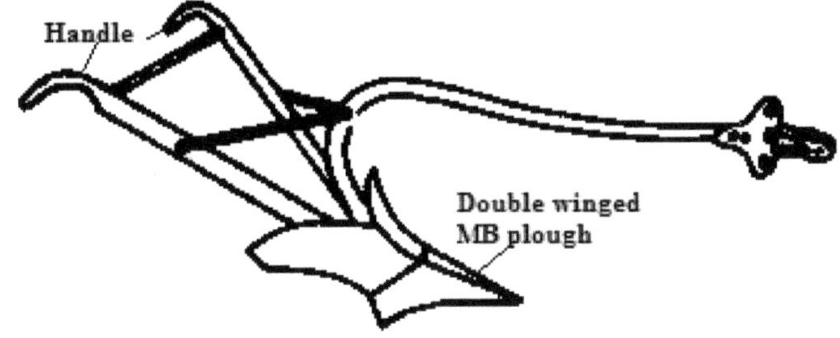

Figure 7.17. An Animal Drawn Ridge Plough.

Chisel and Sub-soil Ploughs

With continuous cultivation over the years land may develop three layers as follows:

☆ Top Layer

☆ Hard or plough pan 7.5 -10 cm

☆ Sub-soil hard pan

Sub-soiling, also called ripping, chiseling or aerating, is the process of breaking up of the dense horizons below the plough layer and to loosen up the soil. To overcome the problems at serial 2, a chisel plough is used while a sub-soiler is used for problems mentioned at serial 3. Both the chisel and sub-soil ploughs are primary tillage implements.

Chisel Plough

The chisel plough is a tool to perform deep tillage with limited soil disruption. Unlike many other ploughs chisel plough does not invert or turn the soil. It is used to break through and shatter compacted or otherwise impermeable soil layers. The main function of the plough is to loosen and aerate the soils, improve infiltration and storage of rainwater in the crop root zone. Since the plough leaves the crop residue at the top of the soil, it can reduce the compaction of the soil. This plough can also be used in no-till and low-till farming, which advocates keeping organic matter and farming residues on the soil surface throughout the year for erosion-prevention. The improved soil structure resulting from chiseling helps in profuse development of root system, which results in high yields. The functional components of the unit include reversible share, tine (chisel), beam, cross shaft and top link connection (Figure 7.18). The shanks of chisel plough are made of nickel alloy and heat treated spring steel. Each shank is typically set from 230 mm to 305 mm apart. Chisel plough standards may be rigidly mounted or spring cushioned or may have spring tips. The chisel plough is typically operated at depths in the range of 300 to 400 mm although some models may run much deeper (450-750 mm). The best results are obtained when soil is dry. Since a chisel plough does not pulverize the soil as much as a MB plough, several post chiseling operations are required to fully prepare the soil. A tractor of sufficient power and good traction is required to pull chisel ploughs as these offer high drag. A chisel plough may require about 10 to 15 horsepower per shank.

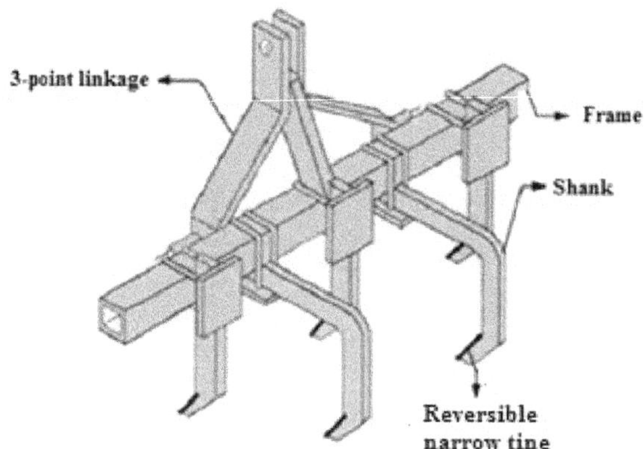

Figure 7.18. A Chisel Plough

Sub-soiler

Sub-soiler (Figure 7.19 top), a single or multi-blade equipment, is used to loosen and break-up the soil at depths below the level of a traditional disk harrow or rotavator. Typically, a sub-soiler is operated at depths of 500-900 mm mainly to improve permeability of the soil. It is designed to break up hard layers or pans without bringing them to the surface. The working depth is controlled by hydraulic system and linkage of the tractor. Since a sub-soiler goes quite deep, only 1 to 2 shanks are used on a sub-soiler. A sub-soiler consists of a beam made of high carbon steel. A hollow steel adaptor is welded to the bottom end of the beam to accommodate hare base, share base having square section, share plate made from high carbon steel and a shank. Share plate is made from high carbon steel, hardened and tempered to suitable hardness. The body of the subsoil plough is wedge shaped and narrow while the share is wide to shatter the hard pan while making only a slot in the top layers. When a sub-soiler is pulled through the soil, it loosens the compacted layers by lifting and cracking them, creating a network of interconnected pores (Figure 7.19 bottom). Many of these pores extend from the depth of loosening up to the soil surface. They can, therefore, act as pathways for root penetration and for transmission of water and air. The best conditions for sub soiling occur during the dry season, when the conditions are optimum for maximum bursting, and it is easier to work with farm implements. It is heavier than chisel plough. Since depth of penetration is more, it is a slow operation as well as requires about 60 to 85 HP to pull one single standard.

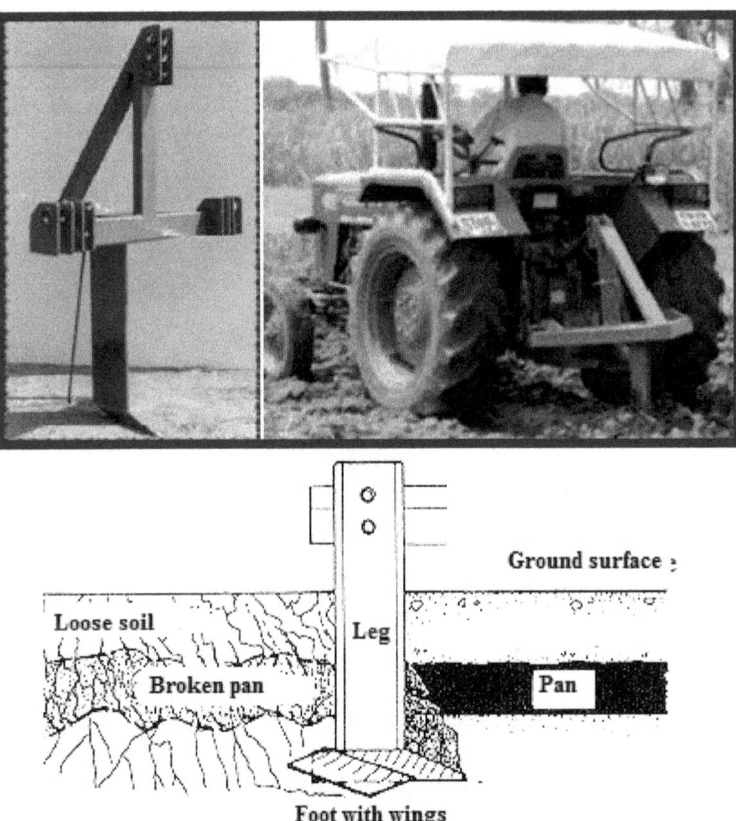

Figure 7.19. A View of a Sub-soiler and Sub-soiling Operation (Top) and Loose Soil with Broken Pan Behind a Sub-soiler (Bottom).

Relative characteristics of a chisel plough and sub-soiler are illustrated in Table 7.4.

Basin Lister

Basin lister is a heavy implement with one or two mould-boards or shovels. These shovels are mounted on a special type of frame on which they act alternately. This implement is used to form listed furrows (broken furrows with small dams and basins) to help prevent runoff from flowing away and/or blowing off the soil in low rainfall areas.

Table 7.4. Relative Characteristics of a Chisel Plough and a Sub-soiler

Chisel Plow	*Sub-soiler*
Chisel ploughs may have a series of standards spaced at 30 cm apart and equipped with replaceable narrow shovels or teeth	Sub-soiler has one or may be two heavy standards
Maximum depth is 45 to 75 cm	It breaks the sub-soil hard pan covering about 90 cm depth
Chisel ploughs may be employed in place of MB ploughs where soil inversion is not required	They cannot be employed for ploughing but can be used to make moles with suitable attachments
The shanks of chisel plough are made of nickel alloy and heat treated spring steel. This can also be used for doing a number of jobs by mounting different types of tines	The standard of sub-soiler is usually long and narrow with heavy wedge like point
It is lighter than a sub-soiler so a relatively low HP tractor can operate the plough	It is heavier than chisel plough. Since depth of penetration is more so a high HP tractor is needed

QUESTIONS (THEORY)

1. What is tillage? List the objectives of tillage operation.

2. Define tool, implement and machine giving specific example of each of them.

3. What do you understand by preparatory tillage? Describe primary and secondary tillage listing the objectives of each of these operations.

4. Name and describe the various methods of ploughing the land.

5. What do you understand by no- till, ridge-till and mulch-till?

6. Write a descriptive note on animal drawn mould-board plough.

7. Name and describe various kinds of plough bottoms of a mould-board plough.

8. Name and describe various components of a mould-board plough.

9. Name and describe various adjustments made for the efficient functioning of a mould-board plough.

10. Describe a disc plough and list its advantages and disadvantages.

11. Explain the terms: unit draft, disc angle, pull, head and land side clearances

12. Describe the standard and vertical disc ploughs clearly bringing out the differences between the two.

13. Name and describe various adjustments required for the good performance of a disc plough.

14. Write short notes on sub-soiling, minimum or reduced tillage, country plough, rotavator, reversible mould-board plough, ridge plough, chisel plough and sub-soiler.

15. Differentiate between conventional and conservation tillage, back furrow and dead furrow, furrow slice and furrow wall, gathering and casting plouging methods and horizontal and vertical suctions in a MB plough.

Chapter 8

Secondary Tillage and Related Equipment

Secondary tillage is defined as the *"tillage operations performed after primary tillage to create proper soil tilth for seeding and planting"*. In fact secondary tillage consists of conditioning the soil to meet various tillage objectives discussed in the previous chapter. Secondary tillage operations are generally performed on the surface soil. These operations do not cause much soil inversion and shifting of soil from one place to another place except in leveling/smoothening operation. The implements used in secondary tillage operations are called secondary tillage implements. They include different types of harrows, cultivators, rollers and pulverizers, rotary tillers, clod crushers, cage wheels and puddlers, tools for mulching and fallowing and similar implements. Compared to primary tillage operations, these operations consume less power per unit area.

HARROWS

Disc harrow is quite suitable for hard ground with full of stalks and grasses. A harrow cuts the soil to a shallow depth for smoothening and pulverizing the soil. It also cuts the weeds/residues and mixes them with soil. Besides, it breaks the clods formed after ploughing, collects the trash from the ploughed land and levels the seedbed. The specific objectives of harrowing are:

- ☆ Used before ploughing to cut residues such as corn and cotton stalk, and weeds and mix them into the soil
- ☆ Used after ploughing to pulverize the soil and to improve its tilth for seeding and transplanting
- ☆ Used to pulverize the topsoil so that the furrow slices make better connection with the bottom of the sole preventing airspace when slices are turned

☆ Io improve aeration, water infiltration and crop growth

☆ Used for summer fallowing

☆ Used to cover the seeds after sowing

Different kinds of harrows such as disc harrow, spike tooth harrow, spring tooth harrow, rotary cross-harrow, soil surgeon harrow, triangular harrow, acme harrow, blade harrow, reciprocating power harrow, zigzag harrow, *etc.* are used in India. Only few of them are discussed in the following sections.

Disc Harrow

This kind of harrow consists of a set (or a number of sets) of rotating steel discs, each set being mounted on a common shaft. Depending upon the sources of power used, disc harrows are characterized as animal drawn and tractor drawn disc harrows.

Animal Drawn Disc Harrow

Animal drawn disc harrow is a mini-version of tractor drawn harrow. It consists of: disc, gang frame, beam, gang angle mechanism, scraper, spacer (spool), clevis, axle, middle tine, and bearings (Figure 8.1).

Disc: Disc is the main part of the harrow. It cuts and pulverizes the soil. It is usually made of steel having the carbon content in the range of 0.80 to 0.90 per cent. The steel sheets at least 3.15 mm thick are used to manufacture the discs. The cutting edge is beveled for easy penetration. The disc has a square opening in the center to allow the axle to pass through it. Discs are arranged in two gangs.

Gang frame: All the gangs are mounted on a frame, called gang frame. The frame is usually a sturdy mild steel structure. The gang frame is bolted to the beam of the implement.

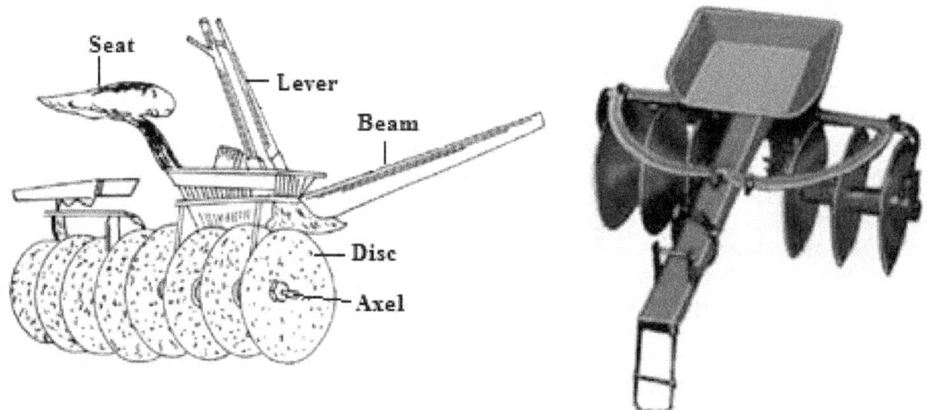

Figure 8.1. Animal Drawn Disc Harrow: Pictorial View (Left) and Photographic View (Right).

Beam: Beam, made of wood or steel, connects the harrow with the yoke. The rear end of the beam has a clevis to adjust the height of hitching to suit the size of animals.

Clevis: It is a part fitted to the beam and the frame that permits vertical hitching of the harrow.

Gang angle mechanism: It is used to adjust the gang angles, which in turn govern the width and depth of cuts of the implement. The lever is usually made of mild steel flat having a wooden handle. The gang angle can be adjusted in the range from 0° to 27°.

Scraper: The scraper is used to scrape the soil from the concave side of the disc so that the disc remains clean for effective operation.

Spacer (spool): Usually made of cast iron, the spacers are used to separate the two adjacent discs and to keep them in position. The spacers have suitable square opening in the middle to allow the axle to pass.

Axle: The axle is usually 20 x 20 mm square section. The length of the axel depends upon the size of the harrow.

Middle tine: This tine is suitably fixed to the rear end of the gang frame so that it breaks the unbroken strip of soil left in between two gangs of the harrow and can easily be replaced (not shown in Figure 8.1).

Bearing: There are one or two bearings, made of cast iron or wood. These are fitted at each end of the gang.

Tractor Drawn Disc Harrow

Discs mounted on one, two or more axles are set at variable angles to the line of motion. As the harrow is pulled ahead, the discs rotate on the ground. Depending upon the disc arrangements, disc harrows have been divided into two classes namely single action, and double action (Figure 8.2).

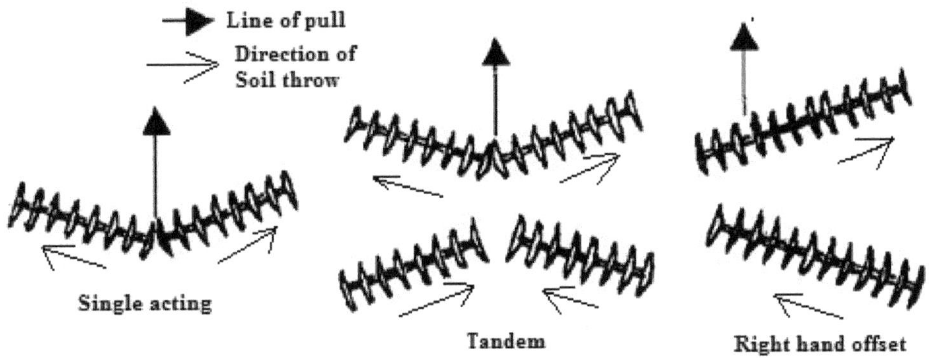

Figure 8.2. Types of Tractor Drawn Disc Harrows as per Disc Arrangement.

Single action disc harrow: In this kind of harrow, two gangs are placed end to end. The discs on the gangs are arranged in such a manner that gangs throw the soil in opposite directions. The right side gang throws the soil towards the right and the left side gang throws the soil towards left (Figure 8.2). Cutting width normally range from 1.2 m to about 6 m.

Double action disc harrow: A disc harrow consisting of two or more gangs, in which a set of one or two gangs follow behind the set of the other one or two, arranged in such a way that the front and back gangs throw the soil in opposite directions (Figure 8.2). Thus, the entire field is worked twice in each trip. Double action harrows are further categorized as tandem, and off-set.

Tandem disc harrow: It consists of four gangs such that each gang can be angled in the opposite direction. It helps to make a much finer and homogenous cut. Unlike the single-action disc harrow, the double-action disc harrow does not leave any particular pattern as it fully scrambles the ground beneath it.

Off-set disc harrow: Unlike the other types of disc harrows, the offset disc harrow has disc gangs that are not in-line with the tractor center line of pull. Since the harrow travels left or right (left hand offset or right hand offset) of the tractor's line of pull, it is why this kind of harrow is called off-set disc harrow (Figure 8.2). It has two gangs in tandem one behind the other, capable of being off-set to either side of the line of pull. The discs on the two gangs are arranged to face in opposite directions such that soil is thrown in both directions. It is widely used in orchards and gardens. These are used as a primary ground tillage farming implement *i.e.* for breaking virgin ground. It produces a much smoother level ground.

Components

A disc harrow mainly consists of discs, gang, gang bolt or arbor bolt, gang angle, gang control lever, spools or spacer, bearings, transport wheels, scraper and weight box. Few of these components are shown in Figure 8.3.

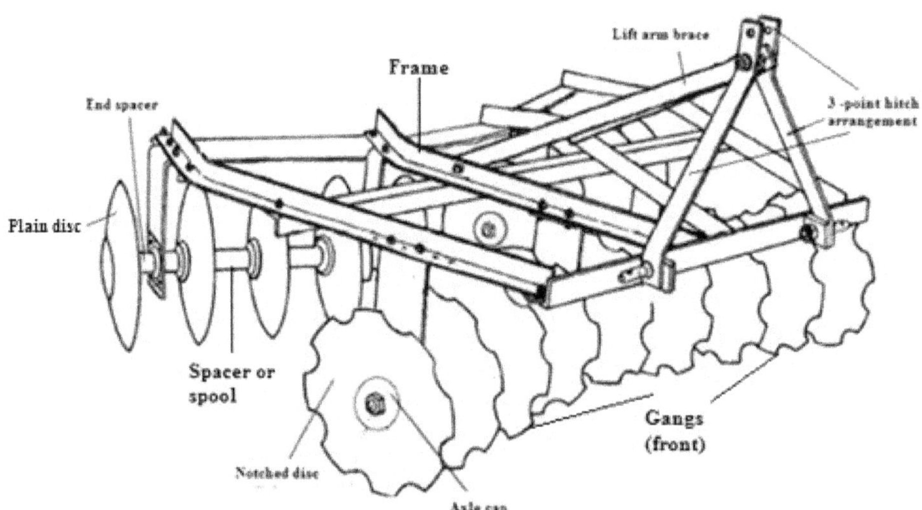

Figure 8.3. Components of a Mounted Harrow.

Disc: It is a circular, concave revolving steel plate used for cutting and inverting the soil. Disc is made of high grade heat treated hardened steel. Tractor drawn disc harrows have concave discs of size varying from 35 to 70 cm diameter spaced 15 to 23 cm for light and 25 to 30 cm apart for heavy duty harrows. Concavity of the disc determines the penetration and pulverization of soil. Two kinds of discs used in disc harrows are plain disc and cut-away disc. Plain discs have plain edges. Most harrows are fitted with plain discs because these are capable of performing all kinds of normal works. Cut-away discs have serrated or notched edges. They cut stalks, grasses and other vegetative matter better than plain discs. Cut-away discs are not very effective for pulverizing the soil, but it is very useful for puddling especially for paddy cultivation. In some cases the front gang is made of cut-away discs and the rear gang of plane discs (Figure 8.3).

Gang: It is an assembly of concave discs mounted on a common shaft separated by spools. The shaft is known as gang axle or arbor axle.

Gang angle: The angle between the axis of the gang and the line perpendicular to the direction of travel is called gang angle.

Spool or spacer: It is made of cast iron and is cast in several shapes and sizes. It is mounted on the gang axle between every two discs to retain them at fixed position laterally on the shaft (Figure 8.3).

Bearing: Since the implement is subjected to heavy radial and thrust loads, ball bearings or tapered roller bearings made of chilled cast iron are used in the harrows.

Transport wheel: In trailing type discs harrows, transport wheels are provided for transporting the harrow on roads to prevent road damage as well as to protect the cutting edges of the discs.

Scraper: It helps to scrape the soil that may stick to the concave side of the discs.

Weight box: A box like frame is provided on the main frame of the harrow to add additional weight on the implement to increase disc penetration in the soil.

Adjustments to Increase Penetration

Following few adjustments are helpful in increasing the disc penetration in the soil.

- ☆ Increase the gang angle, minimum penetration occurs when disc gangs are set perpendicular to the line of draft. Set the gangs such that the forward edges of the disc are parallel to the direction of motion
- ☆ Increase the disc angle
- ☆ Add additional weight on the harrow
- ☆ Lower the hitch point
- ☆ If edges have become blunt, replace the discs to increase penetration. Use sharp edged discs of small diameter and lesser concavity
- ☆ Regulate the speed, penetration is better at low than high speeds

Care and Maintenance of Disc Harrow

Daily Maintenance

☆ Tighten loose or replace missing bolts or parts

☆ Lubricate items that require regular lubrication or those for which lubrication is due

☆ Check all pins to ensure that all retaining hardware is in place

☆ Ensure that all gang components are tight on the axles

Seasonal Maintenance

☆ Coat the outer and inner surfaces of the discs with some oil before the harrow is stored during the slack season

☆ Perform recommended lubrication

☆ Inspect all bearings to ensure that these are tight and are not worn out

☆ Check the tyre pressure and wheel bolts

☆ Make proper operating adjustments for the proposed field activity

☆ Inspect all the oil-bath bearings for leakage

Other Kinds of Harrows

Drag Harrow

Drag harrows have been used since ancient times. These harrows are used to break the clods, to stir the soil, to uproot the early weeds, to level the ground, to break the soil crust and to cover the seeds. Spike tooth and spring type harrows belong to this category.

Spike Tooth Harrow

The spike tooth harrows are available in many styles and sizes. These are either rigid or flexible. While the animal drawn harrows are always of rigid frame type, tractor harrows are both rigid and flexible types. The latter has the advantage that it can be rolled up while transporting. A spike tooth harrow consists of teeth that may be square pointed, drag pointed or round pointed. The teeth, made of wood or steel, are so placed on the tooth bar that one tooth is behind the other. The bars that hold the teeth are also of many kinds such as the "U" bar made of steel, oblong or round wooden bars, or gas pipe bars. The "U" bar is the best form as the material can be used to give the required strength without too much weight. Steel bars may be round, flat or channel shaped. Clamps are used to tightly fasten teeth to the tooth bars so that teeth do not become loose during operation. It is easy to replace the teeth as and when the lower end wears away or to turn it around as the front becomes dull or rounded. The loosening or tightening of the clamp is performed by means of thread and nuts. Besides these 3 parts, other parts include guard, braces, lever and hooks. There may or may not be provision for changing the angle of the spikes while operating the harrow. A particular style is triangle harrow having the basic frame as a triangle. The pegs are rigidly clamped to the crossbars along the three arms of the frame (Figure 8.4 left). In the case of harrows with a wooden frame, the pegs have threaded ends, which are tightened from the top. The pegs of the rigid harrows are fixed slightly tilted so that there is no need to change the angle. The animal drawn harrows are dragged by means of a chain or rope tied to the yoke. These are generally used to stir the soil to a depth of about 5 cm. The harrow covers almost 1 to 1.2 m width. The depth of penetration can be increased by adding weights to the frame. Local make of spike tooth harrow in some regions is known as *Bindha*. The spike teeth as well as the spring tine harrows do not need lubrication.

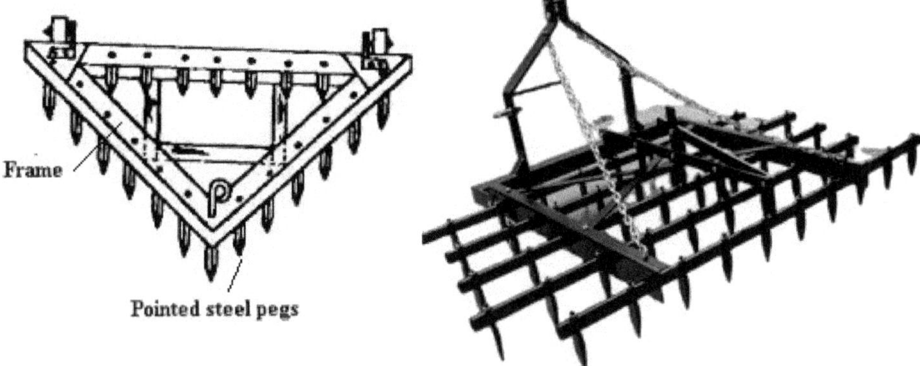

Figure 8.4. Triangular Spike Tooth Drag Harrow (Left) and a Tractor Mounted Spike Tooth Harrow.

Spring Tine/Tooth Harrow

Spring tooth harrow has tough flexible teeth, which are suitable to work in hard and stony soils. The harrow is fitted with springs having loops of elliptical shape. It gives spring action to the teeth during field operation. It is used in soils where obstruction like stone, roots and weeds are hidden below the ground surface. It pulverizes the soil and helps to kill the weeds. The main parts of this type of harrow are: teeth, tooth bar, clamps, frame, lever and links. The basic frame of the harrow is mostly rectangular (Figure 8.5). Guard rails are provided to give rigidity and strength to the harrow. Spring tine tractor drawn harrows have looping, elliptical

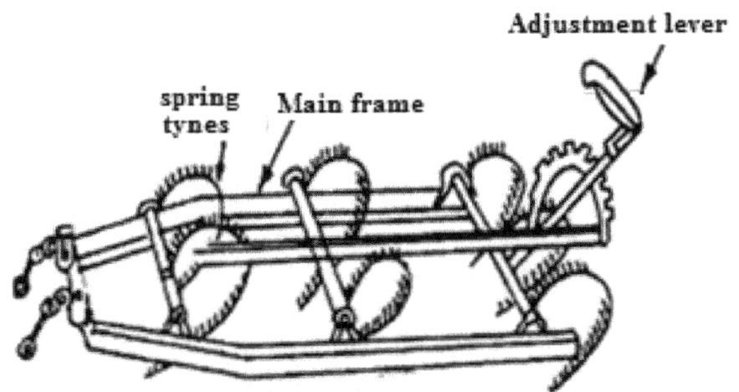

Figure 8.5. Animal Drawn Spring Tine Harrow.

or spring like tines. The animal drawn units have only elliptical tines. The spring tines are bolted staggered on the frame to avoid clogging during operation. Usually the teeth are made of spring steel. Sometimes reversible points are provided so that other end may be used once the one end is worn out. The teeth are fastened to the tooth bar by means of tooth clamps. They provide rigidity and support to the harrow. Teeth may be dismantled for sharpening by grinding. The levers are used for setting the teeth to vary the depth of harrowing (Figure 8.5). The depth of harrowing is 15-18 cm. For light harrowing, the adjustment is done in slanting position. Draft hooks are used for hitching for an animal drawn spring tine harrow.

Patela

The animal drawn version of a *patela* is made of a wooden plank with a number of curved steel hooks bolted to a steel angle section fixed to the rear side of the plank (Figure 8.6). The cutting edge levels and packs the soil and the curved hooks uproot and collect the weeds. The equipment is used for smoothening the soil and crushing and removing the weeds. It is also used for breaking clods, packing and leveling the ploughed soil. *Patela* harrow is 1.2 to about 3 m long and its weight is in the range of 45-55 kg.

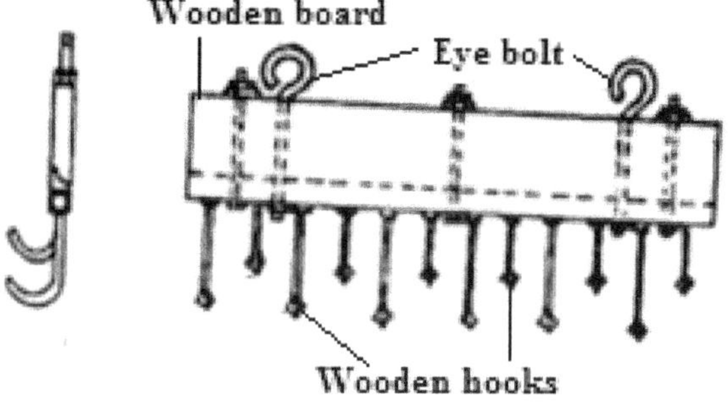

Figure 8.6. A Schematic Diagram of *Patela*.

Blade Harrow

The blade harrow, popularly known as *bakhar*, is quite commonly used harrow by the Indian farmers. Its main body and the beam is made of *Shisham* or *Babool* wood, and the blade is made of steel (Figure 8.7). The action of blade harrow is like that of sweep moving into the top surface of the soil without inverting it. It is generally used in clay soils for preparing seedbeds. It is also used for covering the seed. Sometimes, it is used to chisel out the uncut portion left during ploughing with an indigenous plough. The main functions of course are to pulverize the soil and create soil mulch. The width of cut varies from 38 to 105 cm. *Guntaka* is an improved version of this implement that avoids frequent clogging with the roots and weeds wrapping along the edge of the blade. The improved V-shaped blade of the tool provides relief from clogging. Besides, this version offers advantage in terms of reduced draft, easy penetration and smooth working in the field.

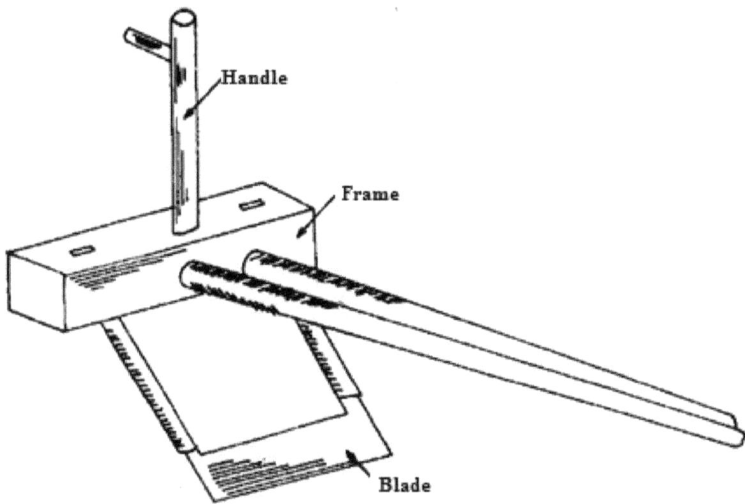

Figure 8.7. Construction of a Blade Harrow.

Acme Harrow

It is also known as knife harrow. It comprises curved knives made of high carbon steel (Figure 8.8). The front part of the knife breaks the soil, crushes the clods and compacts the soil. It is commonly used for good pulverization of the soil. The depth of penetration can be increased by putting additional weight on the implement.

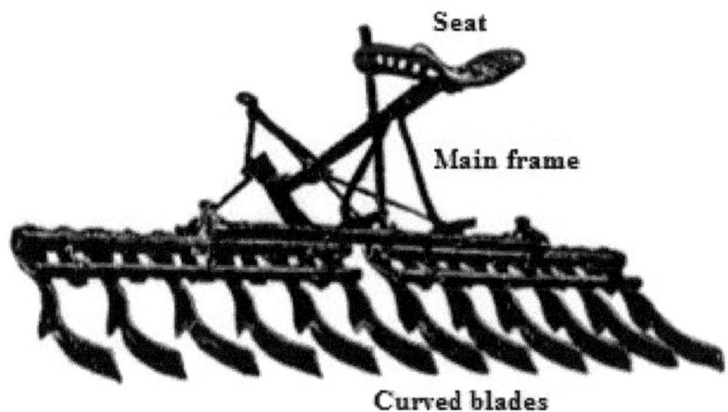

Figure 8.8. An Acme Harrow.

Zigzag Harrow

It is just like a spike tooth harrow. The only difference is that it has a zigzag frame (Sahay, 2019). The teeth are attached at the junctions of the members of the frame (Figure 8.9).

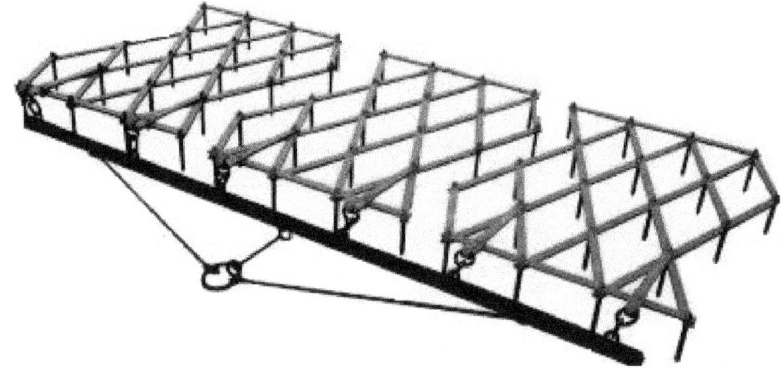

Figure 8.9. A Zigzag Harrow.

POWER HARROW – TRACTOR DRAWN

A power harrow tills the soil maintaining the original profile of the field. It pulverizes the upper and lower layer of soil without turning them upside down. Thus, it not only forms a good seedbed but also makes good soil mulch. It consists of two horizontal crossbars fitted with reciprocating rigid pegs. Power is taken from the PTO of a tractor. The pegs are spaced 200 mm wide and are staggered with respect to each crossbar. The two bars move in opposite directions and hence the implement is dynamically balanced. The oscillating pegs break the clods and pulverize the soil to give a fine tilth. The width of the operation is about 2000 mm having the field capacity of about 1.5 ha/day. Machines with transverse or rotary motions have also been designed.

CULTIVATORS

This implement is used for intercultural operations between crop rows without harming the crop. The cultivator with laterally adjustable tines or discs stirs the soil, and breaks the clods. The tines fitted on the frame of the cultivator comb the soil moderately deep (typically 5 to 10 cm). Its functions are intermediate of those of the plough and the harrow. Few important functions performed by a cultivator are as follows:

✯ Intercultural operations in the fields

✯ Destroys the weeds in the field

✯ Improves soil aeration for proper growth of crops

✯ Conserve moisture by preparing mulch on the surface

✯ To sow seeds when provided with sowing attachments

✯ Reduces surface evaporation and allows rapid infiltration of rain water

The primary function of a cultivator continues to destroy the weeds, even large weeds being completely destroyed during inter-row cultivation. Of course, enough precautions are required to avoid vigorous soil movement that may also bury young crop plants, or deep cultivation that can damage the roots of crops.

Therefore, cultivators are designed to usually work only 50 per cent to 70 per cent of the soil surface, leaving weeds in the crop row unharmed. The cultivators are categorized as 1) Disc cultivator, 2) Rotary cultivator, 3) Tine cultivator.

1. **Disc cultivator:** It is a cultivator fitted with discs.
2. **Rotary cultivator:** In this type of cultivator, tines or blades are mounted on a power driven horizontal shaft.
3. **Tine cultivator:** It is fitted with tines having shovels.

Sweeps/Shovels and Shanks

Sweeps, shovels and shanks are all commonly used components of a cultivator. They are simple, vary greatly in width, shape, and pitch but are quite durable (Figure 8.10). The kind of sweeps/shovel to be used with the equipment depends upon the type of soil, crop and functional requirement. Soil movement away from the shank increases with width and pitch. On the other hand, it decreases with increase in the angle between the leading edges. Sweeps are used to control weeds, as they disturb the whole soil profile (Figure 8.10). Spikes leave much of the soil undisturbed and as such fail to kill weeds. Hoof (Goose foot) style shovels move less soil than standard shovels and sweeps (Figure 8.10). S-shaped (Danish) shanks vibrate more that helps to shake the soil loose and makes the soil free from weed roots (Figure 8.11). On the other hand they are less robust than C-shaped shanks. Cultivators designed for minimum-tillage that operate in high crop residue usually have one shank with a single broad sweep per inter-row. These cultivators have a coulter in front of each shank to cut residue, so it can flow past the shank. In minimum-tillage machines, hilling up around the crop is accomplished with wing attachments or with disc hillers that increases the lateral displacement of soil. The aggressiveness of the operation can be controlled by adjusting the angle relative to the direction of travel. Shovels

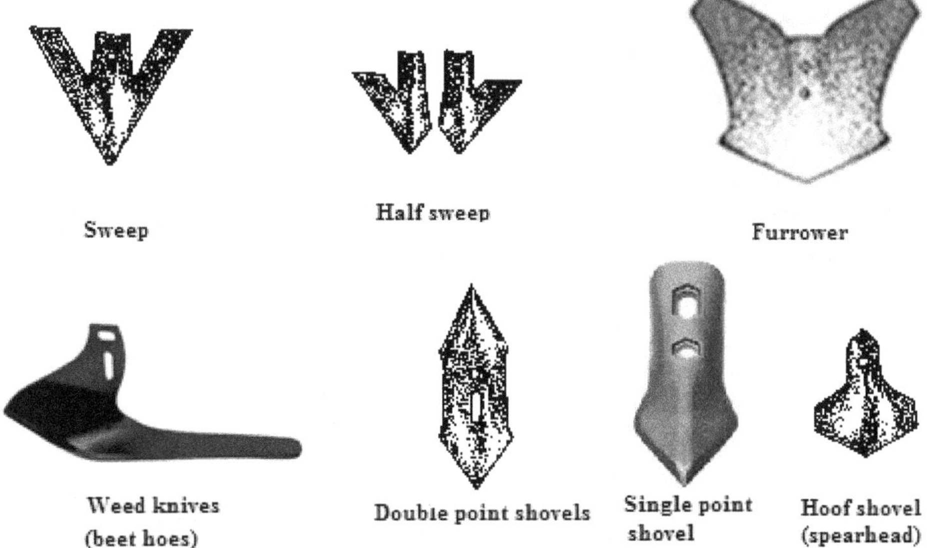

Figure 8.10. selected Shovels and Sweeps Used on Cultivators.

are well hardened and tempered for their effective performance without putting much pressure on the tractor and the implement especially for use in hard soils.

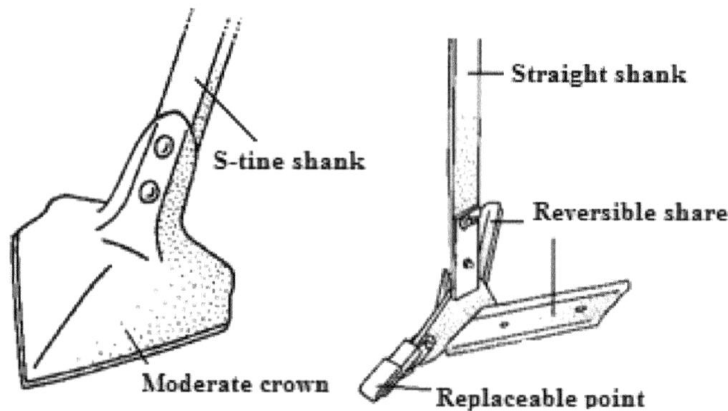

Figure 8.11. Shanks and Share Attachments.

Animal Drawn Cultivators

Numbers of cultivators are available under this category. Most important ones are sweep, junior hoe and cultivator with seeding arrangement.

Sweep

Sweep is used for removing shallow rooted weeds in between crop rows. It consists of V shaped blades having bevel edged wings called sweep. The blades are fitted to the tines by means of counter sunk bolts and nuts. The tines are fitted to a frame. The blades cut the weeds by skimming action under the soil at a shallow depth of 2 to 3 cm. Moreover, cutting action of the blades breaks the capillary passages in the soil. This action results in the formation of good soil mulch for moisture conservation. It is suitable for all row crops and soils and its capacity varies from 1.75 to 2.5 ha/day.

Junior Hoe

It is also intercultural equipment used for weeding between the rows of standing crops. It consists of a frame with hitching arrangement. Six curved tines fitted with reversible shovels are attached to the frame in three rows in a staggered manner. A handle and beam are fixed to the framework for guiding and attaching the unit to the yoke of the animals. The spacing between the shovels can be adjusted as per the row spacing of the crop. The curved nature of tines provides a spring action against stones or roots that helps to release the equipment against such obstacles. The average field capacity is about 1.5 ha/day.

Three-tine Cultivator with Seeding Attachment

It is a combination implement used for intercultural operations and seeding. Seedling attachment is detachable and can be easily detached while performing intercultural operation in crops (Figure 8.12). Combination operations are usually performed with animal or tractor power. The main parts of an animal drawn cultivator-seeder are: frame, seedling attachment, shovel, tine, handle and beam. The rate of seed dropping is controlled manually. A manually operated version of the tool for intercultural operation is also available. It has a lightweight tine for easy operation.

Figure 8.12. Animal Drawn 3-tine Cultivator without Seeding Attachment.

Tractor Drawn Cultivators

Trailed Type Cultivator

It consists of a main frame carrying a number of cross members fitted with tines. A hitching arrangement is provided at the forward end of the cultivator. The height of the hitch is adjusted so that the main frame remains horizontal over a range of depth settings. The tines in each row are spaced widely to allow free passage of the soil and trash around them and staggered to nicely cover the entire width. The working depth is controlled by roughly adjusting the tine in their clamps. Final adjustment is made by a screw lever provided for this purpose. A pair of wheels is provided for easy movement and transport.

Mounted Cultivator

The cultivator comprises a rectangular frame of angle iron mounted on the three point hydraulic linkage. The cross members carry the tines in two staggered lines. Shovels are selected on the basis of type of the soil and crop. Tractor drawn cultivators are categorized as cultivators with spring-loaded tines and with rigid tines depending upon the flexibility and rigidity of tines.

Cultivator with Spring-loaded Tines

These cultivators are fitted with 7, 9, 11, 13 or more tines depending upon the field requirement. The tines are made of high carbon steel and are held in proper alignment on the main frame members. The tines hinged to the frame are provided with two heavy coil springs (Figure 8.13). These springs are pre-tensioned to ensure minimum movement except when the points strike roots or large stones (obstacles). The tines swing back as and when any such obstacle is encountered during the operation. Once the obstruction is crossed, the tines are automatically reset and work continues without interruption. As such, this type of cultivator is recommended for soils embedded with stones or stumps to minimize damage to the implement. A pair of gauge wheels is provided to control the depth of operation.

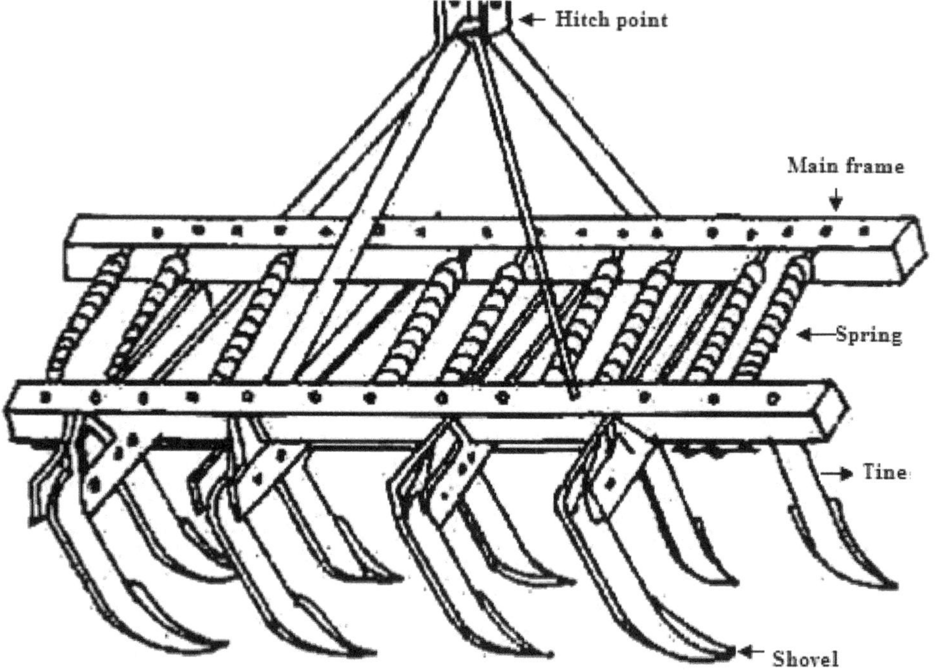

Figure 8.13. Spring-Loaded Tractor Drawn Cultivator.

Cultivator with Rigid Tines

Tines of a rigid tine cultivator are bolted between angle braces fastened to the main bars by sturdy clamps and bolts (Figure 8.14). Since rigid tines are mounted on the front and rear toolbars, the spacing between the tines can be easily adjusted by slackening the bolts and sliding the braces to the desired position. The spacing is so adjusted that the tines are not choked with stubbles of the previous crop or weeds. A pair of gauge wheels is used to control the depth of cultivation.

Duck Foot Cultivator

The name duck foot cultivator is derived from the shape of the sweeps used in this cultivator, being similar to the foot of a duck. It comprises a box type steel rectangular frame, rigid tines and triangular shaped sweeps. The sweeps are made from old leaf spring steel and fixed to tines with fasteners. The sweeps can be replaced once these are worn out or become dull. The tines are made of mild steel flat and forged to shape. Usually this cultivator has 7 sweeps and is about

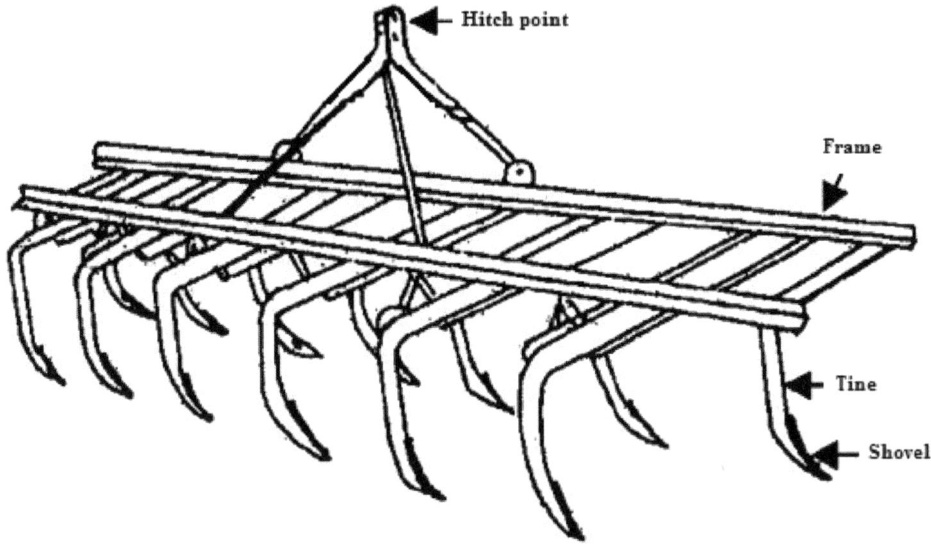

Figure 8.14. A Cultivator with Rigid Tines.

225 cm long and 60 cm wide. The number of sweeps can be reduced according to requirement The depth of operation is controlled by the hydraulic system of the tractor. This cultivator is mostly used in hard soils for shallow ploughing. It is why; it is quite popular in black cotton soils for tillage operation, destruction of weeds and retention of soil moisture.

Rotary Tiller or Rotavator

This implement, also called rotary tiller, cuts and pulverizes the soil by impact forces through a number of rotary tines or knives mounted on a horizontal shaft. It consists of a power driven shaft having a speed of 200-350 rpm on which knives or tines are mounted to cut the soil and trash. Although several types of tines can be fitted on the shaft, but more often sharp edged L-shaped blades are used on the rotor (Figure 8.15). According to source of power used, rotavators are classified as animal drawn, engine operated and tractor-drawn rotavators. A bullock-drawn engine operated rotary tiller is quite useful for timely preparation of seedbed particularly in rice-wheat rotation. Power tiller operated rotary tillers are quite useful for hilly areas and small land holdings. One or two operations of this implement are sufficient for good pulverization of soil depending upon soil and crop condition. It is not recommended for sandy soils. The functional components include, rotor fitted with L shaped steel blades (vary from 36 to 48), gear box, power shaft, sprocket - chain drive, universal joint, leveling board, shield, depth control arrangement, and three point hitching provision. The power from the tractor engine is transmitted to the rotary tiller (rotavator) through PTO of the tractor (Figure 8.15). A clutch is provided in the transmission system for engaging or disengaging power. The speed of rotor is kept at about 350 rpm for rated rpm of 1500 of prime mover. The depth of penetration can be adjusted up to 12.5 cm. A suitable protective cover is provided at the rear to prevent scattering of soil. It can cover about 1.5-2.0 ha/day. A leveling board is attached to the rear side of the unit for leveling the tilled soil (Figure 8.15 bottom). Two adjustable brackets are provided one each on either side of the unit to control the depth of operation. The principal advantages of rotavators are that they completely destroy the above-ground weeds, and chop roots and rhizomes to much smaller fragments than any other implement. In most cases, it completes the operation in one go and requires less energy, it makes huge savings in the cost of cultivation. It helps to improve the soil physical properties resulting in higher crop yield. Moreover, an ideal seedbed is prepared that facilitates the sowing of wheat by seed-cum-fertilizer drill without any operational problems. Nevertheless, excessive pulverization with the implement may result in deterioration of the soil structure.

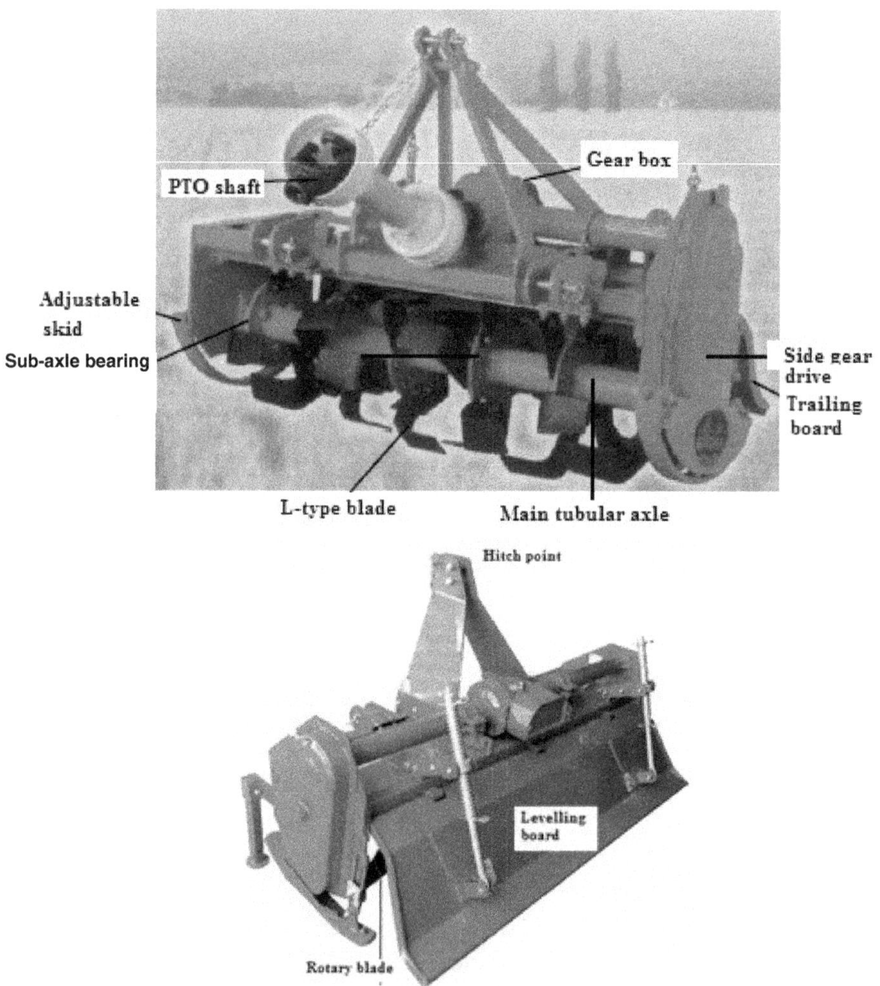

Figure 8.15. Two different Views of a Rotary Tiller.

Types of Blades

1. **L type blades:** Works well in trashy conditions. More effective in cutting weeds but do not pulverize the soil much.
2. **Twisted blades:** Suitable for deep tillage in relatively clean grounds, but needs frequent cleaning due to clogging and wrapping of trash on the tines and shafts.
3. **Straight blades:** It is commonly used on mulchers meant for secondary tillage.

Farmer's feedback on the benefits of rotavator is as follows:

☆ Saves water, time, energy, labour and money

☆ Used successfully for green manuring of *dhaincha*, moong (green gram) and other crops in wet conditions. It can perform puddling and green manuring operations simultaneously provided a heavy planker is also used with the equipment

☆ Incorporates straw and residues of paddy, sugarcane, cotton in the soil in single operation

☆ Better quality of work achieved in less number of operations both in dry and wet land condition compared to traditional practice

☆ No disturbance is caused to the leveled fields

☆ Can be used throughout the year

☆ No injury to finger during transplanting

HOES

A hoe is an ancient and versatile agricultural and horticultural hand tool. A number of traditional and improved hand hoes are used in different parts of country depending upon the nature of soil and crop grown. The shape and size of hoes as well as their names vary from place to place. It is a multi-purpose tool used to shape the soil, remove weeds, clear soil, and harvest root crops. Shaping the soil includes many operations such as piling the soil around the base of plants (hilling), digging narrow furrows and shallow trenches for planting seeds or bulbs. The two commonly used hoes are hand hoes and animal drawn hoes.

Hand hoe is the most popular manually operated weeding tool used on Indian farms. It consists of an iron blade and a wooden handle. The handle may be short or long. The operator holds the handle and cuts the soil with the blade to a shallow depth of 2-3 cm. Besides stirring the soil it also cuts the weeds. The two main problems encountered in the use of hand hoe are the low work rate and uncomfortable operating postures.

Khurpi

Khurpi is a traditional hand tool made by the local artisans for the use of small and marginal farmers. *Khurpi* in India varies widely in size, shape and weight, but they have common basic parts namely a cutting blade and a small wooden handle for the grip. The *khurpi* with a long narrow blade is preferred for weeding around the flower plants, broadcasted crops and vegetable crops. Under normal condition, a man is able to weed out about 0.025 ha/day.

Kodali

Kodali is similar to a *phawda* (broad spade), the difference being that instead of a wide thin cutting blade, a narrow long pointed thicker section blade is attached to the handle. The person working with it has to bend his body. It is used for inter cultivating maize and sugarcane crops, and for earthing-up the potato crops sown in line. The capacity of the tool is about 0.04 ha/day.

Hoe-cum-Rake

The hoe-cum-rake is a multipurpose hand tool consisting of a flat blade on one side like a broad spade and prongs on the other side. The blade and prongs are either made from a single stock with an eye in the center or are joined to an eye by welding. A wooden handle is fitted to the eye for operation. It is a secondary nursery bed preparation tool and is used for light operations. Whereas the flat blade is used for digging, the rake side is used for weeding and/or collection of weeds and trashes. The flat end of the tool is operated with impact action.

Long Handle Weeders

As the name suggests, these hoes are provided with long handles. It allows the operator to perform the operation in standing posture, which is more comfortable than the bent posture required in many hand tools. The two most popular long handle weeders are star type weeder and peg type weeder (http://www.eagri. org/eagri50/FMP211/pdf/lec11.pdf). These weeders are also called as dry land weeders since they are mostly used in dry lands

Star type weeder: A star type weeder is suitable for weeding in garden lands having low soil moisture contents (10-15 per cent). A major limitation of the tool is that it works well only in line sown crops. It consists of a blade for cutting the weeds, a fulcrum wheel for push-pull movement and a long handle to reduce the strain on the operator. The radial arm of the fulcrum wheel is cut into star like projections and hence the name star type weeder (Figure 8.16 left). The operating width of the blade is around 120 mm and the capacity around 0.05 ha/day.

Peg type weeder: The operation and construction of this machine is quite similar to star type weeder. Herein, pegs are welded on the periphery of the wheel hence the name peg type weeder (Figure 8.16 right). The peg type wheel works well in clayey soils. The operating width and capacity is about the same as that of a star weeder.

Improved Hand Hoe

An improved hand hoe has a long handle fitted in the middle of the cutting blade. One end of the blade is about 10 cm wide with sharp edges and the other end is pointed narrow one for making small furrows. It can be used for intercultural operations including weeding close to the individual plants. Being long handle type, it can be operated in the standing position.

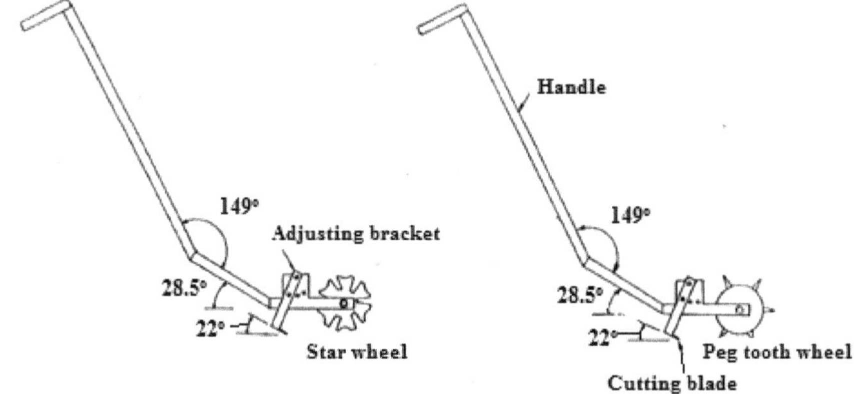

Figure 8.16. Long Handle Weeders: Star Type (Left) and Peg Type (Right).

The Grubber

The grubber is a manually operated pull type hoe suitable for weeding and intercultural operations in upland row crops in black cotton soils. It is provided with three blades having the field capacity of about 0.15 ha/day.

Rotary Paddy Weeder

Rotary paddy weeder is quite suitable for uprooting and burying the weeds into the soil. The operator moves the tool forward and backward in narrow rows of paddy crops. It not only gives higher output but considerably reduces the drudgery of the operator.

The Wheel Hoe

The wheel hoe consists of one or two wheels, two handles and a tine to mount the cutting tool, which is either a reversible shovel or a three prong fork or rake or sweep depending upon the soil moisture conditions and types of weeds. A man operates the hoe in standing position each time pushing it through a short length. The capacity of the machine is around 0.04 ha/day.

Animal Drawn Hoe

Animal drawn weeding implements are pulled either by a single or a pair of animals. These implements may either be single or multi row types. The three tine cultivator or *Triphali*, Akola hoe, Bardole hoe or two blade hoe are some popular hoes for row crop operations in different regions. The crop rows are widely spaced (above 30 cm) to allow the movement of animals and the implement. The main parts of the commonly used blade hoe are: (i) Head piece (ii) Prong (iii) blade (iv) handle and (v) beam. One or more than one cutting blades may be provided on these hoes. The prong makes an angle of about 45° downward with the horizontal plane with a blade attached to it at the end. The hoe width ranges from 25 to 75 cm depending upon the size of the animals/bullocks and types of soil. It only loosens the upper soil surface. It is mostly used for intercultural operations in sorghum, cotton, groundnut and other *kharif* crops.

Cono Weeder

Cono weeder is useful for uprooting and burying weeds in line planted rice fields in wetlands. It has become an essential tool for farmers adopting SRI (System of Rice Intensification) cultivation. It disturbs the topsoil that improves soil aeration and triggers root growth resulting in better growing environment for the crop and higher yields. The weeder consists of a long handle, two truncated conical rollers, and a float. The rollers are fitted at the bottom of the handle in the opposite direction one behind the other (Figure 8.17). The conical rollers have serrated blades on the periphery. When the weeder is operated in between two rows of standing crop, the rollers uproot the weeds and bury them. The float prevents the implement from sinking into the soil. Soil at the time of its use should be moist and little firm. It can cover about 0.05 ha/day.

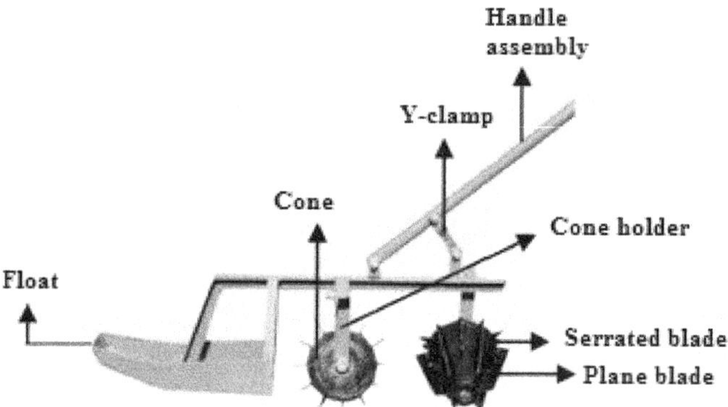

Figure 8.17. Major Components of a Cono Weeder.

PUDDLERS

Puddling is one of the most common farm operations in conventional transplanted rice cultivation. Three main purposes of puddling are:

☆ To reduce deep percolation losses from the rice fields

☆ To control weeds

☆ To create conditions for easy transplanting and proper establishment of seedlings

The processes involved in puddling are ploughing the field, ponding 5-10 cm deep water layer on the soil surface and churning the soil by means of a puddler or any other implement. The puddling creates a semi-pervious hard pan of approximately 10 to 15 cm dense mud. Puddlers are classified as: hand operated puddlers, animal drawn puddlers, and tractor drawn puddlers. As per the construction, puddlers are categorized as straight blade type, helical blade type, paddle type, and cage wheel type. In many areas, even an indigenous plough is used as a puddler. Peg tooth harrow is also used by the farmers to puddle the fields. Since these instruments fail to effectively puddle the fields, rotating blade type puddlers have been developed both for animal and mechanical power sources.

Patela **Puddler**

The unit consists of a mild steel frame, a wooden plank, a gang consisting of pegs for turning the soil, sliding type pegs, handle and hitch system. A pair of bullocks is used for the operation of the implement. A labour generally stands on the plank during the operation to add weight to the plank. It gives additional depth. The tool breaks the soil clods resulting in a smooth puddle. Soil moisture content at the time of operation should be near to the saturation level. It is suitable for shallow puddling with high mechanical dispersion of soil particles. It is capable of preparing homogenized puddled tilth that can suit even for mechanized paddy transplanting.

ANGRAU Open Blade Puddler

The open blade type implement, commonly used for puddling in south India, consists of a series of steel or cast iron blades fastened to a cast iron hub at an angle. For example, the ANGRAU puddler comprises frame, beam, puddling unit, axle, metal-cross and handle (Figure 8.18). Each puddling unit has four paddles (or blades) mounted on an axle. Paddles, made of mild steel sheet, are 3.15 mm thick. The blades are welded to the metal cross made of mild steel rounds. The blades with the metal cross are welded to the axle, at an angle of 10° for 750 mm (30″) puddler and 7½° for 1000 mm (40″) puddler. During the operation, blades (paddles) churn the soil to mix it properly. Besides, it also chops the weeds and mixes them into the soil. A good puddled soil is achieved in 2-3 passes. The animal drawn puddler can cover 0.4-0.5 ha/day.

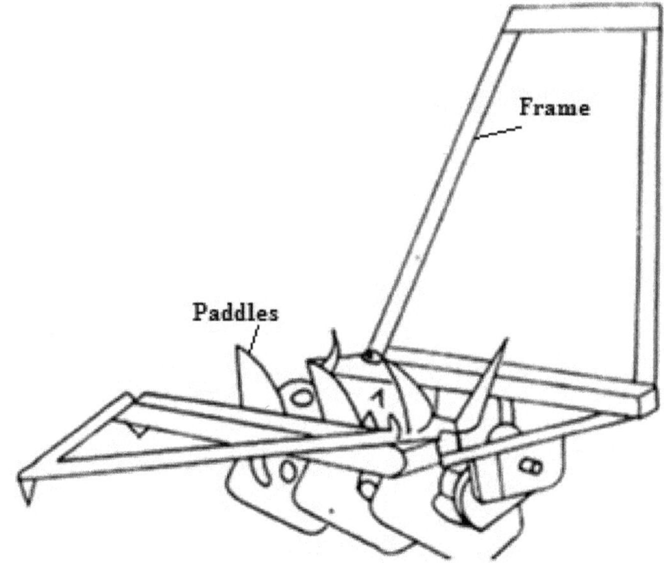

Figure 8.18. Open Blade ANGRAU Puddler.

Tractor Drawn Puddlers

Four classes of tractor drawn puddlers are: (i) tine tiller, (ii) rotating blade puddler, (iii) disc harrow and (iv) power rotary tiller. Among these implements, disc harrow and power rotary tiller are widely used. The 3 point linkage power rotary puddler is very similar in design to a rotavator except that they are less robust and have smaller rotating drums and shoes. They are designed to operate at a maximum of 50-75 mm depth and are normally wider than a conventional rotavator. The tractor drawn puddlers can be used in both wet and dry conditions but standing water in the field is preferred. They leave the field level and work best when laser controlled.

Peg Type Puddler

The peg type puddler consists of a frame made of mild steel angle sections on which three crossbars and the cleats for the three-point linkage are welded. It is operated when the soil moisture content is near the saturation level. It gives a fine tilth and good puddle that facilitates mechanical transplanting. The tractor is fitted with cage wheels to improve traction, and to achieve higher field capacity.

Hydro-tiller Puddlers

A hydro-tiller is a mounted rotavator having a leveling board at its back. It puddles the soil and buries weeds and residue by a powered rotor. The hydro-tiller is propelled forward by a 7-10 kW engine that also drives the rotor attached to a platform that incorporates airtight floats. The speed of operation and degree of puddling is mainly controlled by the weight applied to the rear of the machine. These machines only work in flooded conditions. Transportation to and from the field may pose some problems. Therefore, in most cases wheels are fitted to the outside of the rotor for transportation purposes.

SECONDARY TILLAGE IMPLEMENTS FOR RICE CULTIVATION

Cage Wheels

The performance of a tractor is greatly undermined during puddling the rice fields under water mainly due to slippage of the pneumatic tyres. To overcome this problem, iron/steel wheels, called cage wheels are used both on 2 wheel (power tillers) and 4 wheel tractors for improving the traction of the tractor. A cage wheel is lugged with L angles to provide required grip and facilitate easy movement of the tractor in the rice fields. Besides, cage wheels help to bury weeds. Cage wheels are of two types namely half cage wheel and full cage wheel. Half cage wheels are attachment fitted to the rubber tyre wheels. On the other hand, full cage wheels fully replace the tyre wheels during this operation. The width of the full cage wheel may be about twice that of the half cage wheel.

Paddy Weeder

It is important equipment for intercultural operations in paddy fields. It is used for uprooting weeds and burying them in puddle soil between rows of standing paddy crop. It helps to improve soil aeration. It consists of frame, weeding roll, tines, float and handle (Figure 8.19). The frame is made of mild steel to which float, front and rear weeding rolls and angle regulator are attached. There are two mild steel weeding rolls one in front and the other at the rear. The float, made of mild steel, is placed in front of weeding roll that helps in maintaining easy sliding motion during operation. Weeding rolls with fingerlike projections perfmr the weeding operation in water.

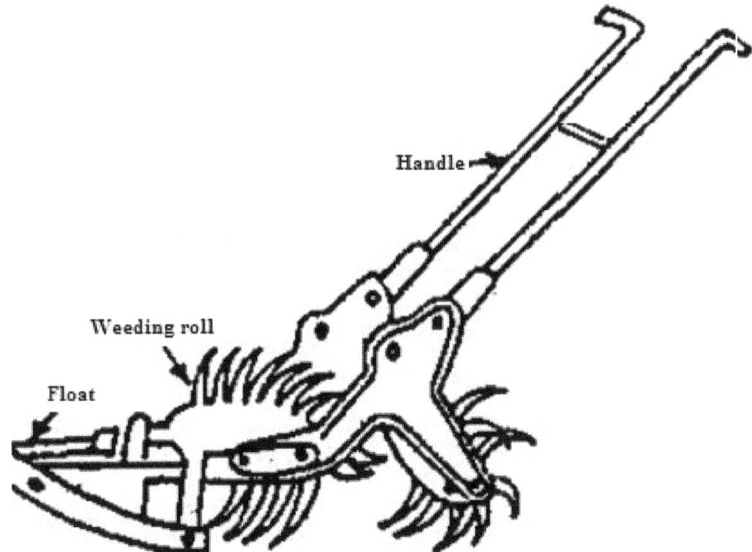

Figure 8.19. A Line Diagram of a Paddy Weeder.

LAND ROLLERS OR PULVERIZERS

Land rollers or pulverizers are used to give a final finish to the operation of seedbed preparation. Several styles of rollers have been invented. Some such rollers are: continental Cambridge roller, corrugated roller, cage roller, Cambridge roller, land packer, and coil roller. The basic principles and function of most of them are almost similar although corrugated or tubular rollers may have certain advantages over the plain smooth rollers (Davidson and Chase, 1908).

Cage Roller (Clod Crusher)

It is made up of a cage of metal bars arranged in the shape of a roller (Figure 8.20). It is used to thoroughly pulverize and firming up the loose soil so that no large air spaces or pockets arc left. The cage roller presses the upper soil against the sub-soil. It makes a continuous seedbed so that moisture is conserved and properly released to the roots of the plants.

Tractor Drawn Spiked Clod Crusher

The clod crusher is used with a tractor drawn harrow or cultivator as a combination tillage tool. It consists of a rectangular frame made from mild steel angle section, hitch frame and a shaft for carrying the drum. The shaft is mounted on bearing pedestals. The drum made of mild steel sheet has pegs welded on its surface. The pegs on the crusher pierce and break the soil clods.

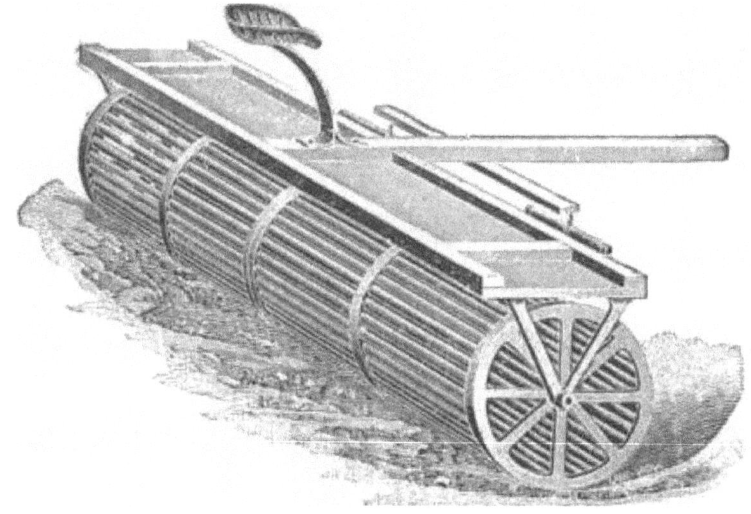

Figure 8.20. Single Gang Cage Roller or Clod Crusher.

Improved *Bakhar*

It is an animal drawn implement used to prepare seedbed in black soils. It consists of a frame made of mild steel angle section in which a V-shaped blade is attached by means of fasteners. The *bakhar* replaces the conventional curved blade with a modified V-blade for reduced draft. Behind the blade a roller made of 150 mm diameter mild steel pipe is provided to crush soil clods. The operator controls the implement through a handle provided on the frame.

V-shaped Roller Pulverizer

The roller pulverizer is constructed of a number of wheel sections arranged on a shaft. The surface of the roller forms a kind of corrugations. It is why; it is also called a corrugated roller. It rolls, pulverizes, packs, levels, cultivates, and mulches the soil.

Clod Breaker

The clod breaker is a simple wooden tool. It is made of a cylindrical wooden log having the iron spikes (Figure 8.21 left). An adult person can pull this implement. More often, it is pulled by animals and has a seating arrangement for the operator's comfort (Figure 8.21 left). By driving the clod breaker over the field in opposite directions, the soil become loose, aerated and is able to absorb more water. A tractor operated clod breaker that can be trailed behind a harrow or cultivator is shown in Figure 8.21 (right).

Figure 8.21. Animal Drawn Clod Breaker (Left) and a Clod Breaker Attachment (Right).

Sheep Foot Roller

It is used to compact the soil and farm roads. The sheep foot roller consists of box frame of heavy angles, hard faced solid forged tamping feet, heavy-duty shaft enclosed in the drum, steel drum on which forged tamping foot are welded, reinforced box section drawbar provided with heavy-duty survival hook, and adjustable cleaner teeth to prevent accumulation of dirt between the teeth. The roller is mounted on a heavy-duty frame. The equipment may have one or two drums and is attached to the drawbar of the tractor. Independent oscillating frame and oscillation stopper are provided in the double drum sheep foot rollers.

Soil Packer

A soil packer, also called culti-packer, is also a kind of clod crusher and packer that ensures better seed and soil contact. It is quite useful for the establishment of small seeded crops like forages (Figure 8.22). It is also available in many other shapes and sizes or as a combination either of two Cambridge rollers off-set by half a disc or Cambridge and cross kill rollers in the same frame (Sahay, 2019).

Figure 8.22. A View of Tractor Drawn Soil Packer.

GOOD AGRICULTURAL PRACTICE IN TILLAGE

- ☆ Inspect the tractor and machinery for proper functioning before venturing into the field

- ☆ Tillage operations should be performed only when the soil moisture is in friable range

- ☆ Avoid over tillage. For example more tillage may be required for root, rhizome and tuber crops but lesser tillage will do for cereal, legume and pulse crops

- ☆ Select and use implements that ensure good soil structure and tilth

- ☆ Use combination tillage equipment for multiple operations to save time and cost of the tillage operation

- ☆ Avoid mechanical tillage operation on lands with slopes exceeding 20°

- ☆ If possible, plough the soil immediately after harvest of the crop especially when there is enough residual moisture in the soil

- ☆ Never walk on the sides of the tractor operated implements to avoid any mishap

- ☆ Never venture to stand or sit on the implement to increase the depth of penetration. Instead put dead weight to achieve the same objective

QUESTIONS (THEORY)

1. Define secondary tillage naming various equipments used for this purpose. List specific objectives of the harrowing operation.

2. What is a disc harrow? Name and describe the various components of a tractor drawn disc harrow.

3. Discuss single action, double action, tandem and off-set harrows clearly bringing out the differences amongst them. Use neat line diagrams to illustrate your answer.

4. Describe a cultivator and list its important functions.

5. Write a descriptive note on rotavator.

6. Differentiate between a star type and peg type weeder.

7. List the functions of a puddler. Explain the construction and working of ANGRAU puddler.

8. Write short notes on animal drawn junior hoe, tractor operated cultivator with spring-loaded tines, and duck foot cultivator.

Chapter 9

Field Capacity Power Requirement and Costing

FIELD CAPACITY

The field capacity of a farm machine is the rate at which it performs its primary function. The most common measure of field capacity of any agricultural machine is ha/hr. The knowledge of this parameter tells the owner as to how many hectares he will be able to cover in an hour of farm operation with a particular machine. This information further helps him in planning field operations, power units, labour, and to estimate machine operating cost. A related issue besides the cost of machine in Rs./hr is the cost of farm operation expressed in Rs./ha. While talking of field capacity of a machine, one is always confronted with the task of matching the power unit to the size and type of machines to carry out field operations on time with minimum cost. While an oversized tractor will increase the cost of field operation, an undersized tractor will compromise on quality and quantity of works. An overloaded power unit will always result in frequent breakdowns. This chapter details all these aspects of farm machine operations.

Theoretical Field Capacity

The theoretical field capacity of a machine depends on the full operating width of the machine and the average travel speed in the field. In other words, it is the rate of field coverage of the implement, based on 100 per cent of time at the rated speed and covering 100 per cent of its rated width.

Theoretical field capacity, $TC = W \times S \times 1000 / 10000 = W \times S / 10$ ha/hr $\qquad$ (9.1)

Here W is the width of the implement, m and S is the speed of operation, km/hr. Theoretical time per hectare accordingly is the time required to complete the operation on 1 ha at the theoretical field capacity. A machine may be capable of operating at 100 per cent field efficiency only for short periods of time. On a long-term basis, the efficiency drops below 100 per cent mainly due to change in speed attributed to slow down at turns, deeper or more intensive tillage, high-yield crop with high residues, or adverse field conditions. The other factor that affects the capacity is working width of the machine being less than the maximum as a result of overlapping to prevent skips.

Calculating Average Field Speed

The travel speed directly depends on the engine speed, and inversely on the transmission ratio. Some speed loss may occur due to the wheel slip. To measure the average field speed, a distance of 100 m is staked in the field. The time taken in seconds to drive between the stakes is noted. If this time is denoted by T, then the average field speed is calculated as:

Speed, km/hr = T x 60 x 60/10 = T x 360

Actual/Effective Field Capacity

The effective field capacity is the rate of field coverage by the machine considering the actual field time and the actual rated width covered by the machine while operating in the field. Actual/effective field capacity is always less than the theoretical field capacity. It is given as:

Effective field capacity = Theoretical field capacity **x** Field efficiency

The equation for effective field capacity is the same as theoretical field capacity (eq. 9.1) with an additional term of field efficiency.

$$C = S \times W \times E/10 \tag{9.2}$$

Such that C is the effective field capacity, ha/hr and E is the field efficiency, fraction. For efficiency in per cent, the above equation changes to

$$C = W \times S \times E_p/1000 \text{ ha/hr} \tag{9.3}$$

Here E_p is the field efficiency, per cent.

Effective operating time per ha, given as 1/C, is the time during which the machine is actually performing its intended function. It is the sum of the theoretical time per hectare plus the time per hectare required for turns plus the time per hectare required for support functions as discussed in the following section.

Field Efficiency

The ratio of actual or effective field capacity (C) to theoretical field capacity (TC) is called the machine's field efficiency (E). It accounts for failure to utilize the full time due to many kinds of time delays and full operating width of the machine (overlapping). The time delays may include turning, filling the seeds, fertilizers or pesticides, emptying grain, traveling to a grain cart, cleaning a plugged machine, checking machine's performance and making adjustments, waiting for trucks, and operator rest stops. Delay activities that occur outside the field, such as daily service, travel to and from the field, and major repairs are not included in field efficiency measurement. Although wider equipment operated at the same speed covers more hectares per hour, but measurements have shown slightly lower field efficiencies of wider equipment. It is because wider implements need longer headlands to turn in raised position and are not performing the operation. It may not be surprising that wider equipment may occasionally result even in a smaller number of hectares covered per hour, if travel speed is slowed due to limited tractor horsepower available or too much wheel slippage. Field efficiencies are known to vary from some 65 per cent to about 90 per cent (Hancock *et al.*, 1991). According to BIS (1979), speed and efficiencies of various implements are given in Table 9.1 and can be used for the calculations.

**Table 9.1. Typical Range of Recommended Speed and
Field Efficiencies of Various Machines**

Machine	*Recommended Average Speed (km/hr)*	*Recommended Field Efficiency (Per cent)*
Plough	4.5	80
Disc harrow	6.0	80
Cultivator	6.0	80
Seed drill	5.0	70
Seed-cum-fertilizer drill	5.0	70
Planter	5.0	70
Ridger	4.5	90
Puddler	5.0	75
Rotavator	2.5	80
Combine (self-propelled)		
Paddy	2.0	75
Wheat	3.5	75
Combine (mounted and drawn)		
Paddy	2.0	70
Wheat	3.0	70

Material Capacity

The working capacity of the harvesting machines is often measured by the quantity of material harvested per hour and is denoted as material capacity, expressed as tons/hr. It is the product of the machine's effective field capacity and the average crop yield per hectare. It is calculated as:

$$M_c = C \times crop\ yield \tag{9.4}$$

Here M_c is the material capacity of the machine. For example, a baler with a C of 5 ha/hr working in a field yielding 7 tons of hay per ha will have M_c of 35 t/hr.

Example 9.1

A 4 x 25 cm double action disc harrow is operated by a tractor having a speed of 5 km/hr. Calculate the actual field capacity, assuming field efficiency as 80 per cent.

Size of the harrow (width) = 4 x 25 = 100 cm = 1 m.

Using eq. (9.3), we have

Actual field capacity, $C = W \times S \times E_p/1000 = (1 \times 5 \times 80)/1000$

$C = 0.4$ ha/hr

Example 9.2

A 3 x 30 cm plough is moving at a speed of 4 km/h. Calculate how much time it takes to plough 500 x 500 m field, if the field efficiency is 70 per cent.

Width of the plough = 3 x 30 = 90 cm = 0.9 m

Effective field capacity, $C = W \times S \times E/1000 = (0.9 \times 4 \times 70)/1000 = 0.25$ ha/hr

Since $C = 0.25$ ha/hr = 2500 m²/hr (Note: 1 ha = 10,000 m²)

Time required to cover the given area = 500 x 500/2500 = 100 hr

Example 9.3

An indigenous plough makes a 20 cm wide furrow at the top having the depth of 10 cm. Calculate the volume of soil handled per day of 8 hours, if the working speed is 2.5 km/h.

Furrow cross-section = 10 x 20/2 = 100 cm² (Bottom width is taken as zero)

Distance traveled in 8 hours = 8 x 2.5 x 1000 = 20000 m

Volume of soil handled = 20000 x 100/10000 = 200 m³

Example 9.4

Calculate the theoretical field capacity of the machine required to combine 1000 ha of corn in 10 field days. Assume that the machine is used for 12 hours per day.

Theoretical field capacity = Area to be covered/time

Theoretical field capacity, ha/hr = 1000/(10 x 12) = 8.33 ha/hr

Example 9.5

A disc harrow 4.5 m wide is operated at a speed of 7.5 km/hr. The disc covered 200 ha area when it was operated for 8 hours a day for two weeks. Calculate the effective field capacity and field efficiency.

Total period, hr = 2 x 7 x 8 = 112

Total area covered = 200 ha

Therefore, effective field capacity = 200/112 = 1.79 ha/hr

Theoretical field capacity (eq. 9.1) = 4.5 x 7.5/10 = 3.38

Field efficiency = 1.79/3.38 = 52.9 per cent

POWER REQUIRED

Tractor engine converts the energy contained in the fuel to a rotating power. In the tractor system, it is further converted to drawbar pull, or PTO output, or hydraulic system output (Figure 9.1). The drawbar pull depends directly on the engine torque, on the transmission ratio considering force losses as a result of rolling resistance. The drawbar power depends directly on the engine power and the losses through the transmission and at the wheel/ground surface.

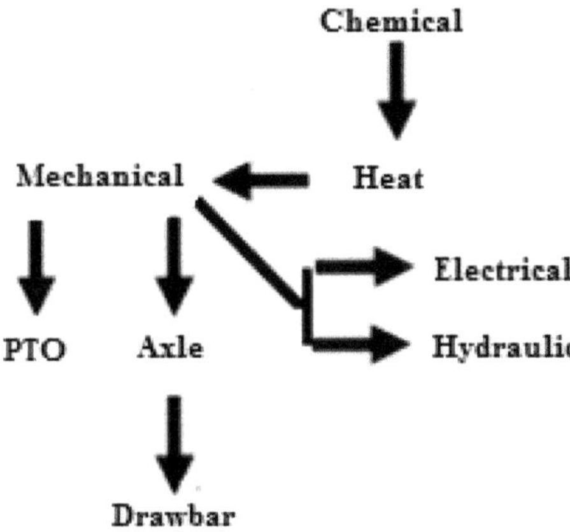

Figure 9.1. Tractor Power Flow Diagram.

The resistance offered by an implement to forward movement determines the draft force needed to perform the required work with the implement. Thus, the draft is the total force parallel to the direction of travel required to pull the implement. It includes both functional draft (soil and crop resistance) and draft required to overcome rolling resistance of the implement. Draft force required pulling many seeding implements and minor tillage tools operated at shallow depths is primarily a function of the width of the implement and the speed at which it is being pulled. For tillage tools operated at deeper depths, draft also depends upon soil texture, depth, and geometry of the tool. Draft force is measured with various kinds of dynamometers such as expanding spring types, hydraulic piston types or load cells. Computerized load cells help to measure draft in the field many times each second so that a mean value can be calculated. Estimated values of draft for several implements are given in Table 9.2. The values given in the Table are multiplied by the actual width of the implement to calculate the total draft exerted by the implement. For example for a 5.5 m wide cultivator, the total draft will be 17.9 kN assuming an average value of draft as 3.25 kN/m per unit width.

Table 9.2. Select Implements and their Draft Requirement

Implement	Draft per Unit Width (kN/m)
Chisel plough	4.5-5.5
Blade plough	4.0-4.5
Disc plough	5.0-6.0
Scarifier	4.0-4.5
Cultivator	3.0-3.5
Planter	2.5-3.5

Estimation of Drawbar Power

Drawbar horsepower is the measure of the pulling power of the engine by way of wheels or tyres or tracks. Since the power is defined as the rate of doing work, it can be written as (Coats, 2002):

$$\text{Power, } P = W/t = F \times d/t \tag{9.5}$$

Such that W is the work done, F is the force, kN and d is the distance travelled, m and t is the time during which distance d is travelled, s. SI units of power are $kg\text{-}m^2/s^3$. If F in the equation is replaced by draft exerted by the implement, then drawbar power required to pull the implement can be calculated as.

$$P_{db} = D \times d/t \text{ kN-m/s (Note that N is } kg\text{-}m/s^2) \tag{9.6}$$

Since d/t is the speed, S, km/hr, eq. (9.6) can also be written as:

$$P_{db}, \text{ kW} = D \times S/3.6 \tag{9.7}$$

In eq. (9.7) D is in kN and S in km/hr. The factor 3.6 has come from the equality: 1 kW = 3600 kN-m/hr = 3.6 kN-km/hr. If S is used in m/s, the factor 3.6 reduces to 1.0 such that P_{db}, kW = D x S.

In terms of HP, eq. (9.7) is written as:

$$P_{db}, \text{ HP} = D \times S/2.68 \approx D \times S/2.65 \tag{9.8}$$

The factor 2.68 or nearly 2.65 has come from the following equalities.

1 Kilowatts (kW) x 1.34 = Horsepower (HP)

Horsepower (HP) x 0.746 = Kilowatts (kW)

Eqs. (9.7 and 9.8) are also applicable for calculation of rotary power (See Example 9.8)

Example 9.6

An implement is being pulled by 2.4 kN force at a speed of 2 km/hr. Find the horsepower required to pull the implement.

Using eq. (9.8) we have

$$P_{db}, \text{ HP} \approx D \times S/2.65 = 2.4 \times 2/2.65 = 1.81 \text{ HP}$$

Example 9.7

A tractor pulls a draught load of 1000 kg while travelling at a speed of 3.6 km/hr. Find the required drawbar horsepower (HP).

$$P_{db}, \text{ HP} = D \times S/2.65 = 1000 \times 9.81 \times 3.6/(2.65 \times 1000) = 13.32$$

Note that 1000 kg is equal to 1000 x 9.81 N or 1000 x 9.81/1000 kN.

Example 9.8

A belt produces a tangential force of 1000 N on a pulley having a radius of 0.30 m running at a speed of 200 rpm. What is the power transmitted?

F = 1000 N or 1 kN, speed = 2π r x rpm x 60

P = F x S/3.6 = F x (2π r x rpm x 60/1000)/3.6 = 1 x (2 x π x 0.30 x 200 x 60/1000)/3.6 = 6.28 kW.

Example 9.9

A tractor of 50 kW is working at a speed of 6 km/hr. What width of cultivator can be used with this tractor, if the draft of the cultivator is 4.5 kN/m?

Using eq. (9.7) with transposition as

D = P_{db} x 3.6/S = 50 x 3.6/6 = 30 kN

Width of the cultivator = 30/4.5 = 6.7 m

PTO Power

The torque output represents the magnitude of the rotational effort developed by an engine or a rotating shaft against a torque load applied to it. It is related as per the following relation.

$$P = 2\pi FRn/60000 \approx FRn/9550 \tag{9.9}$$

$$P = 2\pi Tn/60000 = Tn/9550 \tag{9.10}$$

Such that P is the maximum PTO power, kW, F is the tangential force, kN, R is the radius of force rotation, m, n is revolutions per minute, rpm and T is the torque, a measure of force exerted to rotate a shaft, which is obtained by multiplying the distance from the center of the shaft and the tangential force, R x F with units of N-m. Thus, both the brake power or flywheel power rating (also known as maximum PTO power) and the actual PTO power of the tractor, the power generated at the power-take-off shaft, can be determined with the help of eqs. (9.9) and (9.10).

The PTO should be able to transmit sufficient amount of power to operate specified auxiliary equipment. The drawbar power is always less than PTO power output because of drive wheel slippage, tractor rolling resistance and friction losses in the drive train between the engine and the wheels. The sum of these losses is represented by a tractive and transmission (T&T) coefficient, which is defined as:

$$T\&T \text{ coefficient} = \text{Drawbar power}/\text{PTO power} \tag{9.11}$$

The values of T&T coefficients for various kinds of tractors and soil types are given in Table 9.3 (Coats, 2002). T&T coefficients for four-wheel drive tractors are somewhat higher than those for two wheels drive tractors. Track type tractors have less slip even on soft soils and as such T&T coefficients for these tractors are high even on soft soils. The drawbar power is determined by multiplying the PTO power by the appropriate value from Table 9.3.

Table 9.3. T&T Coefficients of Various Tractor Types as Affected by Floor/Soil Types

Tractor Type	Concrete	Firm Soil	Tilled Soil	Soft Soil
2 wheel drive	0.87	0.72	0.67	0.55
Front wheel assist	0.87	0.77	0.73	0.65
4 wheel drive	0.88	0.78	0.75	0.70
Track type	0.88	0.79	0.80	0.78

Another useful thumb rule is the 86 per cent rule, which states that the PTO power will approximately be 86 per cent of the engine power, and the drawbar power when tested on a concrete track will be approximately 86 per cent of the PTO power. It has been further extended as follows (https://image1.slideserve.com/3430420/bowers-86-method2-n.jpg and https://www. Progressivegardening.com/agricultural-engineering-2/converting-tractor-power-ratings.html).

Engine power = A HP

PTO power = A x 0.86

Max. DBHP concrete = A x 0.86 x 0.86 = A x 0.740

Max. DBHP firm soil = A x 0.86 x 0.86 x 0.86 = A x 0.636

Usable DBHP firm soil = A x 0.86 x 0.86 x 0.86 x 0.86= A x 0.547

Max. DBHP tilled soil = A x 0.86 x 0.86 x 0.86 x 0.86 x 0.86= A x 0.470

Max. DBHP soft soil = A x 0.86 x 0.86 x 0.86 x 0.86 x 0.86 x 0.86= A x 0.405

Example 9.10

A diesel engine produces a torque of 234 N-m at its rated speed of 4200 rpm. What is the power rating of the engine?

Using eq. (9.10), we have

P= Tn/9550 = 234 x 4200/9550 ≈ 103 kW

Example 9.11

What is the torque generated by an engine running at 2200 RPM producing 75 horsepower?

First we convert HP into kW, such that 75 HP = 75 x 0.746 = 56 kW

Using eq. (9.10)

T = P x 9550/n = 56 x 9550/2200 = 243.1 N-m.

Example 9.12

Calculate the power absorbed by the equipment, if the required torque is 320 N-m and the PTO output speed is 1750 rpm.

P= Tn/9550 = 320 x 1750/9550 = 58.63 kW

Example 9.13

Estimate the amount of maximum power available at the PTO and the drawbar of a tractor rated at 45.0 HP.

Using the 86 per cent rule:

PTO HP = Engine HP x 0.86 = 45.0 x 0.86 = 38.7

DBHP = PTO HP x 0.86 = Engine HP x 0.86 x 0.86 = Engine HP x 0.74 = 45 x 0.74 = 33.3

Hydraulic Power

Hydraulic power also known as the fluid power required by the implement from the hydraulic system of the tractor or engine can be computed by the following relation.

$$P_{hyd} = p \times F / 1000$$

Here P_{hyd} is hydraulic power required by the implement, kW, F is fluid flow, L/s and p is fluid pressure or gauge pressure, kPa. Note that $1 \, Pa = 1 \, N/m^2$.

Wheel Slip

Wheel slip (simply termed as slip) represents a loss of forward motion by the tractor and an associated loss of power. It arises because the force at the wheel/surface causes loss of motion, *i.e.*, the tractor does not move forward an amount equal to the amount that the wheel rotates. The slip is given as (Macmillan, 2002):

$$\text{Slip}, I = (m_o - m)/m_o \tag{9.12}$$

Here m is the distance traveled for a number of revolutions with drawbar pull, m_o is the distance traveled for a number of revolutions with zero drawbar pull. Note that slip does not depend to a significant extent on speed, and thus, it is easy to relate slip with drawbar pull.

Selection of Tractor Power

Drawbar horsepower or other kinds of power required for all possible tillage operations is assessed by appropriately considering the time available and field efficiencies. One should not select the largest power that may be required once in a while. Instead, one should take an average view of the overall requirement. Tractor power may be selected as per the average conditions, and appropriate efficiency values. One may add a little extra capacity as a safeguard against overloading.

COST OF FARM MACHINERY

The cost of farm machinery in Rs./hr is calculated in exactly the same manner as that of a tractor. It also comprises annual ownership costs or fixed cost, and operating costs. Since most of the issues have been discussed in Chapter 4, this section only discusses the issue in a brief manner followed by some discussions on calculation of the cost in Rs./ha. To illustrate the procedure, let us rewrite Table 4.4, which summarized the various components of fixed and variable costs for tractor as Table 9.4. An additional column of comments on components of variable cost has been added. All the components and guidelines to calculate fixed cost of a tractor are applicable to calculate the fixed cost of a machine. The only difference is the life of the machine that in terms of hours of use may be much less as compared to the tractor. Use the values reported in Table 4.3 for the respective machine. For the variable cost, many of the components may not either be applicable or may be negligible (Table 9.4). Thus, only repair and maintenance cost has to be included in the variable cost for most machines. Some machines may require additional labour, which may be included in the labour wages while calculating the cost of a tractor or may be included in the cost of the machine. The complete procedure is illustrated through example 9.14.

Table 9.4. Comments on Various Components of Variable Cost while Calculating the Machine Cost

Fixed/ Ownership Cost	Variable/ Operational Cost	Comments on Components of Variable Cost
Initial cost	Fuel consumption	Not applicable
Salvage value	Fuel cost	Not applicable
Life of the machine (Years/hours)	Lubrication cost	Negligible
Interest rate	Repair and maintenance	Consider as per guidelines for tractor
Taxes, insurance and housing	Labour wages	Some machines may require additional labour. It may be considered separately or included in the wages while calculating the tractor cost

Example 9.14

The initial purchase price of 11 tined cultivator is Rs. 24,000/-. The tines are spaced at 20 cm apart. The life of the cultivator is 10 years. The farmer uses the cultivator for 200 hours in a year. Calculate the cost of cultivator in Rs./hr.

Given

Initial purchase price, C = 24000

Life, L = 10 years

Working hours per annum, H = 200 hours

The parameters not given but assumed as per guidelines given in the case of tractor

Salvage vale, S = 10 per cent of C = 10 x 24000/100 = 2400

TIH = 3 per cent (assumed) of average cost

Interest = 10 per cent per annum (Assumed)

(Note: Interest rate for tractor and implement should be same)

RM = 10 per cent of the initial cost

Depreciation, D (Eq. 4.1) = (24,000 - 2400)/(10 x 200) = 10.8 Rs./hr

Average cost A = (24000 + 2400)/2 = Rs. 13200

Interest cost, I = (13200 x 10)/(100 x 200) = 6.6 Rs./hr

Taxes, insurance and housing TIH = 3 x 13200/(100 x 200) =1.98 ≈ 2.0 Rs/hr

Fixed cost, Rs./hr = 10.8 + 6.6 + 2.0 = 19.4

Operational Cost

Fuel cost = Nil

Lubrication cost = Nil

Operator cost = Nil

Repair and maintenance cost, RM = 10 x 24000/(100 x 200) = 12.0 Rs./hr

Total operational cost = 12.0 Rs./hr

Total cost of implement, Rs./hr = 19.4 + 12.0 = 31.4

Note: Since cultivator does not require additional labour, the labour cost is also Nil.

COST OF TRACTOR AND MACHINE

The cost of tractor and machine in Rs./hr is easy to calculate. It is determined by adding the cost of tractor and the machine both in terms of Rs./hr. Besides this cost, most of the time one is interested in knowing the cost of operation per hectare (Rs./ha). This cost can also be easily determined, if the field capacity of the implement discussed previously in this chapter is known or can be calculated.

Operational cost, Rs./ha = Cost of tractor plus machine (Rs./hr)/
Field capacity (ha/hr)   (9.13)

Example 9.15

For the example 9.14, calculate the cost of cultivating 2 ha of land, if the cultivator is operated at a speed of 4 km/hr. Assume that the tractor cost in Rs./hr is 452. Assume any other value required in the calculations.

Cost of tractor plus machine = 452+ 31.4 (Calculated in example 9.14) = 483.4 Rs./hr

Number of tines = 11 (Given in Example 9.14)

Width of the cultivator = 11 x 20 = 220 cm = 2.2 m

Field capacity as per eq. 9.2, ha/hr = 2.2 x 4 x E/10

Assuming the value of E as 85 per cent (slightly higher value than given in Table 9.1 for lower than the recommended average speed can be assumed), we have

Field capacity, ha/hr = 2.2 x 4 x 85/(10 x 100) = 0.748 ≈ 0.75 ha/hr

Cost of operation, Rs./ha = 482.4/0.75 = 644.4

Cost of operation for 2 ha area, Rs. = 644.4 x 2 = 1288.8

TRACTOR AND IMPLEMENT SELECTION

To put together an ideal machinery system on a farm is not an easy task. Equipment that works best one year may not work well the next because of changes in weather conditions or crop production practices. Older equipment may become obsolete because of changes in farming practices. Because of uncertainty in many such variables of unpredictable nature, the goal of a good machinery manager is to have a system that is flexible enough to adapt to a range of weather and crop conditions while minimizing long-run costs and production risks. While selection of tractor has been discussed in chapter 4, following points should form the basis of selection for other farm machinery.

Selection of Farm Machinery

Some fundamental issues related to characteristics of farm machinery and farm operations form the basis of farm machinery selection (Edwards, 2017). Following material from the referenced publication is included with the permission of the University.

Machine Performance

Each piece of machinery must perform reliably under a variety of field conditions. Tillage implements should prepare a satisfactory seedbed while conserving moisture, destroying early weed growth and minimize erosion potential. Planters and seeders should ensure adequate plant population and provide consistent seed placement. Moreover, these should be able to properly apply pesticides and fertilizers. Harvesting equipment must harvest clean, undamaged grain while minimizing field losses. Although the performance of a machine often depends on the skill of the operator, or on weather and soil conditions, nevertheless differences among machines must be given due consideration while selecting a machine. Information contained in field trial reports, test reports and personal experience may provide necessary guidelines.

Machinery Cost

Once a particular type of tillage, planting, weed control, or harvesting machine has been selected, the question of how to minimize machinery costs must be considered? Machinery that is too large for a particular farming situation will cause machinery ownership costs to be unnecessarily high in the long-term. Machinery that is too small may hamper quality operations resulting in lower crop yields. The ownership costs increase in direct proportion to the initial cost of the machine being high for a bigger size machine. Operating costs on the other hand are of minor importance when deciding what size machinery is best suited to a certain farming operation. The following table illustrates some costs associated with the size of the machine. A decision may be taken as per this illustration to arrive at or around the minimum cost point.

Cost/ha	Remark
Total cost	Decrease initially with increasing size but later on increase with size. There is always a minimum cost point
Ownership cost	Increases almost linearly with size
Operating cost	Decreases with increasing size
Labour cost	Decreases with increasing size at a slow rate
Timeliness cost	Initially decreases steeply and later on at a very slow rate

Timeliness Costs

In many cases, crop yields and quality are affected by the dates of planting and harvesting. This is hidden cost associated with farm machinery. The value of the yield losses for not completing the field operation in time is commonly referred to as timeliness costs. As an example, timeliness losses in delaying the harvest are due primarily to more dropped ears, shattering and cracked beans. These losses can be reduced by early harvest of the crop. Here the costs associated with purchase of large machines for harvesting should be balanced against the cost of artificially drying grain harvested at a moisture level higher than that required for safe storage.

Besides what is stated in the previous sections, the following factors influence machinery selection, and are discussed in order of importance.

Number of Crop Hectares

More the acreage, larger the scale of the machinery required to complete all the operations in a timely manner. An alternative way to stretch machinery capacity is to increase the labour force by hiring extra operators or by working longer hours during critical periods.

Matching Machinery to Tractor Power

In putting together a machinery set, it is also important to correctly match machinery sizes and tractor power. As already discussed in this chapter, the horsepower needed to pull a certain implement depends on the width of the implement, the ground speed, draft requirement, and soil condition. Using tractors with horsepower in excess of that required for the implement being pulled results in excessive depreciation and interest costs. On the contrary, using too little horsepower may cause faster engine wear out.

Reduced Tillage/Tillage Practices

The number of field days needed to complete the planting operation depends partly on the number of separate operations to be undertaken on each hectare. Reducing the number of tillage operations to be performed or performing more than one operation in the same trip effectively decreases the amount of machinery capacity needed to complete field operations on time. On the other hand, machinery cost savings from reduced tillage must be compared to possible cost escalation in chemical use and its adverse effects on crop yields, if any.

Crop Mix

Diversification of crops may help to spread out the periods for timely completion of field operations. Growing more than one or two crops reduces the machinery capacity required for a given number of crop hectares. However, it may also require purchasing additional machines especially for harvesting.

Weather

Weather patterns determine the number of days suitable for fieldwork in a given time period each year. Although actual weather conditions cannot be predicted far enough in advance to be used as an aid to machinery selection, past weather records can be used as a guide. Machinery selection should be based on long-run weather patterns even though it results in excess machinery capacity in some years and insufficient capacity in other years.

Risk Management and Timeliness Costs

Numbers of suitable field days for performing field operations fluctuate from one year to another. It results in timeliness costs to vary even when the machinery set, number of crop hectares and labour supply remains the same. Investing in large machinery can overcome this problem by ensuring that crops are planted and harvested on time even in years in which there are few good working days. Frequency of such events vis-à-vis risk involved should be carefully analyzed before going for investing in bigger size machines.

Part Custom Hiring

Some farms may not have enough crop hectares to justify owning a full line of machinery, particularly for harvesting. Custom hiring or leasing certain machinery operations may lower total costs as well as provide more flexibility in the amount of machinery capacity.

Matching Field Capacity

Following formula may be used to estimate the field capacity needed to complete a certain field operation in a set number of field days.

Field capacity required, ha/hr = Area covered, ha/(Days of operation x hours per day)

For example, to combine 100 ha in 10 field days, harvesting 8 hours per day, would require a field capacity of 1.25 ha/hr. The field capacity formula for other implements can be used to calculate the width of the machine required to complete the task in a given duration.

OWNERSHIP OF FARM MACHINERY

One of the most important capital inputs in farm production is tractor and related farm machinery. Since it requires lot of investment, decisions have to be taken on its ownership pattern. The challenge lies whether to own a tractor/machine in sole or joint ownership or hire the same from the market. The issue of interest here are the need and output. The following principles are found useful while taking such a decision (Adegunloye, 2009).

Sole ownership is preferable when the farmer has sufficient cultivated area or the intention is to sufficiently expand the area. It also seems logical if the owner has interest in custom hiring and it is possible for him to arrange the resources. It is all the more worthwhile, if farmer considers that the investment is more paying over the other investment avenues where he can invest this resource.

Hiring of machines should be the preferred choice where the equipment is occasionally required to perform jobs that can be completed in a limited period and where the sole ownership of machinery is expensive. This kind of model known as custom hiring has become quite popular in hiring of harvesting combines, laser land levelers and threshers and to some extent in irrigation pumps.

Joint ownership is another efficient way to use the capital resources on machinery. If the volume of work on a single farm is insufficient to justify sole ownership or cost of the equipment is too high, two or more farmers may jointly buy the equipment. The government of India is also popularizing this kind of models through self-help groups (SHG), cooperative societies and farm machinery banks. Necessary guidelines for joint ownership of the machinery are:

☆ Mutual trust between the partners must exist

☆ Agreement must be entered clearly specifying the priority in case equipment is likely to be used by the owners at about the same time

☆ If circumstances arise to dissolve partnership, procedure to dissolve the same in an orderly manner must exist

The option of co-operative rotations (rotation of crops to suit machines in cooperative ownership) is another way of achieving optimum efficiency in the use of farm machines. Expensive machinery is more nearly utilized to its full capacity with this model.

POINTS TO BE CONSIDERED WHILE MAKING ACTUAL PURCHASE

☆ Choose equipment of well-known and reputed trademark

☆ Buy the latest model with latest features

☆ Go for BIS or ISO mark machines

☆ See whether the machine is repairable or not, how far is the service station or existence of repair shop, how easily the repair can be done?

☆ Whether spare parts of machine are easily available?

☆ Look for the design features such as: points of wear i.e. type of bearing etc. provision for lubrication, ease of adjustments i.e. disc angle, tilt angle, safety provisions i.e. shielding of rotary members, appearance of the machine (eye appeal, aesthetics) and vibrations and noise during operation (should be low)

☆ Methods and provisions for adjustments of various parts should be easy to make and less time-consuming

☆ Adaptability of machine to work under varying environment should be looked into. It must be easy to adopt the equipment to different soil, crop and environmental conditions

☆ If possible purchase multipurpose machines with minimum adjustments. For machines built-in unit packages and design say a multi-crop planter, the time required for the changeover must be minimum

☆ See if the equipment is a trailed or mounted type. Mounted equipments are easy to lift and easy to turn. Trailing equipment, which are attached to the drawbar of a tractor cannot make sharp turns. Extended and swinging drawbars can aid in taking shorter turns

☆ Choose a machine with a proper size of wheels. A small wheel sinks into loose soil, drops into shallow ditches and furrows and makes the turning of the tractor difficult

☆ Human comfort is an important factor in the selection and operation of equipment. The seat must be comfortable, stable and adjustable to suit different sized individuals. There should be proper safety provisions for the operator

QUESTIONS (THEORY)

1. Define field capacity. Differentiate between theoretical and effective field capacities giving appropriate equations for the two capacities.

2. Knowing the engine power, describe 86 per cent thumb rule to calculate PTO and drawbar horsepower.

3. Write a short note on wheel slip.

4. Taking some cue from the cost calculations of tractors, explain the method of calculating the machine cost highlighting the difference between the two calculations.

5. How will you proceed to select the farm machinery for a farm.

Chapter 10

Machinery for Land Leveling and Land Forming

Many fields appear smooth but may be too uneven to uniformly apply irrigation water. For a flat country, it may be almost impossible to say whether a land requires land leveling or not. Some of the following observations made during irrigation of a field may reflect on the need of land leveling.

- ☆ Water does not run down the border strip or a furrow at a uniform speed
- ☆ Water is deeper on one side of a border strip than on the other
- ☆ Water ponds up at places
- ☆ Water breaks over furrows
- ☆ Some parts dry out before or stay wet longer than the rest of the field
- ☆ Water moves in small streams along with soil down slope

As a consequence of one or more of the above mentioned causes, one ends up in an uneven crop growth, reduced crop yields, and loss of water. A properly graded land surface ensures unobstructed smooth flow of water into the land, without eroding the soil and ensuring uniform distribution of water throughout the field. *Land leveling has been defined as a process by which the surface relief of a field is modified to a desired grade and to certain specifications that provide a more suitable surface for efficient application of irrigation water and for improved drainage conditions.* Land leveling is one of the most intensive practices that is applied to agricultural lands and often requires moving a lot of soil from one point to another. The soil movement greatly increases with required change in the land slope and the degree of land leveling. On the contrary, fairly level lands do not require much movement of soil. These lands only require smoothening, an operation less intensive and less costly than land leveling. A related activity is land forming, another process that also requires significant soil movement. For agricultural fields, it may include making dykes or bunds, ditches and drains. All these processes are essentially required for crop cultivation, irrigation water management and drainage improvement. This chapter discusses machinery related to land leveling, dyke/ridge forming and ditch construction.

LAND LEVELING

Land leveling is an essential farm operation to establish good agronomic, water, soil and crop management practices to improve crop productivity. Farmers have recognized the value of land leveling since time immemorial and therefore, considerable attention and resources are devoted to this process even under traditional methods of crop cultivation. Unevenness of the fields and poor farm designs together result in loss of nearly 10-25 per cent of the irrigation water during its application at the farm. Excess water stagnating at low spots not only delays the tillage operations but has a major impact on germination, crop stand and yield of crops through nutrient-water interaction, salt and soil moisture distribution patterns and/or accumulation of salts. All these factors besides reducing yield, impact the product quality which together reduce the potential farm income.

The terms such as leveling, grading, smoothing, smoothening and shaping the field surface are used synonymously, but are practically different. Land leveling is a general term that describes the process to level the land. This term is commonly confused with table-top leveling of the land, which is also not desirable in irrigated lands. Land grading precisely describes the process of establishing a gentle slope in the direction of irrigation. Smoothening operation is performed every cropping season to smoothen the field unevenness caused by cultivation and harvesting processes of the previous season. Smoothening is also applied as a final operation following land leveling since fields even after land leveling may not be quite smooth.

Benefits of Land Leveling

Leveling is performed to modify the existing contours of land to achieve certain objectives of efficient agricultural production system. Some expected benefits of land leveling are as follows:

- ☆ A uniformly graded field helps in the smooth application of irrigation water resulting in its uniform coverage on the field. The higher the difference between the highest and lowest portions of a field, the more the water needed to achieve complete water coverage. Leveled fields receive uniform penetration of irrigation water resulting in high application and distribution efficiencies. Leveling also reduces time to irrigate the fields. Precise leveling may reduce total water requirement to grow the crop up to 25 per cent.

- ☆ The leveled field becomes ready to receive timely agricultural operations like ploughing, seeding and intercultural without any delay

- ☆ Leveling results in reduced requirement of resources such as seeds, fertilizers, chemicals and fuel used in cultural operations

☆ The reduction in weeds up to 40 per cent results in lower cost of crop production as well as less competition for water and nutrients by the crops and weeds. Up to 75 per cent decrease in the labour required for weeding has been reported for properly leveled fields

☆ Good land leveling enables field layouts with larger field sizes. Larger fields reduce area under dykes resulting in increased area under farming. Increasing field sizes from 0.1 to 0.5 ha increases the farming area by 5 per cent to 7 per cent. The increase in farming area gives the farmer the option to reshape the fields that can reduce operating times by 10 per cent to 15 per cent. It also improves operational efficiency.

☆ In rice fields, leveling reduces the time needed for transplanting and allows the farmers to switch over to direct seeding. Change from transplanting to direct seeding reduces labour requirement by 30 person-days per ha

☆ Uniform moisture environment for crops results in more uniform germination, crop stand and maturation. On the contrary, improperly leveled fields have uneven crop stands and uneven maturing of crops resulting in huge penalty on yield. Effective land leveling considerably increases the crops yield on an average of 10-20 per cent. There is a strong correlation between the degree of land leveling and crop yield, crop yield increasing with increasing quality of land leveling. It is why people now prefer laser over traditional methods of leveling

☆ The farm equipment can be operated easily and efficiently on leveled fields. In fact, leveling of field is highly desirable for the proper working of mechanical harvesters

☆ Properly leveled fields help to conserve rainwater especially on dry lands. The rainwater even after heavy rainfall does not stagnate and is transported to the lower end from where it can be easily drained. Thus, the possibility of water logging is reduced considerably because of improved surface drainage. Similarly, excess irrigation water can also be drained through the drainage system

☆ Controlled runoff and reduced velocity of water on properly leveled fields helps to reduce soil erosion

PHASES OF LAND LEVELING OPERATION

Land leveling operation may be grouped into three phases:

1. Land clearing/rough grading
2. Land leveling
3. Land smoothing

Rough grading is the removal of abrupt irregularities such as mounds, dunes and rings, and filling of pits, depressions and gullies. Land leveling or land shaping requires moving large quantities of earth over considerable distance. Since leveling operations leaves irregular surfaces due to dumping of the soil, these irregularities are removed through the process of land smoothing which is the final operation in land leveling. Often these three phases are accomplished together but all of them may not be required in a particular setting. In some setting only one operation may suffice depending upon the requirement. If land smoothing can accomplish the job then why to go for land leveling, being much more expensive.

MACHINERY FOR LAND LEVELING

Depending upon the farm size, source of power and investment capacity, a farmer has a wide choice to choose from an array of implements. Some land leveling machinery is listed as under. The names may vary from one place to another and even within the same area.

Heavy-duty machinery: It is meant for initial cutting and filling of land undulations. Normally, these machines are hired by the farmer because this operation is limited only for the first year.

Leveler: Bullock drawn, power tiller and tractor operated land levelers are used.

Precision land leveler: It is used to provide desired grade in the direction of irrigation.

Leveler with ripper attachment: Rippers loosen the soil prior to earth movement for leveling.

Scrapers: Used after initial leveling using a leveler or even as a normal seasonal operation for land smoothening. Both animal and tractor drawn types scrapers are available.

Scraper plane: It generally creates table-top finish of the land surface.

Drag scrapers: These are used for precision finish and final touch up to cover small areas.

Box scraper: These are used to smoothen and level the ground surface.

Bulldozer/Dozer for Land Clearing

Land clearing is often expensive, time-consuming, and laborious. The common methods of land clearing include: bulldozing/dozing, blasting, pulling and piling, char pitting stumps and forced-draft burning. This section deals only with dozing with a brief mention of the bulldozer.

A bulldozer is a large, heavy, most often a crawler type tractor equipped with a blade. It pushes large quantities of soil, sand, snow, rubble, or similar material during construction or land conversion works. A bulldozer for land-clearing purposes is normally equipped with a land clearing and a leveling blade, a drag for leveling, hooks, and a winch. A claw-like device termed a ripper is also attached at the rear of a bulldozer. In some cases a large brush plough may also be part of various attachments. Compared to the cost of slashing and piling by hand, bulldozing is the cheaper means of removing small trees and bushes.

The ripper, a long claw-like device, is used to tear and split hard ground, weak rocks or old pavements and bases. A ripper can either be a single shank/giant ripper or a multi-shank ripper. A single shank ripper is preferred for heavy ripping.

The ripper shank is fitted with a replaceable tungsten steel alloy tip known as the boot. Ripping breaks the hard ground surface or pavement into small rubble so that it is easy to handle and transport. In agricultural fields, a ripper is commonly used to break-up rocky or very hard earth (such as podzol hard pan), which poses insurmountable difficulties in farming.

A dozer is a tractor unit having a front mounted blade (Figure 10.1 left). The blade is used to push, shear, cut and roll the material ahead of the tractor. It is also used to clear the land of timber and stumps, and moving earth and rocks for short haul. The dozers are effective and versatile earth-moving machines. Dozers blades are either perpendicular to the direction of travel or set at an angle to the direction of travel. Whereas the former pushes the earth forward, the latter pushes it forward but on one side. Small dozers cannot replace the bulldozer needed to perform heavier works, yet these are quite suitable for the following operations in land-clearing operation.

- ☆ Land clearing
- ☆ Dozing (Pushing materials)
- ☆ Ripping (Figure 10.1 right)
- ☆ Towing
- ☆ Assist scrapers in loading
- ☆ Backfilling
- ☆ Spreading earth

Dozers can either be crawler or wheel type tractors. In the latter, these can be single axle drive or two axle drive. Crawler dozers equipped with either bulldozer blades or special clearing blades are excellent machines for land clearing, an operation normally undertaken before land leveling or any earth moving operation.

Figure 10.1. Front Loaded Dozer (Left) and a Ripper on the Rear (Right).

Besides what is stated other modern land clearing equipment may include

- ☆ Tractor-pulled root plough
- ☆ Tractor-mounted tree pusher
- ☆ Tractor- pulled chains

☆ Tractor- mounted rakes

☆ Tractor- mounted special blades

☆ Tractor- powered bulldozer

☆ Tractor- trailed ploughing harrows

These equipment are not being discussed as are beyond the scope of the book.

METHODS OF LAND LEVELING

Manual Leveling

A farmer irrigates the fields season after season. He is able to observe the locations of high and low spots on the field. Pre-sowing irrigation is commonly used to observe the locations where water stagnates and where the land remains dry. Accordingly, the farmers manually move the soil from high to low spots. This operation over the years levels the fields to such an extent that water spreads out relatively uniformly. This method can only be applied to small parcels of land. Moreover, the operation is cumbersome and causes drudgery to the farm workers. Therefore, the operation is now largely performed using animal/tractor power.

Animal or Mechanical Power

Small farmers often rely on animal power such as buffaloes and oxen, or mechanical power such as 2-wheel or 4-wheel tractors and associated equipment. Some animal drawn implements include wooden float, wooden U leveler and bulk scraper, scoop *etc.* The tractor drawn implements include scraper, terrace blade, carrier type scraper, automatic type leveler and land plane *etc.* The operation is carried out on year-to-year basis without any major engineering or technical inputs. Recent addition is the laser land leveling, a high precision land leveler capable of doing both table-top land leveling and grading. Some of these implements are discussed in the following sections.

Animal Drawn Soil Scoop

Soil scoops are used for excavating ditches, clearing drains and doing cut and fill jobs in land leveling (Figure 10.2). It consists of a trough, a blade, hitching loop and handle. Both animal and tractors drawn models are available.

Blade: It is made of light carbon steel with carbon content varying from 0.5-0.6 per cent, is riveted or bolted to the soil trough. The angle of the cutting blade varies from 12° to 15°.

Soil trough: It is made of mild steel sheet and shaped in to a trough having the capacity of 0.15-0.20 m³.

Hitching loop: The loop, made of mild steel, has its ends fitted to the side of the soil trough. The hitching loop is provided with iron rings or pins for connecting the soil scoop to an animal or a tractor.

Handle: In animal drawn version there are two handles made of timber or mild steel plate fitted to the loop and used to control the movement of the implement.

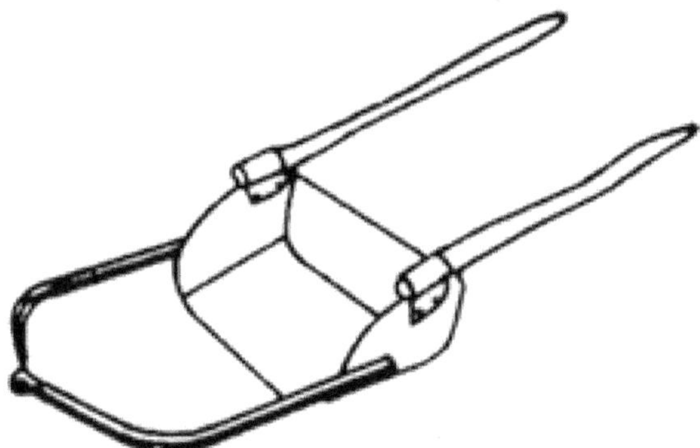

Figure 10.2. A Line Diagram of Animal Drawn Scoop.

Animal drawn scoop is operated by a person walking behind it. It is used for leveling minor irregularities in the field or for filling minor depressions. The terrace blade type has got sitting arrangement and is provided with pneumatic wheels. Initial ploughing and loosening of the soil is necessary before the scoop can be put in operation.

Animal Drawn Leveling Board

Wooden logs or planks known as *patella* are the most common type of land levelers used by Indian farmers (Figure 10.3 left). It is a wooden board 2.0 to 3.0 m long, 0.4 m wide and 0.3 m thick provided with side wings, hitching braces and handle. The shape of the board may vary depending upon the soil condition. Two hooks are provided in the front portion of the board to connect it to the yoke. They are operated in ploughed land to collect loose soil from high spots for dumping into the depressions. For a tractor drawn board, hooks are used to connect the implement to the back of the tractor with a suitable link. It is also used to level paddy fields after puddling. To add weight to the board either extra bags of soil or sand are placed or one or two persons can stand over the board during operation. Similar equipment is the ladder, a rectangular frame made of wood and provided with two steel hooks for hitching at the front (Figure 10.3 right).

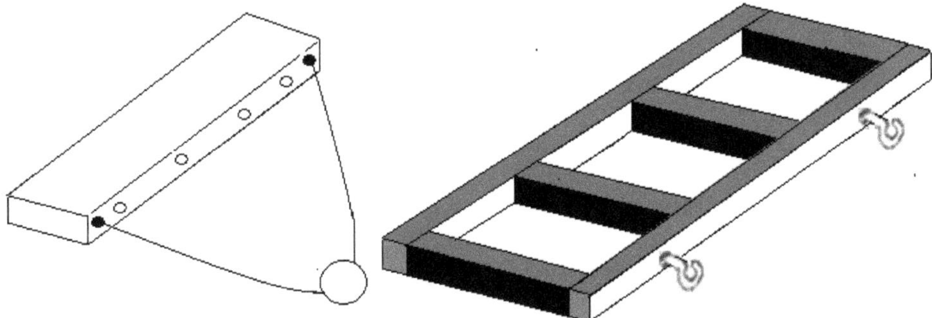

Figure 10.3. A View of *Patella* (Left) and a Ladder (Right).

The other improved version of the land leveler is called the leveling *karaha* or scraper. It is commonly used on medium size farms. It essentially consists of a 'U' shaped steel sheet made of 3 mm thick metal. An extra steel sheet is welded underneath at the center to take care of the wear. Its cutting edge is made of high carbon steel. It is either provided with a wooden handle in the middle or two handles on the sides at the rear end. Provision for hitching is made at the front end. The implement, pulled by a pair of bullocks, requires two persons to operate it. One man controls the bullocks and the other the scraper.

Patella **Harrow**

Patella harrow is an improved design of the *patella* (Figure 8.6). It is a wooden plank normally of *Sal* wood 1500 mm long and 100 mm thick on which a mild steel angle frame is fixed by means of screws. Numbers of curved steel hooks are bolted to a steel angle section, which is fixed to the rear side of the plank. It is used for breaking clods, and packing and leveling the ploughed soil. While the curved hooks uproot and remove the weeds, the angle frame crushes the weeds.

Animal Drawn Buck Scraper

It is a simple tool consisting of a vertical board, tail board, handle and hitch (Figure 10.4). This machine is quite efficient for land grading of small and medium fields using animal power.

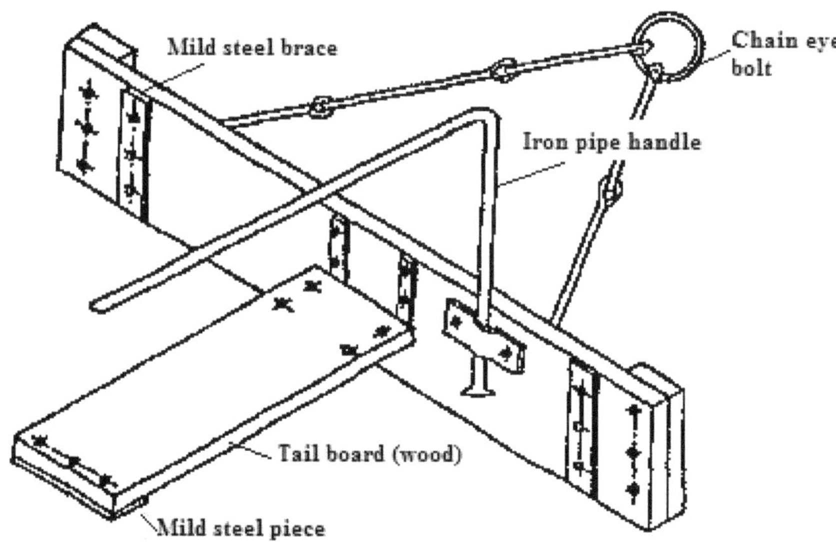

Figure 10.4. Buck Scraper for Leveling.

TRACTOR DRAWN IMPLEMENTS

Tractor Mounted Dozer

Dozer, as previously discussed in this chapter, is an attachment mounted in front of the tractor. It can be easily detached or attached to the tractor. The dozer consists of a thick curved plate and hardened strip (Figure 10.1 left). The strip with a sharp cutting edge is joined to the curved plate of the dozer with the help of fasteners. This strip can be replaced once it wears or becomes blunt. The dozer plate is joined to the tractor with sturdy arms, which help to raise or lower the dozer blade with the hydraulic system of the tractor.

Tractor Drawn Leveler

It is a tractor-mounted implement controlled by tractor hydraulics and three-point linkage. It consists of a hitch system, replaceable cutting blade with sharp edge, and a curved plate with or without side wings (Figure 10.5). Side wings form a bucket. The scraping blade is made of medium carbon steel or low alloy steel, which is hardened and tempered to about 42 HRC. The blade is joined to the curved sheet with fasteners. Thus, it can be replaced once it is worn out or becomes dull. The angle and pitch of the leveler is adjustable. The leveler can also be angled left or right. It can also be used in reverse for back filling operation. During operation, the blade digs into the soil and extra soil is collected in the bucket, which is released in the depressions in the field. Its working depth can be controlled by the hydraulic system of the tractor.

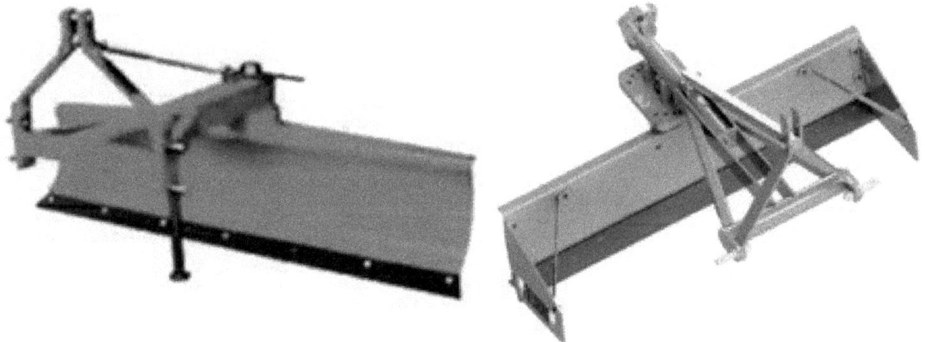

Figure 10.5. Tractor Drawn Land Leveler: Without Side Wings (Left) and with Side Wings (Right).

Tractor Drawn Land Plane

This implement is commonly used for land leveling in irrigated fields (Figure 10.6). The land plane consists of a long frame supported by two wheels and an adjustable blade at some intermediate point. The position of the blade is so adjusted that a high spot cut must fill in a depression before the implement reaches the next high spot resulting in an uninterrupted quick operation. The implement is available in sizes ranging from 1.5 to 4.5 m.

Box and Grading Scrapers

The front of the box scraper is open with a row of teeth that can be adjusted to break the ground as it is pulled. In the rear of the box is a blade facing the tractor which allows soil to collect and be evenly distributed over an uneven surface

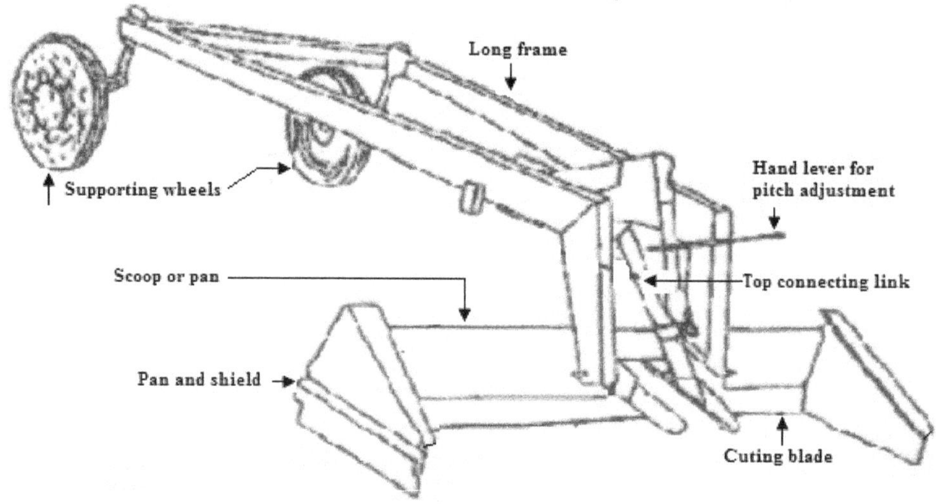

Figure 10.6. A View of a Land Plane.

(Figure 10.7 left). It is used to complete simple tasks such as spreading, leveling, or backfilling. A grading scraper having a row of digging teeth is followed by two small angled blades that run the width of the scraper (Figure 10.7 right). It is used to flatten the rough surfaces. The blades let the material flow around them so that one gets an even distribution of material by dragging the tool over the lose ground.

Figure 10.7. Box Scraper (Left) and Grading Scraper (Right).

LASER LAND LEVELING

The laser controlled land leveling equipment has resulted in one of the most significant advances in surface irrigation technology. It results in a much more level field nearly 50 per cent better than conventional leveling techniques.

Laser Land Leveling Equipment

The laser leveling equipment essentially consists of the following elements (Figure 10.8):

- ✰ Tripod (1) and the laser emitter (2)
- ✰ The laser sensor or receiver (3)
- ✰ The control box [electronic and hydraulic control system (4)]
- ✰ Grading implement, the drag bucket (5) and the tractor and its hydraulic system (6)

Figure 10.8. Components of the Laser Leveling System.

The laser emitter is a battery operated laser beam generator (Figure 10.9 left). It is located on a tripod (1) on or near the field at such an elevation that the laser beam rotates above any obstructions on the field including the leveling equipment. It rotates at a relatively high speed on an axis normal to the field plane. This rotating beam creates a plane of laser light above the field, which acts as the leveling reference. This reference plane is extremely advantageous in the operation since the distance between the laser beam and the earth surface is defined and the deviations from this distance become the cuts and the fills. There is little or no need for the exhaustive engineering calculations as in the conventional approach. On the basis of slope adjusting facility, beam generators are grouped under the following three categories. The cost of equipment accordingly varies with the first kind being the cheapest while the third kind being the costliest.

- ✰ No slope adjustment, resulting in a table-top field (not recommended for field leveling purposes)

☆ Longitudinal slope adjustment, resulting in desired slope in the direction of irrigation

☆ Both longitudinal and transverse slope adjustment

The beam is received by a receiver (Figure 10.9 middle), a light sensor mounted on a mast attached to the land grading implement. In practice, the sensor is a series of detectors situated vertically. With up or down movement of the grading implement, the light is detected above, below or by the central detector. If the detection is made by above or below the central detector, the information through the control system actuates the hydraulic system to raise or lower the implement until the light strikes the central detector. The sensor on the mast is continually aligned with the plane on the laser beam in this manner. This also references the moving equipment with the beam. The sensitivity of the laser sensor system is at least 10 to 50 times more precise than the visual judgment of the manual hydraulic control of an operator. Consequently, the land leveling operation is correspondingly more accurate. The skill of the operator is substantially less critical to the operation.

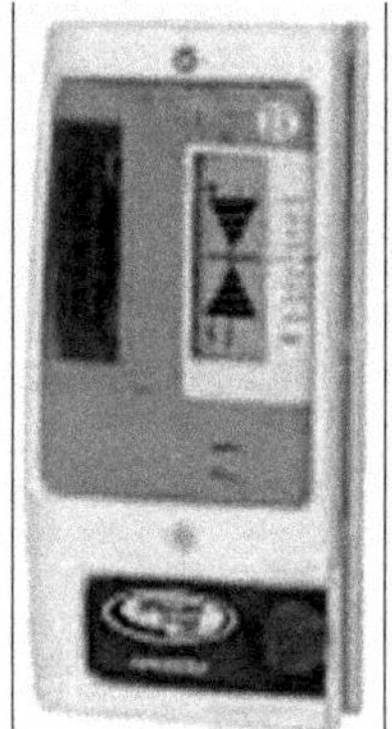

Figure 10.9. Laser Bean Generator (Left), Receiver (Middle) and Control Box (Right).

The control box (Figure 10.9 right), mounted on the tractor within easy reach of the operator, accepts and processes the signals transmitted by the receiver. It displays these signals to indicate the leveler position relative to the finished grade. The three control box switches are On/Off, Auto/Manual, and Manual Raise/ Lower (It allows the operator to manually raise or lower the drag bucket). The control box set to automatic provides the electrical output to control the hydraulic valve. The operator can adjust the settings on the receiver and can override the receiver whenever he needs to pick-up the bucket to transport the soil to another section of the field.

Hydraulic Control System

The hydraulic system of the tractor is used to supply oil to raise or lower the leveling equipment. As the hydraulic pump is a positive displacement pump, a pressure relief valve is provided in the system so that the excess oil pumped is returned to the oil reservoir. The electronic and hydraulic control systems generally have two operational modes. In the observation mode, as the operator drives the equipment, the mast itself moves up or down as per undulations in the field. The monitor provides elevation data from which one can determine the average field elevations and slopes. In this mode, the system operates as a self-contained surveying system. In the planing mode, the mast position is fixed relative to the implement blade, which is then raised or lowered in response to the land topography. The beam plane is located at an appropriate distance above the field and at the desired slopes. The height of the mast sensor is adjusted relative to this plane so that cutting and filling is accomplished simply by driving the tractor over the field.

Tractor and Grading Equipment

The tractor is a standard agricultural tractor. Land grader is also standard equipment, may be as simple as a land plane or a bucket type which loads and carries earth. The HP of the tractor required to operate the system is governed by the size of the land plane or the bucket capacity. Never use under power tractors for this operation. The hydraulic and control systems are appropriately modified so that the whole system operates with the electronic controller.

Steps in Land Leveling Program

Before taking up any land leveling program, it may be necessary to carry out a reconnaissance survey of the area followed by a topographic survey. The reconnaissance survey helps in identifying the location of the farm boundaries, necessity and extent of the land clearing operation, extent and type of problem soils, if any and position of the water table. Topographic survey of the area is carried out to record the high and low spots in the field. It is advisable to take the reading at a regular grid (Say 15 m x 15 m) to ensure high precision. If the laser land leveling equipment has the provision of surveying, it should be used to conduct the survey. It all needs is a laser transmitter, a tripod, a measuring rod and a small laser receiver. A tape, a staff and a compass may also be needed. With one setting, it should be possible to cover a radius of about 300 m. Mean height of the field is determined. Before the actual leveling operation, plough the field preferably from the center of the field outwards. Use rotary tillers, disc harrows or cultivators as may be appropriate. The land should be ploughed when the soil is appropriately moist (*Battar*). In dry conditions, it may result in formation of large sized clods. Besides, it may result in significant increase in tractor power required to plough the fields.

Leveling the Field

During the actual leveling operation, the following sequence of steps is used:

☆ Set the average elevation value of the field in the control box

☆ The laser-controlled bucket is positioned at a point that represents the mean height of the field, normally around the center of the field. The cutting blade is set slightly above ground level (1.0-2.0 cm)

☆ The tractor is then driven in a circular direction from the high areas to the lower areas in the field

☆ When the whole field has been covered in this manner, a final leveling pass is made in long runs from the high end to the lower end of the field

☆ The field is then re-surveyed to make sure that the desired level of precision has been attained. Repeat the process, if necessary

Benefits and Limitations of Laser Land Leveling

Most benefits discussed in the section on benefits of land leveling are achieved with laser land leveling. Since the effectiveness and degree of land leveling is much higher in laser land leveling, the degree of several benefits is also much higher than the conventional leveling methods (Table 10.1, Singh, 2014)). Nonetheless, there are few limitations of this technology as well.

☆ High cost of the equipment/laser instruments

☆ Need for skilled operator to set/adjust laser settings

☆ Less suited to irregularly sized and shaped fields

☆ Limited number of custom hiring operators

☆ Inappropriately selected land leveling equipment by few service providers resulting in less than the projected benefits

Table 10.1. Crops Yield and Water Saving with Laser Compared to Conventional Leveling

Crop	Crop Yield (t/ha)		Water Saving over Conventional Leveling (Per cent)
	Laser Leveling	Conventional Leveling	
Paddy	6.8 (4.6)	6.5	38
Wheat	4.8 (4.3)	4.6	20
Sugarcane	112.0 (13.4)	98.8	24
S. green gram	5.0 (31.6)	3.8	20
Potato	100.0 (11.1)	90.0	25
Onion	100.0 (11.1)	90.0	20
Sunflower	2.3 (15.0)	2.0	20

S. means summer, values in parenthesis are percent increase in yield over the conventional leveling

LAND FORMING EQUIPMENT

Bund Former

Field *bunds* are commonly made to hold water in the field, to prevent runoff and thereby conserve moisture. A *bund* former is used to make *bunds* or ridges by collecting the soil. A *bund* former consists of forming board, beam and a handle (Figure 10.10). The size of the *bund* former is given by the maximum horizontal distance between the two rear ends of the forming boards.

Figure 10.10. An Animal Drawn Bund Former.

Ridger

Ridger, also known as ridging plough, middle buster plough and double mould-board plough, is a machine that cuts and turns the soil in two opposite directions simultaneously for forming ridges. It is also known as furrower because of the furrow formed between the two ridges. It is used to accomplish the following tasks.

☆ To form ridges for sowing seeds and plants of row crops in well tilled soil

☆ Used to make field furrows or channels

☆ Earthing-up and similar other operations

A ridger consists of beam, clevis, frog, handle, mould-boards, braces, share, and sliding shoe. The ridger generally has V-shaped or wedge-shaped share either rigidly fixed or hinged to the mould-boards (Figure 10.11). The nose or the tip of the share penetrates into the soil and cuts and turns the soil in two opposite directions simultaneously to form ridges. During operation the share cuts furrow slice and pulverizes the soil, which moves on to the mould-board and is inverted. While a furrow is formed during the onward movement, ridges are formed in the return pass. The short beam ridger has a gauge wheel, which is attached to the front end of the beam. It facilitates the movement and control of the ridger. It may be operated by one or a pair of bullocks or a tractor. A tractor drawn ridger having a 3-point hitch arrangement is shown on the right side in Figure (10.11).

Irrigation Channel Former

The channel former consists of two inner blades, two outer blades, hitch frame, mainframe and shovel. The front portions of the two inner blades are joined together to form an angle of about 30° in between them. A cultivator shovel is fixed at the junction of two inner blades to penetrate into the soil. The inner blades can be

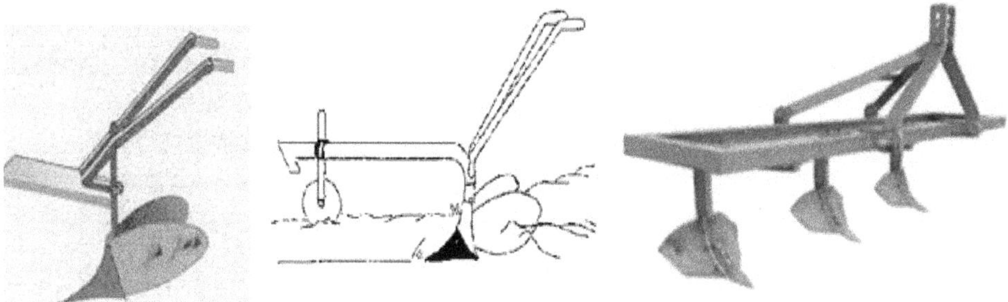

Figure 10.11. Ridger or Ridge Plough: Animal Drawn (Left), Short Beam (Middle) and Tractor Operated (Right).

mounted 50 to 100 mm lower than the outer blades and form a furrow at a lower depth than the surface of the bed for the flow of irrigation water. The two outer blades are placed one on each side of the inner blades and at an angle of 60° to the direction of the travel. The soil collected from the furrow makes *bunds* on both the sides of the irrigation furrow.

Tractor Drawn Ditcher

It is used to make irrigation and drainage ditches. It consists of two curved wings with cutting blades, front cutting point, tie bars for adjusting wing span, and hitch assembly. The sharp cutting blades and cutting point are made of hardened medium carbon or alloy steel. The ditcher depth is controlled by the hydraulic system. The penetration of the ditcher in the soil is aided by its own weight and suction of the cutting point. When the ditcher is drawn forward, it opens the soil in the shape of a ditch with either 'V' bottom or flat bottom. The wings enable the ditcher to slice and roll the tough sod, brush and root sets. The depth and width of the ditch can be adjusted from the operator's seat. It is possible to replace the front cutting points and wings cutting edges once these are worn.

QUESTIONS (THEORY)

1. What do you understand by land leveling for irrigation? List the benefits of land leveling.

2. Describe the various components of animal drawn soil scoop. Explain it with a neat diagram.

3. Discuss the various elements of laser land leveling.

4. Write short notes on tractor drawn leveler, *bund* former, and ridger.

5. Differentiate between box scraper and grading scraper, *patella* and *patella* harrow.

Chapter 11

Sowing and Planting Equipment

Sowing or seeding is an art of placing seeds in the soil in order to ensure good germination. A perfect seeding should ensure the correct amount of seed per unit area (kg/ha), appropriate depth at which seed is placed in the soil and desired row-to-row and plant to plant spacing. Seed drills and planters are used to perform this operation. Whereas a seed drill places the seed in a continuous stream in furrows at uniform rate, a planter places the seeds at a pre-decided distance between the seeds. This chapter dwells upon various methods of sowing, and discusses various aspects of seed drills and planters used in the sowing and planting operations.

METHODS OF SOWING

Commonly used traditional and mechanized methods of sowing are: Broadcasting, Seed dropping behind the plough, Dibbling, Drilling, Transplanting, Hill dropping and Check row planting.

Broadcasting

Broadcasting is the simplest method of seeding. In this method, *seeds are randomly scattered on the surface of the seedbed* (Figure 11.1 left). More often broadcasting is performed manually. Skill of the farmer mainly governs the degree of uniformity of seed distribution in this method. Planking operation follows broadcasting of seeds to properly cover the seeds with soil. A major limitation of this method is that one has to use a somewhat higher seed rate to take care of non-uniform distribution of seeds. The operation can also be performed mechanically. For this purpose, a manually operated centrifugal type unit has been designed that can broadcast seeds as well as granular fertilizers. The unit is made to hang on the belly of the operator. The material is put in the hopper in batches of 3 to 5 kg and the operator operates the unit by rotating the handle as he walks through the field. The machine scatters the seeds on the surface of the seedbed at the controlled rate. It is suitable

Figure 11.1. Traditional Methods: Broadcasting (Left) and *Pora* Method (Right).

to broadcast small seeds like sesame, sorghum, *etc.* The material is spread over a strip 3.5 to 10 m wide. Major advantage of mechanical broadcasting is that large area can be covered in less time. Besides, it ensures relatively high uniformity than the manual broadcasting. The functional requirements of a mechanical broadcaster are illustrated in Figure 11.2. Like manual broadcasting, planking operation is required in the mechanical broadcasting as well. Based on a similar principle, a tractor operated fertilizer or granular insecticide spreader has also been designed. The equipment powered by PTO is mounted on the tractor with the help of 3 point linkage. The power is transmitted to the disc which releases the fertilizer through an oscillating spout.

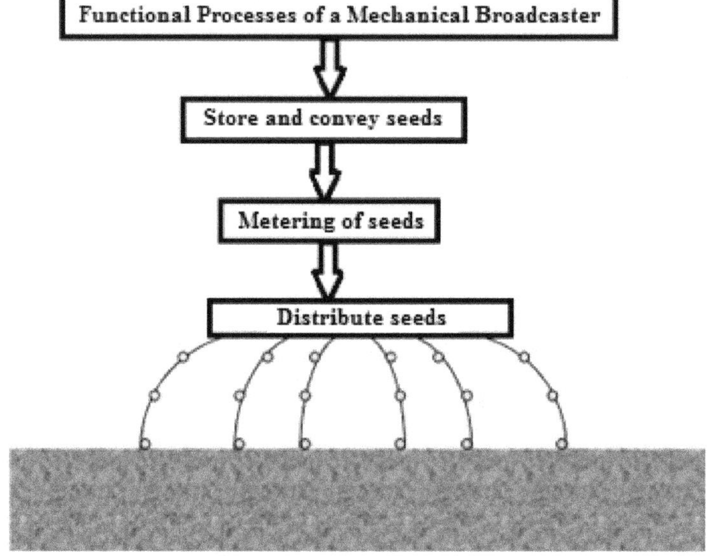

Figure 11.2. Functional Requirements of a Mechanical Broadcaster.

Dibbling

Dibbling is the process of directly placing the seeds in holes made in the seedbed and covering them with soil. In this method, seeds are placed in holes made at definite depth at fixed spacing (Figure 11.3). The equipment used for this purpose is called a dibbler. A dibbling stick is a simple manually operated device consisting of a wooden round stick with one end having a sheet metal cone. The conical end fitted in wooden stick is made from mild steel sheet. The other end has a handgrip. During operation, the dibbling stick is held in vertical position and the conical end is pressed into the seedbed to the desired depth. This action creates a conical cavity in the soil where seed is placed. It is a time-consuming and laborious method. Moreover, it does not suit to sowing of small sized seeds. Dibbling is mostly used to sow vegetable crops.

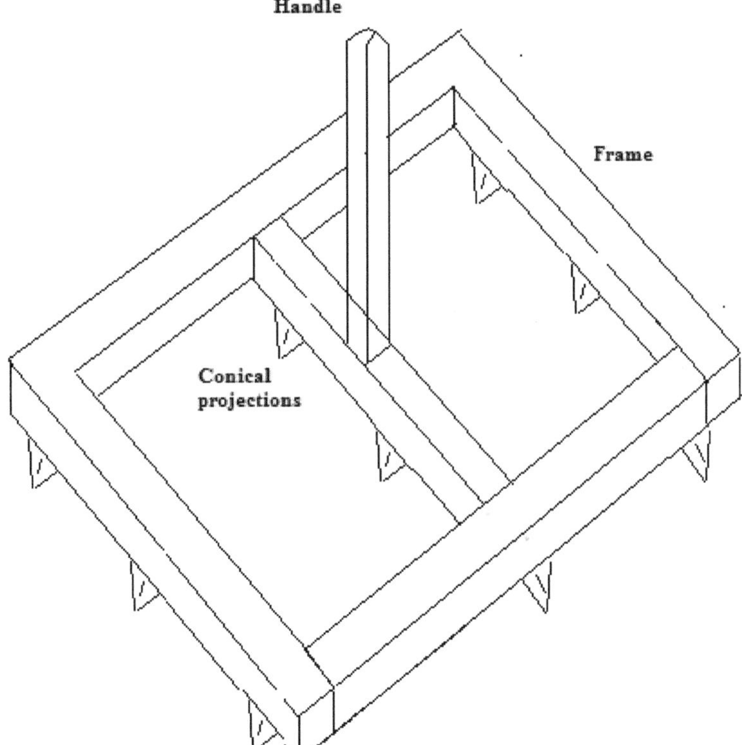

Figure 11.3. Line Diagram of a Hand Dibbler.

Automatic manually operated dibblers of different kinds are now available in the market. In general, a dibbler of this kind consists of a seed hopper, cell type roller for metering of seeds, spring actuated jaws for penetration in the soil, pipe and handle. All parts are made from mild steel except the seed roller, which is fabricated from good quality wood. To operate the dibbler, it is held in both hands to push the jaws into the soil to the desired depth at an angle of 20° with the vertical. The dibbler is then given a jerk at the handle in the forward direction. As a result, roller in the seed hopper rotates and releases one or two seeds depending upon the size of cells. At the same time, the jaws also open and allow the seeds to fall in the cavities. The dibbler is then taken out and shifted to the next point of seeding. As soon as the dibbler is raised, the roller returns to its original position and jaws also close.

Drilling

In the drilling operation seeds are continuously dropped in furrow lines. Drilling operation is performed in one of following three ways.

1. Sowing behind the plough
2. Animal drawn seed drills
3. Tractor drawn seed drills

The operation may cover a single or more than one row depending upon which drilling method and which type of machine is being used. Similarly, metering of seed rate may be manual or mechanical. Seeds may be covered with soil following drilling in a separate operation or may be accomplished simultaneously during drilling. The method is very helpful in placing the seeds at proper depth, maintain proper spacing and ensure that sowing is done with pre-defined seed rate. The last two of three methods listed in this section will be taken up in a more detailed manner in the next section of this chapter.

Sowing Behind the Plough

It is a common method of sowing for crops like wheat, maize, gram, peas, and barley. Two persons are required in this operation. While one person drops the seeds, the other person handles the plough and the bullocks. When seed is dropped in furrows by hand, it is called *kera* method and when it is dropped through a wooden structure known as a *pora* or *nai*, it is called *pora* method (Figure 11.1 right). A *pora* also known as *malobansa* consists of a bamboo tube provided with a funnel shaped mouth. Seeds are placed at a depth of 5-6 cm to ensure satisfactory germination. It is also a slow and laborious method of sowing.

Hill Dropping

In hill-dropping, few seeds are dropped as a hill. In this case, hill to hill spacing in a row is constant. It is unlike drills, where the seeds are dropped in continuous stream and the spacing between plant to plant in a row is variable. The implements used for hill planting are known as planters.

Check Row Planting

Check-rowing is a method of planting where each hill is exactly the same distance away from the adjoining hills. As a result, row-to-row and plant-to-plant distance is uniform and the rows are always perpendicular to each other. A check-row planter is used for check row planting of crops. A field of check-row-planted crop had the appearance of a checkerboard, with a hill of crop at the exact intersection of each line.

Transplanting

Transplanting consists of preparing seedlings in the nursery and then planting the seedlings in the prepared fields. It is commonly used for paddy, vegetable and flowers. It is time-consuming operation. Both manual and mechanical transplanting is possible. Implements used for transplanting are known as transplanters.

Limitations of Traditional Methods

- ☆ Non-uniform distribution of seeds
- ☆ Inter-row and intra-row distribution of seeds is uneven resulting in bunching and gaps in field
- ☆ Poor control over depth of seed placement
- ☆ Need to use high seed rates
- ☆ Thinning may be required to control plant population to desired level
- ☆ Labour requirement is high considering separate application of fertilizer
- ☆ During kharif, placement of seeds at uneven depth may result in poor emergence

SEED DRILL

Seed drill is a machine that places the seeds in a continuous stream in furrows at uniform rate and at controlled depth. It may or may not have the arrangement of covering them with soil. The different functions of seed drill are:

- ☆ To store the seeds
- ☆ Place the seed accurately and uniformly at the desired depth in the soil
- ☆ To meter the seeds at predefined seed rate
- ☆ To place the seed in furrows in an acceptable pattern
- ☆ To cover the seeds and compact the soil around the seed to enhance germination and emergence.

The functional requirements of a seed drill are illustrated in Figure 11.4. A seed drill fitted with fertilizer dropping attachment capable to simultaneously deliver both the seeds and fertilizers in an acceptable pattern is known as seed-cum-fertilizer drill. It carries out all the functions of the seed drill besides storing fertilizer, meter and place it in the desired pattern and depth and cover it with soil. Such a drill may have two separate storage boxes one for seed and another for fertilizer. In some cases, one large box is used, which is divided lengthwise into two compartments, one for seed and another for fertilizer.

Classification of Seed Drills

Based on the power source, seed drills are classified as i) Animal drawn and ii) Tractor drawn. Depending upon the type of seed dropping mechanisms, animal drawn seed drills are further divided into two groups. In one case, seeds are dropped by hand whereas in the others seeds are dropped using mechanical means. In the former, a person drops the seeds in furrows. In the latter, a mechanical device known as seed metering mechanism is used to meter the seeds.

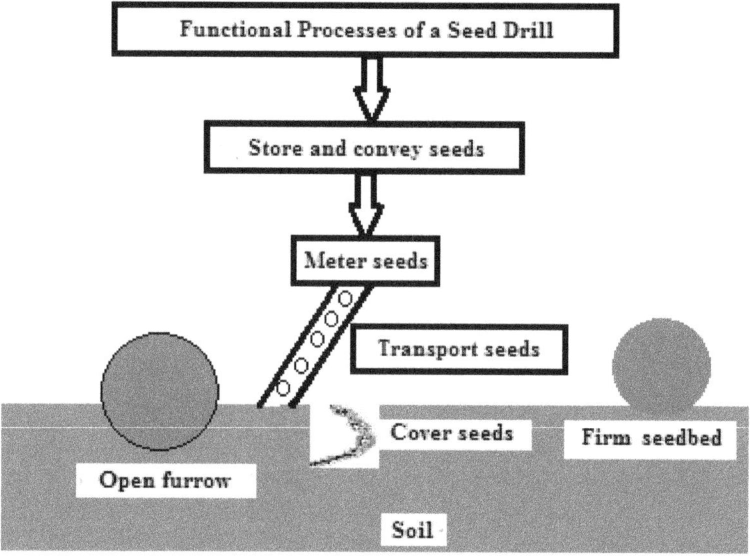

Figure 11.4. Functional Requirement of a Seed Drill.

Components of Seed Drill

A seed drill with mechanical seed metering device essentially consists of (Figure 11.5): (i) Frame (ii) Seed/fertilizer boxes (iii) Seed metering mechanism (iv) Furrow openers (iv) Covering device (vi) Metering control device (vii) Seed rate adjustment lever (viii) Seed tubes and (ix) Transport-cum-power transmitting wheels.

Frame

The frame is the main body of the machine with all parts attached to it. It is made of mild steel angle irons, the size varying with the number of tines. Suitable braces and brackets are used to provide strength and rigidity to the machine so that it is strong enough to withstand all kinds of loads while working in the field.

Seed/Fertilizer Boxes

Trapezoidal shaped seed and fertilizer (only provided in a seed-cum-fertilizer drill) boxes store seeds and fertilizer respectively. These are made of mild steel (2 mm thick) or galvanized iron sheets having a suitable cover. These are mounted side by side on the frame, the fertilizer box being in front. Box dimensions vary depending upon the effective width of the machine and increase with the increase in the number of the furrow openers. A small agitator is sometimes provided to agitate the material to prevent clogging.

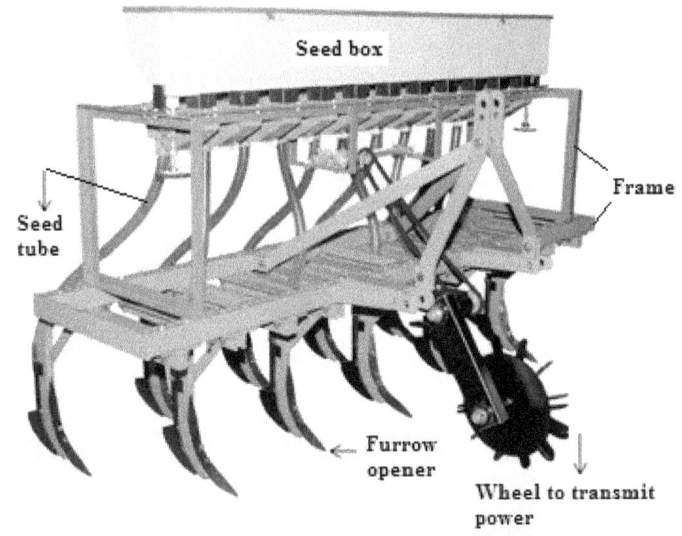

Seed drill

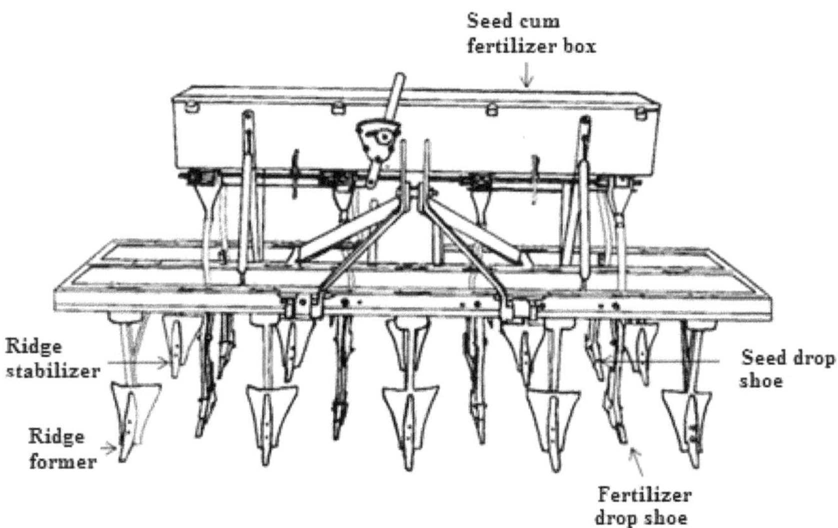

Seed-cum-fertilizer ridger drill

Figure 11.5. A Seed Drill and a Seed-cum-Fertilizer Ridger Drill.

Seed Metering Mechanisms

The mechanisms designed to collect seeds in a singular or group fashion and deliver the same from the hopper to the seed furrow at selected rates is known as seed metering mechanism. It is the key elements of any drill. Numbers of seed metering mechanisms are used few of them being: Fluted roller feed type, internal double run type, cup feed type, cell feed mechanism, brush feed mechanism, auger feed mechanism, picker wheel mechanism and star wheel mechanism. In advanced versions, pneumatic or air system metering units are also being used. Similar fertilizer metering mechanisms are also used to meter fertilizer application rates in a seed-cum-fertilizer drill.

Fluted Roller Feed Type

Fluted roller is a simple, low cost, trouble free device suitable for bulk drilling including the granulated fertilizers. Bulk drilling method is quite common with grass, clover, alfalfa, and small grain seeds wherein a stream of seeds is bulk-fed into the conveying tubes. The fluted roller consists of a fluted wheel, feed roller, feed cut-off and adjustable gate to suit different sizes of grains (Figure 11.6 top). The number of fluted rollers on a drill is the same as the number of furrow openers. Fluted rollers have the longitudinal grooves along the outer periphery such that the length of the groves exposed to seed can be increased or decreased with the sideways shifting of the fluted wheel (Figure 11.6 top). A number of selections are available between the closed position and the full exposure of fluted wheel. It helps to control the seed rate. Both the feed roller and feed cut-off devices are mounted on a shaft that runs through fluted roller. There is an adjustable gate on the discharge side of the fluted wheel. The gate opening can be changed to fit the size of the seed. Fluted rollers, mounted at bottom of seed box, receive seeds into longitudinal grooves. The seeds from the fluted wheel go into the tube through the gate. Mostly, the speed of the square shaft is constant, but on some drills, the speed of the shaft can also be changed to alter the seed rate. This method is quite favourable for sowing small or medium-sized seeds. This mechanism is not suited to bold seeded crops as the seeds are likely to get crushed during metering operation. An improved design of the fluted roller has spiral shaped flutes. This design gives uniform distribution of seeds as compared to straight shaped flutes. On the other hand, most of the low-cost animal drawn seed-cm-fertilizer drills are fitted with straight shaped rollers. The fluted feed mechanism is more positive in its metering action than the internal double run method. The metering device is driven by ground wheel. It is mostly used in sowing wheat crop.

Internal Double Run Type

Internal double run type metering device has a double faced feed wheel having fine and coarse ribbed flanges (Figure 11.6). The face having larger openings is for bold seeds while the face having smaller openings is for smaller seeds. Besides, it has discs mounted on a spindle and housed in a casing fitted below the seed box. A gate at the bottom of the box covers the opening that is not in use at a particular time. Seeds are dispensed as and when the rotating wheel put them towards the corrugations outside the box. At this time, seeds enter the tube. The rate of seeding can be controlled by adjusting the speed of the internal feed wheels by meshing appropriate gears.

Cup Feed Mechanism

It consists of a seed hopper which has two parts namely the grain box, which is the upper part and the feed box, which is the lower part. The seed delivery mechanism is known as the seed barrel, which consists of a spindle, carrying a number of discs with a ring of cups attached to the periphery of each disc (Figure 11.6). The cups have two faces, one for large and the other for small seeds. When the spindle rotates, discs with set of cups rotates and pick-up few seeds and drop them into small hoppers. The speed of the seed barrel controls the seed rate. This mechanism is commonly used on British seed drills.

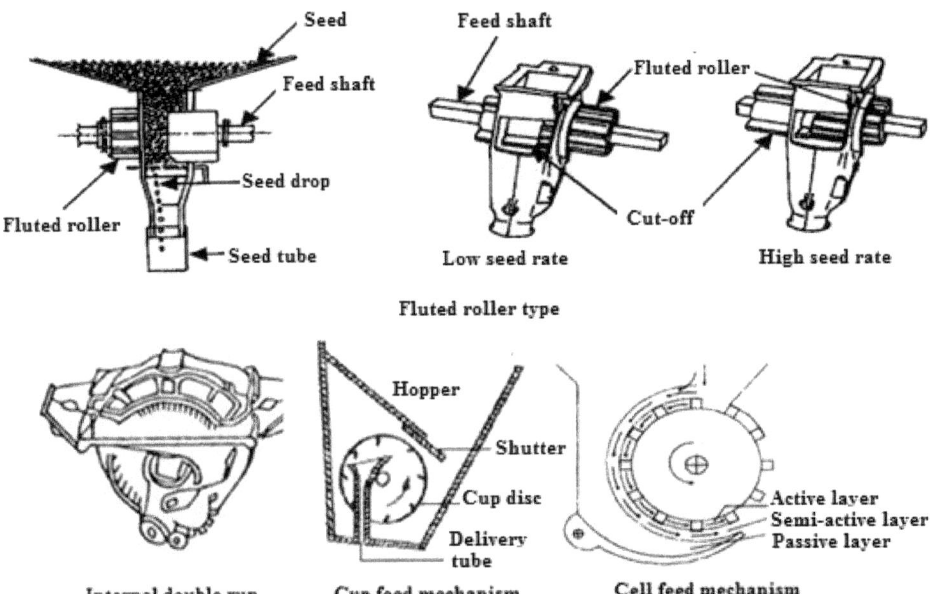

Figure 11.6. Seed Metering Devices.

Cell Feed Mechanism

Seeds are collected and delivered by a series of equally spaced cells on the periphery of a circular plate or wheel (Figure 11.6). The plates can be arranged either horizontally or vertically.

Brush feed mechanism

A rotating brush regulates the flow of seed from the hopper. Many bullock drawn planters in India operate with brush feed mechanism.

Auger Feed Mechanism

It is a distributing mechanism consisting of an auger, which causes a substance to flow evenly in the field through an aperture at the base or on the side of the hopper. Many of the fertilizer drills in India use auger feed mechanism.

Picker Wheel Mechanism

In this mechanism, a vertical plate is provided with radially projected arms. The arms drop the large seeds like those of potato in furrows with the help of suitable jaws.

Star Wheel Mechanism

It consists of a toothed wheel rotating in a horizontal plane and conveying the fertilizer through a feed gate below the star wheel.

Seed Rate Adjusting Lever

This lever is attached to the seed box for increasing or decreasing the seed rate. A scale on the adjusting lever helps to increase or decrease the seed rate.

Fertilizer Metering Systems

The fertilizer metering system controls the rate of fertilizer application as per crop requirement. A fertilizer application and metering system generally consists of fertilizer box, lever, drive shaft, fertilizer metering device, fertilizer delivery pipe and fertilizer boot. The application rate is adjusted with the help of a control lever. The fertilizer goes to the fertilizer delivery pipe which finally reaches into the soil through the fertilizer boot attached to the furrow opener. Some metering devices to meter fertilizers in a seed-cum-fertilizer drill are shown in Figure 11.7.

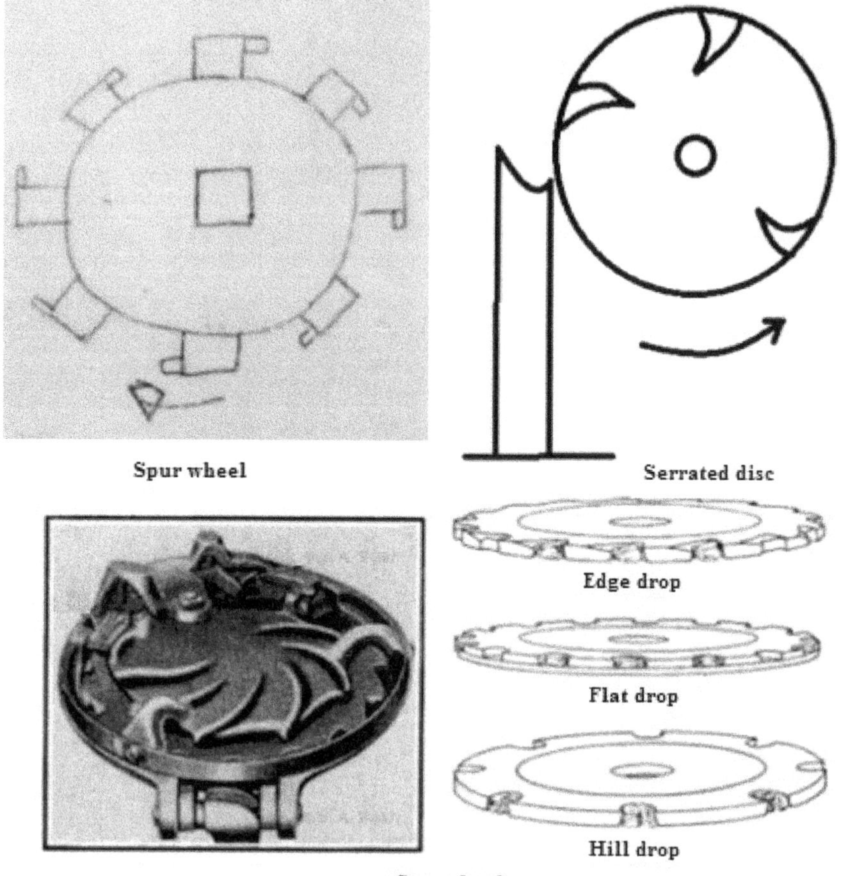

Figure 11.7. Fertilizer Metering Devices.

Spur wheel: Spur wheel, made of aluminium casting, is mounted on the shaft. The wheel has equally spaced 6-12 spurs.

Serrated disc: In this device fertilizer is collected and delivered by the V-shaped serrations on the periphery of the circular disc.

Star wheel: Star wheel is a horizontal disc carrying the stars on the periphery, which meters the fertilizer application (Figure 11.7). The system gets the drive from a bevel gear and pinion.

Furrow Openers and their Types

The furrow openers in a seed drill are used to open a furrow for placing the seeds. Seeds move from the feed mechanism to the seed tube and then into the boot to finally fall into the furrow. The basic requirements of a furrow opener are:

☆ Uniformly open a furrow to stipulated depth

☆ Minimize mixing of soil layers to prevent loss of moisture

☆ Compact the furrow bottom

☆ Minimize disturbance to seed flow

☆ Leave an adequate soil layer between the seeds and the fertilizer

Furrow openers may either be fixed or rotary types. Some of these are shown in Figure 11.8 and described in the following sections.

Shovel Type

Shovel type furrow openers are the most widely used furrow openers. Three kinds of shovels in use are: (a) reversible shovel (b) single point shovel and (c) spear point shovel. Shovel type openers are bolted to the flat iron shanks on the front of the delivery boot, which carry the end of the seed tubes. Springs are used to prevent shock loads resulting from any obstruction encountered during sowing of seeds. As such, these shovels are best suited for stony or root infested fields. Seeds are covered with the soil falling in from the sides behind the delivery boot. It is commonly used for deep placement of seeds, the depth may range from 50 to 100 mm in a well-prepared seedbed. Small shoe shovel type openers are used for shallow (20 to 50 mm deep) placement of seeds in dry farming areas. These are easy to construct, easily repairable besides being cheaper.

Shoe Type

Shoe type openers work well in seedbeds which are either poorly prepared or have trashy condition. They are made of two curved runners with their cutting edges on the ground and meeting at the front. At the rear, they are connected to a common boot through which seeds drop into the furrow. Shoe type openers with single or twin boots are used for sowing in heavy and medium textured soils. Seeds are placed at depths in the range of 20 to 70 mm.

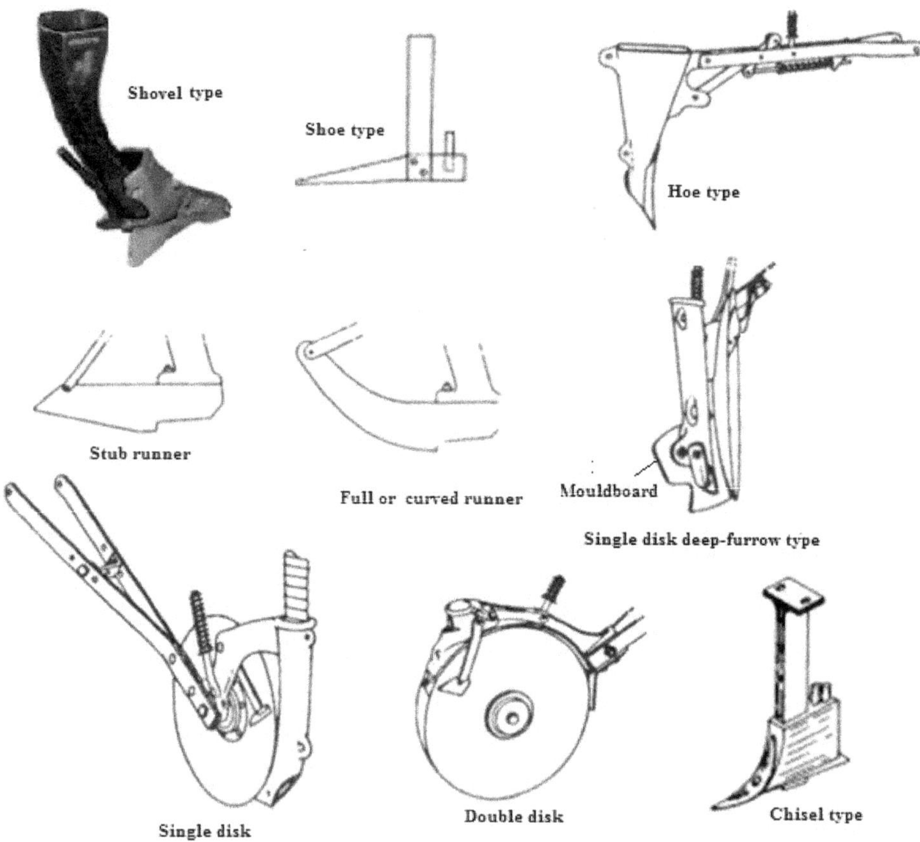

Figure 11.8. Various kinds of Furrow Openers.

Hoe Type

Hoe type furrow openers are made from two flat pieces of steel welded together to form a cutting edge. The shoe is made of carbon steel having minimum carbon content of 0.5 per cent having a minimum thickness of 4 mm. The hoe type openers have good penetration and are used for deep placement of seeds. These work well in trashy soils where the seedbeds are not smoothly prepared and in black cotton soils. These can be equipped with springs for use in stone or root infested soils.

Stub Runner

These are commonly used in rough and trashy fields mostly on corn planters. These are widely used for placement of seeds at shallow depth especially when minimum soil disturbance is desired. These leave minimum soil cover over the seeds.

Full or Curved Runner

It is a simple device that works well at medium depths in mellow soil free of trash and weeds. It is commonly fitted in corn and cotton planters. Horizontal plate type depth gauges are attached to the runner in soft soil.

Pointed Bar Type (Diamond Shaped)

These furrow openers are used to form a narrow slit in heavy soils for placement of seeds at medium depths.

Disc Type

Two types of disc type furrow openers are: single disc type and double disc type. The single disc furrow openers consist of one concave disc. These openers are quite suitable for soils having plant debris and trash mulch. The disc is set at a tilt angle of 5° to form a small ridge. Two scrapers are used to keep the disc clean. A toe shaped scraper is provided on the convex side while a 'T' shaped scarper is provided on the concave side. The scrapper on the concave side also reduces soil throw at high speeds and prevents soil build-up. The seeds are dropped in the boot on the concave side of the disc. It also works well in sticky soils. The main disadvantages are that the discs are costly and maintenance bit difficult.

Double disc furrow openers have two flat discs positioned together at the front and opened at the rear. The included angle between discs is kept at about 10°. The furrow opener consists of tine, shovel, seed tube and boot for seeds and fertilizer. Seeds are placed through the delivery tube placed between the two discs. This type of furrow opener is commonly used on tractor operated seed drills running at high speeds. It is suitable for trashy lands.

Table 11.1. Comparative Suitability of Furrow Openers in different Working Conditions

Furrow Opener Type	Remarks
Rotating	
Double disk	Suitable for high residue conditions
Single disk	Better penetration in hard soils but wider variation in seed placement
Fixed type	
Chisel	Narrow furrow, good for deep sowing and in friable soils
Hoe	Good for stony, harder soil conditions and for deep placement
Inverted 'T'	For reduced till in free flowing soils. Can leave open furrow in wet soils
Runner	For shallow depths in loamy non-smearing soils
Shoe	For placing seed and fertilizer in separate bands
Shovel	Deep placement causing much soil disturbance

Parts of a Furrow Opener

A furrow opener comprises tine, shovel, seed tube and boot. Seed/fertilizer tubes carry the seeds/fertilizer from the metering device to the boot. The boot is a part of the machine bolted or welded to the tine in which lower end of the seed tube is inserted. It conveys the seeds or fertilizers from the delivery tube to the furrow. The most commonly used seed tube has been the steel ribbon type, which is rolled in such a way that the lower edge of the tube is thinner than the upper edge. The main limitation of these tubes has been that they are prone to rust and can be easily damaged. These have now been replaced with polythene or rubber tubes having a minimum diameter of 25 mm. Rubber tubes can easily bend when the furrow openers are lifted. These are also safe against corrosion by the fertilizers.

Coulters and Covering Devices

Coulters are used in front of the furrow openers to cut the trash, if any. Four kinds of coulters are smooth, rippled, rippled with smooth edge and fluted (Figure 11.9).

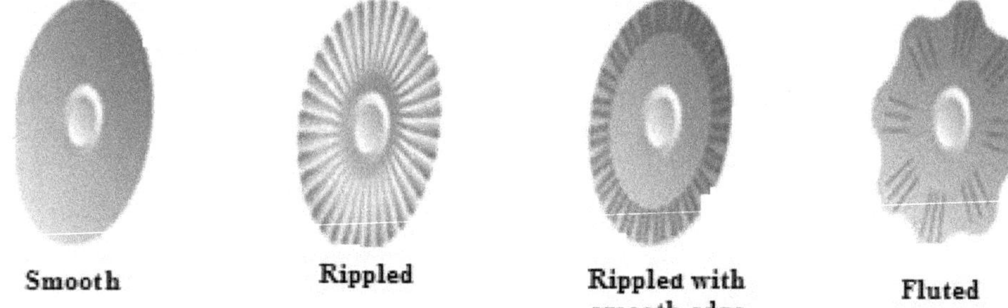

Figure 11.9. Various kinds of Coulters used in Seed Drill for Cutting Trash.

Covering device or furrow closer is a device that refills the furrow once the seed has been placed in it. It performs the following functions:

- ☆ Places moist soil around the seed
- ☆ Covers the seeds to proper depth
- ☆ Leaves the soil directly above the row loose enough to minimize crusting
- ☆ Promotes easy emergence

Covering the seeds is usually done by *patta*, drag chain (Figure 11.10), drag bars, scraper blades, disc hillers, steel press wheels (Figure 11.10), zero- pressure pneumatic press wheels, packers *etc*. Each of these is designed in various sizes and shapes for use under defined soil and crop conditions. For example drag chain is suitable in loose moist soil for grain drill, and steel press wheels for loose sandy soil or furrow drilling in heavy trashy conditions. These wheels increase the crop stand and yield in areas where moisture is a limiting factor. Open center, concave, steel press-wheels are common for corn and other large seeded crops. Zero-pressure pneumatic press-wheels are used for vegetable crops. These are continually flexing and are self-cleaning.

Transport Wheels

Normally two wheels are provided on the main axle to transport the drill on roads. Few seed drills are also fitted with pneumatic wheels. One of the transport wheels is fitted with a suitable attachment to transmit the motion of the wheel to the seed metering mechanism when the drill is in operation.

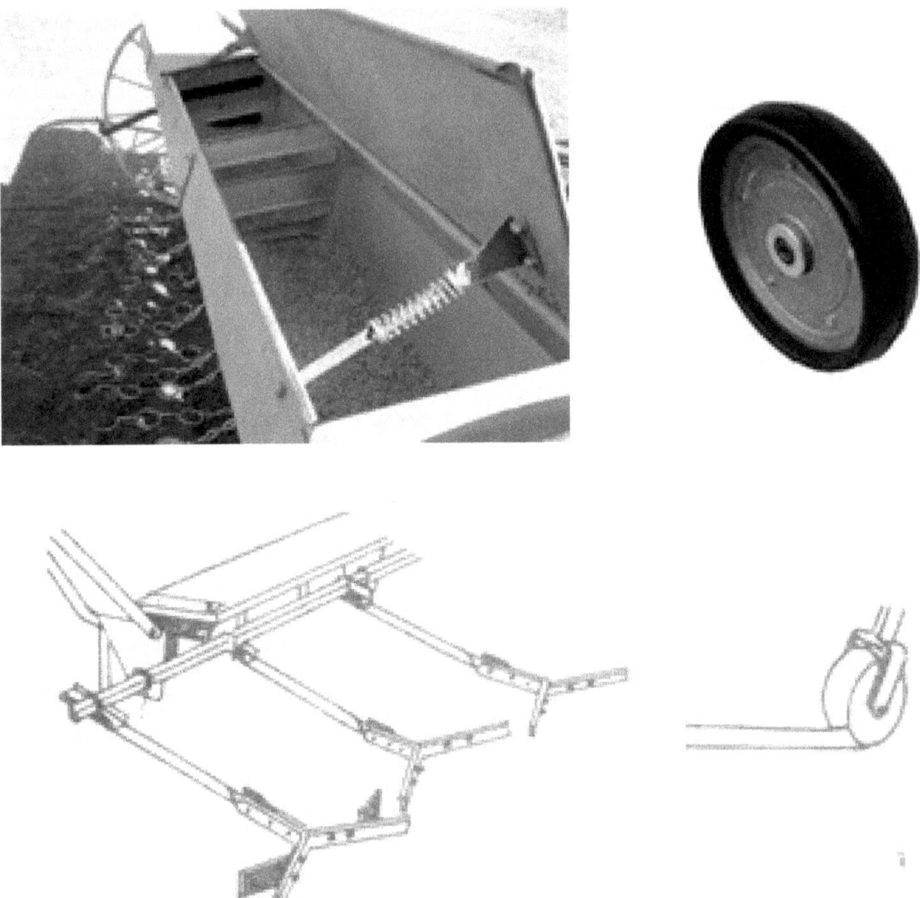

Figure 11.10. Soil Covering Devices: Drag Chain (Top Left), Press Wheel (Top Right), Drag Bar (Bottom Left) and a Disc Hiller (Bottom Right).

SOWING IMPLEMENTS

The Seed-cum-Fertilizer Drill

The machine consists of seed box, fertilizer box, seed/fertilizer metering mechanism, seed tubes, furrow openers, seed rate adjusting lever and transport-cum-power transmitting wheel. Fluted rollers, which are mounted at the bottom of the seed box, are driven by a shaft. These receive the seeds into longitudinal grooves of fluted roller and expel them in the seed tube attached to the furrow openers. By shifting the rollers sideways, the length of the grooves exposed to the seed can be increased/decreased to alter the amount of seed rate. The seed-cum-fertilizer drill is popular because it dispenses with separate application of fertilizer.

Tractor Mounted Ridger Seeder

A tractor drawn implement forms ridges and furrows and carries out sowing either on the flat top ridge, on the side of the ridge or in the furrows as per the crop requirement. It consists of a sturdy frame made of mild steel sections. It supports seed and fertilizer hoppers, seed and fertilizer metering mechanisms, drive system, drive wheels, hitch system, ridger bodies, and furrow openers. The machine is supported on transport wheels. The ridger body consists of two mould boards and share mounted in the front. It is possible to adjust the wingspan of the ridger to vary the width of furrows or ridges. Seed and fertilizer metering units receive power from ground wheel through chain and sprocket system.

Zero Till Drill

No till or zero till drills are now commonly used to sow wheat crop in residue laden fields after the harvest of paddy crop. It consists of a tubular steel section frame on which other parts are mounted. Main parts are: slit/furrow openers, seed and fertilizer boxes, seed and fertilizer metering devices, power transmission unit, depth-control side wheels, hitch points and iron/wooden platform or stand. The main difference between a no till drill machine and conventional drill is that it has narrow shovels known as inverted T-type furrow openers instead of tine type furrow openers. This most important part of the drill is made of medium carbon steel having wear resistant high-speed steel cutting edge. These openers open a narrow slit 3-5 cm wide in untilled soil. The main advantage of narrow shovels is their low draft requirement and easier penetration in the soil. The ground drive wheel having lugs on its circumference acts as a power source to drive the metering mechanism. The zero-till drill is operated in the field when the soil moisture is about 24-27 per cent and the stubble height of the previously harvested crop is not more than 15-20 cm. For its operation, the zero till drill is calibrated like a normal seed drill for predefined seed and fertilizer rates. The depth control wheels are adjusted as per the depth for placing the seed and fertilizer in the soil, which can vary from 5 to 7.5 cm. With the forward movement of the tractor, the ground drive wheel in contact with soil starts rotating and transmits the motion to the seed and fertilizer metering mechanisms through an auxiliary shaft using a set of chains and sprockets. The fluted rollers mounted in seed and fertilizer metering mechanisms expel the desired amount of seeds and fertilizer, which is conveyed through the pipe into the furrow opener boots. The seed sown may remain uncovered under high soil moisture conditions because of inadequate soil backflow into the furrow. In such conditions a lightweight plank is hinged behind the drill to cover the seeds and fertilizer that gently presses the furrow wall to ensure better seed soil contact.

Turbo Happy Seeder

Turbo Happy Seeder has been developed to sow wheat crop. It is capable of direct drilling (ZT) into heavy surface residue loads of combine harvested rice in a single operation (Figure 11.11). It can be converted to a multi-crop seed sowing machine with an additional seed box. It has the following benefits (Jat *et al.*, 2013):

☆ Manages sowing in residues of combine harvested crops including loose and anchored residue

☆ Reduces cost of tillage operations

☆ Provides surface mulch to conserve soil moisture

☆ Saves on irrigation

☆ Buffers soil temperature

☆ Improve soil organic carbon

☆ Helps in overcoming terminal heat effects on the crop

☆ Eliminates residue burning that minimizes air pollution and helps improving environment

☆ Improves crop yields in the long-run

Figure 11.11. A Turbo-Happy Seeder Sowing Wheat in Combine Harvested Field.

The turbo happy seeder has several components; the major ones are listed as under.

Sl.No.	Name of the part	Sl.No.	Name of the part
1	Frame	8	Fertilizer metering device
2	Slit and furrow openers	9	Rate adjustment lever for seed
3	Seed and fertilizer boxes	10	Extra seed box for multi-crop planting
4	PTO drive mechanism	11	Rate adjustment lever for fertilizer
5	Flail	12	Seed boot
6	Drive wheel	13	Depth control wheel
7	Seed metering device	14	Delivery pipes

Most of the system is the same as a regular seed-cum-fertilizer drill with fluted rollers used as metering devices for both the seeds and fertilizer. The major addition is the flails, which are wings made of mild steel and attached to a shaft, called the flail shaft. Flail blades are placed on the rotor and arranged ahead of machine furrow openers. When the flail shaft rotates, the flails clean the residue left in front of tines. It facilitate drilling of seed and fertilizers in the seed rows. During its operation, the machine is mounted on three point linkage and attached to the PTO shaft. It gives drive to the flail with the help of belts. As the tractor moves, the drive wheel also begins to rotate, which gives drive to the seed and fertilizer shaft. As the shafts move, the seed and fertilizer metering system also begins to move to complete the sowing operation.

Pneumatic Multi-crop Drill/Planter

A tractor mounted pneumatic planter consists of a centrifugal blower operated by the PTO of a tractor, which is used to provide the necessary air pressure to pick up and release the seeds in the metering mechanism. It is normally a six-row machine but 2 or 4 rows machines are also available. The machine is suitable to plant single seed at predetermined seed/row spacing. It consists of main frame, aspirator blower, disc with cell type metering plate, individual hopper, furrow openers, PTO driven shaft, ground drive wheel *etc.* It is suitable to plant seeds of grain crops such as sorghum, pigeon pea and maize, oil seed crops such as mustard, groundnut and soybean and cash crop such as cotton and vegetable crop such as okra *etc.*

CALIBRATION OF SEED DRILL

A seed drill requires frequent testing to ensure that the seeding is done at a pre-defined rate. The process of testing the seed drill for this purpose is known as calibration of the seed drill. The calibration in a multi-crop seed drill may also be required when one switches from one crop to another. Therefore, it is always necessary to calibrate the seed drill before operating it in the field. It can either be done in the farm shed or directly in the field. The calibration methods for both the conditions are described in this section.

Farm Shed Calibration Procedure

Determine the nominal width (W) of the seed drill.

W = M x S

Such that M is the number of furrow openers and S is the spacing between the openers, m

Find the length of the strip (L) having nominal width (W) necessary to cover 0.04 ha or 400 m^2 area.

L = 400/W, m

Determine the number of revolutions (N) of the ground wheel of the seed drill having the diameter, D to cover the length, L of the strip.

L = π x D x N = 400/W

$$N = 400/(\pi \times D \times W) \tag{11.1}$$

☆ Jack the seed drill so that the ground wheels turn freely. Make a mark on the drive wheel with a corresponding mark on the body of the drill. These marks will help in counting the revolutions of the wheel.

☆ Fill the seed in the seed hopper. Place containers under each boot to collect the seeds dropped through the boot.

☆ Set the seed rate control adjustment for maximum position and mark this position on the control for reference.

☆ Engage the clutch and rotate the ground wheel for calculated value of N revolutions.

☆ Weigh the quantity of seed collected in the containers and record the observations for each boot separately.

☆ Sum the weights of seed to calculate the seed rate in kg/ha.

$$\text{Seed rate} = \text{Weight of the seeds received} \times 25 \text{ kg/ha} \tag{11.2}$$

☆ If the calculated seed rate is higher or lower than the desired rate of the crop, repeat the process by adjusting the seed rate control adjustment till the desired seed rate is obtained.

While doing the calibration tests, one can also perform two additional tests as follows:

Test for Mechanical Damage

The broken seeds are separated out and their weight is taken. With the known weight of the seeds collected in the containers, percent damaged seeds are calculated as follows:

$$\text{Damaged seeds, per cent} = 100 \times \text{Weight of damaged seeds}/\text{Total weight of seeds} \tag{11.3}$$

Test for Seed Placement Uniformity

Sticky ball method: In this case a 10 m long belt is put under the furrow openers or seed tubes and allowed to travel at the same speed as the desired speed of the seed drill in the field. A sticky layer of grease is applied to the belt so that seeds gets embedded without any displacement. Mount the drill on the stand and operate the drill. Observe the number of seeds dropped and the average distance between the two seeds for each meter of the belt length. Repeat the test several times and take the average value.

Sand bed method: An artificial leveled seedbed made of fine sand is prepared. It is 5 m long having the same width as the seed drill. The sand may be around 25-30 cm deep. Lower the seed drill so that furrow openers are 3-5 cm deep below the sand surface. Allow the drill to travel over the sand bed. Once again observe the seeds dropped and the average distance between two seeds for each meter of sand bed. Repeat the test a few times and take the average value.

Example 11.1

A fluted feed seed drill has 8 furrow openers of single disc type. The furrow openers are spaced 30 cm apart and the main drive wheel has a diameter of 105 cm. How many turns of the main drive wheel of the seed drill are needed to cover an area of one hectare?

Circumference of drive wheel = πD = 3.142 x 105 = 329.9 cm

Total width of seed drill = 8 x 30 = 240 cm

Area covered per revolution = 329.9 x 240 = 79176 cm^2 = 7.92 m^2

Number of turns per ha = 10000/7.92= 1262.6

Example 11.2

Six furrow openers are spaced 20 cm apart and the main drive wheel of the seed drill has a diameter of 1.5 m. Calculate the seed rate per hectare, if seed collected during the test was 20 kg after 600 revolutions of the wheel.

Effective width of seed drill, W = 6 x 20 = 120 cm = 1.2 m

Circumference of drive wheel = 1.5 m

Area covered in one revolution = $\pi D W$ = 3.142 x 1.5 x 1.2 =5.66 m^2

Area covered in 600 revolutions = 5.66 x 600 = 3396 m^2

Seed dropped for 3396 m^2 = 20 kg

Seed rate = 20 x 10000/3396 = 58.9 kg/ha

Calibration in the Field

☆ Fill the seed and fertilizer in the respective boxes. Set the indicator at desired seed and fertilizer rates

☆ Mark a distance of 50 m in the field

☆ Take the seed and fertilizer delivery pipes out from the boots. Put delivery outlets of the pipes in the polythene bags and tight them using rubber rings.

☆ Run the machine and collect the seeds/fertilizers from each delivery pipes after 50 m run of the machine

☆ Weigh the amount of seed and fertilizer collected from each delivery pipe in 50 m run

☆ Calculate the seed rate and fertilizer rate by the following formula

One ha = 10000 m^2

Width of planter = ω m

Distance = 50 m

Weight of seed or fertilizer collected = y kg

Seed or fertilizer rate (kg/ha) is then given as:

(y x 10000)/(ω x distance covered)

If the test has been performed for 50 m run, the above equation is simplified as follows:

Seed or fertilizer rate (kg/ha)= 200 y/ω   (11.4)

If the seed or fertilizer rates are not as per the desired rates, then re-set the indicators or the inclined plates, gears *etc.* in accordance with the desired rates and repeat the whole process of calibration.

Example 11.3

A bullock drawn seed drill is 5 x 22 cm size. The speed of bullocks is 3 km/hr. Calculate the cost of seeding one hectare of land, if the hiring charge of bullocks is Rs.200/- per pair/day, hiring charges of seed drill is Rs.50/- per day and wage of operator is Rs.150/- per day of 8 hours.

Width of seed drill = 5 x 22 = 110 cm = 1.1 m

Area covered/hr = width x speed = 1.1 x 3 x 1000 m = 3300 m or 0.33 ha

Or use the formula of field capacity

Field capacity, ha/hr = 1.1 x 3/10 = 0.33

Time taken/ha = 1/0.33 = 3.03 hrs

Hiring cost of per hour = (200+50+150)/8 = Rs 50 per hour

Cost of seeding per hectare = 3.03 x 50= Rs. 151.50

Example 11.4

A 4 tined seed drill with a spacing of 25 cm between the tines is used to sow an area of 4 ha. The seeds are placed 5 cm deep with each furrow opener opening a 5 cm wide furrow. If the seed drill is operated at a speed of 3 km/hr, calculate the time and power required to sow the crop. Assume that the pressure exerted by the soil is 0.45 kg/cm^2 and field efficiency is 90 per cent.

Effective width of seed drill = 25 x 4=100 cm = 1 m

Time taken to cover 4 ha area = 4 x 10000/(3 x 1000) = 13.3 hr

Gross time taken = 13.3/0.9 = 14.8 hr

Furrow section = 5 x 5 = 25 cm^2

Area covered by 5 tines = 5 x 25 = 125 cm^2

Total draft = 0.45 x 125 = 56.25 kg

Power required = 56.25 x 3 x 1000/(75 x 60 x 60) = 0.625 HP

(Note: 3 km/hr is converted to m/s)

RICE TRANSPLANTING

Manual rice transplanting is a very tedious, expensive and labour consuming operation. To overcome these problems, several attempts have been made to develop walk behind (manual), tractor operated and self-propelled rice transplanters in rice growing countries such as Japan, China, Korea and India. Some major requirements for the effective use of these transplanters are:

☆ The fields are puddled well and soil is allowed to consolidate for one or two days to avoid sinking of the transplanter

☆ Special kind of mat type rice seedlings are prepared for use with the tansplanters. Mat nursery minimizes root damage during separation and planting of seedlings. Besides, it uses less land, can be installed closer to the farm house, and minimizes labour requirement for transporting and replanting.

Walk Behind/Manual Rice Planter

The manual rice transplanter consists of frame, movable tray, handle, skids and seed picking fingers (Figure 11.12). As already stated, it needs mat type nursery so that it is able to transplant the seedling uniformly without any damage to them during the operation of transplanting (Yadav *et al.*, 2014). It has provision to adjust both the planting depth and the hill-to-hill spacing. Manual rice transplanters are available in 5-6 rows with comb type fingers. By pressing the handle, the forks pick-up the seedlings, push them downward and place them in the prepared soil. For every stroke of the handle the seedling tray moves sideways so that seedlings are uniform picked by the forks. The operator has to pull the machine after finishing planting in a row. The machine has a safety clutch mechanism, which prevents breakdown of the planting device from the impact against stones in the field. Its working capacity varies from 0.3-0.4 ha/day and requires two persons, one for operating the transplanter and the other for filling the tray with mat seedlings. It is desirable to check the performance of the transplanter after covering 2-3 m for the condition of transplanted seedlings, hill-to-hill distance, depth of placement and number of seedling per hill. If all these parameters are in order, the machine can be operated for further transplanting operation. The machine saves time and money compared to purely manual transplanting.

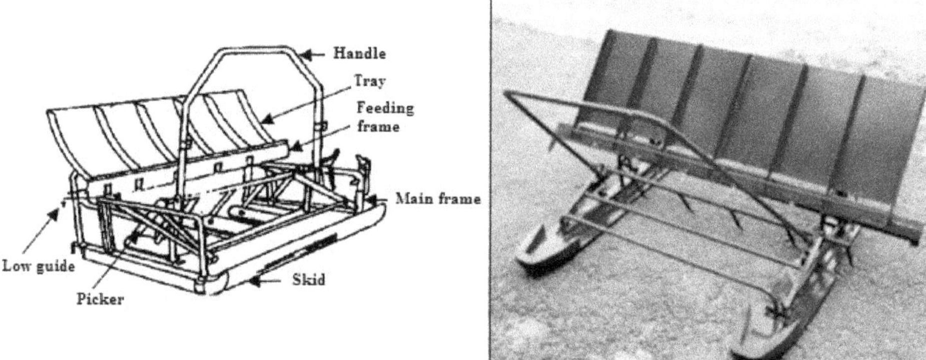

Figure 11.12. Labeled Diagram (Left) and a View of Six Row Manual Rice Transplanter.

Engine Operated Paddy Transplanter

A self-propelled paddy transplanter consists of i) Air cooled petrol/diesel engine, ii) Main clutch, iii) Running clutch, iv) Planting clutch, v) Seedling table, vi) Float, vii) Star wheel, viii) Accelerator lever, ix) Ground wheel, x) Handle and xi) Row bar linkage mechanism *etc.* The four row machine having four bar linkage mechanism is a riding type and employs a double acting transplanting mechanism for enhanced transplanting speed that ensures high field capacity. The mat type 15-20 days old seedlings are used for planting. Mats are placed on the seedling table. Power from the engine goes to the main clutch. Here it follows two routes, one going to planting clutch and the other to running clutch. The planting clutch controls the operation of four bar linkage mechanism. The machine has provision for adjustments of number of seedlings per hill, depth of transplanting and hill-to-hill distance. Whereas the row-to-row spacing is 300 mm, hill-to-hill spacing can be varied from 120 to 220 mm in five settings. Thus, one can vary the plant population. Depth of transplanting can be set from 15 to 45 mm. Once adjusted, the depth of transplanting is automatically maintained during the whole period of transplanting. A four bar linkage mechanism picks up 3 or 4 seedlings in each fork at a time from the mat and plant them in the puddled soil. A fiber glass float supports the machine. Two ground wheels are used to move the unit using a suitable gear combination. A marker demarcates the transplanting width. The planting capacity of the machine is about 0.05 to 0.1 ha/hr. An 8-row engine operated semi-automatic transplanter is shown in Figure 11.13.

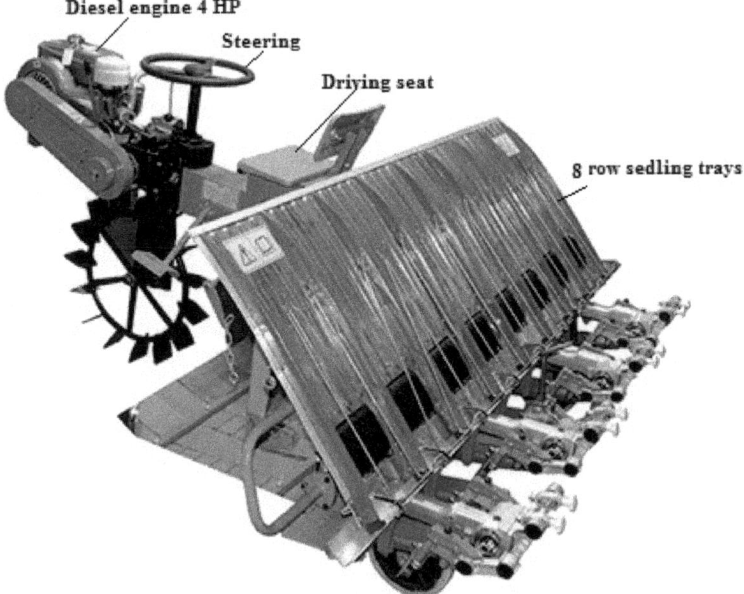

Figure 11.13. 8-Row Semi-Automatic Ride type Transplanter.

Preparation of Mat-type Nursery

The modified mat nursery is prepared in a layer of soil mix on a firm surface, may be concrete floor/polythene sheet/seedling trays *etc.* The basic steps to prepare the mat type nursery are listed as under:

- ☆ Select a location having fertile soil where irrigation water is available and is near to the field to be transplanted. The field should be away from the tube well and trees to avoid shade, debris and damage by birds. The field should be properly leveled. To plant 1 ha (with 2 seedlings/hill at 20 x 20 cm spacing), use 18–25 kg good quality seeds (>80 per cent germination and establishment)

- ☆ Soak the seeds for at least 24 hr or more depending upon the varieties as some varieties may take longer to bud. Drain the soaked seeds and cover them to keep moist for another 24 hr. By this time, the seeds sprout (bud) and the first root is about 2–3 mm long. Prepare 100 m^2 of nursery area for each ha

- ☆ Prepare the soil mix comprising 70–80 per cent soil, 15–20 per cent well-decomposed organic manure and 5–10 per cent rice hull or rice hull ash. Incorporate around 1.5 kg powdered di-ammonium phosphate or 2.0 kg 15-15-15 powdered NPK fertilizer for every 100 m^2 of nursery area

- ☆ Spread 50-60 gauge, 90 cm wide polythene sheet with 1-2 mm diameter perforations over the surface. Place one or more iron frames having compartments of size 40 x 20 x 2 cm for manually operated transplanter, 45 x 21 x 2 cm for engine operated transplanter and 58 x 28 x 2 cm for self-propelled transplanter over the polythene sheet. Number and size of compartments vary from one machine to another depending upon the machine specifications

- ☆ Fill the frame almost to the top with the soil mixture. Spread about 50-60 g of pre-germinated seed evenly in each compartment to achieve uniform density of 2 or 3 seedlings/cm^2. One should possibly use nursery sowing seeder to ensure uniform seed distribution. Sprinkle water immediately to soak the bed

- ☆ Remove the wooden frame and continue the process until the required nursery area is completed

- ☆ Irrigate the area after sowing. The flow of water for first 2-3 irrigation should be very mild. The water level should be uniform so that there is no damage to newly formed mats. Care must be taken that the seedling mats are always wet. If the nursery can be flooded then at 7 days after seeding (DAS), maintain 1 cm water level around the mats. Protect the nursery from heavy rains during the first few days after seeding

- ☆ After an interval of about 10 days, spray 300 g urea for 200 mats. If the temperature and water are adequate, but the seedlings show yellowing (N deficiency), then sprinkle seedlings with 0.5 per cent urea (1.5 kg urea in 300 L water/100 m^2)

☆ Drain the water from the nursery a few hours before the nursery is uprooted for transplanting

☆ Seedlings reach sufficient height for planting in 15–20 DAS. Give a cut with a sharp blade along the nursery boundaries of the mat. The uprooted nursery is now ready for use in the transplanting machine.

Preparation of Field for Transplanting

Plough the field using a mould-board or wooden plough to a depth of 20-25 cm. It should expose most of the weeds. Irrigate the field to keep the fields flooded or saturated with water. After two or three days, puddle the soil in standing water. Apply uniformly half of nitrogen and total quantities of phosphorus and potash on drained surface at the time of last puddling to incorporate in the top 10-15 cm deep soil. Transplant the seedlings one or two days thereafter.

DIRECT SEEDED RICE

Direct seeded rice (DSR) seed drill is normally fitted with inclined plates seed metering system and inverted T-type tines. The seeding operation is carried out under the following two conditions.

Sowing after application of pre-sowing irrigation: Apply irrigation and cultivate the moist field 2-3 times followed by planking. Sow the seeds 3-4 cm deep with seed drill. Use *suhaga* (planking) immediately after sowing. Both the operations of field preparation and sowing may preferably be done in the evening to avoid moisture loss.

Sowing in dry field (Dry sowing followed by light irrigation): Cultivate the field 2-3 times followed by planking. Sow the seeds 2-3 cm deep with seed drill. Do not use *suhaga* after sowing. Irrigate the field immediately after sowing.

Tractor Mounted Direct Rice Seeder

It is an attachment to the commercially available tractor mounted cultivator for direct sowing of rice in dry land conditions. The direct rice seeder comprises seed box, seed metering disc fitted with cups mounted close to its periphery, ground wheel, cultivator shovel, furrow closer, clutch lever, power drive system, and frame. During operation, discs rotate, the cups pick-up rice seeds from the hopper and delivers to the outlet attached to furrow opener through the PVC tubes.

Advantage of Direct Seeding of Rice

☆ Helps to save water

☆ Helps to save fuel

☆ Minimizes environmental pollution

☆ Saves labour

☆ Decreases soil disturbance compared to transplanted rice

☆ Reduces methane emissions due to a shorter flooding period

Drawbacks of Direct Seeding of Rice

- ☆ Weed control may pose problem
- ☆ Relatively high seed requirement 20-25 kg/ha as against 10-12.5 kg/ha in transplanting
- ☆ Proper leveling of land is necessary in DSR
- ☆ Timely sowing is necessary so that the plants come out properly before the onset of monsoon rains

PLANTER

A planter is a farm implement that sows (plants) seeds in a precise manner at equal intervals (same distance between seed to seed) along the rows. A planter is normally used for large sized seeds, which cannot be sown by usual seed drills. The main functions of a planter are:

- ☆ To open the furrow
- ☆ To meter the seed
- ☆ To deposit the seed in the furrow
- ☆ To cover the seeds and compact the soil over it

A planter consists of the following components.

Hopper: A planter has a seed hopper for each row. It is usually made of mild steel, although other suitable material can also be used.

Seed metering device: Several devices such as horizontal rotating plate, vertical rotors, inclined rotors, pneumatic (air-pressure) metering devices, vacuum pick-up devices and others are used for seed metering in planters. The most common device is a horizontal rotating plate at the bottom of seed hoppers. The rotating plate receives the seeds from the hopper. It has suitable notches or holes called cells. Depending upon the type of notches on the plates, it is categorized as i) Edge drop, ii) Flat drop and iii) Hill drop. The edge drop carries the seed in the cell on the edge of the plate. In a flat drop, seed is carried on a flat in the cell of the plate. Each time only one seed is allowed in the cell. In a hill drop, the round or oval cells around the edge of the plate are large enough to admit several seeds at a time. The plate moves under an arrangement called spring-loaded cut-off device or knock out arrangement.

Knock out arrangement: This is a spring-loaded device which knocks out the seeds from the cells or picker heads of the mechanism. It consists of rollers, star wheels or rounded points which are forced into the cells by the pressure of a spring to eject seeds out of the cells. The spacing of seeds or hills in the row is determined by the ratio of linear or peripheral speed of the cells to the forward speed of the planter and by the distance between the cells in the metering unit.

As per the working mechanism, the rotating plate receives seeds from the hopper. The plates move under an arrangement called cut-off, which allows only those seeds that are accommodated in the cells. Knock out mechanism knocks out the seeds from the cells or picker heads of the mechanism. The accuracy of the planter depends upon several factors such as:

- ☆ Speed of seed plate
- ☆ Shape and size of cells
- ☆ Shape of hopper bottom
- ☆ Uniformity of seed size

Some planters with varying features are shown in Figure 11.14. Besides, few of the planters are discussed in the following sections.

Tractor Mounted Modular Planter

Tractor mounted planters are extremely well adapted for intercropping operations. A modular planter has been developed to plant seeds of different crops in adjacent rows or alternate rows. All the planting units have an individual hopper with seed and fertilizer chamber, ground wheel and seed metering mechanism, *etc.*, all mounted on a common frame and hitched to the tractor with the three point linkage. It is suitable for planting groundnut, soybean, Bengal gram *etc.*

Inclined plate planter

Maize planter

Cotton and maize planter

Potato planter

Figure 11.14. Commonly used Planters for Various Crops.

Bed Planter

A bed planter makes beds and sows crops simultaneously. It is suitable for sowing crops like wheat, maize, peas *etc*. In the case of wheat, bed planter can sow two or three rows of wheat on each bed. The bed planter consists of a frame made of mild steel sections, planting hoppers, fertilizer box, furrow openers, bed shaper and power transmitting wheel. The furrow openers are ridger type and have mould-board and share point. The wing span of the mould-board can be adjusted. The share is made of medium carbon or alloy steel tempered to suitable hardness. A vertical disc type seed metering unit is used in the machine. To reduce the draft requirement of the machine, a roller type bed shaper is used.

Tractor Mounted Ridge Planter for Winter Maize

As the name indicates, it is a tractor mounted implement consisting of seed hopper, metering plate, chain drive system, seed tubes, ground wheel, furrow opener, ground wheel tension spring, ridger bottom and ridger beam. The inclined plate type seed-metering device having cells on its periphery is used to meter seed rate. The seed metering mechanism receives power from the ground wheel. The planting unit is mounted on a commercially available ridger consisting of two mould-boards. The share mounted in the front is made of medium carbon or alloy steel. During operation, the cells of the inclined plate pick-up a seed or two to drop them on the ridge in the furrow through the seed tube. The furrow opener can be set to sow seeds on the side of the ridge. Thus, the machine is suitable for planting maize or other similar crops on one side of the ridges.

Potato Planter

Traditional methods of potato planting require huge labour force, may be around 200 man days/ha. In order to mechanize potato planting, several machines have been developed and commercialized. While general features of a potato planter are shown in Figure 11.15 (left), a seed metering device for cut potatoes is shown in Figure 11.15 (right). In this device, a picker wheel with arms at the periphery moves through the tuber supply. The sharp pins pierce into the supply, pick-up a piece of potato and leave it at the delivery point. As against it, cups are used when whole potatoes are used as seed material. The tractor operated vertical belt, paired row automatic potato planter consists of shovel type furrow openers, two vertical rubber belts fitted with metal cups for picking potato tubers (Figure 11.15, AICRP, 2016). The cups on the belt are arranged in a zigzag manner. Number of cups per belt are 40. The planter is equipped with multi-speed gears to adjust the seed spacing with the help of chain and sprockets. Provision has been made to adjust row to row spacing from 600 to 710 mm and plant to plant spacing from 200 to 280 mm. The power to the metering mechanism is provided through the ground wheel having a diameter of 640 mm and power transmission gear ratio of 9:2. Fertilizer can also be applied along with sowing of potatoes.

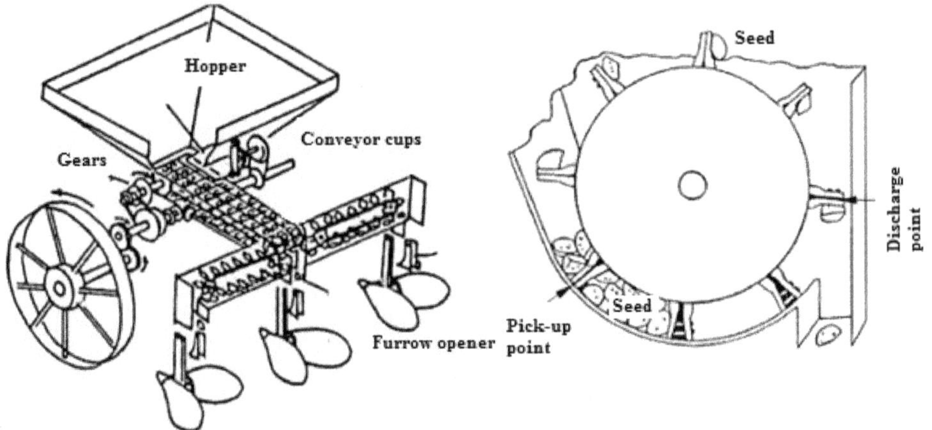

Figure 11.15. A Potato Planter (Left) and Seed Metering Device for Cut Potatoes (Right).

GENERAL PRECAUTIONS DURING SOWING

- ☆ Level the seed/seed-cum-fertilizer drill with respect to the ground surface
- ☆ Select and lock the metering mechanism levers at the calibrated values
- ☆ Always operate the drill at optimum speed to ensure uniformity of seeding and fertilizer application
- ☆ Lock the position control lever and set the depth control wheel as required to maintain uniform depth of sowing
- ☆ Keep watch on the seed drill marker position as it may get easily distorted at turns. Adjust as and when required
- ☆ Check seed and fertilizer boxes at regular intervals lest the drill is run with empty boxes
- ☆ Always engage a skilled operator to perform the operation

QUESTIONS (THEORY)

1. What do you understand by broadcasting? Discuss manual and mechanical broadcasting methods.

2. What do you understand by seed metering mechanism? Name and describe various seed metering devices used in the seed drills.

3. What is the purpose of a furrow opener in a seed drill? Name and describe at least 5 kinds of furrow openers used for this purpose.

4. Describe the need of a turbo-happy seeder. Discuss its construction, operation and benefits.

5. Describe the step wise procedure for calibration of seed drill in the farm shed.

6. Describe the step wise procedure for calibration of seed drill in the field.

7. Discuss the step-wise procedure to prepare the mat type nursery for rice transplanting.

8. What do you understand by direct seeded rice? What are the advantages and disadvantages of direct seeding?

9. What do you understand by a planter? How it differs from a seed drill? Briefly explain a potato planter.

10. Write short notes on: dibbling, seed-cum fertilizer drill, functions of a covering device, zero seed drill, tractor mounted ridger seeder, walk behind rice transplanter, Engine operated paddy transplanter, bed planter

Irrigation Pumps

Tools and equipment required for an irrigation system cover a wide range of instruments such as pumps to pump groundwater, diversion structures in water delivery systems, notches and weirs to measure irrigation water, sprinkler and drip systems to apply irrigation water and sensors to measure soil moisture. Similarly, ditchers are commonly used to construct ditches for efficiently functioning irrigation and drainage systems. Efficient irrigation and drainage practices help to improve crop yields, crop quality, minimize water use, and save and protect natural resources. This chapter is not intended to be a full-fledged treatise on irrigation and drainage systems management. For the limited purpose of this chapter, discussion only revolves around the most commonly used basic equipment, the pumps and drainage ditchers. Even under the head pumps, discussion is restricted to centrifugal, turbine, submersible and solar pumps commonly used by the farmers.

WHAT IS A PUMP?

Turbo-machines are mechanical devices that either extract energy from a fluid or add energy to a fluid because of dynamic interactions between the device and the fluid, either a gas or a liquid. The machines that add energy to the fluid are called pumps. A fluid pump or hydraulic pump is a device or machine having moving mechanical components. It makes use of mechanical energy supplied to it from some basic sources of energy like diesel/electricity/solar/wind. This device then converts the mechanical energy into hydraulic energy that appears as pressure and velocity of the fluid. Thus, *"a hydraulic pump is any device that uses input force to create pressure or kinetic energy, which in turn creates flow"*. The most common applications of pumps include transfer of liquids from one place to another (*e.g.* water from an underground source to a water storage tank) and distribute/circulate liquids around a system as cooling water in automobiles or lubricants through machines and equipment.

To understand the working of a pump, let us consider it as a black box. Fluid enters the pump at certain velocity and pressure, which may be even zero, and leaves it with increased energy, that is velocity and pressure. It is possible because the black box (hydraulic pump) has performed two functions:

1. First, its mechanical action created a vacuum at the pump inlet, which allowed atmospheric pressure to force liquid from the reservoir into the inlet line to the pump

2. The black box houses the rotating components, which moves the fluid either by confining it in definite volume and then displacing it or by imparting energy to the fluid by dynamic action of the moving parts that in turn increases velocity and pressure of the fluid

CLASSIFICATION OF PUMPS

Although a number of classification criteria classify the pumps in various groups, but normally, the pumps are classified into two groups.

Positive displacement pumps, also called hydrostatic pumps or variable delivery pumps *etc.*, are all discontinuous or intermittent flow pumps (Miller 1995). On the basis of motion, these pumps are further categorized as:

☆ Reciprocating pumps

☆ Rotary pumps

A hydrostatic pump converts mechanical energy to hydraulic energy with comparatively small quantity and velocity of liquid. The pressure developed and the work done is a result of essentially static forces rather than dynamic effects.

Non-positive displacement pumps, also called rotodynamic, hydrodynamic or dynamic pressure pumps, are all constant delivery pumps (Gülich, 2007). In a hydrodynamic pump, liquid velocity and movement are large; output pressure actually depends on the liquid flow velocity.

Although not very exhaustive, Figure 12.1 presents the classification of the pumps.

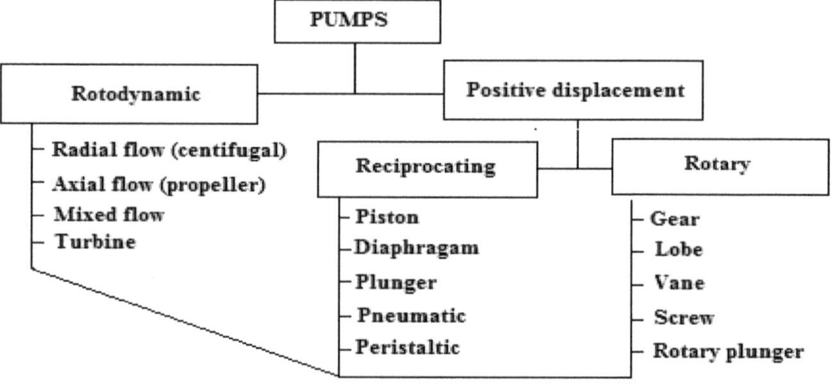

Figure 12.1. Classification of Pumps.

Rotodynamic or hydrodynamic pumps are further categorized as follows:

- ✩ Centrifugal pump
- ✩ Turbine pump
  - ❖ Deep well turbine pump
  - ❖ Submersible pump
- ✩ Propeller Pump
- ✩ Jet pump
- ✩ Air lift pump

This chapter is only devoted to centrifugal, deep well turbine, and submersible pumps because mostly these pumps are used in the on-farm irrigation sector. Besides, solar pump is also included because it has recently been introduced in this sector. Solar pumps are normal pumps, but the source of power is renewable energy from the sun.

Advantages and Disadvantages of Hydrodynamic Pumps

The advantages of hydrodynamic pumps are:

- ✩ These pumps provide smooth continuous flow
- ✩ Have fewer moving parts
- ✩ Wide working range in size and capacity
- ✩ Initial and maintenance cost is low
- ✩ They are capable of handling almost all kinds of fluids including slurries and sludge
- ✩ The operation is simple and reliable
- ✩ No damage to pump occurs in the event of closure of discharge pipe during pump operation
- ✩ There are no drive seals, therefore the risk of leaks is completely eradicated

The disadvantages are:

- ✩ These pumps are not self-priming and hence must be positioned below the fluid level or some arrangement for priming needs to be made
- ✩ Discharge is a function of output resistance
- ✩ Volumetric efficiency is low
- ✩ Primarily used to transport fluids but have little industrial application

Working Principle of Centrifugal Pumps

The centrifugal pump operates on the principle of transfer of energy (or angular momentum) to the fluid from a rotating impeller housed inside a casing. As the impeller rotates, a partial vacuum is created that helps to suck the liquid from the reservoir. The fluid entering the axis or eye of the impeller is discharged from the impeller periphery. The fluid gains kinetic energy and momentum because of the angular momentum imparted by the high-speed impeller. A part of the kinetic energy of the fluid changes to pressure energy (head) while the fluid passes through a diverging area between the impeller periphery and the casing. The fluid flows radially outward or axially into a diffuser or volute chamber before exiting the pump. It may be noted that the maximum possible suction lift at the mean sea level is 10.33 m of water (corresponding to 1 atmosphere). The suction lift in general is much less than this theoretical value. Maximum suction at any location depends upon pressure applied at the liquid surface and decreases as this pressure decreases with altitude of the location. It also depends upon the vapour pressure of liquid at pumping temperature. Maximum suction decreases as vapour pressure increases, varying inversely with temperature. Maximum suction also varies inversely with pump capacity.

The head developed by the pumps depends upon the pump design and the size, shape, and speed of the impeller. The rotational speed of the impeller and the flow resistance of the system govern the flow capacity. The flow output decreases with increase in system resistance (load), as some fluid slips back at higher resistance. These pumps, within limits, operate at approximately constant head and variable flow rate. Centrifugal pumps can be operated in a closed discharge line condition because the liquid can re-circulate within the pump without causing damage. It is always better to avoid such a condition, as it will have adverse consequences on the system and pump health. These adverse effects arise from unstable operation, excessive heating of the fluid and because of energy dissipation within the pump.

KINDS OF PUMPS

Centrifugal pumps may be classified into the following types.

1. According to flow direction in the pump: radial, axial and mixed flow types

2. According to casing design or energy conversion: Volute type, vortex type or diffuser type

3. According to number of impellers or stages: Single stage pump or multistage or multi impeller pump

4. According to number of entrances to the impeller: Single suction pump or double suction pump

5. According to disposition of shaft (axis of rotation): Horizontal shaft pump or vertical shaft pump

6. According to impeller type or liquid handled: open impeller, semi open impeller or closed impeller types

7. According to specific speed: Low specific speed or radial flow impeller pump (further grouped as relatively low, medium and high specific speed), medium specific speed or mixed flow impeller pump, high specific speed or axial flow type or propeller pump.

8. Based on the working head, H: Low head if H < 15 m, medium head if 15 < H < 40 m and high head if H > 40 m

9. On the basis of method of drive: Direct connected- mono-blocks, geared and belt or chain driven, PTO driven

Only three important classifications listed at serial no. 1, 2 and 6 are discussed in the following sections.

Based on Flow Direction in the Pump

Radial Flow Pump

In these pumps, *'the fluid entering the pump along the axial plane is accelerated by the impeller and exits at right angle to the shaft (radially)'*. A spiral casing known as volute surrounds the impeller of the radial flow pump. It is why these pumps are also known as volute type pumps. The water leaving the vanes is directed to circumferentially flow in the volute chamber. As the area of the volute chamber gradually increases in the direction of flow, the fluid velocity reduces while pressure increases. As the water reaches the delivery pipe a considerable part of kinetic energy has already been converted into pressure energy. Since it is not possible to completely avoid eddies, some loss of energy takes place due to the continually increasing quantity of water through the volute chamber. Radial flow pumps operate at higher pressures and lower flow rates than axial and mixed flow pumps (Figure 12.2 top left).

The name centrifugal pump most often applies to the radial flow type of pumps (Figure 12.3 top), although the term is widely used to describe all impeller type rotodynamic pumps including the radial, axial and mixed flow types. Centrifugal pumps are most appropriate for liquids having low to moderate viscosity under a wide variety of flow conditions. These are commonly used in systems with low viscosity fluids such as handling water or milk (dairy plants). Much of the discussions in this chapter primarily apply to centrifugal pumps.

Axial Flow or Propeller Pump

In many applications such as those associated with drainage and irrigation, one requires high discharge at low head. Axial-flow pumps are commonly used under such conditions. In the axial flow pumps *'the fluid enters and exits along the same direction parallel to the rotating shaft'* (Figure 12.2 Top right). The impeller rotates within a cylindrical casing with fine clearance between the blade tips and the casing walls. Fluid particles, in course of their flow through the pump, retain their radial locations. The inlet guide vanes properly direct the fluid to the rotor. The outlet guide vanes eliminate the whirling component of velocity at the discharge port. The number of impeller blades usually varies from 2 to 8, and the ratio of hub diameter to impeller diameter ranges from 0.3 to 0.6. The impeller action lifts the fluid instead of accelerating it. The propeller pump can be oriented in vertical, horizontal, or angled position. Except at design capacity, the efficiency of the axial flow pump is less than that of the centrifugal pump. The axial flow pump can result in overloading of the drive motor when flow rate is significantly less than the design capacity. The head curve for the axial flow pump is much steeper than that of the centrifugal pump.

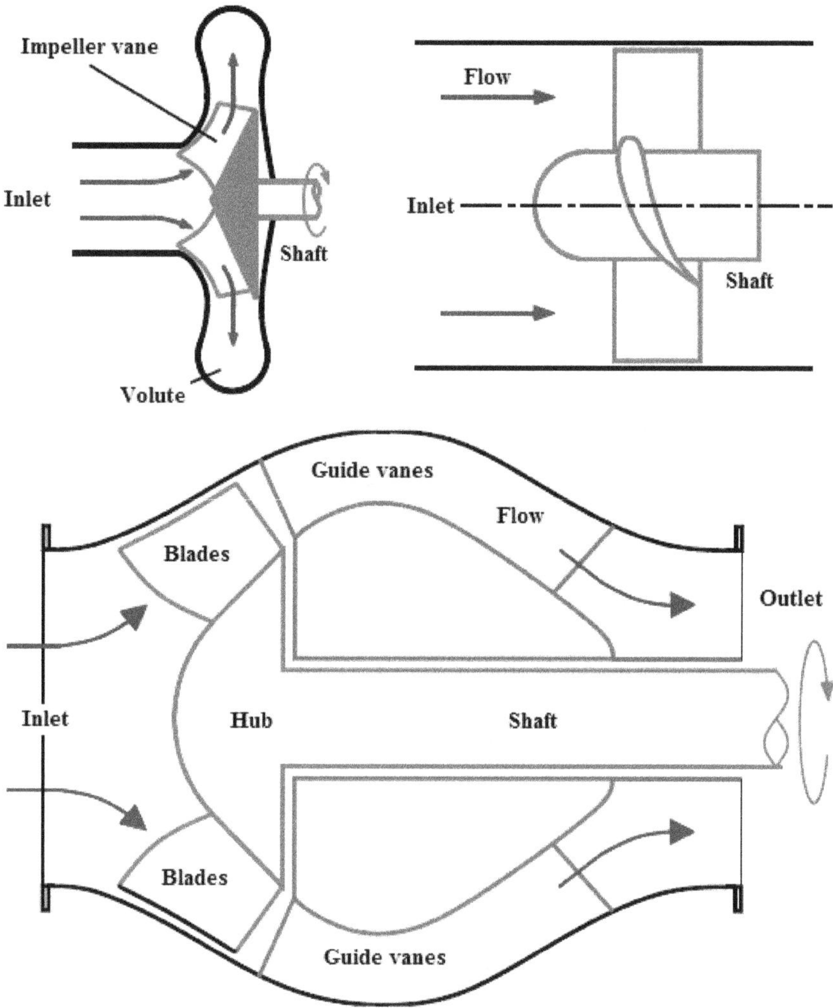

Figure 12.2. Categorization of Pumps on the Basis of Flow: Radial Flow (Top Left), Axial Flow (Top Right) and Mixed Flow (Bottom).

Mixed Flow Pump

Mixed flow pumps, as the name suggests, function as a compromise between radial and axial flow pumps, *'the fluid experiencing both radial acceleration and lift and exits the impeller somewhere between 0-90° from the axial direction'* (Figure 12.2 Bottom). Therefore, mixed flow pumps operate at higher pressures than axial flow pumps while delivering higher discharge than radial flow pumps. The exit angle of the flow dictates the pressure head-discharge characteristic in relation to radial and mixed flow.

Casing Design or Energy Conversion

Volute Casing

The name is derived from the spiral shape of the casing, which is so constructed to collect the liquid as it leaves the outer edge of the impeller vanes. In this case, offsetting the impeller effectively creates a curved funnel such that cross-sectional area increases towards the pump outlet. The increase in area of flow decreases the exit velocity and therefore, pressure increases in the casing as the fluid moves towards the outlet (Figure 12.3 bottom). Volute flow pumps are same as the radial flow pumps as discussed previously.

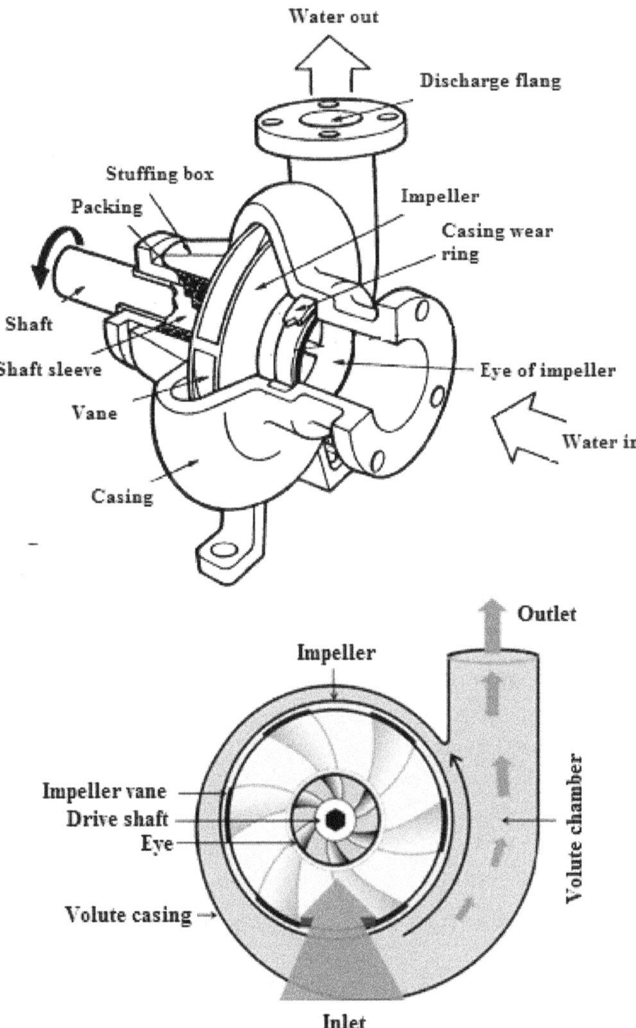

Figure 12.3. Parts of a Centrifugal Pump (Top) and a Schematic of Volute Casing (Bottom).

Vortex

A circular chamber between the casing and the impeller is used in a vortex casing (Figure 12.4 left). Vortex casing reduces eddies formation to a considerable extent resulting in increased pump efficiency.

Diffuser

In diffuser casing, a diffuser is provided as an additional component. The diffuser has many vanes providing gradually enlarging passages surrounding the impeller periphery. The fluid leaving the impeller passes through these guide vanes or diffuser (Figure 12.4 right). The fluid pressure increases as the fluid is expelled between a set of stationary vanes that diffuse the liquid. This provides a more controlled flow and a more efficient conversion of velocity head into pressure head. Provision of fixed diffuser increases the efficiency of the pump up to 90 per cent. Diffuser pumps, because of their resemblance to a reaction turbine, are also named as turbine pumps.

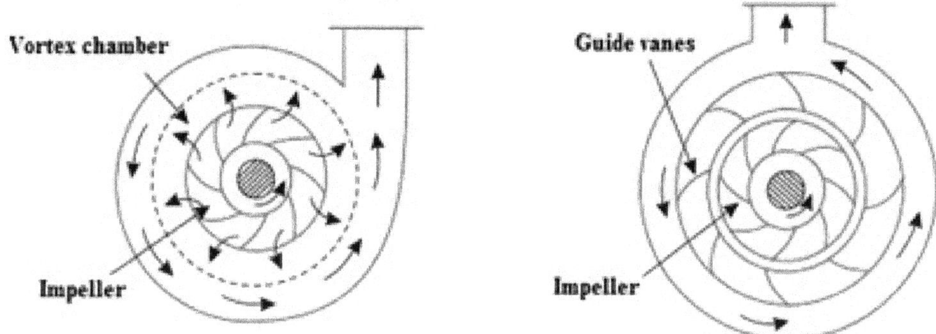

Figure 12.4. Vortex (Left) and Diffuser (Right) type of Casings.

Types of Impellers

The impellers are made of iron, steel, bronze, brass or aluminium. The vanes of the rotating impeller impart a radial and rotary motion to the liquid forcing it to the outer periphery of the pump casing (Figure 12.5 right). The liquid is collected in the outer part of the pump casing called the volute. As previously discussed, based on the direction of flow in the impellers with reference to the axis of rotation, pumps are radial, axial and mixed flow types. According to another classification, impellers are open, semi-open and closed types.

Open Impeller

An open impeller has vanes attached to a center hub mounted directly on the shaft. There is no wall surrounding the vanes (Figure 12.6 left). It makes open impellers weaker than semi-closed or closed impellers. On the other hand, open impellers are easier to clean and repair. Open impellers are mostly used in small pumps or pumps that are required to handle suspended solids. The number of vanes can vary from 2 to 8 but can be more.

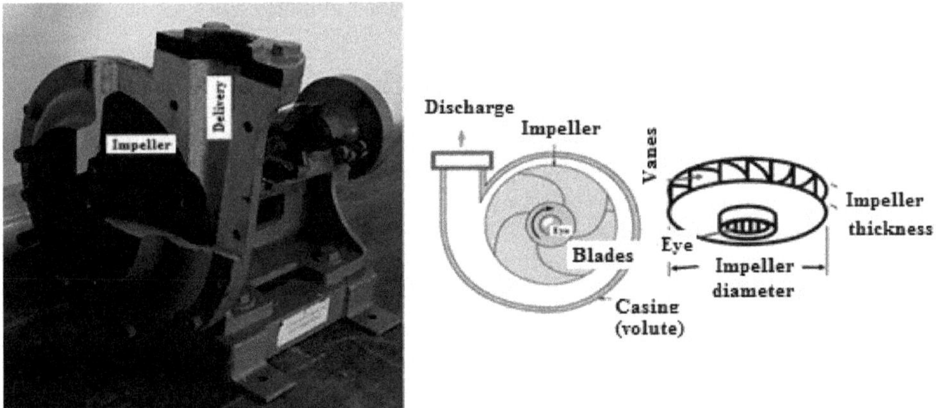

Figure 12.5. Details of an Impeller of a Centrifugal Pump.

Semi-open Impeller

Semi-open or semi-closed impellers have a back wall that adds strength to the impeller (Figure 12.6 middle). A common problem with such impellers is their reduced efficiency, but the ability to pass solids is an important trade-off.

Closed Impeller

Closed impellers have a back and a front wall (plate) around the vanes. It adds to its strength (Figure 12.6 right). The liquid moves in the cavity between the vanes and plates. Closed impellers are primarily used in large pumps. Their performance is greatly affected when fluids being pumped have solid particles. These are also difficult to clean, if clogged.

Figure 12.6. Open Impeller (Left), Semi-open Impeller (Middle) and a Closed Impeller (Right).

COMPONENTS OF A PUMP

Rotating Parts

A pump comprises some rotating and some stationary components. The rotating components are mainly the shaft (Figure 12.3 top) and the impeller (Figure 12.3 and 12.5 left).

Shaft and Impeller

The impeller discussed in the previous section is mounted on a shaft usually made of steel or stainless steel. It is carefully sized to support the impeller. An undersized shaft can result in increased pump vibrations, shorter bearing life, the potential for shaft breakage, and an overall reduced pump life. On the other hand, an oversized shaft will unnecessarily increase the cost of the pump. The shaft also couples the pump to the motor, which drives the pump (Figure 12.3 top). In most pumps, the portion of the shaft that is under the sealing arrangement is covered with a shaft sleeve. The shaft sleeve is usually made of bronze or stainless steel. The shaft sleeve is used to correctly position the impeller on the shaft. It also protects the shaft from wear and tear. The location where the shaft passes through the casing is called the stuffing box, discussed in the section on stationary parts. The pump shaft is supported and held in place by the bearings which have to be designed to handle all kinds of the loads created by the rotation of the impeller, and sized to provide a reasonable service life. Bearing failures are one of the most common causes of pump downtime.

Stationary Components

Important stationary components are: the casing, stuffing box, suction and delivery pipes, and reflux and/or foot valve.

Casing

The casing functions as a cover housing the impeller. Its main function is to enclose the impeller at suction and delivery ends and thereby form a pressure vessel. While the rotating impeller increases the kinetic energy of the liquid, the casing helps to covert this kinetic energy into pressure energy. The pressure at suction end of a single-stage pump may be as little as one-tenth of atmospheric pressure and at delivery end may be twenty times the atmospheric pressure. The fluid is then discharged through the discharge pipe. Note that casing of a centrifugal pump is also known as volute. Purpose of the volute is to collect the liquid discharged from the periphery of the impeller at high velocity and gradually reduce its velocity as the fluid passes through the increasing area of the volute. This design helps to convert the velocity head to static pressure.

Stuffing Box

As already stated, stuffing box is provided where the shaft passes through the pump casing. A sealing arrangement is used to seal the gap between the shaft and the wall of the stuffing box. A stuffing box is an assembly used to house a gland seal, packing or other type of sealing element such as a mechanical seal to prevent leakage of fluid between rotating and stationary parts of the pump.

Suction and Delivery Pipes

One end of the suction pipe is connected to the pump inlet *i.e.* the eye of the impeller, and the other end dips into water source from where water is to be pumped.

Discharge pipe is provided on the pump casing and attached to the delivery side (Also see installation section).

Foot Valve with a Screen (Optional)

A foot value is a non- return value fitted at the lower end of the suction pipe, which retains the water after the pumping to avoid frequent priming (Figure 12.7). It also helps to prevent the trash from entering the suction side.

Figure 12.7. A View of the Foot Valve.

Reflux Valve

Its main purpose is to retain water in the suction pipe section between the pump and the reflux valve to minimize need of priming. When water is pumped to overhead tanks or is conveyed long distances, a reflux valve is also used on the delivery side to permit flow only in the forward direction. It not only checks the return flow but also prevents the water hammer in case the pump suddenly stops to operate.

Self-priming Device

This device is an optional attachment to achieve self-priming of the pump. This device eliminates the need for an external priming mechanism. These kinds of pumps are self-priming centrifugal pumps.

PUMP PERFORMANCE OR CHARACTERISTIC CURVES

The pump performance curves are of great importance to the engineer who is responsible for selecting pumps to suit a particular flow system. These curves comprise (Sahu, 2017).

✰ Head, H and capacity, Q curve shows how much water a given pump will deliver at a given head

✰ Overall efficiency, η curve represents the relationship between the pump efficiency and the discharge

✰ Brake horsepower curve gives the relation between the discharge and horsepower

✰ Net positive suction head (NPSH) required over the capacity range

The typical characteristic curves showing the total dynamic head, brake horsepower, efficiency and NPSH all plotted over the capacity range of the pump at a constant impeller speed are shown in Figure (12.8).

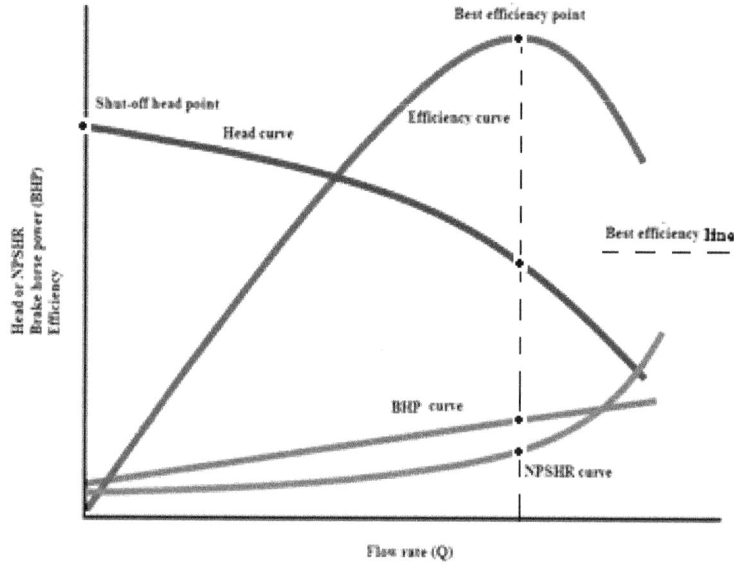

Figure 12.8. Characteristic Curves of a Centrifugal Pump.

Rise head curve: The head curve rises with decreasing flow rate.

Shutoff head: It indicates the head developed by the pump at zero discharge. In other words, it represents the rise in pressure head when the discharge valve is closed.

Brake horsepower: Brake horsepower increases initially with increasing discharge but subsequently there is a fall as one approaches the maximum discharge.

Efficiency: The efficiency is a function of the flow rate. Efficiency is zero when the discharge valve is closed (shutoff head) and flow is zero. As the discharge is gradually increased, flow and efficiency also increase gradually reaching a maximum value at some flow rate. The point at which it attains a maximum value is commonly referred as the normal or design flow rate or capacity of the pump. Later on, flow through the pump becomes turbulent and turbulence increases with increasing discharge. At that point, efficiency starts dropping and thereafter continues to drop as the pump approaches run out condition reaching a Q_{max} just before such a situation.

The Best efficiency points (BEP): These are the points on the various curves corresponding to the maximum efficiency.

Following is the list of some general features of characteristic curves of a centrifugal pump:

- ✩ H is almost constant at low flow rates
- ✩ Maximum H(shutoff head) is at zero flow rate
- ✩ Head drops to zero as Q approaches run out condition
- ✩ Efficiency is always zero at $Q = 0$ and $Q = Q_{max}$ or at run out condition
- ✩ η is not an independent parameter
- ✩ $\eta = \eta_{max}$ at roughly $Q=0.6\ Q_{max}$ to $0.93\ Q_{max}$
- ✩ $\eta = \eta_{max}$ is called the best efficiency point (BEP)
- ✩ All the parameters corresponding to η_{max} are called the design points, Q, H, and P
- ✩ Pumps design should be such that the efficiency curve should be as flat as possible around η_{max}

Pump Selection Criteria

Large numbers of pumps are being manufactured and are available in the market. Yet, not all of them will suit a given setting. Therefore, for a given set-up, one need to choose an appropriate pump such that it operates with maximum efficiency at a given head and discharge. Following step-wise procedure is used to select an appropriate pump for a particular field application.

- ✩ First and foremost consideration is the required flow rate. Assess the required discharge depending upon the cropping pattern, peak requirement including that of animals and household needs, if living in a farm house
- ✩ Determine the static head knowing the depth to water table and the required height to which water is to be delivered
- ✩ Determine the friction head, which depends on the flow rate, the pipe size and the pipe length. Normally, a fraction say 10 per cent of the total static head may be added as frictional head loss
- ✩ Calculate the total head, the sum of the static head plus the friction head
- ✩ Select the pump using information provided in the pump manufacturer's catalogue using the total head, flow required and the desired efficiency

INSTALLATION OF THE PUMP

The most crucial factor to ensure that the pumping system runs for years without causing frequent problems is to install it right. Even high-quality pumps but without proper installation are likely to give numerous operational challenges.

All centrifugal pumps come with an operation and maintenance manual, which also lists the installation procedure. Thoroughly read the manual and install the pump as per procedure prescribed by the manufacturer. Following steps provide some guiding principles.

- ☆ Location
- ☆ Foundation
- ☆ Alignment
- ☆ Piping
- ☆ Wiring

Location: As far as possible, locate the pump close to the water source to reduce the suction lift. The suction lift should be around 4-5 m but should be reduced with increasing altitude of the site. The place should be easily accessible for inspection and maintenance. Do not locate the pumps in areas likely to be flooded. Adequate drainage should be provided if it is not possible to avoid such an area. The main purpose is to keep the pump and motor dry and clean. Leave adequate space for operation, maintenance and inspection of the pump and equipment. For submersible and turbine pumps enough vertical space should be available.

Foundation/base plate/skid: Permanent foundation consists of the base plate fixed to foundation bolts embedded in concrete. The concrete foundation should be rigid enough to prevent vibrations. Make the concrete foundation well in advance of installation of pump and other accessories including motor/engine. It will allow time for drying and curing. The foundation should be provided with anchor bolts for attaching the base plate. The location of foundation bolts should be marked as per the outline drawing supplied by the manufacturer. A pipe sleeve of higher diameter than the diameter of the bolt may be used so that final positioning of the bolts can be easily and accurately made. If the pump is proposed to be mounted on a steel frame or any such similar structure, it should be mounted directly over the supporting beams. The beams and the structure should be rigid enough to prevent distortion and potential misalignment. The pump base must be leveled in such a way that the pump shaft remains in vertical alignment. The pump suction and discharge flanges should be in vertical and horizontal plane respectively.

Alignment of pump and motor: The pump shaft and the drive shaft should be aligned straight to have good performance from the system. The alignment should be repeatedly checked as and when a component is moved. The faces of the coupling halves should be spaced far enough apart so that they do not strike each other when the driver rotor is moved toward the pump. The alignment can be checked with a straight edge and an outside caliper, taper thickness gauge, or dials indicators. To check the parallel alignment, place a straight edge across both coupling rims at the top, bottom and at both sides. The unit is in parallel alignment, if the straight edge rests evenly on the coupling rim at all positions. Allowance may be given for temperature changes and for coupling halves that are not of the same outside diameter. Take care that the straight edge is parallel to the axis of the shafts. The outside caliper or the taper thickness gauge is used to correct any angular misalignment and to verify the correct gap between the coupling flanges. Angular alignment check is made by inserting a taper gauge or feelers between the coupling faces at 90-degree intervals around the coupling. The unit is in angular alignment when the coupling faces are exactly the same distance apart at all points. Install coupling guard after the final adjustment.

Installing pump: Ensure that all foreign material has been removed from the pump before mounting. Remove all shipping protection material prior to the operation of the pump. The orientation of the rotation of the centrifugal pump is indicated on the casing. Ensure that the pump is fixed in the correct rotational direction, before assembling the coupling on the base.

Suction and discharge pipe flanges/piping: The suction piping should be at least as large as the pump flange. Use an eccentric reducer at the suction flange with the straight side up. Avoid using flow-retarding fittings. If necessary, place them at least four times the pipe diameter away from the pump suction. The suction and discharge pipes should be supported independently with the help of pipe hangers and support blocks. This helps to prevent strain on the centrifugal pump joints and casing. The pipe alignment needs to be checked before a connection is made. A flexible fitting may be used on both suction and discharge sides. It helps to eliminate misalignment loads or stresses being transmitted to the pump. Use a concentric taper on the discharge side to increase discharge pipe diameter. All valves and additional fittings should be of the same size as the discharge mainline. The discharge pipe size should be adequate to maintain reasonable velocities and reduce friction losses.

Screening: Make adequate provision for the installation of a suction screen or strainer to prevent any debris from clogging the impeller. The open area of any such strainer should be equal to at least four times the area of the pipe. The screen should be rigid enough so that it does not collapse when flow is reduced due to clogging.

Wiring: All wiring should be of adequate size to prevent voltage drop. Ground wire should be connected as per the approved standard.

Figure (12.9) shows schematically a correctly installed pump and prime mover. Various components of the dynamic head against which the pump has to operate are also shown in this figure.

Operation

Since most pumps are not self-priming, adequate precautions should be exercised to ensure that there is enough liquid in the pump casing when the pump is being put into service for the first time. Besides, liquid in the pump casing needs to be replenished as and when the liquid in the pump casing is low or the pump has not been used for a long period of time. Moreover, most centrifugal pumps lose prime due to ingestion of air or vapours into the pump. When vapours or gases fill the pump casing, the pump impeller becomes gas-bound and is incapable of pumping. It is because the clearances and displacement ratios in pumps used for water and other more viscous fluids cannot displace the air as its density is much lower than the liquids. Under this situation also pump requires priming. *"The*

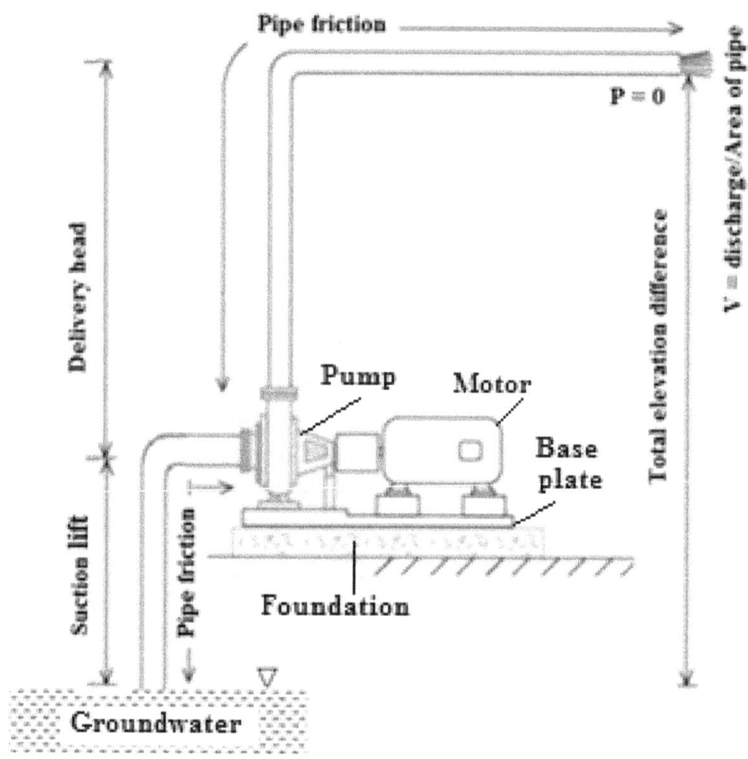

Figure 12.9. A Correctly Installed Pump showing Total Dynamic Head.

operation in which suction pipe, casing of the pump and a portion of delivery pipe is filled by the fluid from outside is known as priming". This process helps to remove air from the impeller and casing necessary to pump the liquid. To do priming, open the priming cock as well as air vent on the pump casing. Pour water into the priming funnel. Continue pouring water until water flows out uniformly from the air vent and is free from air bubbles. A reflux valve or a foot valve at the end of the suction pipeline helps to minimize priming or a self-priming gadget can be used instead. Note that extended operation of a dry pump will destroy the seal assembly. While starting the pump, the gate valve on the discharge line should be closed. As the motor approaches full speed and the pressure gauge on the discharge side is stable, gradually open the discharge valve of the centrifugal pump.

Operator should follow the following safety guidelines before operating the pump.

☆ Go through the operations manual furnished with the power source. Carefully read all safety information given in the safety information section of the manual

☆ Verify that rotation is correct and that the shaft rotates freely

☆ Check all piping connections for tightness and leaks

☆ Inspect all accessories and make sure they are appropriate for the installation

☆ Verify that the driver and coupling are aligned correctly and that all guards are in place

☆ Ensure that all bearings and grease seals are lubricated

☆ Oil levels should be checked and also maintained during pump operation

☆ Follow the instruction on all tags and labels attached to the equipment

☆ Never run the pump with the discharge valve closed for extended periods of time

To stop the pump, the operator should follow the guidelines as listed:

☆ Close the delivery valve

☆ Close the vacuum gauge cock on suction side, if present

☆ Shut down the power unit

☆ Stop the supply of cooling water in case of engine operated system

COST OF PUMPING

The cost of pumping is grouped under the heads listed in Table 12.1.

Table 12.1. Cost Components for Irrigation Pumps

Fixed Cost	Variable Cost
Depreciation (Based on initial cost, salvage value and life of the pump)	Fuel or electric energy consumption
Interest on investment	Lubrication (negligible)
Housing, and taxes if any	Repair and maintenance
	Operator wages

Most of the issues related to cost calculations have been discussed in Chapter 4 for the tractor and Chapter 9 for farm machines. Same procedure applies even in this case except for the fuel or electricity consumption that forms the subject of this section.

Power Required and Energy Consumption

Water power is the theoretical power required to pump any fluid. *It is the power required by a pump in lifting a given quantity of fluid in a unit time assuming no losses of power in the pump* (Miller, 1995). If Q is the discharge of the pump and H the total head against which the pump is working, then the water power generated by the pump is given as:

$$WP = \gamma Q H \ \text{Newton-m/s} \tag{12.1}$$

Here, WP is the water power, γ is the specific weight of the fluid (N/m^3), Q is the discharge of the pump (m^3/s), and H is the total head or total dynamic head (m) as illustrated in Figure 12.9. Since 1 Newton-m/s is equal to 1 W, the water power in eq. (12.1) is in Watts. Since γ (N/m²) is also written as ρg where ρ is the density of the liquid (Water ~ 1000 kg/m³) and g is acceleration due to gravity (9.81 m/s²), we can write

WP =1000 x 9.81 Q H = 9810 QH					(12.2)

Since 1 Watt is also equal to 0.00134 horsepower, we have

WHP = 9810 QH x 0.00134 = QH/0.076					(12.3)

Writing Q in L/s, we have:

WHP = QH/76 ~ QH/75					(12.4)

Eq. (12.1) is valid for all fluids including water but eq. (12.2) onwards are specific to water.

Shaft Horsepower

It is the power required at the pump shaft and can be derived as:

Shaft Horsepower = Water horsepower/η_p					(12.5)

Here η_p is the pump efficiency.

Brake Horsepower (BHP)

It is the actual horsepower an engine or electric motor must supply to drive the pump.

For direct driven pump, BHP = Shaft Horsepower					(12.6)

With belt or indirect drives, BHP = Water horsepower/$(\eta_p \times \eta_d)$					(12.7)

Here η_d is the drive efficiency.

Horsepower Input to Electric Motor (IHP)

IHP = Water horsepower/$(\eta_p \times \eta_d \times \eta_m)$					(12.8)

Here η_m is the motor efficiency.

Kilowatt input to electric motor or energy consumption in kilowatt-hours is given as:

Energy in Kilowatts = IHP x 0.746 or BHP x 0.746/η_m					(12.9)

Cost of operation of electric motor = Energy in Kilowatts x
						Hours of pumping x Cost per Kilowatt Hour					(12.10)

Example 12.1

A centrifugal pump delivers 900 L/min to fill a square tank having an area of 0.5 ha. If the pump operates for 8 hours a day, what should be depth of water in the tank at the end of the day?

Water delivered per day (m³) = (900 x 60 x 8)/1000 = 432 m³

Depth of water in the tank (mm) = 432/(0.5 x 10000) = 0.0864 m = 86.4 mm

Example 12.2

For the example of 12.1, what should be the pump capacity to fill the tank to a height of 50 mm in 15 hr?

If Q is the capacity of the pump, then

The quantity of water pumped in a day = Q x 15 = 15 Q m³

Water required = 50 x 0.5 x 10000/1000 = 250 m³

Therefore, Q x 15 = 250

Q = 16.66 m³/hr = 4.66 L/s

Example 12.3

A pump delivers 20000 L of water to a height of 25 m in every 2 min. Calculate the horsepower required to drive the pump, if the efficiency of the pump is 80 per cent. Assume any value or condition required to solve the problem.

Water delivered per second = 20000/(2 x 60) = 166.6 L/s

Using eq. (12.4)

WHP = 166.6 x 25/75 = 53.5

Assuming that the pump is direct driven, we have

SHP = BHP = 53.5/0.8 = 69.4 HP ≈ 70 HP

Example 12.4

A pump lifts water @ 80 m³/hr against a total head of 30 m. Calculate the water power of the pump? If the pump efficiency is 80 per cent, what size of prime mover is required to operate the pump? If a direct driven electric motor with an efficiency of 80 per cent drives the pump, compute the cost of electric energy at the end of 50 days. Assume that the pump operates 10 hours a day. Assume the cost of electricity as Rs 3.50 per unit.

Discharge of the pump (Q) = 80/60 x 60 = 0.022 m³/s, Total head (H) = 30 m

Using. Eq. (12.1)

Water power, WP = γ x Q x H = 9810 x 0.022 x 30 = 6474.6 W or 6.5 kW

Since the efficiency of the pump is 70 per cent, the shaft power required is

Shaft power = 6.5/0.8 = 8.13 kW

For a direct driven pump, the shaft power determines the required size of the prime mover. Input power is given as:

Input power = 8.13/0.80 = 10.2 kW

Total energy consumption in 50 days = Input power x Hours of pump operation x Number of days = 10.2 x 10 x 50= 5100 kW-h

Since 1 kW-h is equal to 1 unit, therefore, the cost of electric energy = Rs. (5100 x 3.5) = Rs. 17850.

Example 12.5

Calculate the cost of pumping water per day of 8 hr from a well with a centrifugal pump with the following data.

Suction head = 4 m; Delivery head = 8 m; Friction head = 1.5 m

Output of the pump = 30 L/s; Pump efficiency = 70 per cent

Motor efficiency = 90 per cent ; Cost of electricity = Rs. 3.0 per unit.

Assume any other value required in the calculations.

Total head = 4+8+1.5 +0.1 x 1.5 = 13.65 m (Assuming minor friction losses as 10 per cent of major friction losses)

Water horsepower = 30 x 13.65/75 = 5.46 (Eq. 12.4)

Brake horsepower = WHP/pump efficiency = 5.46/0.7 = 7.8

Since 1 HP is equal to 0.746 kW, we have

Motor size in kW = 7.8 x 0.746 = 5.82

Therefore, units consumed = 5.82 x 8/0.9 = 69.33 ≈ 51.73 kW-h

The cost of pumping = 51.73 x 3.0 = Rs. 155

Example 12.6

What should be the capacity of the pump to command an area of 5.5 ha? Take the irrigation interval as 22 days and depth of irrigation as 5 cm. Assume that the pump will be operated 15 hr a day.

If Q is the capacity of the tube well, then

The quantity of water pumped in a day = Q x 15 = 15 Q m^3

Area to be irrigated per day = 5.5/22 = 0.25 ha

Water required to irrigate 0.25 ha = 10000 x 0.25 x 5/100 = 125 m^3

Therefore, Q x 15 = 125

Q = 125/15 = 8.33 m^3/hr = 2.33 L/s

TROUBLE SHOOTING

Some of the common troubles that might occur during operation of the centrifugal pumps and major remedial actions to resolve these problems are listed below (https://www.sameng.co.za/full-service-pump-manufacturers/centrifugal-pumps-troubleshooting-guide).

No Water is Delivered by the Pump

Sl.No.	Causes	Remedies
1.	Lack of prime	Fill pump and suction pipe with liquid to be pumped
2.	Loss of prime	Check for leaks in suction pipe joints and fittings, and mechanical seal. Vent casing to remove accumulated air
3.	Suction lift too high	If no obstruction at the inlet, static lift may be too high. Raise the liquid level to be pumped or lower the pump. Manage friction losses in suction pipe
4.	Discharge side head too high	Check that valves on discharge side are open. Check pipe friction losses. If high, larger diameter discharge pipe may reduce the losses
5.	Speed too low	Check whether the motor is directly across-the-line and is receiving full voltage. Alternatively, frequency may be too low; the motor may have an open phase
6.	Wrong direction of rotation	Check motor rotation. Rotation in the wrong direction will damage the pump
7.	Impeller completely plugged	Dismantle pump to clean the impeller with appropriate tools

The Pump Discharge is Low

Sl.No.	Causes	Remedies
8.	Air leaks in suction piping or gaskets at the inlet and outlet of the pump admitting air	Threaded connections are a common source of leakage in the pipes. The connections should be coated with white lead or pipe cement and tightened, the gaskets should be tightened
9.	Air leaks in stuffing box	Increase seal lubricant pressure to above atmosphere
10.	Speed too low	See item 5
11.	Discharge system head too high	See item 4
12.	Suction lift too high	See item 3
13.	Impeller partially plugged	See item 7
14.	Cavitation due to insufficient NPSH	Increase positive suction head on pump by lowering pump or increasing suction pipe size or raising fluid level. Cool suction piping at inlet to lower liquid temperature or pressurize suction vessel
15.	Defective impeller	Inspect impeller and replace, if damaged
16.	Defective packing	Replace worn out packing and sleeves
17.	Foot valve too small or partially obstructed, the foot valve or the reflux valve may be defective, so that water is not retained in the suction side	Area of valve should be at least as large as area of suction pipe- preferably 1½ times. If strainer is used, net clear area should be 3 to 4 times the area of suction pipe. Replace, if necessary. Check for any obstruction and clean it. Often the flap in the foot valve does not operate properly. This should be checked and replaced, if necessary
18.	Suction inlet not immersed deep enough	Immerse the suction inlet to proper depth
19.	Wrong direction of rotation	See item 6
20.	Impeller diameter too small (probable cause if none of the above)	Check whether a larger impeller can be used; otherwise, cut pipe losses or increase speed, or both, without overload drive

Not Enough Pressure is Developed

Sl.No.	Causes	Remedies
21.	Speed too low	See item 5
22.	Air leaks in suction piping	See item 8
23.	Mechanical defects	See item 15, 16, 17
24.	Obstruction in liquid passages	Dismantle pump and inspect passages of impeller and casing. Remove obstructions
25.	Air in liquid	See item 14
26.	Excessive impeller clearance	Adjust impeller clearance
27.	Impeller diameter too small	See item 20.

Pump Operates for Short Time, then Stops

Sl.No.	Causes	Remedies
28.	Incomplete priming	Free pump, piping and valves of all air. If high points in the suction line prevent this, they need to be corrected. Also see item 5
29.	Suction lift too high	See item 3
30.	Air leaks in suction piping	See item 8
31.	Air leaks in stuffing box	See item 9
32.	Air or gases in liquid	See item 14

Pump takes Too Much Power

Sl.No.	Causes	Remedies
33.	Head lower than rating resulting in pumping of too much water	Impeller's diameter needs to be changed
34.	Cavitation	See item 14
35.	Mechanical defects	See item 15, 16, 17
36.	Suction inlet not immersed deep enough	See item 18
37.	Liquid heavier (either viscosity or specific gravity) than allowed by the manufacturer	Use larger driver
38.	Wrong direction of rotation	See item 6
39.	Stuffing box too tight (Packing)	Release gland pressure and reasonably tighten it. If sealing liquid does not flow while pump operates, replace packing. If packing is wearing too quickly, replace shaft sleeves and keep liquid seeping for lubrication
40.	Casing distorted by excessive strains from suction or discharge piping	Check alignment. Examine pump for friction between impeller and casing. Replace damaged parts
41.	Shaft bent due to damage - through shipment, operation, or overhaul	Dismantle pump and inspect shaft

Sl.No.	Causes	Remedies
42.	Mechanical failure of critical pump parts	Check bearings and impeller for damage. Any problem in these parts will cause a drag on shaft
43.	Misalignment	Realign pump and driver
44.	Speed may be too high	Check voltage on motor
45.	Electrical defects	The voltage and frequency of the electrical current may be lower than that for which motor was built or there may be defects in motor. The motor may not be ventilated properly due to a poor location

Pump Leaks Excessively at the Stuffing Box

Sl.No.	Causes	Remedies
46.	The packing material used may be worn out, incorrectly inserted or may not be of the right kind	Examine and take appropriate action
47.	The shaft may be worn out	Replace the shaft

Pump is Noisy

Sl.No.	Causes	Remedies
48.	The suction lift may be too high	Examine and take appropriate action to reduce the suction lift
49.	Mechanical defects such as bent shaft, improper alignment between the pumps and the driving unit, broken or worn-out bearing	Examine to identify the cause and take appropriate action

TURBINE PUMP

Turbine pumps are used when water level in the tube wells or in the open well or the sump is below the practical suction lift limit of the centrifugal pump. Although the turbine pump also operates on the principle of centrifugal pump, it differs from volute type as stationary guide vanes guide upward the water thrown by the impeller to the periphery. The gradually enlarging vanes guide the water to the casing. As a result, some kinetic energy is converted into potential energy. Therefore, the turbine pump generates heads several times that of a volute pump. A vertical turbine pump consists of electric motor, head shaft, column pipe, line shaft, impeller shaft, impeller, bowl, suction case and suction pipe. All these components are grouped under five basic elements, namely Pump bowl assembly, Column pipe assembly, Line shafting, Surface discharge head and the Driver. It has a shaft at the bottom end of which is the turbine pump and on the top a driving motor, which is placed above the ground level (Figure 12.10 left).

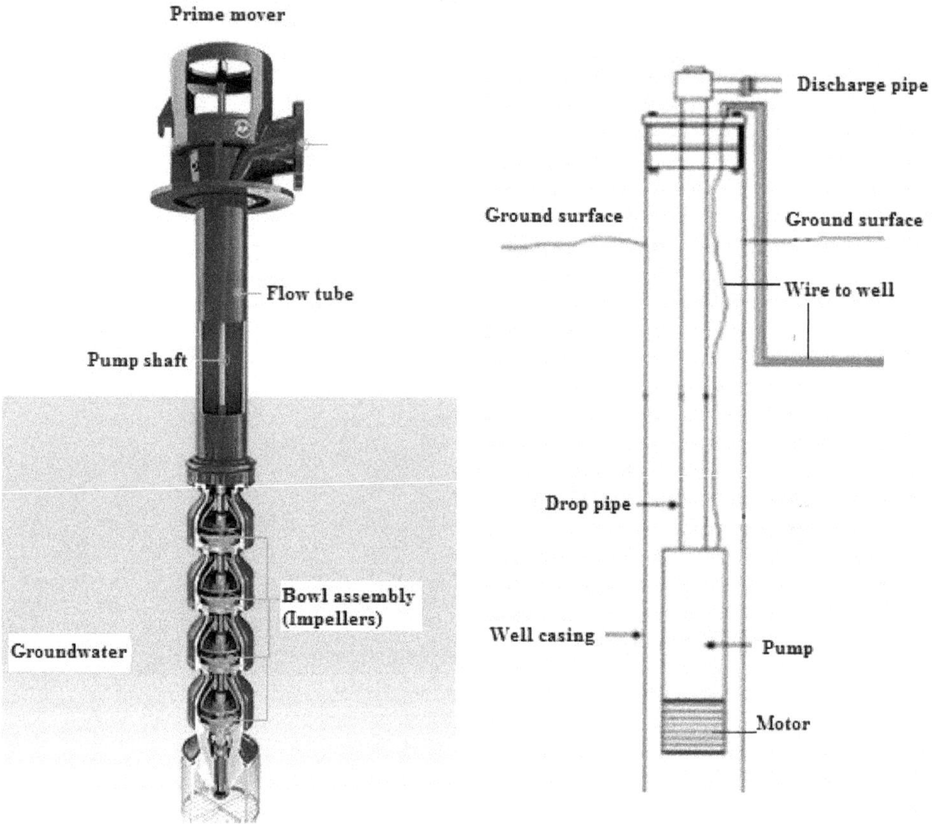

Figure 12.10. A Schematic Diagram of a Vertical Turbine Pump (Left) and a Submersible Pump (Right).

Pump Bowl Assembly

Bowl assembly is the main pumping element of a vertical turbine pump. Number of bowls or stages in an assembly depends upon the head requirement of an individual installation. A single stage pump bowl assembly consists of a suction case or bell, an enclosed or semi-open type impeller, a top intermediate bowl and a discharge case. Impellers are connected to the pump shaft by split cone-shaped impeller lock collets or by key. The suction case/bell serve as the intake of the pump bowl assembly. Liquid moved by the impeller flows through the pump bowls. The top inter bowl acts as a diffuser and guides the flow into the discharge case, which in turn delivers it into the column pipe. For units of two or more stages, intermediate bowls and impellers are added for each stage. Every intermediate bowl acts as an inter stage diffuser that directs the fluid vertically to the next stage impeller. Since the pump unit remains submerged in water, no priming is required in this case as in a centrifugal pump. On the other hand, some minimum submergence is needed at the beginning of the suction. It is why; the pumping unit is located below the drawdown level of the water source. Pump shaft bearings are located in the suction bell, top inter bowl/discharge case, for multi-stage pumps in each inter bowl between stage impellers. Basic design features of a bowl assembly differ as different types of lubrication arrangement are made to lubricate the shaft *i.e.,* self water lubrication or oil lubrication. From a sanitary point of view, it is always desirable that pump bearings are water lubricated because any leaking lubricating oil may contaminate the water.

Column Pipe Assembly

The column pipe assembly conveys water from the bowl assembly, located deep in the well or sump, to the discharge head at ground level. Either threaded or flanged column pipe assembly is used depending upon the pump features, size or customer's requirements. Whereas a threaded column has straight threads on both the column and column couplings, flanged column pipes have male and female spigots to ensure proper alignment of the two adjacent column pipes. The flanged column pipes are bolted together with a gasket between the flanges to eliminate leakage.

Line Shafting

The line shaft assembly transmits torque from the driver to the bowl pump assembly. It rotates within the column pipe. A lead washer inserted between the two shaft ends facilitates dismantling the joint when necessary. Bearings at specific intervals support the shafting.

Surface Discharge Head

Surface discharge head performs multiple functions. Firstly, it provides a base from which the pump is suspended. Secondly, it directs the fluid flow from the pump column to the piping system. Thirdly, it also provides a way to seal the line shaft.

Driver

It is generally a three-phase induction motor. A well-maintained turbine pump provides trouble free service for several years.

SUBMERSIBLE PUMPS

These are a special kind of centrifugal pumps where a long vertical shaft connecting the motor and pump unit (as in the case of turbine pump) is replaced by a short shaft. The prime mover and pump are closely coupled and both are submerged in water (Figure 12.10 right). Thus, *"a submersible pump is a centrifugal pump driven by a closely coupled electric motor constructed for submerged operation as a single unit"*. It consists of a pump-motor assembly, discharge column pipe, head assembly and waterproof cable. All submersibles are multi-stage pumps mounted on a single shaft, and all rotating with the same speed. Each impeller and matching diffuser is called a stage. The impeller of one stage passes the water to the eye of the impeller of the next stage through a diffuser. The shape of the diffuser is such that it helps to slow down the flow of water and convert the velocity head to pressure head. The water inlet into the submersible pump is between the pump and the motor. Submersible pumps are suitable for tube wells having a bore diameter equal to or exceeding 100 mm. The pipe should have enough strength so that there is no possibility of breakage. The waterproof electrical wiring is used. Use of a properly grounded electrical system minimizes the possibility of shorting. Shorting may damage the entire unit. The discharge pipe supports the pump and motor assembly. For the same output, submersible pumps consume less power besides requiring less space. These pumps are ideal for lifting water from deep tube wells for irrigation, domestic and industrial applications.

SOLAR WATER PUMPING SYSTEM

India is blessed with an abundant solar resource with clear sunny weather for more than 300 days in a year. The annual range of solar radiation is from 4 to 6 KWh/m^2-day (ICID, 2019). These together with high utilization of groundwater for agriculture in the country make solar-powered irrigation systems an attractive alternative to run groundwater pumps in India. Out of total of 4 30 million pumps operating in India, 21 million are grid operated, 8.8 million are diesel powered and only 0.13 million are solar-powered (https://www.ceew.in/sites/default/files/CEEW-Solar-for-Irrigation-Deployment-Report-17Jan18_0.pdf). It is heartening to note that more than 181,000 solar power pumps have been installed in the country over the past three years under the *Pradhan-Mantri Kisan Urja Suraksha evam Utthaan Mahabhiyaan* (PM-KUSUM) scheme. A solar power system is essentially a normal pump except that it derives renewable power from the sun through a solar system set-up. A solar system comprises solar panels that absorb sunlight, inverter to convert direct current into alternating current, mounting structure to hold the panels in place, batteries to store the extra power generated, grid box and other parts such as cabling or wiring, nuts *etc.* The size of the solar system is designated in kW such as 1 kW, 3 kW, 5 kW, 10 kW *etc.* Besides, some systems are equipped with solar tracking system and integrated battery solution.

The pump: This is the heart of the solar water pumping system. Solar pumps are rated according to the volts that need to be supplied for its operation. A 12 volt pump is a small one, 24 volts is more a norm, while 48 volts and upwards will require more power and are capable of pumping more water. Wiring being costly, smaller wire sizes are used even in higher voltage systems without sacrificing power output from panels to pump. Some pumps require additional accessories such as filters, float valves, switches, *etc.* to function optimally. The manufacturer or distributor usually specifies as to what is required.

Solar Panel: Photovoltaic (PV) panels (Figure 12.11) are rated in watts of power produced. Panels have now become much more affordable than before (ICID, 2019). A pump only requires a certain amount of power to produce a certain amount of pressure and flow. Thus, one should always size the PV array by the amount of power needed. It may be noted that most solar pumps actually require about 20 per cent more wattage than specified when wiring the panel directly to the pump. Also, a larger panel will allow the pump to turn on earlier and later in the day and also in relatively lower light conditions (Waterman, 2013).

Figure 12.11. A Solar Panel on a Groundwater Well.

Inverter: The actual power from the solar panel to the pump depends upon the power of the inverter and its efficiency. Most inverter comes with an efficiency of 80 per cent. Therefore, a 3 kW inverter system will give up to 2.4 kW of peak power for use at a given point of time.

Linear current booster: A small device installed between the panels and the pump allows the pump to switch on during low light conditions. Linear current boosters are sized in accordance with the pump voltage and panel output.

Wiring accessories: These include minor components that will connect the panels to the pump. A combiner box is used to make wiring safe and simple when more than one panel is used. Circuit breakers are installed in the box to enable safe and quick shut-off of the panels as and when the system needs servicing. The circuit breakers can also serve as a switch for turning the pump on and off at the discretion of the user.

Batteries: Many types of batteries are in use, the lead-acid type being the most common ones. Currently, lithium-ion batteries are becoming highly popular. A motive power (traction battery) is a lead-acid battery commonly used in stand-alone PV systems. It differs from deep discharge batteries because they use heavier, thicker plates and strong inter-cell connections to withstand the mechanical stresses from deep discharges. Battery storage systems often have power ratings in kilowatts (kW), typically in the range of 1-7 kW. The power rating of the battery is its capability to supply power. The battery storage capacity is measured in ampere-hours (Ah) or kilowatt-hours (kWh). A 12-volt battery rated at 480 Ah stores 2.25 kWh of energy.

Solar tracking system: Trackers track the sun movement and direct the solar panels or modules toward the sun. These devices change the orientation of the solar panels throughout the day following the sun's path to capture maximum energy. Single-axis solar trackers as the name suggests rotate on one axis moving back and forth in a single direction. Some single-axis trackers include horizontal, vertical, tilted, and polar aligned. Dual-axis trackers on the other hand make the solar panel to continually face the sun. A dual-axis tracker is able to track sun rays by switching the solar panel in various directions. These types include tip-tilt and azimuth-altitude consisting of servo motor, stepper motor, rain drop sensor, temperature and humidity sensor and an LCD.

A set-up of a solar power pumping system on canal irrigation in Haryana is shown in Figure 12.12. On the right is a rain gun being operated in an adjoining field for irrigation (Figure 12.12). Some advantages and disadvantages of solar power pumping systems are as follows (Hartung and Pluschke, 2018).

Figure 12.12. A Solar-Powered Grid Connected System on Canal Irrigation in Haryana (Left) and a Rain Gun in Operation (Right).

Advantages of Solar-Powered Irrigation

☆ Rugged construction with long life

☆ PV technology constitutes a reliable source of energy for pumping of irrigation water in remote areas, particularly in areas that are not connected to the electricity grid or where regular supply of liquid fuels and maintenance services is not guaranteed

☆ Solar power irrigation system can help to stabilize, increase and diversify production through improved access to energy and water

☆ Reduced dependency on fossil fuel can play an important role in climate change mitigation by reducing green house gases emissions in irrigated agriculture. Energy subsidies for fossil fuels can be reduced while offering an alternative to farmers and rural communities whose livelihoods may otherwise be negatively affected

☆ Cost of solar panels continues to drop resulting in reduced cost of water pumping in the long run. It has made solar systems economically viable and competitive with other sources of energy

☆ Potential for job creation in the renewable energy sector (producers, suppliers, services)

☆ Potentially more efficient use of water when combined with drip or other water- efficient irrigation technologies

☆ Lower hourly yields over more hours per day results in gentler abstraction of sensitive groundwater resource, reducing risk of borehole collapse

☆ Easy to operate and maintain

Disadvantages

☆ Relatively high initial investment need innovative financing models (or subsidies) especially for small and marginal farmers

☆ Since optimal operation and maintenance requires a certain degree of technical knowledge and skill, farmers need to be trained and adequate services (extension services or private service suppliers) for this purpose need to be arranged

☆ If not adequately managed and regulated, there is a risk of over-abstraction of groundwater and unsustainable water use as low energy costs can lead to wasteful use of water using irrigation methods that have low field water application efficiencies

☆ Taxes on imported equipment may distort and artificially keep prices high

☆ Production of PV panels requires some toxins and rare minerals; mining and production of these tends to produce environmentally harmful waste

☆ Panels need to be disposed of appropriately to avoid environmental harm

☆ Illegal pumping might increase and may lead to depletion and degradation of groundwater resources

☆ Lack of codes and standards to guarantee quality of solar power irrigation system may pose few problems

☆ Systems are vulnerable to theft and hence often not covered by insurance agencies, a pre-requisite for financing by the banks

Relative merits and demerits of solar power irrigation system and diesel pumps have been listed in the following Table (Abu-Aligah, 2011).

Type	Advantages	Disadvantages
PV pump	Unattended operation	High investments
	Low maintenance cost	Water storage may be required as the pump operates at sub-optimal level during cloudy times and during night
	Long life	Skill required for maintenance
Diesel pump	Low investment cost	High operational cost
	Fast and easy installation	High maintenance costs
	Familiarity of the farmers to technology	Short life
		Environmental issues with noise and air pollution

QUESTIONS (THEORY)

1. Explain the working principles of a centrifugal pump.

2. Discuss the pump characterization based on the nature of the flow through the impeller. Draw neat diagrams to explain the flow process in each case.

3. List the advantages and disadvantages of hydrodynamic pumps.

4. Name and describe the various types of casings used in hydrodynamic pumps.

5. Discuss open, semi-open and closed impellers indicating their use under particular conditions.

6. Draw a neat diagram showing suction head, delivery head, and friction head in a pump set-up.

7. What do you understand by pump's characteristics curves? Show various curves using a neat diagram briefly explaining the various characteristics.

8. Write descriptive notes on turbine and submersible pumps.

9. How will you proceed to calculate the monthly expenditure on electricity consumed to pump a given quantity of water to a given height?

10. What is priming of a centrifugal pump? Why you need to prime a centrifugal pump before its start?

11. Write a descriptive note on pump installation.

12. List the causes and remedies of the 2 problems of a pumping system namely no water is delivered by the pump and pump takes too much power during pumping.

13. Discuss the salient features of a solar water pumping system. List its advantages and disadvantages.

Chapter 13

Plant Protection Equipment

Advancement in agricultural sciences, intensive farming and expanded irrigation facilities has tremendously increased the cropping intensity, even going up to 300 per cent in many cases. As such, most fields remain covered under crops around the year. It has significantly increased the incidences of plant pests and diseases. Moreover, climate change has added a new dimension to this scenario leading to proliferation of crop insects, pests and diseases. As such, use of pesticide and fungicide to control pests and diseases has become a routine practice at Indian farms. A pest management exercise requires thorough knowledge of the pest problems, agro-chemicals to be used and their proper application to cover the infested area. While the first two issues will be discussed in other subjects, this chapter is devoted to the equipment used for applying pesticides for their proper application. This issue is vital to the success of pest control operations because inadequate and improper application of pesticide will not yield desirable results. Sprayers and dusters are commonly used for pesticides application. The sprayer or duster used for a specific situation should be able to apply proper dosage to the target area in an even manner. Moreover, the chemical should reach the target in proper droplet size. This chapter discusses commonly used kinds of sprayers and dusters in pest control operations.

SPRAYERS/DUSTERS

Agro-chemicals are commonly used on plants to achieve one or more than one of the following objectives:

- ☆ Application of pesticides to control insects, pests and diseases
- ☆ Application of herbicides to remove weeds
- ☆ Application of fungicides to minimize fungus diseases
- ☆ Foliar application of micro-nutrients on the plants to overcome their deficiencies to enhance plant growth

☆ In modern agriculture chemicals are also used to defoliate or condition crops for mechanical harvesting, growth regulating hormones to increase fruit set or prevent early dropping of fruits, thinning of fruit blossoms and application of biological materials such as viruses and bacteria in sprays to control pests

The agro-chemicals can be applied either in liquid formulation or in solid forms. Whereas liquid forms are applied using sprayers, solid powder forms are applied using dusters.

Kinds of Sprayers/Dusters

Based on the kind of energy used and kinds of applications, sprayers/dusters are categorized in many groups. Some of these are the followings.

☆ Manually operated sprayers/dusters

☆ Power operated sprayers/dusters, which include high-pressure orchard and general purpose sprayers and air-blast sprayers which utilize an air stream as a carrier for sprays

☆ Boom-type field sprayers

☆ Air-craft sprayers/dusters

☆ Granular applicators

☆ Ground- rig dusters

Aerosol generators, which atomize liquids by thermal or mechanical means and are widely used for control of mosquitoes and other diseases transmitting vectors but have limited application for agricultural pest control

Manually operated sprayers and dusters: Suitable for small holdings operated in pressure range of 1-7 kg/cm^2.

Engine operated sprayers and dusters: The sprayer pump or the duster fan is driven by an engine and therefore, there is little variation in output, pressure and performance of the sprayer/duster. These power sprayers/dusters may also be operated by the PTO shaft of the tractor. These are suitable for large scale works with higher working pressure of 20-50 kg/cm^2.

Air-craft operated sprayers and dusters: It consists of gear pump or centrifugal pump to spray liquid through the nozzle in the pressure range of 3-8.5 kg/cm^2. The pump is driven by a wind driven propeller. Lightweight aluminium tank of capacity of 450- 2200 L is used to store the liquid depending upon the size of the plane. Lower capacity tanks are used in case of unmanned drone spraying. These are suitable for large scale commercial farming.

On the basis of volume of spray, the spraying is classified in four groups.

1. High volume spraying
2. Low volume spraying
3. Ultra low volume (ULV) spraying
4. Foam spraying

The range of volume of spray mix in each of the above cases is arbitrary. Following spray volume ranges are taken as a guide for field crop spraying (https://niphm.gov.in/Recruitments/ASO-PHE-Manual-NIPHM-03102013.pdf)

High volume spraying: Spraying @ 300 - 500 L/ha or on an average 400 L/ha

Low volume spraying: Spraying @5 - 300 L/ha or on an average 150 L/ha

Ultra low volume spraying: Spraying @ < 5 L/ha

Foam spraying: Special nozzles are used to perform foam spraying. Besides, a foaming agent (chemical additive) is added to the spray solution.

The selection of a particular kind depends upon the type of vegetation, kind of pests and approach to the field. There is distinct advantage in low volume over high volume application. The higher the volume applied more the time, more the labour and more the cost of application. However, the lower volume applications need concentrated solutions and uniform sprays.

SPRAYERS

A sprayer is a machine used to apply liquid chemicals on plants. The main functions of a sprayer are:

- ☆ Breaking the chemical solution into fine droplets of effective size
- ☆ Distributing the droplets uniformly over the plants
- ☆ Applying the chemicals with sufficient pressure so that it positively reaches the plants
- ☆ Regulating the amount of spray to minimize excessive application

The liquid formulations of pesticide are applied in small drops to the crop using different types of sprayers either in diluted (with water, oil) or undiluted forms. Normally, the emulsifiable concentrate (EC) formulations, and wettable powder formulations are suitably diluted with water, a common carriers of pesticides. In some cases, oil is also used as diluent or carrier of pesticides. The volume of liquid spray required besides other things depends upon the spray type and coverage, total target area, size of spray droplets and number of spray droplets. The required spray volume will be more for coarse spray droplets than for smaller size droplets. Moreover, the spray volume required will be more, if spray has to be made on both the sides of leaves than only on one side of the leaves.

Considering all these issues, a good sprayer should possess the following qualities

- ☆ It should produce a steady stream of spray material in desired droplet size so that the plants are treated uniformly
- ☆ Deliver the liquid at sufficient pressure so that the spray solution reaches whole of the foliage and uniformly spreads over the plant body
- ☆ It is light weight but strong enough to withstand rough handling
- ☆ It is easy to work with
- ☆ Easily repairable

Basic Components of a Sprayer

A sprayer essentially comprises various parts such as: Pump, Chemical tank, Agitator, Air chamber, Pressure gauge, Pressure regulator, Valves, Strainer, Suction line, Delivery line and Nozzle(s). It is a general list although some components may vary from one kind of sprayer to another. Some of these parts for a knapsack sprayer are shown in Figure 13.1.

Pump: A pump is a device is used to pressurize and move the fluids from one point to another. Most hydraulic sprayers are equipped with a positive displacement pump capable of developing pressure needed for the spraying operation. A simple piston-type pump consists of a plunger operated inside a cylinder, which displaces a volume of fluid by physical or mechanical action. The discharge capacity of the pump is approximately proportional to the speed of the pump. A pressure relief valve or by-pass valve is provided to protect the positive displacement pump from any damage that may occur, if the discharge line is closed for some reason or the other and for the convenience of the operator.

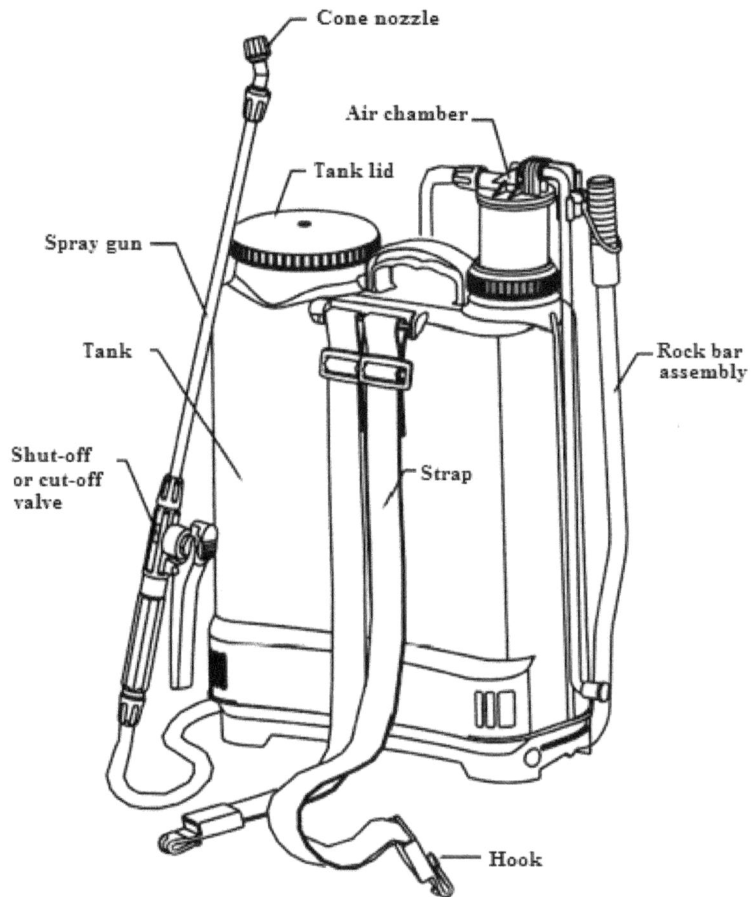

Figure 13.1. Components of a Knapsack Sprayer.

Tank: The container to hold the chemical solution is made up of metal sheet such as brass or steel or synthetic rubber or PVC having good resistance against corrosion and similar actions. The size of the tank varies according to the pump capacity and the volume requirements. An opening is provided at the top of the container to pour the solution.

Agitator: It is used to stir the chemical solution so that the solution contents remain in homogenous condition. Normally, the propeller or paddle type mechanical agitators or hydraulic agitators are used. The agitator is positioned at the bottom of the tank. Complete agitation of spray material in the tank allows the use of all kinds of spray materials such as powdery emulsions, fungicides, cold water paints or other spray material.

Air chamber: In a reciprocating type pump where discharge is made in one of the two strokes, an air chamber is provided on the discharge line of the pump to level out the pulsations of the pump.

Pressure gauge: The dial gauge on the discharge line indicates the pressure at which liquid is delivered from the pump. A properly calibrated pressure gauge guides the operator to make proper adjustment of the pressure.

Pressure regulator: The pressure regulator helps to adjust the pressure required for a particular spray job within the pressure range of the pump. Besides in a positive displacement pump, it also serves as a safety device in automatically unloading the excess pressure by directing the unused flow from the pump back to the tank.

Valves: Valves are used to regulate the flow of a fluid by opening, closing, or partially obstructing various passageways. Three types of valves are commonly used.

Cut-off valve is provided in the delivery line to control the flow of liquid from the sprayer. It is operated by hand.

By-pass valve is provided in the delivery line to by-pass the flow from pump to tank when flow in delivery line is less than the pump capacity.

Relief valve is an automatic device to regulate the pressure of fluid or gas within a predetermined pressure range.

Pressure regulator: It is an automatic device to control the pressure of fluid or gas within a range of settings.

Strainer: A strainer is used to filter any dust particles present in the chemical solution. It is a small circular plastic ring having nylon wire mesh. It is located in the suction line between the tank and the check valve. In some sprayers strainers are provided at the mouth of the chemical tank as in knapsack sprayers

Nozzle: It is the most important functional component, which breaks the fluid into fine droplets. Atomization of spray fluid is usually achieved by discharging the liquid under pressure through an orifice called nozzle. In some cases, a blast of air is used to atomize the jet of liquid. Kinds of nozzles have been discussed in a separate section of this chapter.

Spray gun: The delivery hose is connected to the spray gun. It is the most important part of the low pressure sprayers *e.g.* knapsack sprayers, rocker sprayer *etc.* A spray gun essentially consists of a cut-off valve, an extension rod-straight or goose-neck and an appropriate nozzle (Figure 13.2). The cut-off valve upon actuation opens or closes the system. When the cut-off valve is open, it allows the liquid to reach the nozzle assembly through a metallic or PVC extension rod. Spray from the spray gun is readily adjustable during the operation (Figure 13.2)

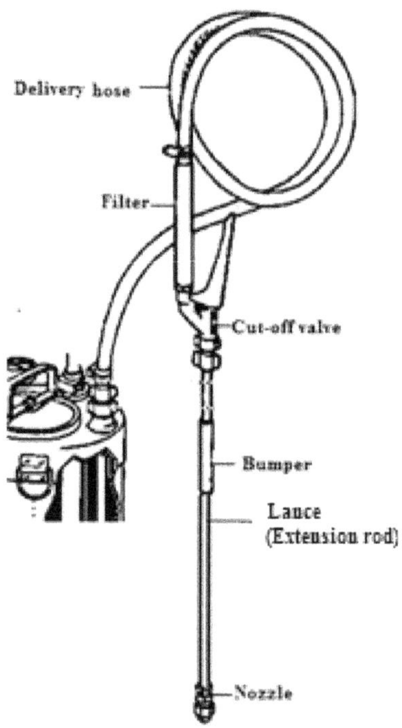

Figure 13.2. A View of a Spray Gun for a Hand Operated Sprayer.

Spray lance: Spray lance is a pipe (extension rod) through which liquid reaches the nozzle.

Spray boom: It is a long metallic or PVC pipe on which several nozzles are fitted. In fact, it is also a spray gun wherein spray nozzles are fitted on the pipe mounted at right angle to the lance. Ideally suited for row crops, can be used with foot/rocker/knapsack/power operated sprayers. The most useful application is on high power - high capacity sprayers such as tractor operated and/or power tiller operated sprayers. In such cases, booms are mounted on suitable structure. The delivery hose is connected to the spray boom. The area covered by spray boom is much larger compared to spray guns.

Over-flow pipe: It is a conduit pipe through which excess fluid in the delivery line is by-passed into the chemical tank by the action of a relief valve or pressure regulator.

The shoulder strap: It is adjustable strap to put the sprayer around the shoulders. Other ends are fastened to the tank with steel buckles. It should be wide enough to prevent any inconvenience or pain to the shoulder of the person using the sprayer.

Components of a Nozzle

Spraying nozzle is a device for emitting spray liquid, breaking it up into small droplets and throwing the droplets away from the nozzle orifice. Nozzles of different designs are used to produce appropriate spectrum of droplets sizes. Although parts may vary from one kind of nozzle to another, yet some important parts especially referring to disc nozzle are given as under.

Nozzle body: It is the main component which encloses all other components of a nozzle (Figure 13.3).

Swirl plate: It is a metal disc with two tangential holes which imparts a swirl or rotation to the liquid passing through it (not shown).

Nozzle disc: It is the component which breaks the fluid into fine droplets. It is a flat disc with an orifice at the center. As the spray solution from the swirl plate reaches the disc, the disc builds-up further pressure on the fluid. When the fluid passes out of the orifice, it breaks into fine droplets. The disc has a specific design to impart a specific type of discharge to the outgoing fluid. The popular nozzles are a) hollow cone b) solid cone c) fan or flat type (not shown). This issue is discussed in a separate section of this chapter.

Nozzle boss: It is a lug on the spray boom or the spray lance to which the nozzle body or cap is screwed.

Filter/Strainer: It is a small circular plastic ring with nylon wire mesh to filter any dust particle coming with the chemical solution (Figure 13.3).

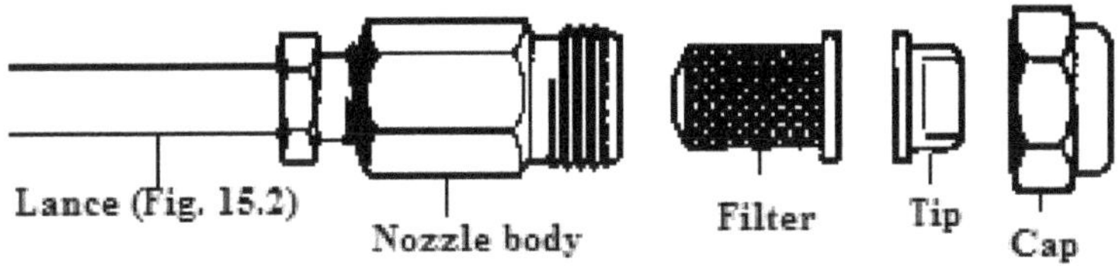

Figure 13.3. Major Components of a Nozzle.

Spacer: Two runners/plastic rings one placed in between nozzle and swirl plates and the other between swirl plate and strainer for effective travel of the solution (not shown).

Tip and cap: While the tip is the functional component, cap usually protects the tip (Figure 13.3). These often include items such as seals, washers and gaskets that provide a more secure fit between nozzle tip and cap units to reduce leaks. These items help the tip to perform at its optimum capacity.

The performance of a nozzle depends upon the following three characteristics.

Exit velocity: For a particular nozzle, it is proportional to the square root of pressure.

Discharge: It is proportional to orifice area and the exit velocity. Since the exit velocity is proportional to the square root of the pressure, discharge is also proportional to the square root of the pressure.

Operating pressure: Operating pressure determines both the velocity and discharge. The pressure also controls the droplet size, the higher the pressure the smaller the droplet size. Low pressures below $1.5 \, \text{kg}/\text{cm}^2$ are undesirable as nozzle fails to work satisfactorily at such pressures.

SPRAYER TYPES

Based on the kind of energy to break the liquid into droplets, the spray nozzles used in agriculture are classified as hydraulic energy nozzles, gaseous energy nozzles and centrifugal energy nozzles. Almost all sprayers for high volume spraying make use of hydraulic nozzles. The knapsack type low volume sprayers are generally fitted with air blast nozzle or gaseous energy nozzle. The hand held battery operated sprayers are fitted with a spinning disc type nozzle which works on centrifugal energy. Based on the kinds of nozzles used, sprayers are categorized as per the following list.

Manually Operated	*Power Operated*
Hydraulic energy	
Syringes, slide pump	High pressure sprayer (hand carried type)
Stirrup pumps	High pressure trolley/barrow mounted
Knapsack or shoulder-slung: Lever operated knapsack sprayer, Piston pump type, Diaphragm pump type	Tractor mounted/trailed sprayer
Compression sprayer: Hand compression sprayer, Conventional type, Pressure retaining type	High pressure knapsack sprayer
Stationary type: Foot operated sprayer, Rocker sprayer	Aircraft, aerial spraying (Fixed wing, helicopter)
Gaseous energy	
Hand held type	Knapsack, motorized type
	Hand/Stretcher carried type
	Tractor mounted
Centrifugal energy	
Hand held battery operated ULV sprayer	
Knapsack motorized type	
Tractor/vehicle mounted ULV sprayer	
Aircraft ULV sprayer	

Manually Operated Sprayers

These are common/utility types of sprayers. They either work on hand compression or foot compression. Manually operated hydraulic sprayers *viz.* hand atomizer, knapsack sprayers, twin knapsack sprayers, foot sprayers, hand compression sprayers; air carrier sprayers such as motorized knapsack mist blower-cum-duster and centrifugal rotary disc type sprayers are specially suitable for spray applications on crops.

Hand Atomizer

The hand sprayer is a small capacity pneumatic sprayer that comes in various types and models. Two main types are: the single-action and continuous-action atomizer.

Single action type is made of an external plunger pump, container and nozzle (Figure 13.4 left). The containers are made of tin, glass, brass, copper-bearing sheet steel or plastic. To operate, the spray solution is filled in the container. Held by both the hands, the piston pump is worked by sliding action. A dip-tube draws liquid from the tank due to hand actuation of the plunger. Hydraulic pressure so created forces spray solution through the nozzle. Since there is no pressure chamber, it does not retain pressure and therefore the operator has to pump continuously without any break. It is suitable only for small scale application in nursery or kitchen gardens *etc.* A common problem associated with this kind of sprayer is that it fails to ensure thorough coverage. Moreover, pressure fluctuation results in poor performance of the nozzle causing variations in discharge rate, changes in spray angle and fluctuations in spray droplets sizes.

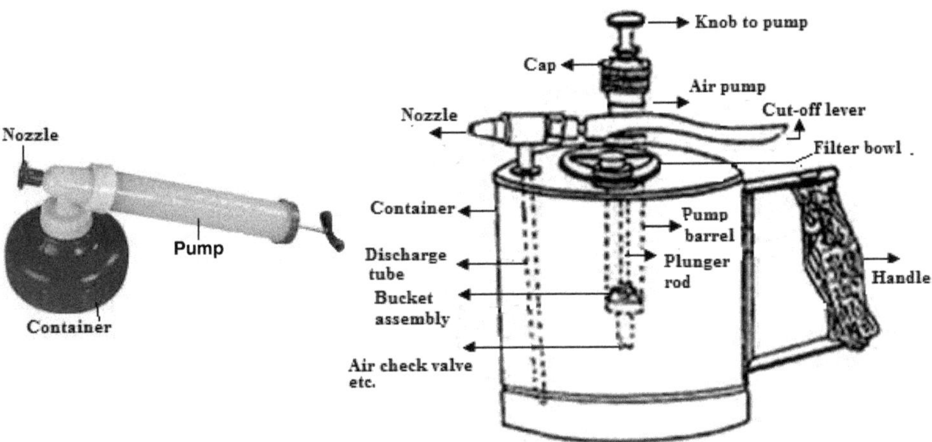

Figure 13.4. Construction of a Hand Atomizer: External Pump (Left) and Internal Pump (Right).

The continuous action type consists of liquid container of 0.5 to 3.5 L capacity made of tin, glass, brass, copper-bearing sheet steel and PVC. It has a built-in plunger pump, pressure gauge (optional), nozzle and flow cut-off lever (Figure 13.4

right). The tank is filled to approximately 3/4 of its capacity. Air is compressed in the remaining space by means of a knob which actuates the plunger pump. The pump builds-up a pressure of about 0.15-0.3 kg/cm^2, enough to force a continuous stream of spray for sometime before it is required to pressurize again. When the spring actuated cut-off lever is pressed, compressed air passes over the end of the tube of which the other end dips into the spray material to perform the spraying operation. The application rate ranges from 45 to 100 L/ha. It is ideally suited for home gardens and small fields. Chemicals with suspension characteristics are not effectively sprayed with this type of sprayer because there is no agitation mechanism.

Hand Compression Sprayer

It is similar to continuous action type hand atomizer but is carried on the shoulder of the operator. It consists of a tank of 10-12 L capacity for storing spray material, a vertical air pump, pressure gauge, filling port, spray lance, nozzle and a flow control lever (Figure 13.5). All the parts are made from brass alloy and the tank is fabricated to withstand high pressure up to the order of 18 kg/cm^2. The chemical tank is filled to 75-80 per cent of the tank volume. The pump is operated to pump air into the tank to build pressure up to 2.0-3.5 kg/cm^2. The fluid passes through the nozzle on pressing the flow cut-off lever. At this time the operator sprays the fluid on the crops. The tank has to be frequently pressurized to maintain proper atomization of the spray liquid. The application rate ranges from 45 to 100 L/ha. It is suitable for spraying chemicals on field crops and the lawns. Pressure retaining hand compression sprayers are now available that ensure that air charged once may last for weeks. Such pumps require sturdy tank to withstand high pressure. Therefore, these have not become popular.

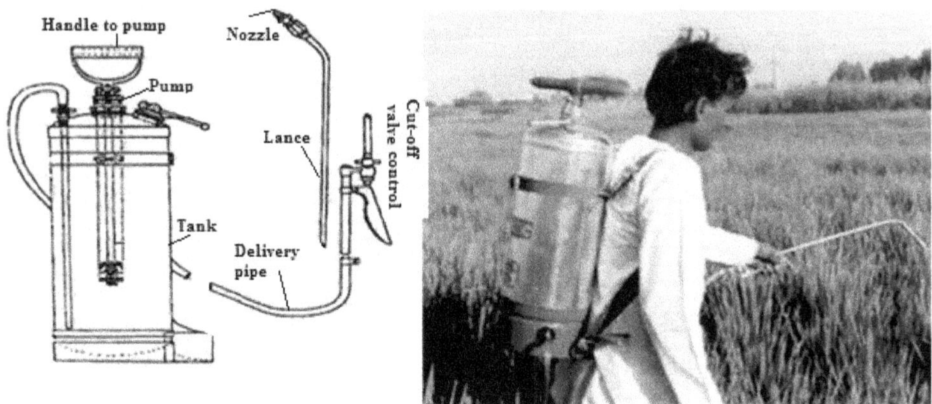

Figure 13.5. A Hand Compression Sprayer: Line Diagram (Left) and Pictorial View (Right).

Knapsack Sprayer

The spraying apparatus consists of a knapsack flat or bean-shaped tank of 9 to 22.5 L capacity. The tank is made of galvanized iron, brass, steel or PVC to ensure high resistance to corrosion from acid or bases (Figure 13.6). The tank is ergonomically designed imitating the curve of human back. It ensures close contact of the apparatus with the back so that it is comfortable to carry it to the field as well as during the spraying operation. A hydraulic pump and an air chamber are permanently installed inside the tank. Besides, it consists of a handle to operate the pump, agitator, filter, delivery hose, and has a spray lance having the nozzle and flow control lever (Figure 13.6). It has two straps for mounting the sprayer at the back of the operator. The handle of the pump extends over the shoulder or under the arm of the operator. Thus, it is possible to pump with one hand while spraying with the other. Uniform pressure can be maintained by continuous operation of the pump. When the pump is operated, it draws the fluid in the tank through the suction hole and delivers it to the spray gun. As soon as the cut-off lever is pressed, spray in the form of fine droplets comes out of the nozzle. The pressure varies from 3 to 12 kg/cm^2. It mostly depending upon the kind of pump used. However, a pressure of 3-4 kg/cm^2 can be maintained in most cases without much effort. At an application rate of 500 L/ha, the coverage is about 0.5-1.0 ha/day. The sprayer can be used for spraying row crops, vegetables and nursery stocks and 2.0-2.5 m high shrubs and trees.

Figure 13.6. A Photographic view of a Knapsack Sprayer.

The salient features of knapsack sprayers are given as under:

- ☆ Light in weight and so easy to carry on the back of the operator
- ☆ Useful to develop high pressure with less effort
- ☆ High work rate, easy and economical operation
- ☆ Robust and simple to maintain
- ☆ Both left and right-hand operation are possible

✫ A variety of easy to replace nozzles are available depending on the practical demand for optimum spraying results

Rocker Sprayer

The rocker sprayer is a long lever high-pressure sprayer designed for operation with one or two lances. The complete assembly is mounted on a wooden board, which is held to the ground by the foot of the operator. Major components of the sprayer are: pump assembly, platform with frame and fork, operating lever, pressure chamber, suction hose with strainer, delivery hose, extension rod with spray nozzles, *etc.* (Figure 13.7). Since there is no built-in chemical tank, a separate container is used to store the chemical. The pump is operated with a long lever in a rocking to and fro motion which sucks the liquid from the inlet pipe submerged in the spray liquid. During operation, the pump draws the fluid through the suction hose and delivers it to the delivery hose via the pressure chamber. The other person holds the lance and manages the spray through the flow control knob. He directs the spray chemical to the target. The sprayer is capable of building-up a high pressure of 14-18 kg/cm^2, which may even go up to 36 kg/cm^2. With a high jet spray gun or bamboo lance the chemical can be sprayed to a height of up to 10 m. Long hoses up to 30 m with either one or two outlets can be used. The output of the sprayer is 70-90 L/hr with one nozzle. Coverage is about 1.5 ha/day. The sprayer is popular in coconut and areca nut fields, flower gardens, cotton and tapioca areas.

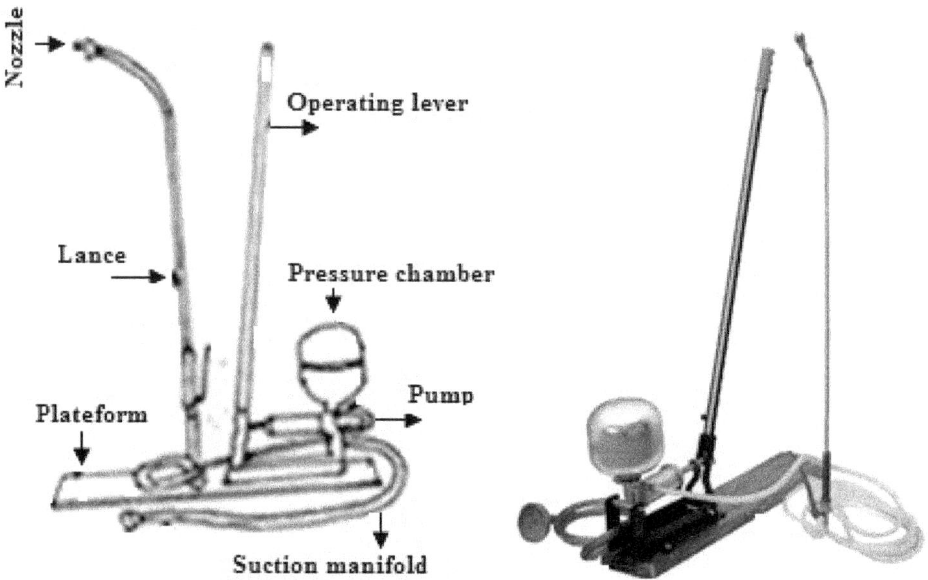

Figure 13.7. Line Diagram (Left) and Photographic view (Right) of a Rocker Sprayer.

Foot or Pedal Dprayer

The foot or pedal sprayer as the name suggests consists of a pump operated by the foot lever instead of hand (Figure 13.8 left and middle). The pump barrel is mounted on a steel frame that provides it stability when placed on the ground. The pump has a pedal attached to the plunger rod, which by its upward and downward movement operates the sprayer. It consists of plunger assembly, stand, suction hose, delivery hose, extension rod with a spray nozzle *etc.* It has a provision of two strong springs, which retract the foot lever to its original position after each pumping stroke. One end of the suction hose is fitted with a strainer and the other with a flexible coupling. Similarly, the delivery hose has one end fitted with a shut-off pistol valve and the other with a flexible coupling. A separate tank is used to store the chemical as there is no built-in tank in this sprayer. The plunger pump, a positive displacement pump, builds-up a high pressure of 17-21 kg/cm² and throws spray liquid to large distances with a suitable boom. A minimum of two-person team is required to perform the task of spraying. The principle of operation is similar to that of a rocker sprayer. The suction hose is kept in the chemical container. When the pump is operated by the foot, it draws the fluid through the suction hose and delivers it to the delivery pipe. Upon pressing the flow control lever, the fluid is released from the nozzle in the form of spray. Constant pedaling is required for continuous spray. The discharge rate with one nozzle is in the range of 110-135 L/hr covering about 1.0 ha/day. The foot sprayer, an all-purpose sprayer, is suitable for both small and large scale spraying on field crops, in orchards, vegetable gardens, tea and coffee plantations, rubber estates, flower crops and nurseries *etc.*

Figure 13.8. A Foot/Pedal Sprayer (Left and Middle) and a Stirrup Sprayer (Right).

Stirrup Sprayer

The stirrup pump sprayer is commonly used for mosquito control, but because of its simplicity, low cost and ease of operation, it is equally popular amongst the small farmers for spraying in orchards, nurseries, flower crops, vegetable gardens *etc.* The pump of the sprayer always remains submerged in a bucket containing the spray liquid. It is why; it is also named as bucket sprayer. It consists of a double action pump, seamless brass tube barrel, adjustable stirrup foot rest, angular delivery spout, relief valve, foot valve, plunger, shaft fitted with travel limiter, D type handle, and brass balls. The spray lance has a detachable gooseneck bend, a nozzle, and trigger cut- off valve with strainer (Figure 13.8 right). All the major parts are made of brass. Usually a flat fan spray nozzle is used with the sprayer. For operation, the barrel fitted with the pump is placed in a bucket containing spray liquid. One person operates the pump by placing his foot on the foot stirrup. The other person holds the spray lance and directs it towards the target. The field capacity of the sprayer is around 0.3 ha/day.

Power Sprayers

Most of these sprayers are hydraulic sprayers with a power unit to drive the pump. These are capable of developing high pressure and releasing high discharge to cover large areas. A power sprayer essentially consists of: (i) Prime mover and pump (ii) Tank (iii) Agitator (iv) Air chamber (v) Pressure gauge (vi) Pressure regulator (vii) Strainer (vii) Boom (ix) Nozzles. Most of the parts are similar to what has been discussed in the manually operated sprayers. The other parts are described as follows:

Prime mover and pump: Prime mover, usually a combustion engine, supplies power to the power sprayer. Horsepower of these units generally varies from 1 to 5 HP. The portable sprayers use a petrol engine so that the equipment can be easily taken to the spray sites being lightweight. The pump units employs mostly piston or plunger pumps but in several types even centrifugal pumps are used. In that case an air chamber is not required.

Tank: PVC or steel tanks of relatively large capacity are widely used to prevent corrosion. An opening with a cap, and fitted with a removable strainer is provided for filling, inspection and cleaning. A drain plug is provided at the bottom to drain the liquid, whenever required and for cleaning.

Agitator: Propeller or paddle type mechanical agitators are used for agitation. Normally, a horizontal shaft with flat blades attached to it is used to agitate the liquid in the tank. It rotates at about 100 to 120 RPM. Tip speed of the paddle should not exceed 2.5 m/s as it may cause foaming.

Air chamber: An air chamber is provided on the discharge line of the pump to level out the pulsations of the reciprocating pumps thereby providing a constant nozzle pressure.

Boom: Field sprayer mounted on a tractor has a long boom on which nozzles are fixed at specified spacing. The boom can be adjusted vertically to suit the height of plants in the fields.

Nozzle: It is used to break the liquid into the desired spray and deliver it to the plants. The flow rate of a particular nozzle is proportional to the square root of the pressure and the orifice area of the nozzle.

The complete assembly is mounted on the stretcher type frame or on wheelbarrow for easy transportation. The number of lances may vary from 1 to 6 depending upon the model. In some models there is a built-in storage tank of fiber glass having a capacity of 100 L, while in others a separate storage tank is required. To operate the machine, the engine/electric motor is started to actuate the pump. The shut-off trigger valve of the lance at this time is closed. The pump draws the spray liquid from the tank, imparts pressure energy and sends it to the delivery line/lines. The operator operates the trigger/shut-off valve and directs the lance towards the target. Spray pattern is adjusted by adjusting the nozzle or selecting an appropriate nozzle. A bamboo lance can also be used to deliver the spray liquid to large distances/heights. These sprayers are suitable for spraying in orchards even with tall trees up to a height of 15 m, tea and coffee plantations, rubber plantations, vineyards and field crops.

Motorized Knapsack Sprayer

Motorized knapsack sprayer is one of the most versatile and simple amongst the power operated machines. It is lightweight (12-20 kg including accessories), is powered by 1.2-3.0 HP petrol engine with a shock-proof cushioned frame, which comfortably sits on the back of the operator. It is capable of delivering 6.8 to 42.5 m^3 of air per minute at a velocity of 200-420 km/hr at the nozzle. The tank, having the capacity of 10-12 L, is mostly made of high density polyethylene (Figure 13.9 Top Left and Right). Another tank of 10-15 L capacity is provided for the fuel. Some manufacturers also provide diffuser and deflector accessories with the delivery hose for adjusting the swath as per requirement. To operate the machine, the tank is filled with the spray liquid leaving about half a liter space for the air cushion. When the engine is operated, a part of the air generated by the blower is directed into the tank to form an air cushion over the liquid within the tank. It is why the tank is not filled to its full capacity. Liquid from the tank passes through a tube to the nozzle on the spray lance by gravity, partly helped by the air pressure exerted over the liquid within the tank. The spray liquid flows out of the nozzle by means of an air current generated in the machine. The machine fitted with a rotary pump can spray on trees about 8 m high. The discharge rate varies in different makes, but can be adjusted. Effective width is 7-8 m horizontally and 5-6 m vertically. It is suitable to cover large areas and orchards with tall trees.

Motorized Knapsack Mist Blower-cum-Duster

The motorized knapsack mist blower has a small 2-stroke petrol/kerosene engine to which a centrifugal fan is vertically mounted. The spray tank made of plastic has also a compartment for fuel. The engine-fan unit is mounted on a common frame, which fits well at the back of the operator. The flow of spray liquid is due to gravity aided by suction created at the tip of the nozzle. The fan produces a high velocity air stream, which is diverted through a 90-degree elbow to a flexible (plastic) discharge hose having a divergent outlet. For operation, the tank is filled with the spray liquid. As the engine is operated, the fan produces a high velocity air stream. The control valve for the spray liquid is opened gradually to adjust the flow rate. The operator directs the discharge hose towards the target. The air stream shears the spray liquid producing mist as it comes in contact with the atmosphere. It is used for spraying tall crops, orchard trees and coffee gardens. The sprayer can be converted into a duster by changing few parts. The accessories for this purpose are supplied by the manufacturer. The conversion is achieved through the following actions:

☆ Replace the liquid delivery hose by a higher diameter pleated hose that can easily carry the powdery chemical from the tank into the air stream

☆ Provide a air distributor at the bottom of the tank for stirring and keeping the chemical in suspended form

Figure 13.9. A Motorized Knapsack Sprayer (Top Left), Major Parts of the Sprayer (Top Right) and a Tree/Blower Sprayer (Bottom).

Tree Sprayer or Blower Sprayer

These sprayers are tractor mounted or trailed carrier types. A tractor mounted sprayer can be attached to 3-point linkage of the tractor. It utilizes PTO power to run the pump. The trailer type has a 4-stroke petrol/kerosene engine to drive the centrifugal fan to produce an air stream of high volume and velocity. A micronized nozzle is used to produce uniform and fine droplets of spray liquid in the range of 150-200 microns. Besides, it has a plastic tank for storage of spray liquid, rotary pump to draw the spray liquid from the tank to feed it to the nozzle and a fiber glass casing. All these components are mounted on the stretcher type of frame. The sprayer can be carried by two persons to the place of spraying or on the tractor as the case may be. For operation, the tank is filled with spray liquid and the PTO or engine is started to drive the fan and the rotary pump (Figure 13.9 Bottom). The flow rate is adjusted by slowly opening the control valve. The trailed carrier type sprayer is placed under the tree and manually moved around to complete the spraying operation.

Aero Blast Sprayer

The machine consists of a spray tank of about 400 L capacity, pump, fan, control valve, filling unit, spout, adjustable handle and spraying nozzles to release the pesticide solution into a stream of air blast produced by a centrifugal blower. The air blast distributes the chemical in very fine droplets throughout the whole swath on one side of the tractor (Figure 13.10). The major portion of swath is through the main spout while the auxiliary nozzles cover the swath area near the tractor (Figure 13.10). The sprayer is mounted on the tractor 3-point linkage and is operated by tractor PTO. The orientation of the air outlet can be adjusted for direction and width of coverage. It is commonly used to spray horticultural trees and crops like cotton, sunflower *etc.*

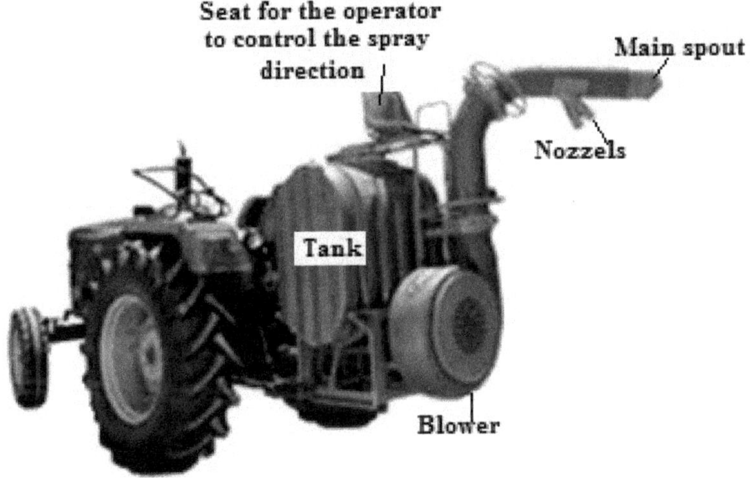

Figure 13.10. An Aero Blast Sprayer.

Tractor Drawn Boom Sprayer

The sprayer essentially consists of a tank made of fiber glass or plastic, pump assembly, suction pipe with strainer, pressure gauge, pressure regulator, air chamber, delivery pipe, and spray boom fitted with nozzles (Figure 13.11). It is categorized under the hydraulic energy sprayers. The complete sprayer is mounted on 3-point linkages of the tractor. The PTO power of the tractor is used to operate the pump of the sprayer. The spray boom can be arranged as overhead spray boom or ground spray boom. The overhead spray boom is used to spray tall crops. In this case, planting is done in such a way that for each of about 18-20 m wide planted strip, an unplanted strip of about 2.5 m width is left for the movement of tractor. For ground spray boom the planting is done in rows keeping in view the track width of the tractor. Ground spray boom is used when the crop height is still small.

Figure 13.11. A View of the Tractor Drawn Boom Sprayer.

Battery Operated Ultra Low Volume (ULV) Sprayer

It is a hand-held sprayer with a spinning/rotating disc having grooves or teeth designed for ultra-low volume (ULV) and controlled droplet application of insecticides, fungicides, pesticides, herbicides and other liquid formulations. The sprayer powered by a motor of 6 to 12 volt battery rotates the spinning disc at 4000-9000 revolutions per minute. The spinning disc receives the concentrated chemical from a plastic container having a capacity of about 1 L. Average droplet size varies between 35-100 microns. The handle of this sprayer has provision to adjust the angle of spray head. For operation, the spray liquid is filled in the container. It flows in the form of drops at the center of spinning disc. The disc, attached to the motor, rotates at a very high speed. The spray liquid travels in the grooves of the disc and gets fragmented into very fine droplets while leaving at the periphery of the disc. The droplets are thrown outwards due to centrifugal force. A cut-off valve is provided in the hose line to stop the flow of spray as and when desired. ULV formulations are applied at only 2.5-7.5 L/ha. One hectare of crop can be treated in around 2.5 hr. ULV sprayer is used to spray chemicals on row crops like cotton, cow pea, groundnut, tobacco and vegetables. It is ideally suited for home gardens.

Self-Propelled Light Weight Boom Sprayer

The machine consists of a lightweight power unit and a spraying unit. It is normally operated by a 5 HP diesel engine. A handle is provided to control the machine. The spray pump and two narrow pneumatic wheels get power from the engine through gears, chains and sprockets. The whole system is supported on a frame. Sometimes, a caster wheel is provided at the rear, which acts as a supporting wheel. The spray boom is mounted on the power unit through a canopy frame. The machine is suitable for chemical applications on wheat, vegetable and other crops

Self-Propelled High Clearance Sprayer

The machine designed at PAU, Ludhiana features a chassis with 1200-mm ground clearance, four wheels, and 20 HP diesel engine, gearbox, tank, seat for the operator, spray pump and boom having 18 nozzles with a boom width of 10.80 m. It has four forward and 1 backward speeds. The boom height is adjustable from 31.5 cm to 168.5 cm to suit different crops and can be folded during transport. The front two wheels are narrow in width (20 cm) and are driven wheels, while the rear wheels are steering wheels. Fenders are provided in front of the drive wheels to deflect the crop branches away from the wheels to minimize mechanical damage. The wheel track is 135 cm. Two rows of cotton crop come under the machine chassis during its operation. It is most suited to spraying tall crops like cotton, sunflower, wheat and similar crops.

Fogging Machine

When droplets smaller than 15 μm fill a volume of air to such an extent that visibility is reduced, it is termed as fog. Usually, thermal fogging machines are used to create fog in which pesticide dissolved in oil of suitable flash point is injected in the hot gases produced by burning the fuel. The liquid is vapourized, which condenses on coming in contact with air. A fogging machine consists of fuel tank/reservoir, formulation tank, pump, fogging nozzle, fogging coil, water pump and other controls. These machines are used for spraying in glass and green houses. For field crops, fogging operation is carried out early in the morning so that fog created remains close to ground. The fogging machines are normally used to control flying insects.

Air-Craft Spraying

Conventional methods of pesticide spray application are known to create many problems such as:

- ☆ Excessive application of chemicals
- ☆ Low spray uniformity, deposition, and coverage
- ☆ Higher cost of pesticide application
- ☆ Water and soil pollution
- ☆ Field conditions may prevent access to wheeled tractors that delays the spraying operation and may cause soil compaction
- ☆ Difficult to cover very large areas
- ☆ The process is time-consuming and is quite drudgery and results in reduced effectiveness in controlling the pests and diseases

To overcome these problems, aerial spraying came into operation being faster and avoiding travelling inside the fields. In this system, a specific pesticide is carried in the aircraft that follows a pre-mapped route to spray crops. Earlier, the aircraft employed for spraying were manned, resulting in severe harm to the pilots in case of any mishap. An alternative to reduce the risk of fatal accidents is to use unmanned aircraft or drone mounted sprayers, which improves coverage, boosts chemical effectiveness and makes spraying job easier and faster without undue risks. Reports have emerged to show that pesticide spraying with drones is speedier, results in even distribution of spray, gives better efficiency, reduces the pesticide use by 30 per cent, saves the farmers from inhalation and drudgery and has automatic controls to cruise back to the initial position once the pesticides is consumed. Some disadvantages associated with aircraft spraying are as follows:

☆ High wind speed and temperature inversion may limit treatment application

☆ Some fields may not be amenable to treatment such as those in the vicinity of tall trees, waterways, and overhead power lines

☆ Precise application in dense crop canopies is difficult

☆ Volatility and spray drift can be a problem (Figure 13.12)

☆ Significant environmental contamination can occur, if spraying operation is incorrectly executed

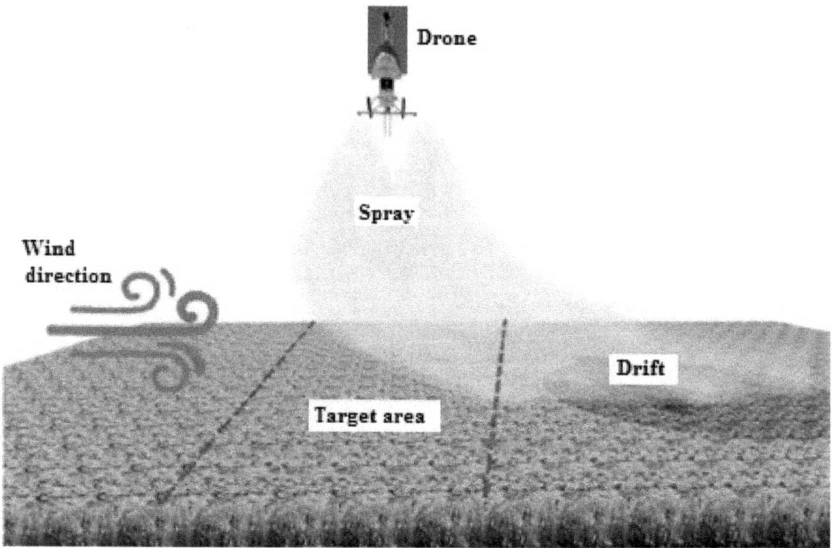

Figure 13.12. Application of a Drone for Crop Spraying and likely Pattern of Spray Drift.

It may not be out of place to mention that drone-spray has been illegal so far. Insecticides Act 1968, mandates that approval/permission of the Central Insecticides Board (CIB) must be sought before undertaking any aerial application of pesticides. No permission/approval so far has been granted by the CIB in the past for the use of drones to spray pesticides.

TYPES OF NOZZLES

Various nozzles types are used as per their suitability to accomplish different tasks. According to their construction nozzles are categorized as: disk, regular vermorel, modified vermorel, Bordeaux, cap, solid stream and self-cleaner types

(Goodwin, 1912). The disk and Bordeaux nozzles have larger capacity than the vermorel and self-cleaner types. On the basis of shapes or patterns of spray, nozzles are further categorized in 5 classes.

Types Based on Construction

Disc nozzle: The disc nozzle is larger in capacity than most of the vermorel or self-cleaner types. It is compact, lightweight, less liable to clogging, and do not have any projecting parts that may cause trouble while moving around the equipment. The type of spray, whether fine or coarse, can be regulated to some extent by size, number, and angle of the holes in the disk.

Vermorel nozzle: Capacity of the vermorel nozzle is normally small. It is largely because the liquid must pass through small orifices and abruptly change direction that reduces its speed.

Modified vermorel nozzle: In this nozzle, liquid enters the whirl chamber through a hole at one side of the chamber, which gives it the whirling motion around the needle of the degorger. Packing under the nut prevents any leakage around the needle stem and a spring holds the needle back leaving the way to the hole in the cap unobstructed. The angle of the head of the nozzle can be set by turning it around on the threaded stem.

Bordeaux nozzle: It is commonly advocated and used by western orchardists in the USA. It makes a flat fan-shaped spray, which is coarse and much heavier in the center of the fan than at the edges. These are also of large capacity and can be adjusted to throw a solid stream of liquid.

Cap nozzle: This nozzle is of small capacity, a miniature type of disc nozzle, and is suitable for bucket pumps and small hand sprayers. Large capacity nozzles are preferable to the vermorels.

Solid stream nozzle: It is best suited for spraying tall trees. Because of their extremely large capacity, it is not possible to use them with small power machines.

Self-cleaner nozzle: A self-cleaner nozzle has narrow passage-ways. A needle cleans the hole in the cap when the head of the nozzle is pushed back.

Types Based on Shape of the Spray

Spray nozzles on the basis of the shapes of spray pattern are classified as: solid stream, flat fan type, hollow cone, solid cone and adjustable nozzles. Such classification is at best arbitrary because many variations or intermediate forms of sprays are given by some nozzles.

Solid stream: It sends a simple jet of focused fluid that has no true droplets. A solid stream is formed by forcing the fluid through a shaped orifice that forms the spray into a jet (Figure 13.13).

Flat fan nozzle: They are also called fan nozzles. The spray liquid is thrown from an elliptical orifice to give a flat shaped sheet of spray (Figure 13.13). These are used for band spraying. These nozzles are generally used on booms. Proper distance is maintained in between the two nozzles to ensure some overlapping to give even distribution. The normal working pressure is about 2.8 kg/cm². However, these fan nozzles can also be used for herbicide application at low pressure of around 1.5 kg/cm² to avoid drift of the fine droplets.

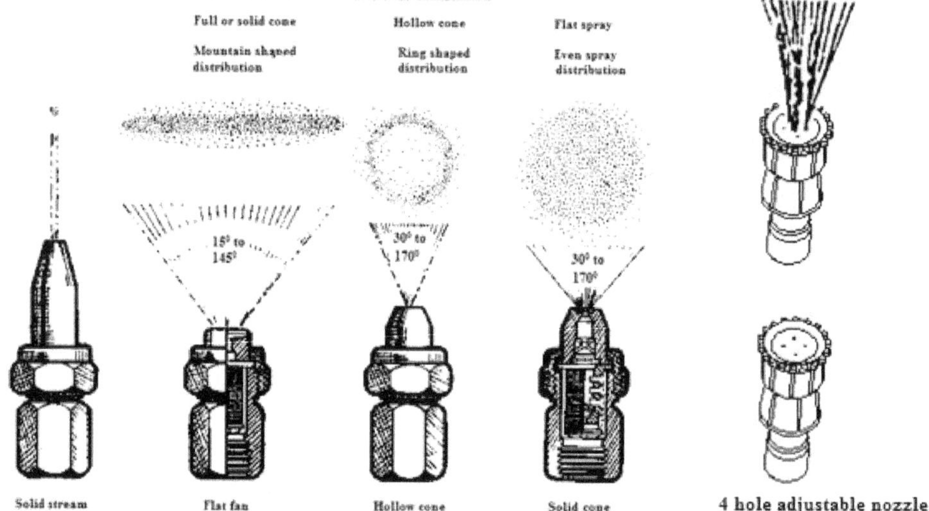

Figure 13.13. Various kinds of Nozzles used in Agriculture.

Hollow cone nozzle: It is quite popular hydraulic nozzle for spraying insecticides and fungicide. It produces a hollow cone pattern of spray consisting of a mixture of different sized droplets (Figure 13.13). The nozzles of different designs are made from brass, stainless steel and plastic materials. The simplest design is made of brass metal having an orifice hole drilled in it and a rotral with tangential cut grooves providing swirl motion to spray liquid. It breaks down into droplets while emerging from the nozzle under pressure. This kind of simple brass nozzle can be screwed onto a hand lance/boom. The other designs of nozzles consist of a stainless steel disc having a central circular hole through which spray emerges from a swirl chamber behind it. The disc and the swirl plate are suitably fitted in the body of the nozzle. The body of the nozzle has threads for screwing (fitting) it to the lance/boom. The normal working pressure of a hollow cone nozzle is about 2.8 kg/cm².

Solid cone nozzle: These nozzles are also known as deflector nozzles or flood jet nozzles. These nozzles are commonly used to spray herbicides. Having low pressure (≈1.5 kg/cm²), the spray essentially consists of coarse droplets but covers the entire area over a small range (Figure 13.13). The construction is more or less similar to a hollow cone nozzle except for the addition of an internal jet, which strikes the rotating liquid just within the orifice of discharge. The breaking of the drops is mainly due to impact. In these nozzles, spray liquid emerging from a circular hole strikes an inclined smooth face and is deflected at an angle. Thus, the liquid spreads as a sheet in a wide angled fan pattern.

Adjustable nozzle: These are also called triple action nozzles because of varying patterns of sprays that can be obtained by manipulating the swirl velocity of spray liquid in the eddy chamber. The hollow cone spray pattern consisting of fine spray particles, or a jet spray for orchard/tree spraying and a medium coarse spray pattern can be obtained by simple adjustments (Figure 13.13). Adjustable nozzles are generally used with foot operated sprayers, rocker sprayers or high pressure hydraulic sprayers for spraying trees.

Droplet Size

The size of droplet is important as it affects drift and penetration distance of droplets towards the target. Fine droplets are required to control insects, pests or diseases and bigger size droplets for application of herbicides, *etc.* The higher the number of fine droplets produced by a device, the better is the deposition on the target area. A compromise is usually made to prevent drift, achieve wide coverage of plant or target area and more penetration. The optimum droplet sizes for various target groups are indicated in Table 13.1.

Table 13.1. Optimum Droplet Sizes for different Target Groups

Target Group	Droplet Size (microns)
Flying insects (drift)	10-15
Crawling and sucking insects (drift)	30-50
Plant surfaces (limited drift)	60-150
Soil applications (no drift) as in case of herbicide application	250-500

CALIBRATION OF SPRAYER

The rate of application of pesticide should be adequate and uniform over the whole area. Too much application or too less application of pesticide is undesirable. Too much application results in wastage, crop injury, and may entail additional costs. On the other hand, too less application results in poor pest control, wastage of pesticide, time and money.

The pesticide application by any sprayer is regulated by 3 factors namely nozzle discharge rate, swath width and walking speed of operator. The flow rate of the nozzle is determined by the following formula.

$$F = S \times D \times A / 10000 \tag{13.1}$$

Here F is flow rate in L/min (This represents flow rate from all the nozzles of the sprayer, if more than one nozzle is used), S is the swath width, m, D is operator's walking speed or tractor speed, m/min and A is the application rate in L/ha. Note that if any of the 3 values are known the fourth one can be calculated. Equipment manufacturers normally provide tables that give the capacity of their equipment. More often, it is difficult to rely on such tables. It is because the pump and the nozzle may wear out as the sprayer gets old. In that event, besides the performance of the sprayer, rate of application may also change. The frequent calibration of the sprayer therefore, is essential to make sure that the pesticide is applied in adequate amounts.

There are several ways to calibrate the sprayer. The sprayer can be calibrated theoretically with the help of eq. (13.1) and practically in the field. It is desirable to frequently verify the theoretical calibration with field practical calibration. For the practical field calibration a small area is demarcated, say 100 m² (10 x 10 m). The sprayer is filled with a known volume of water, say 5 L. Then the operator sprays this area as he will do in the actual field ensuring uniform and even distribution of spray. Afterwards the quantity of water still remaining in the sprayer is measured by a jar, say it is 2 L. The quantity of water sprayed in such a case is 3 L on an area of 100 m². It means to cover a unit area (one hectare or 10,000 m²); 300 L of formulation will be needed. For this case, the application rate is 300 L/ha. We can also use the following formula (http://www.ikisan.com/sprayers.html)

Application rate, L/ha = (1 hectare/Marked area) x

$$\text{Spray volume used for marked area} \qquad (13.2)$$

For the illustrated example, we have

Application rate, L/ha = (10000/100) x 3 = 300 L/hr

Example 15.1

A knapsack sprayer discharges 600 mL spray solution every minute to cover one meter swath. If the operator is walking at a speed of 30 m/min, what is the rate of application in L/ha?

F = 0.6 L/min; S = 1 m, D = 30 m/min

Using eq. (13.1), we have

F = S x D x A/10000

A = F x 10000/(S x D)

A = 0.6 x 10000/(1 x 30) = 200 L/ha

Example 15.2

A battery operated ULV sprayer has to spray at the rate of 10 L/ha. What should be the flow rate of the sprayer, if the operator walks at a speed of 1 m/s and the swath width is 3 m?

F = ? L/min, S = 3 m, D = 60 m/min (1 m/s = 60 m/min), A = 10 L/ha

Using eq. (13.1), we have

F = 3 x 60 x 10/10000 = 0.18 L/min or 180 mL/min

PRECAUTIONS IN SPRAYING

Spray materials are highly poisonous. These may prove to be highly dangerous to the operators or any other person whosoever unintentionally inhales the spray. Although, little amount of toxic material taken into the body may not produce ill-effects at the first instance but as the spraying proceeds, the cumulative material on and in the body may suddenly give rise to serious poisoning. It requires great care and handling skill to avoid any mishap during the whole operation of spraying. Following safety precautions before, during and after the spraying should be observed to avoid/minimize the ill-effects.

Before the Operation

- ☆ Identify the pest and ascertain the damage done. Undertake the spraying operation only if the crop damage has exceeded the economical injury level
- ☆ Use only the recommended least toxic pesticides
- ☆ Carefully read instruction manual of the equipment and label on the pesticide
- ☆ Check the spraying equipment and accessories to be used. All components especially strainers, sprayer tank, cut-off device and nozzles should be clean and operational
- ☆ Check parts such as 'O' ring, seal and gasket, worn out nozzle tips, hose clamps and valves. Replace the worn out parts
- ☆ Test the sprayer to ascertain whether it gives the required liquid output at rated pressure. Nozzle spray pattern and discharge rate should also be checked
- ☆ Calibrate the sprayer. Set spraying speed and nozzle swath by adjusting spray height and nozzle spacing
- ☆ Make sure that appropriate protective clothing and gadgets are available for use. These should include coveralls, protective boots, gloves, glasses/goggles, hat and a respirator.
- ☆ Ensure that soap, towel and plenty of water is available at the site
- ☆ Never leave pesticides unattended in the field

During Dilutions

The pesticide must be handled very carefully being quite poisonous. Any dilution exercise should be carried out observing the following precautions.

- ☆ Carry out the dilution exercise in the open
- ☆ Exercise enough care to avoid inhaling the fumes
- ☆ Put on appropriate protective clothing and gloves
- ☆ Mix water and concentrate in a large clean container. The container and any measuring cup used for this purpose should not be used for any other purpose. It is good to mark them with an appropriate label such as 'DANGER - POISON: DO NOT TOUCH'. They should be carefully washed and safely stored in the equipment shed for future use
- ☆ Put a small amount of water into the bucket first. Put the required amount of pesticide into the water. Rinse the measuring cup with clean water and add this solution also to the bucket. Stir to thoroughly mix the pesticide into the water. Pour this solution into the sprayer tank and then add the rest of the water to the tank
- ☆ Stir the solution carefully with a plastic, aluminium or steel stirrer avoiding any splashing. Never leave the paddles unattended after use as they can

be dangerous to small children and animals. They should also be carefully washed and stored for future use along with the containers

☆ Avoid using wooden paddles as these can soak up the pesticide and must be carefully disposed immediately after use. It is best to bury them along with the empty pesticide containers

☆ Pesticide solution enough to fill the sprayer should be mixed at any one time

During the Operation

☆ In case of any doubt, recheck the pesticide use instructions and the equipment

☆ Do not transfer pesticides from original container and/or packing into another container

☆ Make sure pesticides are diluted/mixed in correct quantities. The formulation should be poured carefully to avoid splashing

☆ Wear appropriate clothing to avoid skin contamination

☆ Protect eyes and mouth from exposure to the spray. As an example, clogged nozzles or hoses should never be blown out with mouth

☆ Do not spray under the conditions of high wind, high temperature and rain. Suspend spraying whenever windy weather is encountered

☆ Select proper direction of spraying to avoid drift. Always proceed upwind so that operator moves into uncontaminated area. Hold the nozzle and boom at proper height. Operate the sprayer at correct speed and correct pressure

☆ Never spray if the wind is blowing towards grazing livestock or pastures in regular use

☆ Never eat, drink or smoke when mixing or applying pesticides. Follow correct spray techniques

☆ Never allow children or other unauthorized persons to stand nearby during mixing/spraying

After the Operation

☆ Never leave unused pesticides in the sprayer. Leftover pesticides should be disposed of in pits on some wasteland nearby

☆ Do not empty the spray container into irrigation canals or ponds

☆ Empty pesticide containers should be crushed and buried. Do not use them for any other purpose

☆ Clean buckets, sticks, measuring jars, *etc.* used in preparing the spray solution

☆ Remove and wash protective clothing and footwear. Wash yourself well and put on clean clothes

☆ Keep an accurate record of pesticide use

☆ Ensure that humans or animal do not enter the treated areas until safe to do so. Identify the sprayed plots using some kind of flag or similar other indication

☆ Properly clean the sprayer at the end. After use, oil it and then safely keep away in store room

DUSTER

Many insects and plant diseases can be controlled more effectively and economically with dusts rather than with liquid sprays. Cotton, potatoes and other field crops, as well as garden and orchard crops are commonly dusted with various kinds of insecticides and fungicides. *The dusting powders are low concentration ready to use type, dry formulations containing 2 to 10 per cent pesticide.* The non-toxic inert material or dry diluents added to the dry formulations are talc, soapstone (made from talc-schist, a type of metamorphic rock) and attapulgite, a naturally mined clay *etc.* Note that sulphur dust is not diluted with inert material. The major advantages of pesticide dusting are:

☆ Ready to use products with no need for further dilution

☆ Reduced risks of handling concentrate

☆ Useful in areas where water is scarce especially in dry land agriculture

Major disadvantages are:

☆ Serious drift problems may be encountered because of fine dust particles

☆ The operator and field labour is exposed to dermal (skin) and inhalation hazards

☆ Pesticides may be carried to neighbouring field/area and in the air causing pollution

☆ Precise metering and even distribution of dusting powders in field conditions is difficult and may prove to be a limitation

Dusters are used for dusting pesticides. An air stream through the dusters carries the pesticides on to the plants. A duster essentially consists of hopper, agitator, feed control, fan or blower and delivery nozzle.

Hopper: Like a tank in a sprayer, hopper in a duster stores the pesticide.

Agitators: Agitators prevent dust preparations from caking and packing in the hopper. Most agitators have projections on horizontally mounted shafts, which revolve and keep the dust broken up.

Fan/blower: A centrifugal type of fan is commonly used in dusting machines. Fans operate in 2,500 to 5,000 RPM range. The high velocity of the air aids in breaking the dust into a fine fog. Fans usually have one discharge opening but may have openings for each hose.

Feeds: The dust is uniformly fed either directly into the fan or into the air stream coming from the fan. Brushes or curved blades with or without leather strips revolve over the opening. The best results are obtained when dust is broken-up into a fine fog-like dust. If small pellet-like particles are blown out, they do not adhere to the plants and so is the very fine dust, which may drift away from the target even under light wind conditions.

Delivery nozzle: It delivers the pesticide to the target point.

TYPES OF DUSTERS

Dusters, like sprayers, are also categorized as manually operated and power operated. Some manually operated dusters are: bulb type, plunger type, bellow type, and crank rotary type. In the power operated type we have: traction duster, motorized knapsack sprayer-cum-duster, power take off duster, and wet duster. Besides, aerial dusters are also used to cover large areas.

Manually Operated Dusters

Bulb Type Duster

A bulb duster consists of a rubber bulb filled with pesticide dust. When the bulb is hand squeezed, the pesticide gets dusted at the target spot through the delivery nozzle (Figure 13.14 Top).

Bulb duster

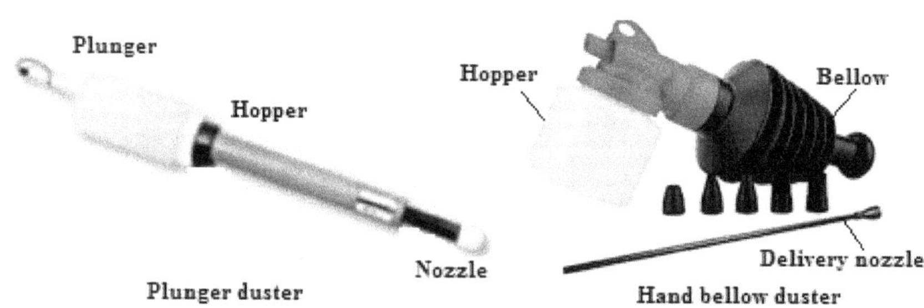

Figure 13.14. Manually Operated Bulb, Plunger and Bellow Type Dusters.

Plunger Type Duster

It consists of a chamber for the dust, an outlet, a cylinder with piston and handle. By moving the piston back and forward in the cylinder, a current of air flows over the dust in the hopper. The dust is carried away through a delivery spout (Figure 13.14 Bottom Left). Small hand pump dusters of this type are suitable for small areas like vegetable gardens.

Bellow Type Duster

It is also a low cost dusting machine having a simple design. On squeezing, the collapsible bellows puffs and pushes the air into the dust hopper that expels the dust from the outlet in the form of a small cloud (Figure 13.14 Bottom Right). It has a capacity of 2.5 to 5.0 kg and is used for spot treatments.

Knapsack Type Duster

Knapsack dusters have a hopper through which a current of air is blown to pick up the dust. Lever operated leather bellows produce the air current required for dusting. Shoulder straps are used to carry the duster on the back of the operator. These dusters are also suitable for small areas.

Rotary Duster

The hand rotary duster, a commonly used duster by the farmers, may be hand carried type or shoulder mounted type. In the latter case, it is available in two models, shoulder mounted (Figure 13.15 left) and belly mounted (Figure 13.15 right). It consists of a hopper, a fan, gear box, handle, delivery hose and a deflector plate. The hopper is made of plastic, aluminium or mild steel sheet. Hoppers made of mild steel sheets are coated with anti-corrosive material for longer life. The duster has a mechanical agitator, which is connected to the gearbox. It churns the chemical and prevents the clogging of the outlet. The adjustable orifice plate mounted below the hopper outlet helps to control the application rate. For operation, the hopper is filled to 1/2 to 3/4 of its capacity. It is mounted on the shoulder/belly with the help of adjustable straps. When the handle is rotated, the fan rotates at high speed and draws air from outside. The chemical from the hopper is fed into the air stream on the suction side of the fan. From here it passes through the delivery line fitted with a spoon type deflector. It is directed towards the target while continuously rotating the handle. It is used to dust nurseries, vegetable gardens, field crops, tea and coffee plantations, green houses, glasshouses and godowns. A hand rotary duster can discharge dust powder up to 150 g/min and displace air about one m^3/min at 35 RPM. These machines are capable to treat about 1 to 1.5 ha/day.

Power Operated Dusters

Traction Dusters

It is the simplest duster deriving power from the turning of the land wheels. These are bellow type and two wheel trailer types. Later are used to cover large areas.

Motorized Knapsack Duster

The motorized knapsack duster is used to cover large areas. It consists of a power driven fan, a hopper to store about 9 kg of powdered dust and a delivery spout. The fan creates strong air flow which causes the dust to blow off from the hopper to a considerable distance vertically or horizontally. A movable spout suitably fitted with the unit regulates the direction of dust application. Agitation is achieved by directing a part of air from the discharge. The discharge of the dust is controlled by rotating the pleated hose on the blower elbow, which carries it to the discharge hose. For a spraying-cum-dusting unit, it is easy to convert the sprayer to a duster or vice-versa. It is achieved by replacing the liquid feed tubes by appropriate dust feeds. Parts required for conversion of a sprayer into duster can be purchased at nominal cost from the suppliers of knapsack sprayers-cum-dusters.

Figure 13.15. Rotary Type Dusters: Shoulder Type (Left) and Belly Mounted (Right).

Tree Duster

The tree duster consists of petrol or petrol/kerosene run engine, blower/fan directly coupled to the engine that produces a high volume high velocity air stream, hopper for the storage of the chemical, discharge chute and other control attachments. The assembly is mounted on the stretcher frame. The equipment can be conveniently carried by two persons to the dusting place. For operation, the hopper is filled with the chemical and the metering mechanism is adjusted for the desired application rate. The equipment is placed under the tree to be dusted. Once the engine is started, the chemical drops in the discharge chute where it comes in contact with the high velocity air stream and is carried to the target in the form of cloud. The air velocity imparts enough energy to the dust particles so that these can easily penetrate the canopy of the tree. The equipment is moved around the field to complete the dusting operation. Tree duster is used for dusting the orchards and tall trees. The dust can be discharged at a rate of 1 to 8 kg/min. A power duster can cover about 10 ha/day

Wet Dusters

A normal duster is used in a wet dusting system. On the top of the duster's outlet, a spray nozzle is fitted to wet the dust. For this purpose, a small hand operated sprayer of 2 L capacity is mounted on the lid of the shoulder mounted rotary duster. In this process, the dust particles released from the duster are wetted resulting in improved dust deposition. It also minimizes losses due to drift. Therefore, wet dusting is more effective and economical and is particularly suitable for dry land crops. A village artisan can easily convert a normal duster into a wet dusting system.

Aerial Duster

An aircraft is used for dusting large acreages with pesticides. Airplanes have been successfully used for dusting field crops and orchards. A V-shaped hopper near the front seat stores the dust. The dust in the hopper is stirred just above the outlet by an agitator. The feed consists of an opening across the width of the fuselage. A slide covering the opening is operated by the pilot, which regulates the flow of dust as per pre-decided application rate. A venturi nozzle is mounted underneath the fuselage and slightly in front of the dust outlet. The rear end of the nozzle is tipped slightly downward. The blast of air generated by the plane's propeller rushes through the venturi nozzle. The high velocity of air through the nozzle creates a partial vacuum in the feed opening, which also aids in the flow of dust. The air blast catches the dust and discharges it in a whirling cylindrical column. The dust while coming down spreads and settles on the plants.

PRECAUTIONS AND MAINTENANCE

- ☆ The dry dust powder should be loose filled in hopper. Do not compact it by hand or any other means

- ☆ Some clod formation is unavoidable as the dust powders absorb moisture from the atmosphere. Such clods should be crushed before filling the dust into the hopper

- ☆ The dusting powders are very finely divided particles and can remain air-borne for long time and can drift far away distances. The fine particles can very easily be inhaled and enter the body system. Therefore, the operator should wear protective clothing. Besides, he must cover his nose and mouth so as to avoid inhalation of pesticide drift

- ☆ The operator should never operate against the wind direction

- ☆ The dusting should be avoided under high wind conditions. It is better to apply the dust in early morning or in late evening hours, avoiding mid-day and afternoons

- ☆ After the completion of the work, carefully remove the dust powder from the hopper. Even after this, some dust material may remain in the hopper, feeders, and discharge tube. The dust should be removed by briskly cranking and blowing action. A dry brush may be used to remove the dust from inside the hopper, *etc.*

- ☆ The lubricating oil should be applied on moving parts *e.g.*, gearbox, crank handle, agitator, fan bearing, *etc.*

SELECTION OF APPLIANCES

Purchase of any appliance is a long-term investment; sprayer/duster being no exception to it. Keeping in view the type of job, one should check the appliance for the following:

- ☆ Suitability for the job
- ☆ Ease of operation and maintenance

☆ Good performance

☆ Good serviceability

☆ Availability of parts and after-sales service

☆ Available at reasonable cost. Price apart, be sure that the equipment is most suitable for the job you intend to perform

Check the equipment for the following while making actual purchase.

☆ Examine moving parts for smoothness and finish

☆ Make sure that the detachable parts fit securely and are easy to assemble and disassemble

☆ Select the nozzle which is most suited for the kind of application. If possible, test the nozzle for discharge rate and degree of spray angle

☆ Ask for demonstration and make sure you understand completely how it works, before payment is made

IMPORTANT FORMULAE AND CALCULATIONS

Theoretical suction capacity of power sprayer of plunger type is given as:

$$Q = (\pi/4) \times D^2 \times L \times n \times 10^{-6} \tag{13.3}$$

Here Q is theoretical suction capacity in L/min, D is the diameter of plunger, mm, n is rev/min, and L is the stroke length, mm.

Therefore the volumetric efficiency is given as:

$$\text{Volumetric efficiency} = 100 \times \text{Actual suction capacity} / \text{Theoretical suction capacity} \tag{13.4}$$

$$\text{Water horsepower, kW} = Q \times \rho \times H/(6116 \times 1000) \tag{13.5}$$

Here Q is the discharge, L/min, H is the pressure in m of water, and ρ is the density of spray in kg/m^3. [Note: $9.81/(1000 \times 60) = 1/6116$]

$$\text{Pump efficiency, per cent} = 100 \times \text{Water horsepower} / \text{Shaft horsepower} \tag{13.6}$$

$$\text{Spraying efficiency (per cent)} = 100 \, (\text{Spray volume required} / \text{Actual spray volume consumed}) \tag{13.7}$$

Example 13.3

Suction volume of a pump is 25 L/min and its efficiency is 85 per cent. Calculate the shaft power at a pressure of 32 kg/cm^2.

Using eq. (13.5), we have

Water power, kW = 25 x 1000 x 320/(6116 x 1000) = 1.3 kW

Note: 1 kg/cm^2 = 10 m of water

Using eq. (13.6)

Shaft power = 100 x 1.3/85 = 1.53 kW

AMOUNT OF INSECTICIDE NEEDED FOR SPRAYING

The dose of pesticides application is generally reported in terms of per acre or hectare *e.g.* kg/ha or L/ha or gm ai/ha. The latter unit stands for gram of active ingredient per ha. Since the pesticide must be applied in exact quantity as per recommendations, following procedures are used to calculate the amount of chemical required.

Case 1

Pesticides come in the form of pesticide concentrate. Being very strong, these must be diluted before use wherein a small volume (amount) of the pesticide is mixed with relatively large volume of water. The application rate of a particular pesticide is the amount of mixed pesticide solution (chemical plus water), which is needed to treat an area of a particular size. If you check the pesticide label, many times this information is directly given in the following forms.

☆ Mix 5 g powder with 5 L water to cover 30 m^2 area

☆ Use 1 packet of powder per 10 L water to treat 10 m^2 area

In such cases, it is easy to work out the pesticide required to cover an area say 1 ha or 10000 m^2.

☆ Using the application rate, calculate the amount of pesticide concentrate needed for the size of the area to be sprayed

☆ Calculate the amount of water needed to dilute the pesticide to the correct strength

Case 2

Sometimes, the recommendations is made in terms of active ingredient (a.i.) or per cent concentration (0.1 per cent, 0.05 per cent concentration) and sometimes in parts per million (100 ppm, 200 ppm). The total amount of insecticide (T) needed, kg depends on the average surface area (S) to be sprayed m^2, the target dosage (Y) of insecticide, g/m^2 and the concentration of active ingredient (C) in the formulation, per cent. Thus

$$T = 100 \times N \times S \times Y/C \tag{13.8}$$

We can also directly calculate the amounts as shown in the following examples.

Example 13.4

A fungicide abc 40 WP (means 40 per cent by weight of active ingredient in the wettable powder) is used for spraying at 250 ppm. It is proposed to treat an area of 2 ha @ 300 L/ha. How much of the formulation is required to treat the area?

Quantity of solution required for treating 2 ha @ 300 L/ha = 300 × 2 = 600 L

In terms of ppm means parts per million, 250 ppm means 250 parts (a.i) per 1000000 parts of solution. So, 600000 mL (600 L) should contain 250 × 0.6 = 150 mL a.i.

Since 40 WP means 40 gm a.i. in a 100 gm formulation, 150 gm a.i. will be available in

(100/40) 150 = 375 gm

Example 13.5

2.5 D xyz dust formulation is proposed to be applied to 1 ha at 300 gm a.i. per ha. How much dust formulation should be applied to treat the area?

a.i. to be applied in 1 ha = 300 x 1 = 300 gm

2.5 D formulation has 2.5 parts a.i. in 100 parts of formulation. So 300 gm a.i. will be in 100 x 300/2.5 gm formulation = 12000 gm or 12 kg.

QUESTIONS (THEORY)

1. Write a short note on need of spraying/dusting of agro-chemicals in agriculture.

2. Discuss the classification of spraying equipment on the basis of volume of spray.

3. What do you understand by a sprayer? List various qualities of a good sprayer.

4. Discuss the various components of a sprayer.

5. Draw a labeled neat diagram of a knapsack sprayer showing its various parts.

6. Name the various manually operated sprayers. Discuss the various types of hand atomizers.

7. Write short notes on hand compression sprayer, rocker sprayer, salient features of knapsack sprayer, foot or pedal sprayer, stirrup sprayer, aero blast sprayer, battery operated ultra low volume sprayer, fogging machine, rotary duster.

8. Write a descriptive note on motorized knapsack mist blower-cum-duster.

9. Discuss the limitations of conventional spraying. Write a short note on aircraft spraying and its limitations.

10. Name and describe the kinds of nozzles classified on the basis of construction.

11. Name and describe the kinds of nozzles classified on the basis of shape of the spray pattern.

12. Name and discuss few important parts of a duster. List the advantages and disadvantages of dusters.

13. Write short notes on tree duster and wet duster.

Chapter 14

Harvesting and Threshing Equipment

HARVESTING DEFINED

Harvesting is defined *as the operation of cutting, picking, digging or any combination of these operations for removing the whole crop or edible part of the crop from either under or above the ground including plucking the useful parts or fruits from the plants.* Some examples of harvesting are: grain harvesting, root crop digging and harvesting, vegetable harvesting, forage harvesting and fruit plucking *etc.* Harvesting is the most important farm operation after the crops have ripened and have reached maturity. Harvesting at appropriate time and moisture content helps to minimize losses, achieve higher yield and better quality products. Any undue delay or early harvesting results in lower yield. Following four actions or their combinations perform the actual operation of the harvesting. For example, manual harvesting usually involves a combination of slicing and tearing actions.

☆ Slicing action with a sharp tool

☆ Tearing action with a rough serrated edge

☆ High velocity single element impact with sharp or dull edge

☆ Scissors type action with two elements

Few important equipment and related terms in connection with harvesting are as follows:

Mower: A machine used to cut herbage crops and leave them in swath.

Reaper: A machine used to cut grain crops.

Reaper binder: As the name suggests, it is a reaper, which cuts the crops and ties them into neat and uniform bundles/sheaves.

Swath: The path made by the width of a scythe or a mowing-machine blade during mowing or the mown grass or grain lying on such a path while harvesting operation is in progress is called a swath.

Windrow: A windrow is a row of cut (mown) grass or small grain crop. It may naturally form as the machine mows the grass or may be formed by combining two or more swaths.

Windrower: A machine used to cut crops and deliver them in a uniform manner in a row. A windrower comprises a cutter bar driven by an engine or powered by a tractor, a reel to sweep the crop onto a platform, and a conveyor that carries and deposits it on one side in a windrow for drying.

Harvesting operation can be performed by manually operated tools, animal drawn machines or mechanically operated machines including self-propelled combine harvesters. Most harvesting and threshing operations are now performed using mechanically operated machines.

CLASSIFICATION OF HARVESTING EQUIPMENT

Harvesting machines may be classified in a number of ways. Commonly the following two criteria are used.

On the basis of source of power used, harvesting equipment are characterized as manual, animal and power driven. Manual tools are often referred to as traditional methods of harvesting.

On the basis of the crops that are to be harvested, harvesting equipment are classified as follows: It may not be treated as a complete classification as other tools/machines depending upon the specific crop requirements are available.

Cereal harvesting machines: These include reaper, reaper-cum-binder and combine harvester.

Fodder crops: Fodder crops including fresh grass is harvested with mowers, which are of two types namely cutter bar mowers and rotary mowers. Dry grass or hay is harvested with the forage harvester, which are cylinder and flywheel type or the flail-type and hay baler.

Root crop harvesting equipment: These include diggers such as potato diggers, elevator diggers and the spinners. Besides, we have equipment for crops that require topping such as sugar beet harvester and tapioca puller.

Cotton harvesting equipment: These include cotton stripper and cotton picker.

Fruit harvesters: These are used to harvest fruits normally on tall trees. Some of these are discussed in the next chapter.

TRADITIONAL HARVESTING METHODS

Most traditional methods rely on manual harvesting. Commonly used tools are the sickle, scythe and *dao*.

Sickle

Sickle is the simplest harvesting tool used to harvest crops and cutting other vegetation types. It essentially consists of a curved metallic blade and a wooden handle (Figure 14.1). Whereas the blade, the main metallic part of the sickle, is made of carbon steel, the handle is made of well-seasoned wood. The forged end of the blade to fix the handle is called tang. Protective metallic bush fitted at the junction of the blade and the handle to keep the tang tight in the handle is called ferrule. The blades of the sickles are either plain or serrated. Accordingly, sickles are classified as plain (Figure 14.1 left) or serrated sickles (Figure 14.1 right). The plain or serrated edge in the inner side of the blade is called cutting edge. Serrated blade sickles cut the crop by principle of friction operating like a saw blade. The cutting operation is performed by holding the crop in one hand and pulling the sickle along an arc with the other hand. The teeth of the serrated sickle are regularly sharpened for its efficient working. The Punjab sickle, an improved version of the traditional sickle, has a wooden handle with a bend at the rear for better grip and to prevent hand injury during operation. This sickle with self-sharpening blade saves 25 per cent labour, operating time and 35 per cent cost of operation compared to conventional method of harvesting with local sickle. Sickle is commonly used for harvesting of crops like wheat, pearl millet, teff, paddy, sorghum and forages *etc.* Harvesting with a sickle is quite slow and labour consuming process requiring 80-110 man-h/ha.

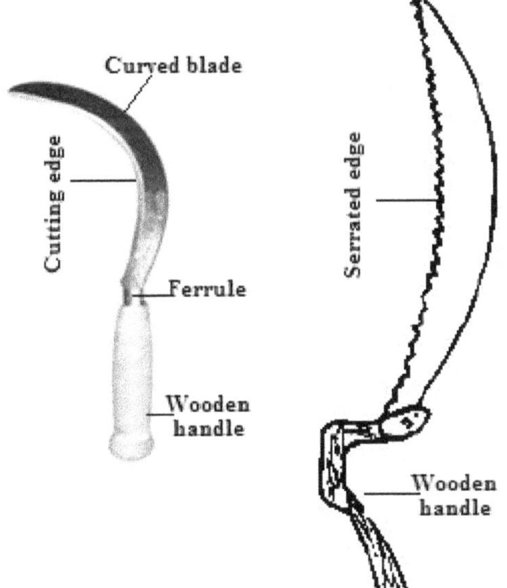

Figure 14.1. A Plain Sickle (Left) and a Serrated Sickle (Right).

Scythe

A scythe, used to harvest fodder crops, consists of a curved blade and a long pipe handle. The handle and blade makes an angle of about 90° with each other. The blade is made from medium carbon steel or low alloy steel and forged to shape.

The cutting edge needs regular sharpening for smooth cutting of the crop. A person operates the scythe in standing posture by gripping the handle at suitable positions and swings the blade in a curvilinear motion. Blade is kept close to the ground during the operation. It is possible to cut an area of about 1.2 m wide and 0.6 m long in one stroke. The cut crop is swept and windrowed in the second stroke of the blade.

Dao

Dao, a multi-purpose tool, is available in various shapes and sizes depending upon the purpose of its use. Normally, it has a long blade curved at the tip. The blade is forged from old leaf spring steel. It is hardened and tempered to suitable hardness for long service life. The tang end of the *dao* is fitted to a bamboo or wooden handle. The tool is operated by striking the blade on the stems or twigs or by dragging the blade on the grass or weeds just like a sickle. The tall weeds are cut by swinging action. It is also used for cutting wood and bamboo, splitting wood for fuel purposes, and ripping of bamboo for making bamboo based products.

The major limitations of traditional methods of harvesting are:

☆ A time-consuming and tedious operation

☆ It is difficult to ensure timeliness of harvest. It may cause damage to standing crops in case of rains and storms

☆ It is difficult to ensure timely land preparation and planting of the next crop

☆ High labour requirement

In order to overcome these limitations, most farmers are now adopting mechanical harvesting and/or combine harvesting, which also simultaneously performs the threshing operation. As per Bureau of Indian Standards the cutting and conveying losses in any harvesting system should not be more than 2 per cent.

MOWERS AND REAPERS

Both the mower and reapers are harvesting equipment. Whereas mower is mostly used to cut grasses as for lawn mowing and fodder such as alfalfa, *berseem etc.*, a reaper is used to harvest grain crops. The mower cuts the stems at about 3-10 cm above the ground. The basic difference between a mower and a reaper is that the mower only cuts the crop and lays it behind the mower in a swath (which afterwards is collected by labour or raking machinery), the reaper makes a windrow of the crop.

Mowers

Mowers have been classified in a number of ways. Several types of mower are available in the market differing in the cutting mechanism. These are: (i) Cylinder mower (ii) Reciprocating mower (iii) Horizontal rotary mower (iv) Gang mower and (v) Flail mower. Besides this main grouping, several sub-grouping of cutting mechanisms have also been made but are not included for brevity. Based on power sources, mowers are classified as push mowers, animal drawn mowers, tractor drawn mowers, self-propelled mowers, electric mowers, battery powered cordless mowers and petrol mowers. According to the mode of hitching, mowers are classified as trailed type, semi-mounted and integral mounted type. Semi-mounted and integral-mounted mowers can be further classified as rear, mid and front-mounted. According to drive used, mowers are classified as ground-driven, engine-driven and PTO driven. This section only deals with the mowers as per the classification based on the cutting mechanisms.

Cylinder Mower

Also known as reel mower, cylinder mower is the most common kind of mower used to cut grass in the lawns. It consists of a horizontal rotating cylinder with helical blades that rotate vertically at the front of the mower to produce a continuous scissor-like cutting (Figure 14.2). Forage or grasses are harvested (cut) continuously with the rotation of blades. It can be manually operated or may be self-propelled lawn mower (See chapter 15). These mowers may be ganged into sets of three or more to form a gang mower. The gang mower is pulled by a tractor to mow large fields.

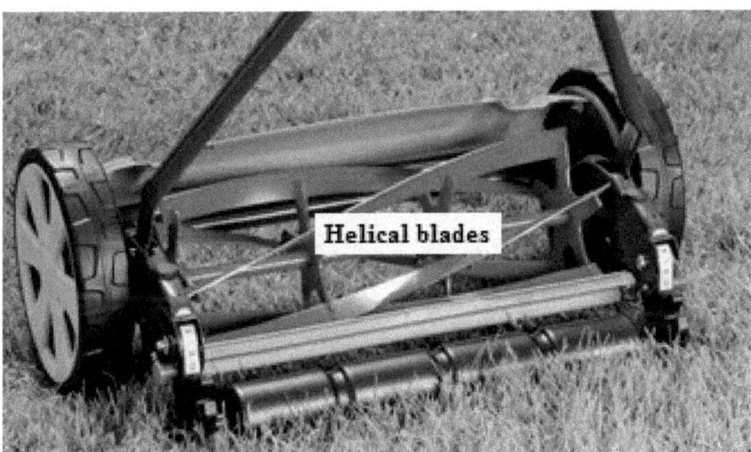

Figure 14.2. A Pictorial View of a Cylinder Mower.

Reciprocating Mower

It is the most commonly used mower, which has a knife having sections that reciprocate against stationary fingers. Here the plant is pressed by the active knife against the fixed ledger-plate of the finger guard and sheared. It is also known as finger-type cutter bar mower.

Horizontal Rotary Mower

It consists of rapidly rotating blades in horizontal plane (60-70 m/s) under the mower like the propellers of a plane. It relies primarily upon the inertia of the material being cut to furnish the opposing force required for shear. The ground also acts as one of the shearing elements. When the revolving blade hits the plant, it is sheared and carried with the blade. Rotary mowers are designed to work on medium to high cut, although they can work on any type of grass.

Gang Mower

A gang mower is an assembly of two or more cylinder mowers.

Flail Mower

The flail mower has numbers of small blades on the end of chains attached to a horizontal axis. The cutting is carried out by axe-like heads striking the grass at high speed. These mowers are used on rough ground, where the blades may frequently be fouled by other objects, or on tougher vegetation such as brush (scrub). The chopped lengths of the stalks can be varied from fine-medium to coarse by increasing or decreasing the clearance between the shear-bar and the flail- tip. A flail mower becomes a hedge-cutter when used in an upright position for trimming the sides of hedges as a boom mower.

Conventional Mower/Reaper

The conventional mower/reaper mainly comprises: i) Frame, ii) Power transmitting unit, iii) Cutting Bar, iv) Shoes, v) Ledger plate, vi) Wearing plate, vii) Knife, viii) Grass board and ix) Pitman (Figure 14.3). Cutting operation utilizes both impact and shear.

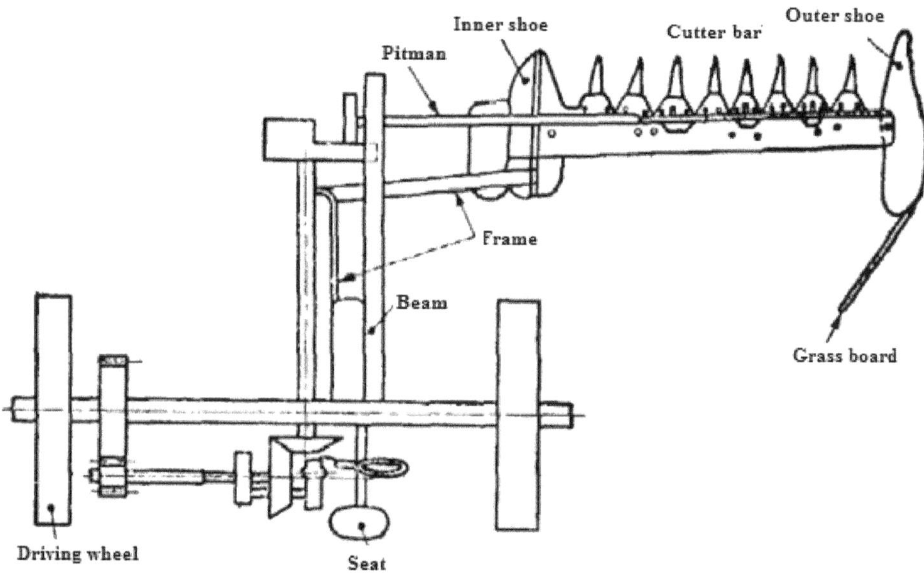

Figure 14.3. Line Diagram of a Tractor Mounted Mower showing some Major Parts.

Frame

The frame lends strength to the machine besides providing space for locating gears, clutch and bearings. The lever for lifting the cutting bar is attached to the frame. A flywheel is used to store energy that helps to maintain the steady speed of the cutting mechanism.

Power Transmitting Unit

The power-transmitting unit consists of axle, gears, crank wheel, crankshaft and pitman. The cutting mechanism is driven independently of the forward speed of the mower. Tractor drawn semi-mounted or mounted type mowers gets power from the PTO shaft. A shaft connected with the PTO shaft drives a pulley with the help of a universal joint. This V pulley rotates another smaller pulley on the crankshaft of the machine and reciprocating motion is transmitted to the cutter bar, the knife making about 1600 cutting strokes per minute.

Cutter Bar

The cutter bar is made of high grade steel, which works like a knife. It is an assembly comprising fingers, knife guides, wearing plates and shoes (Figure 14.3 and 14.4). The knife is a metal bar, on which triangular sections are mounted. The knife section makes reciprocating motion within the knife guards on the cutter bar (BIS, 1981). The knife moves on the ledger plate provided with the knife guards. A properly registered knife stops at the center of the guard on each stroke. Knife clips/guide, made of malleable iron or steel; hold the knife section down against the holder plates. Knife clips placed with wearing plates are spaced three or four guards apart. The cutter bar is normally designed to make cut 25-50 mm above the ground, but it can even go up to 100 mm or more.

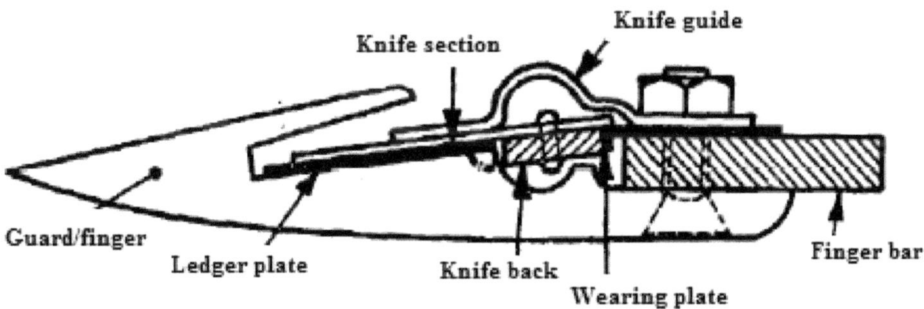

Figure 14.4. Major Parts of a Cutter Bar.

Shoe

A shoe on each end of the cutter bar is provided to support and regulate the height of the cut above the ground (Figure 14.3). The inner shoe supports the inner end of the cutter bar. It has a adjustable and replaceable sole underneath, and is adjusted to regulate the height of cut. The outer end of the cutter bar is supported by the outside shoe also having an adjustable sole to regulate the height of cut. The pointed front part of the outer shoe acts as a divider that separates the cut crop/grass from the standing uncut crop/grass.

Ledger Plate

It is a hardened metal inserted in the guard (finger) over which knife sections move to give a scissor like cutting action.

Wearing Plate

It is a hardened steel plate attached to the finger bar that forms a bearing surface for the back of the knife.

Knife

It is the reciprocating part of the cutter bar, comprising knife head, knife back and knife sections. The knife head is the portion of the knife connected to the pitman. Knife back is the strip of steel to which knife sections are riveted and the knife head is attached. The knife section is a flat steel plate (triangular shape) with two cutting edges, which are either plain or serrated.

Grass Board and Stick

The grass board clears the cut material from a narrow strip next to the standing crop. It provides a place for the inner shoe to operate on its next turn. A stick is also provided that helps the forage to fall correctly. Its height can be adjusted being high for tall and low for short grass. Being spring-loaded, the grass board is free to move to some extent as and when it meets an obstruction.

Pitman

Pitman is a type of connecting rod pinned to the crankshaft with the help of a pin. It transmits reciprocating motion to the knife head. It also supports weight and absorbs vibrations. Although pitman is commonly made of seasoned wood, yet steel pitman is also used.

Problems and Adjustments

Breaking of Knives

The breaking of knives is the major trouble in the operation of a mower. It may be caused due to play in bearings and worn knife head holders. Besides, poor alignment due to which the mower works at a certain angle may also cause breaking of the knife.

Alignment of Cutter Bar

Under normal working conditions, the standing crops exert pressure on the cutter bar of the mower tending to push it backward. For the correct operating position, the crank pin, knife head and the outer end of the knife should be in a straight line during the cutting operation. Moreover, this line should be at 90° to the direction of travel of the mower. For achieving this, the cutter bar is set at about 88° to the direction of motion *i.e.* an inward lead of 2° is given to it in order to take care of the back pushing action of the crops. The controlling factors for a properly aligned cutter bar are: cutter bar lead, the register of the knives (see next section) and the angle of the pitman. The last feature makes the knife and the pitman run in a straight line resulting in a proper cutting action. To adjust the mower when it is not in operation, the outer end of the cutter bar is given a lead of about 2-cm per meter of cutter bar. To measure the alignment, a cord or piece of string is tied to the pitman box or bearing. It is pulled along the top of the cutter bar, so it lines up with the pitman or over the middle of the knife head. The direction of the cord is then compared with that of the line of rivets or the back of the knife bar. It is not compared with the back of cutter bar as it is tapered. If the cord and the line of rivets are parallel, the cutter bar has no lead or lag (Huber, 1948). It means the mower requires adjustment. It is adjusted by rotating an eccentric bushing, which will push forward the outer end of the cutter bar to increase the lead. Besides what is stated, the pitman should operate at right angles to the crank pin and parallel to the face of the flywheel.

Registration of Cutter Bar

A knife is said to be in proper registration, if midpoint of the knife section during operation stops at the center of guard on each stroke (Figure 14.5 left). Improper registration (Figure 14.5 right) may be a result of cutter bar out of alignment, incorrect length of pitman, bent drag bar or brace bar or both, drag bar or brace bar too short or too long or too much looseness in the pitman connections (Huber, 1948). To adjust an improperly registered cutting bar, identify and remove the cause or move the entire cutter bar in or out with respect to the pitman crankshaft. Failure to properly register the cutter bar results in an uneven cutting and an uneven loading of the mower (Figure 14.5 right). It also results in high draft and clogging of the knife.

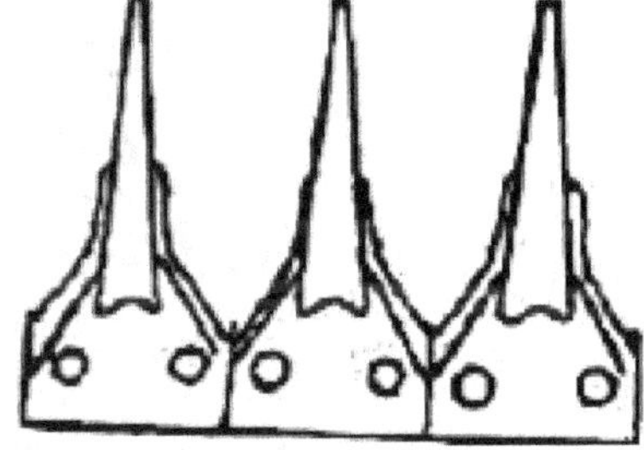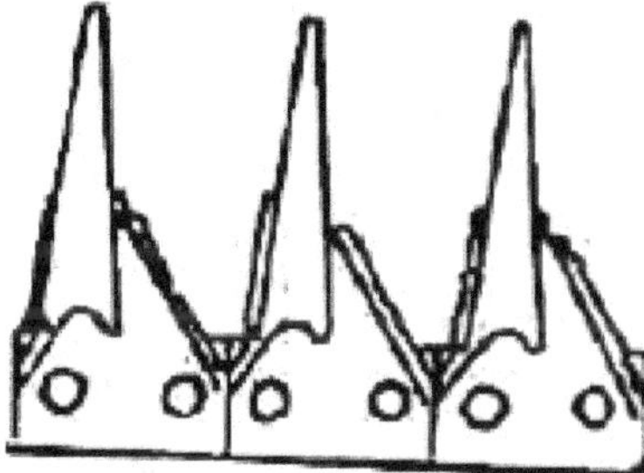

Figure 14.5. Registration of Mower: Proper Position of the Cutter Bar at the End of Stroke (Left) and Improper Position (Right).

Vertical Conveyor Reapers

These types of reapers are mostly used to harvest wheat and paddy crops. They are available in various models and can be operated either by tractors or power tillers. Even self-propelled reapers are available. A front mounted power tiller operated reaper is powered by the engine fly wheel either through V belt or by providing gear box and propeller shafts. The crop is guided by the star wheels to the cutter bar and held in vertical position by the springs. The capacity varies in the range of 0.25-0.35 ha/hr depending upon the cutter bar length, which may range from 100 to 160 cm. The crop is conveyed to the side with the help of a conveyor belt. A front mounted tractor reaper is powered by the PTO shaft (Figure14.6 right). It can be lowered and raised by the hydraulic control. Operation is almost similar to the power tiller operated reaper except that its capacity is in the range of 0.4-0.6 ha/hr.

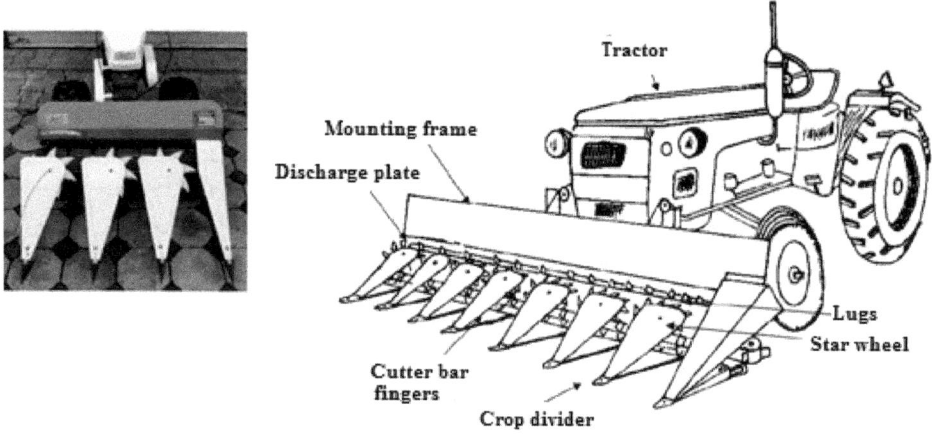

Figure 14.6. Front Mounted Power Tiller Operated Reaper (Left) and Tractor Operated Reaper (Right).

Reaper-cum-Binder

It is a normal reaper having the additional attachments of gathering, and knotting units. Thus, it simultaneously performs the operations of cutting and binding the crop. The cutter bar cuts the crop at the height of about 10 cm from the ground level. It is then accumulated with the help of guards and fingers at the center for feeding in the knotting unit. Here bundles are made automatically. The harvesting capacity is 0.25-0.35 ha/h.

ROOT CROPS HARVESTERS

Commonly known as digger/harvesters, root crops diggers are now available for almost all kinds of crops such as potato, sugar beet, groundnut, carrot, onion, tapioca *etc.* Basic principles as well as the operations performed by most of these machines are similar. Some of the required operations are:

☆ Cutting of the unwanted portion at desired height

☆ Appropriate disposition of the tops to prevent interference with other steps

☆ Digging or loosening the crops just below the tuber zone

☆ Elevating the crop and separating them from clods and other foreign materials

☆ Depositing the clean material at appropriate location

This section describes 4 equipment namely potato digger, groundnut digger, turmeric harvester and tapioca digger.

Potato Digger

Potato digger, a rear mounted tractor PTO driven machine, is used for digging and windrowing the potatoes. The machine consists of a cutting blade and an elevator roller chain of iron bars (Figure 14.7 left). The potatoes dug by the blade are lifted to the conveyor, which is under periodic shaking. The potatoes delivered at the rear of the machine are manually collected. Its capacity ranges from 0.15-0.2 ha/hr. It is normally operated by a 20-25 HP tractor.

Figure 14.7. Potato Digger (Left) and Groundnut Digger (Right).

Groundnut Digger

As the name suggests, groundnut digger is used for digging the groundnut crop. It is a tractor mounted PTO operated machine. The machine is suited to harvest both erect and spreading varieties of groundnut grown in most soil types. It consists of a digging blade and a spike tooth conveyor (Figure 14.7 right). The loosened groundnut vines are lifted and shaken by a conveyor chain that removes the soil attached to the vines. The vines, free of soil, are dropped and windrowed behind the machine. Harvested crop is collected manually.

Tractor Operated Turmeric Harvester

The unit consists of a main frame, a hitch frame and a blade with five bar points for easy penetration into the soil. The blade is fixed at an inclination of 20° to a cultivator frame with straight tines at both ends for easy penetration into the soil. It is hitched to the tractor at the hitch frame. For digging of turmeric, the blade with bar points penetrates into the soil, lifts the turmeric rhizomes along with the soil and conveys them to the lift rods. These rods aid in separation of the rhizomes from the soil. The soil slip back to the ground and the dugout rhizomes deposited at the center of the unit. The field capacity of the unit is 1.6 ha/day. Machine harvesting saves 70 per cent of cost and 90 per cent of time compared to manual digging. Besides, the extent of damage caused to the rhizomes and non-harvested rhizomes left in the field are much less at 2.83 per cent and 2.42 per cent respectively.

Tapioca Puller

Tapioca (*Manihot esculenta* Crantz) commonly known as cassava or Brazilian arrowroot is a perennial shrub grown primarily for its roots for food, feed and industrial products. Tapioca is mostly harvested by hand. The roots of tapioca are pulled out of the ground by holding and lifting the lower part of the stem. Both manual and power operated tools are now available to harvest tapioca. The manual harvesting tool operates on grip and lift principle with the help of a lever (Amponsah *et al.*, 2017). It consists of a frame to which an immovable griping jaw is attached (Figure 14.8 left). A chisel tip serves as the base for lifting cassava roots even from hard and dry soils, where the grip and lift principle alone fails to uproot the crop.

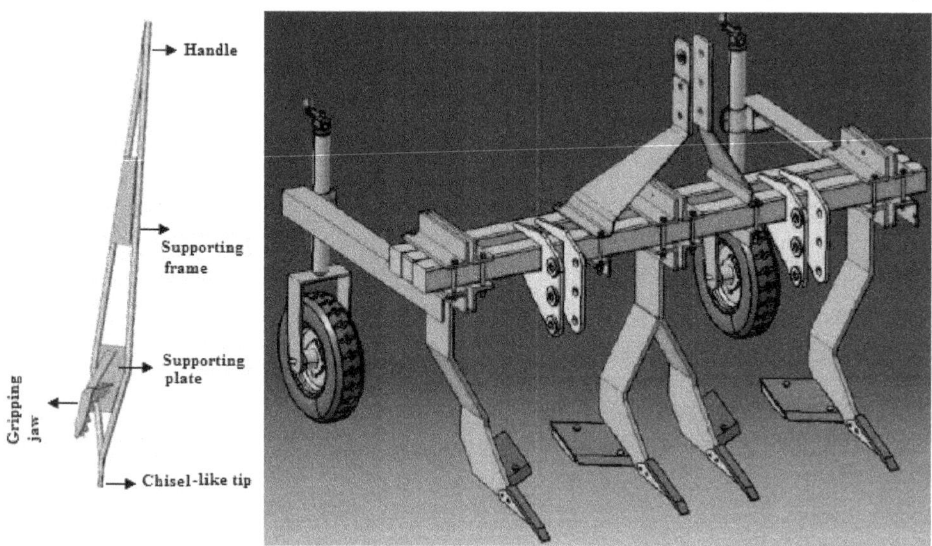

Figure 14.8. Tapioca Pullers: Manually Operated (Left) and Two Row Tractor Operated Puller (Right).

The power operated tapioca harvester consists of a mainframe, shank with depth adjustment, digging blade with pegs and a three-point hitch system (Figure 14.8 right). The shank made up of mild steel has a digger blade attached to it. The shank is designed as a bent leg plough with an angle of 150° to accommodate the dug cassava tubers. The trapezoidal digging blade is made of 13 mm thick mild steel plate. The blade angle is around 20° that helps in easy penetration into the soil. The depth wheels also help to adjust the depth of operation. In hard soils, it is possible to add dead weight to the main frame to further ease the penetration of the blades. Pegs are provided at the front end of trapezoidal digging blade to reduce the draft requirement of the unit and for easy penetration. The row spacing can be altered by moving the shanks on the main frame. The base cutting to remove the above soil surface portion of the plant needs to be carried out before the digging operation.

THRESHING

Threshing is the process of loosening the edible part of the cereal grain from the scaly, inedible chaff that surrounds it. In other words, it is the operation of detaching the grains from the spikes, panicles, ear heads, cobs or pods. A mechanical thresher commonly used for threshing is a versatile machine that separates grains from the harvested crop, provides clean grain without causing much loss and damage. The grain loss or damage during threshing may occur in terms of broken grain, un-threshed grain, blown grain, and spilled grain. In order to reduce these losses, threshing is usually done after the grain moisture content has reached 15-17 per cent. Clean un-bruised grain fetch good price in the market and has long storage life. According to the Bureau of Indian Standards, the total grain loss should not be more than 5 per cent, in which broken grain should be less than 2 per cent.

Threshing is normally achieved by

- ✰ Rubbing
- ✰ Impact
- ✰ Stripping

Traditional Threshing Methods

The early methods of threshing includes hand threshing, bullock threshing, and to some extent human powered thresher (Zakiuddin *et al.*, 2012). Hand beating with a stick or a flail has been quite common method for rice threshing since rice is easily shattered. A flail is a hand threshing implement consisting of a wooden handle at the end of which a stouter and shorter stick is so hung as to swing freely. Pigeon pea and many other pulse crops are quite amenable to this method of threshing. With this method, a worker is able to obtain 15-40 kg of product per hour. Beating sheaves of crops on a hard slanted surface, commonly a curved surface such as a drum (Figure 14.9 left), is also practiced in many areas. The other common method in India is bullock treading. In this method, crop in sheaves of about 30 cm thick are spread on the threshing floor in a circle and bullocks are made to walk over it

Figure 14.9. A view of Traditional Method of Rice Threshing (Left) and Traditional Winnowing Operation (Right).

in a circular path. The repeated trampling under the bullock feet helps separation of grains from straw. The process continues till all the grains are separated. It is followed by winnowing (Figure 14.9 right). On an average a pair of bullocks can thresh about 140 kg of grains per hour. Now many farmers use a tractor to achieve the same objective. The sheaves in this case are regularly turned over between the two passages of the tractor.

Olpad Thresher

Olpad thresher has its origin at a small village named Olpad in Gujarat State. Olpad thresher is a simple, bullock drawn implement operated by the man. It consists of notched discs placed on three axles, fixed on a wooden or iron frame. The operator can sit on the seat on the frame to control the movement of animals. Although the number and size of discs may vary, mostly about 20 circular grooved discs each of 45-cm diameter and 3-mm thick placed 15 cm apart in three rows are used. All discs are staggered to give effective cutting of the straw (Figure 14.10 left). Besides, it has three or four wheels to facilitate its movement from one place to another. A pair of bullocks pulls it around over the dried crop spread in a circular path on the threshing floor (Figure 14.10 right). Threshing is continued till the entire material becomes a homogeneous mixture of grain and *bhusa* (chaff). Although mainly a wheat thresher, it can also be used for threshing barley and few other crops. Threshing is fairly efficient and cheap. It reduces the drudgery of the operator and gives comparatively higher output per unit time although overall it is also quite slow with low output.

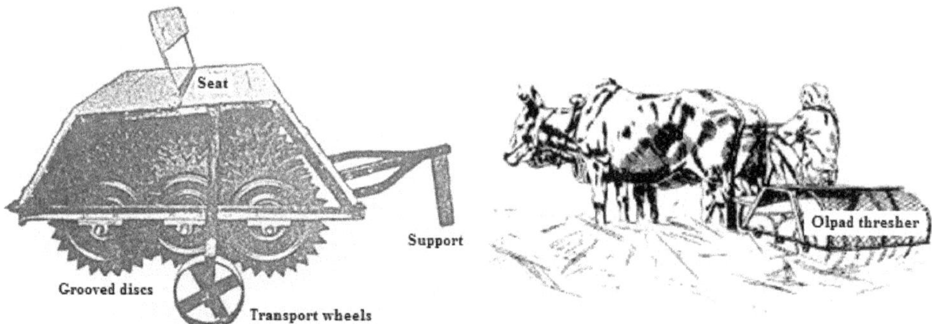

Figure 14.10. A view of an Olpad Thresher (Left) and an Olpad Thresher in Operation (Right).

Hand Operated Drum Thresher

A hand-operated drum thresher is used to thresh paddy or cut stocks of pigeon pea. It basically consists of a big drum fitted with metal rings or teeth. A handle is used to rotate the drum. The paddy or cut-stocks of pigeon pea are threshed by hand-holding the sheaves and pressing the upper portion of dried plants against the rotating drum. An improved version is pedal operated paddy thresher discussed in a subsequent section of this chapter.

In all traditional or semi-mechanical methods the threshed material is subjected to winnowing to separate grain from straw either with natural wind flow (Figure 14.9 right) or blast from a winnowing fan.

A Word on Manual Winnowing

'Winnowing is the process of separation of heavier components of a mixture from its lighter substances with the help of wind'. It is used for separating grains from the mixture of grain and chaff or straw or *bhusa* after the process of threshing. In the manual winnowing, a mixture of grains and husk is dropped from a height (Figure 14.9 right). The heavier particles (grains) form a heap below as they fall vertically downward. The husk being lighter is carried away by the wind and forms a different heap at some distance away from the grain heap. No machine is needed to perform the operation. It is easy to adopt and costs less. On the other hand, it requires blowing wind to perform the task. To overcome this limitation, many farmers use winnowing fans to create artificial wind to complete the task even in the absence of naturally blowing wind. Note that this operation is incapable of separating out materials heavier than grains such as stones, incompletely threshed or completely un-threshed grains, which also form the part of grain heap.

Limitations of Traditional Methods

- ☆ Traditional methods are laborious, slow and time-consuming with limited output
- ☆ They are only suitable for small scale farming, because it is difficult to handle large volume of crop
- ☆ Threshing by traditional methods involves drudgery and takes more time to obtain required quality of *bhusa*
- ☆ Winnowing operation is essentially required following the threshing of grains, which is again a weather dependent operation
- ☆ Task cannot be performed during inclement weather
- ☆ Availability of labour during peak season may pose problems
- ☆ Since it is a time-consuming process, pilferage, rotting of stacked crops, damage by rodents or other animals or birds and also the damage to crop through blowing winds, storms, rain and fire cannot be ruled out
- ☆ The efficiency of the methods is low and consequently there is a tremendous loss of the produce besides human/animal power loss
- ☆ The cost of threshing with traditional methods is high. The labour requirements for manual threshing and cleaning is 250 man-hours/ha as compared to mechanical threshing and cleaning, which is 20 man-hours/ha
- ☆ Quality of produce may not be up to the mark and therefore is likely to fetch less price in the market
- ☆ Higher losses due to undetached grains
- ☆ Great concern on account of musculoskeletal disorders and injuries to different parts of the human/animal body

MECHANICAL THRESHERS

Threshers have become the most important component of farm mechanization in India. Historically speaking, the famous Ludhiana thresher was first introduced in India in 1956-57. The tractor operated thresher was used mainly for threshing wheat. It was a good machine in the sense that it threshed, cleaned and bagged the grains in one go. Besides, it made good quality straw (*bhusa*). Further development works continued especially after 1965 with the main objective to develop threshers that required low horsepower. As a result of these efforts, spike tooth cylinder thresher was commercially marketed in the country in 1970. This simple design helped to reduce the weight of machine and hence the cost was greatly reduced. The output capacity also improved. Currently, mechanical or motorized threshers or threshers operated by tractor power have become quite common for threshing of cereal crops. However, by some simple replacement of a few accessories and the appropriate changes in settings, these machines can be used for threshing other crops as well. Various kinds of threshers available in the market are categorized as follows:

According to the crop being threshed these are categorized as single crop or multi-crop threshers.

Based on the type of feeding, threshers are divided into throw-in type in which the whole crop is thrown into the cylinder and hold-on type in which the panicle end is only pushed into the cylinder while the remaining straw is manually or mechanically held outside the thresher.

According to functional components these are categorized as drummy, regular or through flow and axial flow threshers. In the through flow threshers, threshed straw and separated grain flow in a direction perpendicular to the axis of the threshing cylinder. On the other hand, the threshed straw and separated grain flow in a direction parallel to the axis of the threshing cylinder in the axial flow threshers.

According to types of threshing cylinder these are categorized as peg type, loop type, syndicator, beater or hammer mill type, spike tooth type and rasp bar type.

On the basis of source of power threshers are categorized as manually operated like pedal threshers, engine/tractor operated (power range of 5-40 HP and capacity range of 200-500 kg/hr) and self-propelled engine type as combine harvesters

Parts of a Thresher

A mechanical thresher consists of several components such as drive pulley, fan/blower, feeding chute, spikes, cylinder, concave, flywheel, frame, towing hook, upper sieve, lower sieve, transport wheel, suspension lever, cam pulley and shutter plate. All these parts can be functionally grouped under the following 6 heads (Figure 14.11).

Main frame

- ✰ Feeding unit (chute/tray/trough/hopper/conveyor)
- ✰ Threshing unit/cylinder (hammers/spikes/rasp-bars/wire-loops/ syndicator)

☆ Cleaning unit (concave woven wire mesh/punched sheet/welded square bars)

☆ Blower/aspirator and sieve-shaker/straw-walker

☆ Power transmission unit

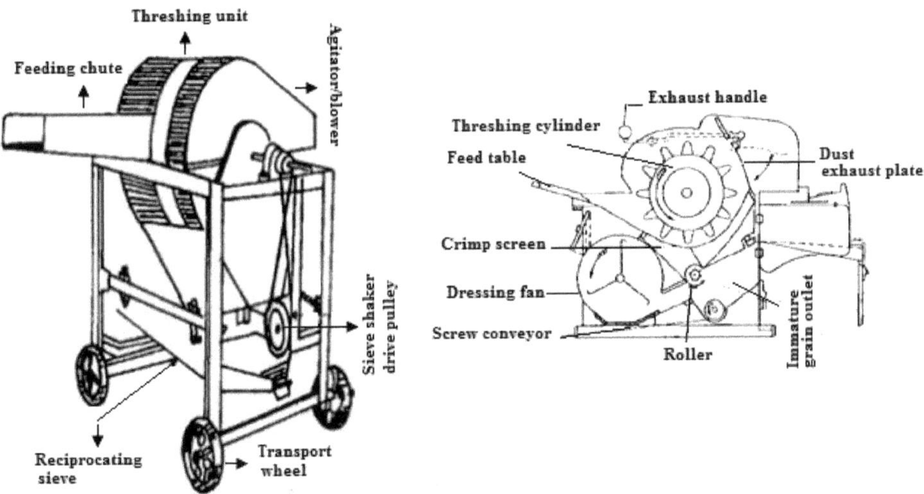

Figure 14.11. Major Components of a Thresher External view (Left), Internal view (Right).

Main Frame

A strong frame made of heavy angle iron sections provides strength to the thresher besides providing space to mount all functional parts. The main frame should be strong enough to withstand vibrations encountered during field operation.

Feeding Unit

The feeding unit performs the operation of conveying the crop into the threshing unit. As stated earlier, two kinds of feeding units are throw-in-type or 'hold-on-type'. The former type of feeding is quite common in the threshers. The feeding mechanism may be a feeding chute (Figure 14.12 left) or a feeding hopper (Figure 14.12 right). Feeding chute is generally used with hammer mill type threshing cylinder. The cut crop is fed perpendicular to the direction of motion of rotating beaters. The chute is inclined to the horizontal at an angle of 5-10 degrees. To ensure safety of the labour, the minimum length of feeding chute should be about 90 cm, out of which a minimum of 45 cm should be covered (Figure 14.12 left). Farmers should only buy threshers that are fitted with chutes as per BIS standards. In the case of feeding hopper, a hopper is placed on the top of the threshing cylinder. A rotating star wheel mechanism is placed between the hopper and threshing drum that allows uniform feeding of crop to the drum. Since the initial cost of such a system is high, it is mainly used on large threshers. Besides these two mechanisms, conveyors fed threshers are also available in which feeding of the crop is done using a conveyor (BIS, 1981).

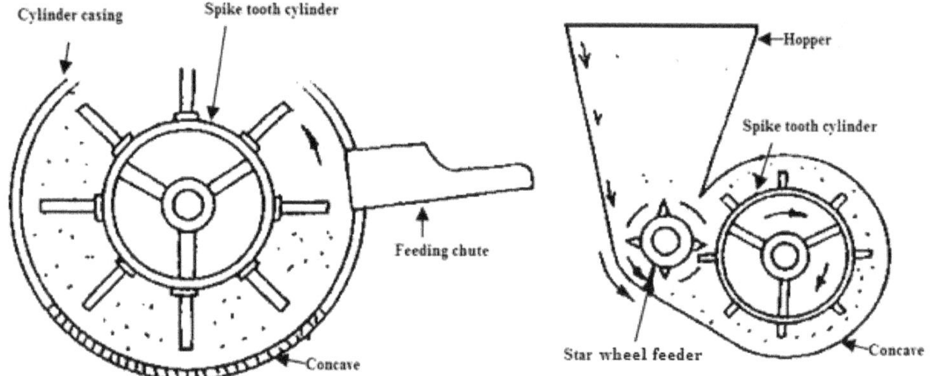

Figure 14.12. A Chute Type (Left) and a Hopper type Feeding Mechanism (Right).

Cylinders

Various kinds of cylinders are used in the threshers. Each has its own merits and demerits. Some cylinder types are discussed in the following sections (Figure 14.13).

Drummy type: It consists of beaters mounted on a shaft rotating inside a closed casing and concave. It has no separation and cleaning systems, although a centrifugal blower is provided for partial separation and cleaning of grains.

Hammer mill type: A hammer mill type cylinder uses beaters to do the threshing. The beaters are made of flat iron pieces and are fixed radially on the rotor shaft. The cut crop is fed perpendicular to the direction of motion of rotating beaters. This type of thresher requires more power as compared to spike tooth type of thresher. It is similar to drummy type but is provided with aspirator type blower and sieve shaker assembly for cleaning grains.

Spike-tooth type: Spikes are mounted on the periphery of a cylinder that rotates inside a closed casing and concave. The teeth on the concave and cylinder are arranged in such a manner that the cylinder teeth pass midway between the teeth on the concave. It is provided with cleaning sieves and aspirator type blower. Most farmers prefer spike tooth type threshers because of their simple design, low cost and their ability to make fine *bhusa*. A spike tooth cylinder with spikes of flat front and streamlined back has lower energy consumption. If pegs instead of spikes are provided, then it is known as peg type cylinder.

Raspbar type: In this type of cylinder, there are slotted plates, which are fitted over to the cylinder rings in such a way that the direction of slot of one plate is opposite to another plate. It is fitted with an upper-casing and an open type concave at the bottom of the cylinder. The cleaning system is provided with a blower fan and straw walker. It gives better quality of *bhusa*. It can be used for a wide varieties of crops such as wheat, paddy and soybean.

Angle bar type: Angle iron bars with rubber pads on their faces are helically fitted on the cylinder.

Wire-loop type: In this type of drum, number of wooden or MS plates are fitted on a hollow cylinder. On these plates, numbers of wire loops are fixed for threshing purpose. A woven wire mesh type concave is provided at the bottom. This type of cylinder is common in the manually operated paddy threshers. Holding the bundle against the loops of revolving cylinder does the threshing of paddy crop.

Syndicator type: It is essentially an adoption of chaff cutter for threshing. It has a set of rollers, which cuts the crop into pieces. The rims of the flywheel are fitted with three to four serrated chopping blades. Threshing is done mainly due to cutting helped by rubbing and impact. This kind of thresher is able to handle crops having high moisture content. The main limitations are: i) chopping knives need to be sharpened every 3-5 hours of operation and ii) use of positive feed rollers makes the machine more prone to accidents.

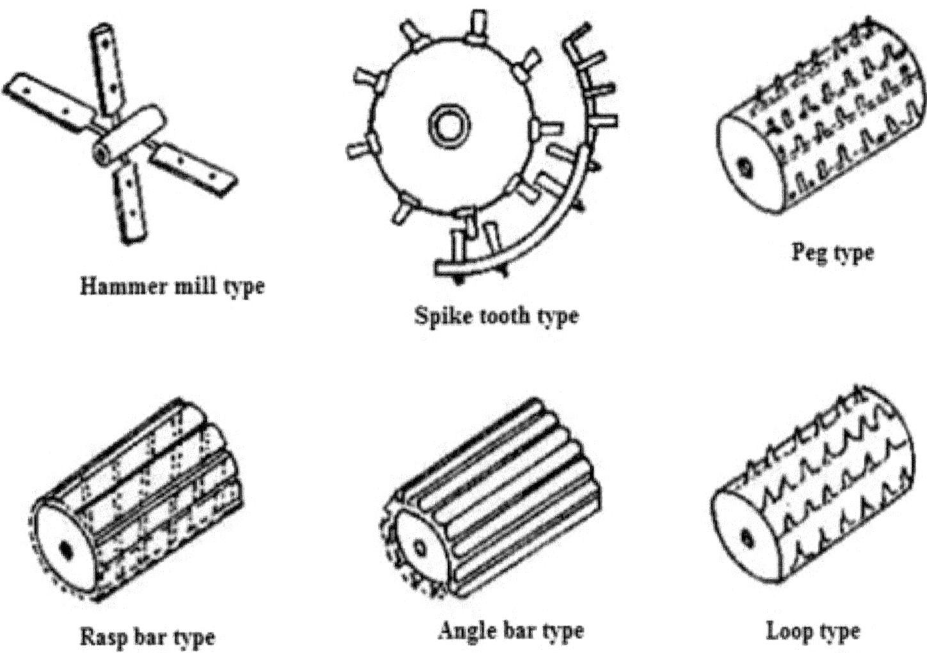

Figure 14.13. Various kinds of Cylinders used in Threshers.

Concave

Cylinder and concave together makes a threshing unit as the threshing takes place only in the space between the cylinder and the concave. This space is adjustable varying from 5-13 mm for wheat and 5-10 mm for paddy. The threshing efficiency increases with decreasing concave clearance. On the other hand, losses increase with decreasing concave clearance. The concave clearance at the inlet is less as compared to outlet. Concave is a curved mostly a semi-circular unit (Figure 14.14). It is of two kinds namely screen type made of MS rods or wires or perforated with perforations made in the steel sheet. Concave is provided in the thresher to hold the fed crop inside the threshing chamber and allows only grain and small amount of chaff to pass through it. Threshed grains fall on the sieve through the concave.

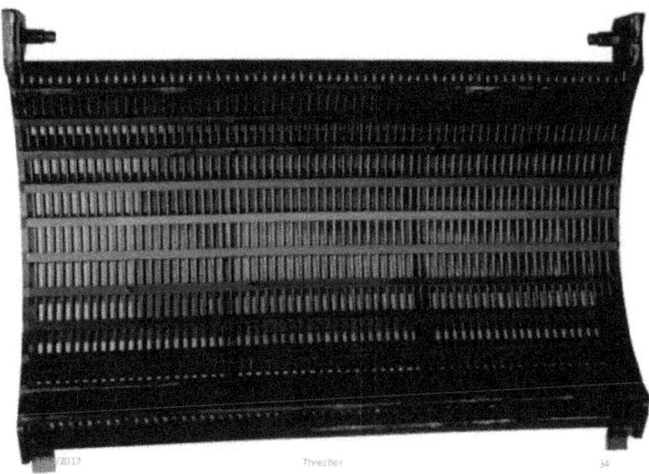

Figure 14.14. A view of a Screen type Concave.

Cleaning Unit

This unit separates the grain from chaff. It consists of aspirator or blower and sieves and sieve shaking mechanisms. Whereas a thresher provided with an aspirator unit is called an aspirator type thresher, those fitted with blower blowing the air in the horizontal direction is called drummy threshers.

Blower and aspirator: Blower or fan is generally installed on the main shaft. Air sucking duct known as aspirator comprises primary and secondary ducts. Whereas primary duct removes a major portion of straw, dust and other foreign matter, secondary duct does the final cleaning of the grains. Rest of the separation-cum-cleaning is done by screens.

Screens: Most power threshers are equipped with two screens. The top screen allows the grains to pass to the second screen while taking the chaff *etc.* out from it. The bottom screen sieves out the smaller grain or weeds seeds to deliver the cleaned grain out of the grain outlet. The hole size of the screen is mainly governed by the crop to be threshed.

Shaking mechanism: The screens are made to oscillate to improve the effectiveness of screening. A crank attached to the screen helps in shaking the screens. The crank receiving power from the main axle converts its circular motion into oscillating motion of the screen. The frequency of oscillation *i.e.* strokes of crank per unit time can be is adjusted.

Power Transmission Unit

Threshers are usually powered with tractors and sometimes with electric motors or diesel engine. In a tractor operated thresher, tractor PTO shaft is coupled with a flat pulley. A corresponding matching pulley of appropriate size is provided on the thresher main shaft. These pulleys are connected with flat belt to operate the thresher. Blower fan is provided on to the main shaft of the thresher, which rotates to accomplish its required function. The screens are oscillated with the help of a V-belt and a crank wheel, powered by the main shaft of thresher. A heavy flywheel made up of cast iron is provided on one end of the main shaft of thresher. It stores the energy to release it continuously and equally to the entire threshing cylinder.

Transport Wheels

Transport wheels are provided so that thresher can be easily transported from one place to another. Earlier, wheels were mostly made of cast iron. Now pneumatic wheels are used especially in new and large capacity threshers for easy transportation.

Threshers as already discussed are categorized on the basis of various mechanisms used and their relative placement. Following two kinds of threshers are discussed for illustration purpose.

Axial Flow Type

In axial flow type, the crop is fed from one end, which moves axially and the straw is thrown out from the other end after complete threshing of the crop. It consists of a spike tooth cylinder or a peg type cylinder, woven-wire mesh concave and upper-casing provided with helical louvers (angled slats).

Multi-crop Thresher

The thresher consists of raspbar threshing cylinder, oscillating sieves, concave and winnowing and cleaning attachment. Whole crop in some cases such as wheat and rice and only ear heads in other cases such as pearl millet, sorghum and maize are fed into the machine. It is possible to adjust the cylinder and concave clearance and blower speed to make the machines suitable for threshing various crops.

Principle and Working

- ☆ The crop is fed from the feeding tray into the working slit between the circumference of the revolving drum having attached spikes and the upper-casing

- ☆ The speed of the spikes is higher than the plant mass due to which they strike the latter resulting in part of the grain being separated from straw

- ☆ Simultaneously, the drum pulls the mass through the gap between the spikes and the upper-casing with varying speed. Due to this the spikes move in the working slit with varying speed. The material layer is struck several times by the spikes against the ribs, causing threshing of the major amount of grains and breaking stalks into pieces

- ☆ Since the system is closed, the thicker stalk which cannot be sieved through the concave, joins the fresh stalk and the same process is repeated until the stalk size is reduced to the extent that it can pass through the concave apertures

☆ Threshed crop falls on the concave and from there to the top sieve of the cleaning unit. Due to reciprocating motion of the sieve lighter material accumulates at the top and grain falls on the bottom

☆ In threshers with aspirator blower, blower sucks the lighter material from top of the screen and throws it out of the blower outlet

The motorized threshers usually require two or three workers for their operation. The capacity of the machine depends on the type of machine, the nature and maturity of the grain, the skill of the workers and other organizational inputs.

Factors Affecting Threshing Performance

An effective threshing process should ensure that the un-threshed kernels getting ejected with the straw through the concave is less (http://ecoursesonline. iasri.res.in/course/view.php?id=59). Besides, the amount of the material passing through the concave is high and the loss due to grain damage is low. The factors affecting the quality and efficiency of threshing are broadly classified in three groups:

Crop factors: Variety of crop, moisture content of the crop material at the time of threshing.

Machine factors: Feeding chute angle, cylinder type, cylinder diameter, spike shape, size and number, concave size, shape and clearance

Operational factors: Cylinder speed, feed rate, method of feeding, machine adjustments.

Threshing efficiency is expressed as the threshed grain received from all outlets with respect to total grain input expressed in mass percentage as follows:

Efficiency, per cent = 100[Threshed grain from all outlets/total grain input]

It can also be expressed as:

Efficiency, per cent = 100[1 – Un-threshed grain from all outlets/total grain input]

For good performance of the machine, it should be installed on clean level ground and various adjustments made according to crop and crop conditions. One must set i) concave clearance, ii) sieve clearance, iii) sieve slope, iv) stroke length and v) blower suction opening. Besides these, in a multi-crop thresher cylinder concave grate, top sieve hole size and cylinder speeds for threshing different crops needs to be changed/adjusted.

Precautions during Operation of a Thresher

☆ Install the machine at a level site

☆ Orient the machine in the direction of wind

☆ Do not stand on unstable platform during threshing

☆ Do not wear loose clothes, wristwatch and bangles while working on the thresher

☆ Operate the machine at the recommended speed

☆ As far as possible feed the thresher continuously and uniformly avoiding overfeeding of crop materials in the hopper

☆ Feed should be dried to appropriate moisture level before the threshing operation

☆ The feed should be free of any extraneous material such as wood or iron

☆ To avoid any accident/mishap, employ trained labour and give proper instructions before the threshing work begins

☆ Don't permit smoking or lighting the fire near the threshing yard

☆ Rest the machine after a continuous operation of 8-10 hours

☆ Do not cross over the flat belt or any moving component of the threshing machine

☆ Ensure proper lighting in case the machine is operated at night as poor visibility may lead to accidents

☆ At the end of day inspect all the sieves for clogging, grease or oil the various bearings and parts as may be required

☆ At the end of the season, remove all belts for proper storage and keep the thresher under a covered place

☆ Always keep a first aid box and a fire extinguisher near the place of work

Manually Operated Drum Thresher

Foot operated manual paddy thresher consists of wire-loop type threshing cylinder, power transmission system, mild steel sheet body and foot pedal. The threshing cylinder consists of wire-loops of 'U' shape embedded in wooden or metallic strips joined to two discs. A shaft carries the threshing cylinder and is connected to the transmission system (Figure 14.15 left). The transmission system consists of meshed gears or sprocket-chain mechanism with the larger gear or sprocket connected to the foot pedal/bar. The foot pedal/bar is in raised position when the machine is not in operation. The threshing cylinder begins to rotate upon pressing the pedal. For threshing, the paddy bundle is held in hands and the ear head portion of the crop is placed on the rotating cylinder. The wire-loops hit the ear heads and grains are detached from the rest of the crop. The machine can be easily matched and run with an engine (Figure 14.15 right).

Maize Dehusker-cum-Sheller

Maize de-husker-cum-shellers of various kinds in different sizes are available in the market. Three kinds shown in Figure 14.16 are tubular maize sheller (left), handle operated maize sheller (middle) and motorized maize sheller (right). An octagonal hand operated maize sheller (fin type) is used for shelling maize cobs, especially for seed purposes. It has tapered fins fitted inside a pipe of 60-65 mm diameter. To operate, the sheller is held in left hand and the dehusked maize cob in right hand (for right hand person). The cob is inserted in the sheller and is repeatedly given forward and backward twists or clockwise and anticlockwise strokes. The tapered edges of the fins dig into the space between the rows of the grains in the cob and

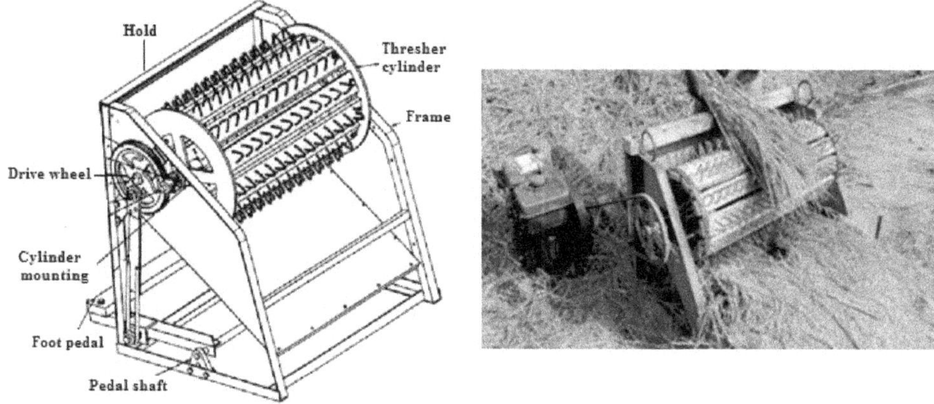

Figure 14.15. A Pedal Operated Paddy Thresher (Left) and an Engine Operated Model (Right).

the grains are released from the cob with forward and backward strokes. After grains are separated from one end of cob, the other end is inserted in the sheller to complete the removal of grains from the whole cob. Due to the taper edges of the fins, one end of the sheller has larger opening while the other smaller. For shelling the larger end of the cob it is inserted in the larger opening and the smaller end of the cob in the smaller opening. Weight of the machine is about 200-250 gm. With 100 per cent shelling efficiency, the capacity of the tool is around 20 kg/hr.

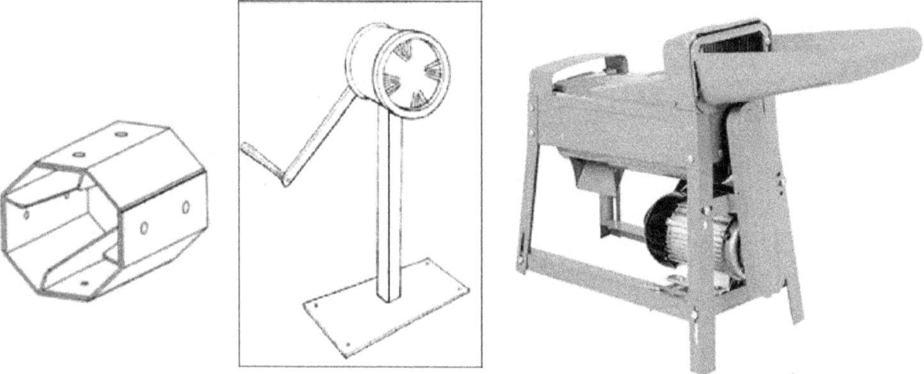

Figure 14.16. Maize Sheller: Hand Operated (Left), Hand Operated with Handle (Middle) and Power Operated (Right).

Handle operated corn thresher vary in features, may be as simple as the hand operated sheller or as technical as a power operated sheller. Of the latter category, the power operated shellers are of two types namely spring type and cylinder type. Let us discuss one kind in this section and the other in the following paragraph. The handle operated model is constructed of heavy-duty cast iron. The spring-type sheller consists of a rotating fluted cylinder, a rotating disc and a spring pressure plate. The cobs are fed to rotating fluted cylinder and kernels are removed from the cobs as they move in between cylinder and disc. Spring tension on the hinged pressure plate is easily adjustable with a thumb-screw as per the size of the corn cob. A hand crank is used to operate the machine. Some machines are also provided with a pulley groove for powered operation. The threshing rate > 98 per cent can be achieved.

A power operated maize sheller uses 30-36 cm diameter cylinder of 80-100 cm length. Pegs/spikes staggered at varying heights are provided on the periphery of the cylinder for better shelling efficiency. These remove the grains from cobs using axial flow movement. The cylinder speed is maintained in between 500-600 rpm. The cob while passing through the concave from feeding side toward the end of sheller is rubbed against drum resulting in removal of grains. Blower is provided to remove lighter material. Concave clearance and cylinder speed can vary and adjusted as per recommendations. A 5-10 HP electric motor or tractor can operate the machine. It can give output of 500-1500 kg/hr depending upon the size of power source and machine.

COMBINE HARVESTER

A combine harvester performs multifarious activities such as harvesting, threshing, separating, cleaning and collecting grains all in one go while moving through the standing crops. A bagging arrangement may also be provided with a pickup attachment. Mostly two kinds of combines are in operation namely self-propelled type and PTO driven type. The leading manufacturers of self-propelled combine harvesters in India especially in Punjab and Haryana are Kartar Agro, Vishal Agriculture Works, Preet Agro Industries, Claas India, John Deere, Dashmesh Group, Indo Farm, Swaraj, Malkit Agro Tech Pvt. Ltd., Hira Agri Industries, and New Hind Agro Pvt. Ltd. The size of the combine is given by its width of cut. The combine harvesters of various manufacturers have cutter bar width varying from 2500 mm to 4500 mm having an engine power of 45 kW to 95 kW. The main functions of a combine are:

- ✰ Cutting the standing crop
- ✰ Feeding the cut crop to threshing unit
- ✰ Threshing the crop
- ✰ Cleaning the grains from straw
- ✰ Collecting the grains in a container

A self-propelled combine may have more than 17000 parts. The major components are:

Header, Reel, Cutter bar, Elevator canvas, Feeder canvas, Feeding drum, Threshing drum, Concave unit, Fan, Chauffer sieve, Grain sieve, Grain auger, Tailing auger, Tail board, Straw spreader, Return conveyor, Shaker, Grain elevator, and Grain container. These components are grouped in 5 units. These units are described as follows (Figure 14.17).

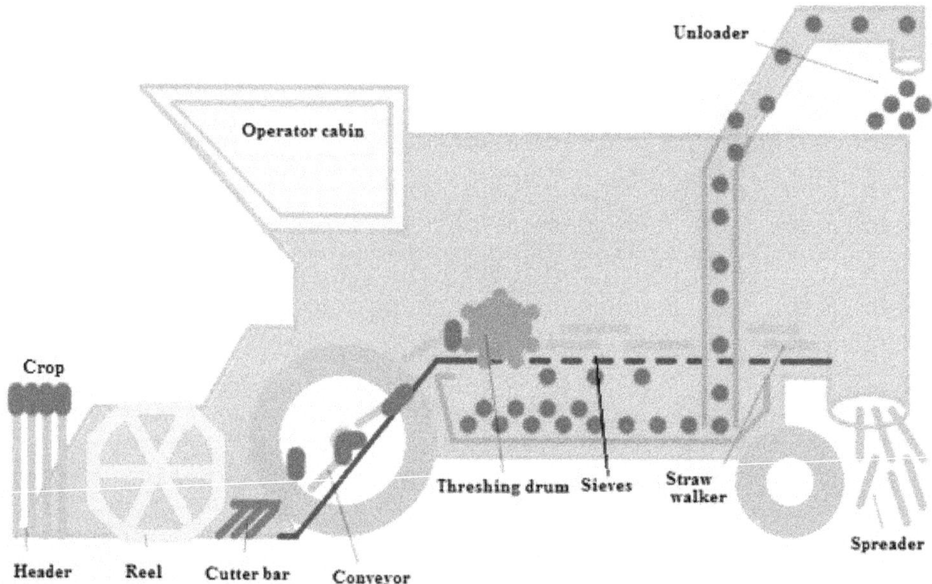

Figure 14.17. A Schematic view of a Combine Harvester.

Header and Gathering Unit

Header unit cuts, gathers and feeds the crops to the threshing unit. It comprises headers at the front, which has a pair of sharp pincers called crop dividers at either end. Generally speaking, the wider the header, the faster and more efficient is cutting operation. There are two kinds of header namely scoop type and T type, the former type being used in small and latter type in large combines. The header is hydraulically powered and can be raised, lowered, and angled in different ways from the cabin. The header can be dismantled and towed behind the harvester lengthwise so that it is easier to transport the combine through narrow lanes. A slowly rotating wheel made of wooden slats revolves in front of the cutter bar. It is also called the reel (or pick-up reel) that pushes the crops down toward the cutter. The reel gets power through suitable gears and shafts and is adjustable up and down.

Cutting and Feeding Unit

Harvesting is done by the cutting unit. It uses a cutter bar similar to that of a mower/reaper. The serrated edges of the knife prevent the straw from slipping during cutting operation. It cuts the standing crop and pushes it towards the conveyor. Chain and slat type of feeding unit receives the harvested crop from cross auger (pushes the harvested crop to conveyor) and delivers it into the threshing unit.

Threshing Drum and Concave Unit

Threshing takes place between the cylinder and concave unit of the combine. The basic components of the threshing unit of the combine are similar to a power thresher. As soon as the crop is threshed, the threshed material moves to a straw walker, which oscillates and separates the grains, which move to the cleaning unit.

Separating Unit/Cleaning Unit

Separation of grain from crop stubble begins at the concave. The remaining grains and straw then moves to a straw walker, which oscillate at 200-300 stokes/min. It drops the grain in the cleaning unit by tossing action. The cleaning unit consists of two sieves and a blower. While the sieves oscillates for better cleaning efficiency, the chaff is winnowed by the blower. The blower speed is adjusted in such a way that the chaff is blown off to the rear side of the machine. Finally, clean grain is delivered to the clean grain auger and an elevator takes the clean grain to the grain hopper. The un-threshed grains pass through tailing augur and go for re-threshing.

Handling Unit

It comprises clean grain auger where threshed grains are collected; an elevator for conveying grain to the tank and grain unloading auger for bagging.

To operate the machine and these units, the machines have several primary and secondary controls (Table 14.1, Shukla *et al.*, 2021). Name and functions of these controls are similar in all machines although there location with respect to the operator's seat may vary from one machine to another considering the operator's comfort and ease of operation.

Table 14.1. Some Selected Controls in Self-propelled Combine Harvesters and their Functions

Name of Control	Location of Control	Function
Primary control		
Steering wheel	In center and front of operators' seat	To control vehicle direction while in motion
Header assembly control	R.H.S. and in front of operators' seat	To lower or raise the header assembly
Ground speed control	R.H.S. and in front of operators' seat	To control speed of combine harvester
Gear shift lever control	R.H.S. and parallel to operators' seat	To change ground wheel speed
Brake pedal	R.H.S. and below the operators' seat (near to steering wheel column)	To stop combine harvester
Clutch pedal	L.H.S. and below the operators' seat (near to steering wheel column)	To engage and disengage engine with transmission system
Secondary control		
Reel position control	R.H.S. and forward of operators' seat	To raise and lower reel
Engine speed control	R.H.S. and forward or abreast to operators' seat	To control the speed of engine
Whole assembly operation control lever	L.H.S. and in the back of operators' seat	To transmit power from guide drum to cutter bar
Grain unloading augur control	L.H.S. and in the back of operators' seat	To discharge grain
Header engagement control	L.H.S. and in the back of operators' seat	To operate feeding
Thresher clutch control	L.H.S. and in the back of operators' seat	To transmit power from engine to guide drum

Requirements for Efficient Harvesting

For the successful use of combine harvesters or any other mechanical harvesting equipment, following field and crop conditions must be ensured.

★ Field must be fairly level without undulations to ensure smooth operation of the machine and uniform stubble length

★ Field should be as large as possible since the efficiency of harvesting machines increases with increasing size of the fields

★ Water must be drained in advance so that fields are relatively dry at harvest time of rice

★ For efficient functioning of small reapers and binders, crops should be grown in rows

Safety Associated with Combine Harvesters

★ Never attempt to dislodge stalks and sheaves with feet or hands while the machine is in operation. Always shut down the combine, turn off the ignition before removing plugged or lodged material

★ All combine adjustment should be made with machine shut off except those that are to be made while the machine is in operation

★ While working under the header, don't rely on hydraulic cylinder alone. Use locks and solid blocks while working beneath a header

During transport of the combine harvester, following additional points may prove useful.

★ Empty the grain tank to reduce the weight of the combine. It will also lower its center of gravity

★ Move the unloading auger to transport position

★ Remove the header as it will reduce the width of the machine

★ Make sure that slow moving vehicle signage, lights and reflectors are in good condition

Stubble Managment for Combine Harvester

Super Straw Management System (SSMS)

Stubble is defined as the crop residue left in the field after harvest, including stem, leaf and glumes of cereals etc. When crops are harvested using a combine harvester, the crop is not cut close enough to the ground, thereby leaving large quantities of stubble behind. To manage this stubble is a problem especially under pressure to get the field ready for next sowing season in the shortest possible time. As such, farmers burn the stubble on the field. Super straw management system (SSMS) attached at the rear of the combine (Figure 14.18 left) chops the paddy straw into minute pieces. Its blades, revolving at high speed of >1500 rpm, ensure that it is chopped and distributed evenly over the entire width of the cutter bar (Figure 14.18

Figure 14.18. SSMS on the Rear of the Combine (Left) and Inside View of Rotor and Blades (Right).

right). It facilitate the operations of other *in-situ* straw management machinery like happy seeder, no till drill or mould- board plough *etc.* in the fields. The SSMS has the option of switching on or off using a small metal sheet. This enables the combine harvester with SSMS attachment to be used as a traditional combine harvester without dismantling the SSMS. The only problem with the equipment arises when the moisture content in the paddy straw is on the higher side. Some specifications of SSMS approved by Government of India are as follows (Ministry of Agriculture and Farmer's Welfare, 2018). According to these guidelines the size of chopped paddy straw should not be more than 20 cm.

Sl.No.	Item	Specification
	Rotor	
1	Rotor diameter	165-170 mm
2	No. of lugs on rotor in a row	6
3	No. of rows in periphery	4
4	Length of pivotal flail	172 mm
5	Width of flail	50.8 mm
6	Thickness of flail	5 mm
7	Number of flails in one set	2
8	Spacing between flails of one set	38-40 mm
9	Distance between adjacent flail units	201.4 mm
10	Number of rows/bars of serrated blades	1
11	Number of serrated blades in a row	24
12	Spacing between serrated blades	50 mm
13	Clearance between pivotal blade and concave	22.5 mm
14	Overlapping of pivotal blade on serrated blade	60 mm (Adjustable)
15	Rotor speed	1600-1800 rpm

Sl.No.	Item	Specification
	Spreader	
1	Spreader curved width	1676.4 mm
2	Total number of flaps	6+2 (side)
3	Length of flap	470 mm
4	Distance between flaps	Adjustable
5	Spreader angle with horizontal	9° (adjustable)
6	Spreader angle with line of travel	15°
7	Spreader sheet thickness	2.5-3.0 mm
8	SMS sheet thickness	4-5 mm

Straw Balers

Balers have become quite popular because of the problems associated with rice straw burning. A straw baler is used to make bales, or bundles of the straw. It also tightly binds the bundles with twine, wire, or string. Balers use whole straw and therefore, no pre-processing is required (AICRP, 2008). Thus, the straw balers can be operated in the combine harvested fields where loose straw is picked up and baled. The standing stubble remains untouched. However, the best performance of the baler is obtained where the stubble shaver is also used after combine harvesting the paddy field. For example, a self-driven New Holland baler cuts the stubble to ground level using a rotary slasher.

Balers are classified in several ways.

- ☆ Based on the shape of the bales these are classified as round, rectangular or square balers. The rectangular balers are essentially plunger type balers

- ☆ Based on mobility bales are field or stationary balers

- ☆ Based on mode of mobility, these are classified as tractor mounted or self-propelled field balers

- ☆ Based on feeding, tying and bale-length controlling mechanism these are manual, semi-automatic or automatic balers

A field baler essentially has a unit to pick-up the material to be baled in the windrow using tines in the baler's reel. The straw is elevated and a conveyor moves it to the bale-chamber. A reciprocating plunger compresses the straw and moves it through the bale chamber. This movement of straw is resisted by applying force, which together controls the degree of compaction and bale density. In the compression unit straw is compacted with plunger at a speed of 540 to 560 rpm in bales of rectangular shapes. A metering device is used to control bale length forming bales of varying length from 40 to 110 cm. The height and width of bales are generally fixed at 45 cm. Other operations such as separating the bales, placing wire or string around the bale and tying and cutting are also performed in this section. The weight of bales varies from 15 to 45 kg depending on moisture content of straw and length of bales. Once these processes are over, the back of the baler swings open, and the bale is discharged. A tractor operated field baler is shown in Figure 14.19. Stationary balers do not have the first two components but have a feeding chute for manual feeding. The energy requirements vary widely from 0.57 to 1.05 kWh/tonne. Generally, balers require a tractor of 40 HP or more.

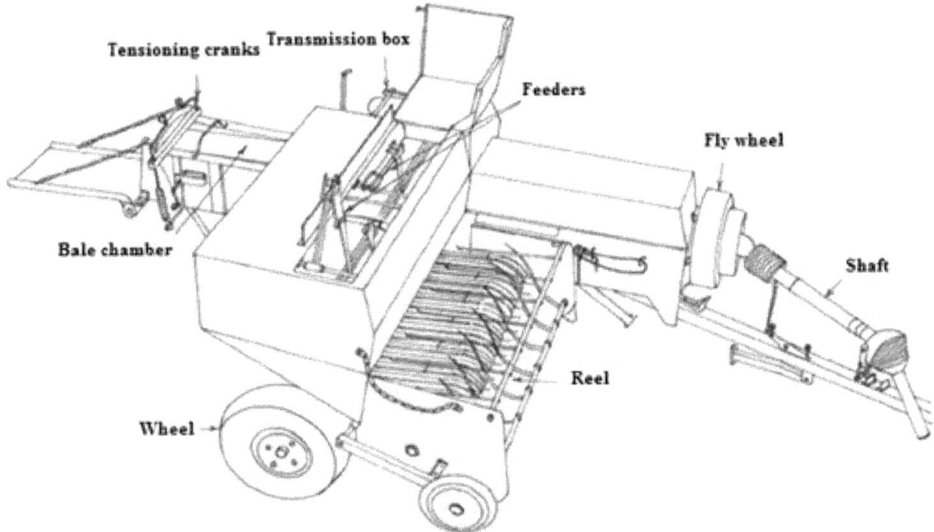

Figure 14.19. Labeled Diagram of a Tractor Operated Baler.

QUESTIONS (THEORY)

1. Name and describe the traditional harvesting tools. What are the limitations of traditional methods of harvesting?

2. Name and describe the various components of conventional mower/reaper.

3. Draw a neat sketch showing the various parts of a cutter bar.

4. What do you understand by the alignment and registration of cutter bar in a conventional reaper/mower.

5. Briefly describe the potato digger, turmeric harvester and tapioca puller.

6. What do you understand by threshing? Discuss at least two methods of traditional harvesting. List the limitations of the traditional methods of harvesting.

7. Name and describe the various components of a mechanical thresher.

8. Discuss the various kinds of cylinders and concaves used in threshers.

9. Discuss the principle and working of a thresher.

10. List the various precautions needed while operating a thresher.

11. What is a combine harvester? Name and describe its various units.

12. Briefly describe the super straw management system.

13. Briefly describe the straw baler.

14 Write brief notes on reaper-cum-binder, Olpad thresher, winnowing, manually operated drum thresher for paddy, maize dehusker-cum-sheller.

15. Differentiate between reaper and mower, horizontal rotary mower and flail mower.

Chapter 15

Garden Tools

A gardener often requires different kind of tools and equipment for carrying out various horticultural operations. Thus, *a garden tool, may be a hand tool or power tool, is any tool used to make gardening easy.* Accordingly, these tools and equipment are categorized as 'hand tools' and power equipment. Hand tools are less expensive, have multiple functions, and are easier to use in small spaces. Power equipment on the other hand are costly, require fuel, electricity or battery for functioning. On the other side, labour-intensive tasks are much easier to perform with the power operated tools. This chapter discusses some hand and power tools such as general purpose tools, pruning knife and secateurs, budding and grafting equipment, fruit harvesting equipment and other accessories.

GENERAL PURPOSE TOOLS

Most garden crops are transplanted. Some of the field operations required to accomplish a perfect transplanting are:

- ☆ Prepare the soil to make it friable
- ☆ Gently remove transplant from the original container without disturbing its roots
- ☆ Set the plant into the soil
- ☆ Firm the soil gently around the plant
- ☆ Apply irrigation

One may need one or more than one of the following tools to accomplish the aforementioned tasks.

1. Spade or *Kassi*
2. *Khurpa* or *Khurpi*
3. Shovel

4. Small trowel

5. Axe

6. Wheelbarrow, wooden board, or plastic sheet to transport plant

7. Watering can

Spade

Spade, known as *Kassi,* is a versatile tool to perform multifarious activities. It is used for digging or turning over the soil, cleaning of irrigation channels, making bunds in the field and small plots, *etc.* A spade is made of a blade fixed on a wooden handle (Figure 15.1 left). The blade is made of tempered steel or in some cases even of stainless steel. During operation, the blade is kept vertical, as it requires minimum efforts. A similar but fork type spade (Figure 15.1 right), commonly used as horticultural tool, is available in different types depending upon the type of job to be done. The digging fork is used for digging and managing the soil already turned by spade. The tines allow the implement to be pushed more easily into the ground. It can rake out stones and weeds and break up clods. A fork normally has four prongs. Prongs are either round or square in section. Forks for weeding the border areas around the fields have lesser number of forks normally 3-4 and are narrower and lighter than digging forks. Forks are also used to manually handle the harvested root crop such as potatoes. The soil attached to the tubers automatically drops through the forks while not much damage is done to the tubers. The prongs in such forks are much broader than digging fork and their number is also more, being 4 or 5.

Figure 15.1. A view of a Spade and a Fork Spade.

Khurpa

Khurpa or *Khurpi* (a smaller version of *Khurpa*) is quite a versatile tool (Figure 15.2 right). It is made of cast iron with a wooden handle attached to one side, It is used for digging soil and for weeding in small gardens or vegetable farms. It is traditionally used in a squatting posture for operations like tilling, bed preparation, weeding and digging at small scale especially in ornamental beds and pots.

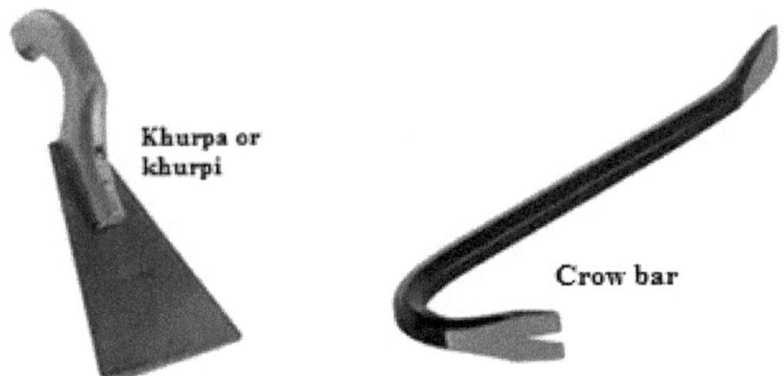

Figure 15.2. Views of a *Khurpa* and a Crow Bar.

Crow Bar

This too is used for digging holes or pits for planting and fencing (Figure 15.2 right). It is a hand tool fabricated from an octagonal bar. One of the ends is pointed while the other is spoon or chiseled shape. It is made either from the structural steel or from medium carbon steel. It is commonly used in hard or stony soils. *Kodali* is also a hand tool consisting of a wide blade and a wooden handle used for loosening and cutting the soil.

Shovel

There are different kinds of shovels. The variations in shapes of the blade and its tip change the type of shovel such as round, square, scoop *etc*. Any one of them is selected depending upon the task in hand. A shovel comprises a grip, a handle, a collar, where the blade is fastened to the handle and a blade, typically made of metal or, in some cases of plastic (Figure 15.3 left). The flat portion at the top of the blade is called the step. It is used to put the foot and body weight to push the shovel into the soil.

Figure 15.3. Photographic view of a Shovel (Left) and a Wheelbarrow (Right).

Trowel

Trowel, or also called mason trowel, is a pointed, scoop-shaped metal blade with a handle. It is also a multi-purpose tool for breaking up earth, digging small holes for planting, weeding, mixing fertilizer or other additives, and transferring plants to pots. Besides, a scoop-shaped tool with handle used for digging and lifting small plants without damaging the root zone for transplanting is known as transplanting trowel. Garden shovels and trowels are useful multi-purpose tools for general gardening tasks.

Axe

The axe is a simple hand tool available in various sizes and shapes consisting of a cutting edge and an eye for fixing the handle. It is forged to shape from a single piece of iron. The operator holds the handle with both hands at convenient position and the tool is raised to suitable height and struck with force against the material normally the wood. The penetration is caused through impact action, which shears the slice of wood. The axes are made from high carbon steel and the cutting edge is hardened to 550-650 HB. Axe is a multipurpose cutting tool used for felling and lopping of trees, splitting of logs for firewood and dressing of logs for timber conversion. Small axes are also used for clearing the bushes. Garden hatchet is also an axe-like tool used for cutting branches. It has a sharp edge on one side and is provided with a long handle.

Garden Line

A garden line is 6-20 m nylon rope having a pin at one and reel at the other end. The reel is used to wound up the nylon string for storage. A garden line is used to create perfectly straight rows and edges. It is used during seed sowing, trenching, lawn edging and transplanting operations.

Wheelbarrow

A wheelbarrow consists of a tray placed suitably on a pipe frame. A wheel in the front facilitates its easy movement. It is manually pushed using the handles. It is used to move heavy materials from one place to another, and also to collect trash in the garden (Figure 15.3 right).

Watering Can

A watering can is a portable container varying in size from 0.5 L to 10 L. It has a handle and a funnel used to water the plants by hand. It is made of aluminium metal or plastic. At the end of the funnel any device, like a cap, with small holes (sprinkler) can be attached to break up the stream of water into droplets. It is used for watering seedbeds, nursery beds and potted plants to avoid washing off the soil and damage to young seedlings (Figure 15.4).

Garden Hose

A garden hose is a flexible tube used to carry water attached to a hose spigot (tap). The common attachments available for use at the end of the hose are sprayers and sprinklers. These can be used to concentrate water application at one point or spread it over a large area.

Figure 15.4. A Watering Can with Sprinkler Attachment.

TOOLS FOR SPECIALIZED TASKS

Rake or Garden Rake

A garden rake is used for breaking up the soil surface into a fine tilth to make it ready for sowing. It is also used to collect weeds and stones, remove dead grass from the lawns, and to perform few other tasks often performed by the harrow in agriculture (Figure 15.5 left).

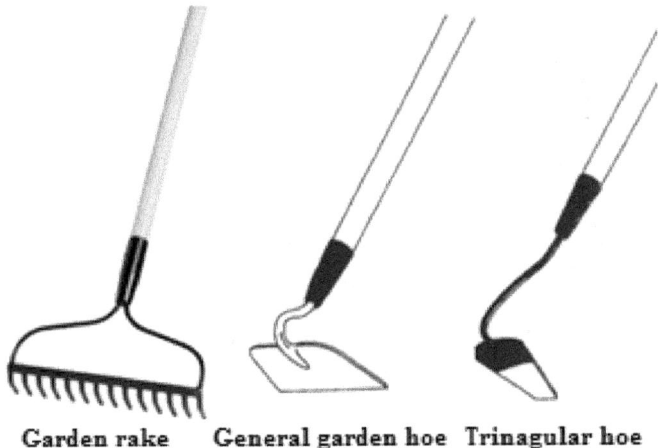

Figure 15.5. Garden Rake, General Garden Hoe and Triangular Hoe.

Garden Hoe

It has a long handle with a paddle and blade at the end. It is used to cultivate small fields and for carrying out the weeding operations. Several kinds of garden hoes are available to serve specific purposes. Two of these kinds are shown in Figure 15.5 (middle and right). Triangular headed hoes are also used for making shallow drill in conjunction with a garden line.

Hedge Shear

Hedge shear is used to perform various gardening operations like pruning, cutting and trimming of hedges and shrubs to give them desired shape by removing haphazard growth. It makes them to look attractive. Moreover, it can also be used for cutting of shrubs. It consists of two cutting blades with tongs. Each cutting blade is made of high carbon steel in a single piece. The tongs are inserted in wooden handles for better grip. The size of the shear is according to the size of the blades, varying from 15 to 30 cm in length and 0.8 cm in thickness. The hedge shears is manually operated (Figure 15.6 left), but diesel operated or electricity operated models are also available.

Grass Shear

Although hedge shear is also termed as grass shear and can be used as such but other kinds of grass shear, slightly different from hedge shear, are also available. Grass shear is used for trimming and side dressing of the lawn to maintain it in a proper shape. The most important part of a grass shear is its cutting blades, which are made of high carbon or alloy steel. The blades are sharp at the cutting edges. These are joined to a V shaped spring steel handle, which always keeps the shearing blades open. Cutting is due to the shearing action of the blades. The length of the blade varies in the range of 15 to 20 cm (Figure 15.6 right). A grass shear is operated manually but gasoline operated or electricity operated machines are also available.

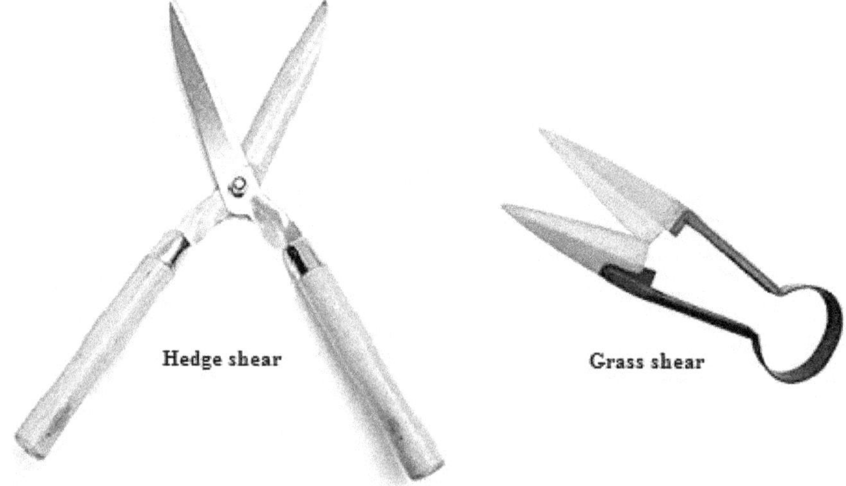

Figure 15.6. Manual Hedge Shear (Left) and Grass Shear (Right).

Dibber

Dibbers are used to make holes in seedbed so that seeds, seedlings or small bulbs can be planted. They are made of stainless steel sharp dibber and an ergonomically designed hard wood handle (Figure 15.7). The two are joined together using a ferrule. Variety of dibbers in various shapes and sizes such as straight dibber, T-handled dibber, trowel dibber, and L-shaped dibber are available and used depending upon the type of task and ease of doing work. A sharp dibber glides effortlessly into all kinds of soils.

Figure 15.7. A Straight Dibber for making Holes in a Seedbed.

Garden Sword

A garden sword is used to cut grass. The garden sword consists of a metal strip or blade with one of the edges sharpened to cut the grass. It is a simple manually operated hand tool used in a sitting or squatting position. The blade has a tang, which is inserted in the wooden handle and joined to it by riveting.

PRUNING AND LOPPING TOOLS

Plants differ from other living things in one important aspect that is their vigour can be improved by removing unhealthy or even healthy parts of the plants. This removal of plant parts is known as pruning or lopping. There is always some confusion between the terms tree pruning and tree lopping as both are associated with tree trimming. But the terms have a thin line of difference. Tree pruning is a *'process of carefully removing the parts of the plant including the healthy parts - to make it still healthier or more attractive or to improve other qualities, such as its fruit-bearing ability'*.

Note that when we talk of pruning, emphasis is on improvement, not removal. On the other hand, tree lopping is *'basically the trimming of branches of trees in order to reduce its size'*. Moreover, tree pruning and tree lopping also differ on the basis of season and species. While pruning for trees and shrubs may be required regularly, tree lopping can be done at infrequent intervals or when the trees reach a certain height beyond which it may pose problems. The tools and equipment used for pruning and lopping may be similar or quite different as both require cutting and trimming of trees. Tree pruning is often done making use of lopping hand sheers having the ability of cutting thick stems. Manual sheers, gasoline powered shears and electric sheers are used for tree lopping.

Pruning or Slashing Knife

Pruning and slashing knife or the billhook is a manually operated hand tool, which consists of a blade and tang joined rigidly to the handle (Figure 15.8). The tip of the blade is either hooked or curved in order to cut or slash the small branches or twigs of a plant or tree by pulling action. The blade is made from high carbon steel, tool steel or alloy steel and hardened to 45-55 HB. The blade is forged to shape with a sharpened cutting edge. The handle is made from good quality wood or plastic. To perform the cutting operation, the blade is engaged with the thin branch or twigs and the knife is pulled towards the operator. It is used for cutting and slashing of thin branches and twigs of plantation crops and orchards. It is mostly used for heavy pruning operations to remove unwanted and dense branches or twigs of plants.

Figure 15.8. A Pruning or Slashing Knife.

Pruners, Clippers or Secateurs

Pruners, clippers or secateurs are the most widely used tools for pruning shrubs, flowers, vines, and small growth on trees. They have applications in gardening, arboriculture, farming, flower arranging, and habitat management in nature conservation. Some important activities in which pruning shears find application are: cutting of branches, de shooting, disbudding, cutting of scion sticks, defoliation of leaves from the sticks and to 'top off small trees. Pruning shears on the basis of their action and shapes are named as single cut, double cut, parrot nose cut, roll cut, *etc.* A pruning shear has two blades, one sharp made of high carbon or alloy steel, while the other is blunt, made of copper. The blunt blade is used to support the branch to be cut. Both of these are fitted to a handle each usually made of mild steel covered with plastic tubes. On the basis of functional action, pruning tools are of three kinds namely Anvil, Bypass and Rachet type. Anvil loppers and pruners tend to crush the material being cut. Therefore, they are used where cleanliness of the cut is not as important as the removal. A bypass pruner or lopper is reserved for cuts that might affect the health of the plant and therefore, it is always kept razor sharp. Ratchet pruners are similar to anvil pruners, but they feature a mechanism that cuts in stages and causes much less stress to the wrists. Pruning shear are strong enough to prune 2 cm thick branches and making cuttings for propagation (Figure 15.9 left). The secateur is a handy tool to cut the branches or twigs, which are difficult to cut by pruning knives. Pruning secateur is used by gardeners for comparatively quick and easy pruning. The blade of the secateur is made from high carbon steel, alloy steel or tool steel. This tool has been found especially useful in pruning those plants which are thorny. The size of the secateurs is 150, 175, 200 or 225 mm (BIS, 1999). It is an essential tool of the gardener in plant propagation.

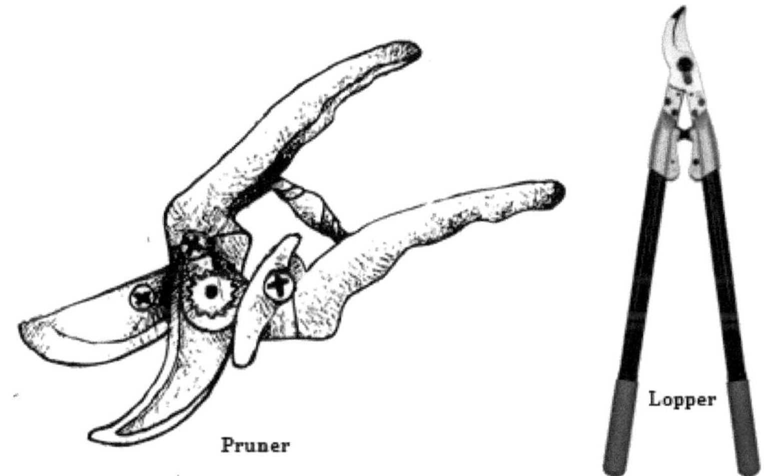

Figure 15.9. View of a Secateur or Pruner (Left) and a Lopper (Right).

Loppers

Pruning tools used with one hand are known as pruners (or hand shears) while large tools that require use of both the hands are known as loppers. These are used to cut branches up to 6.5 cm thick. This tool is similar to a pair of hand shears, but the blades are thicker and the handle is much longer (Figure 15.9 right). Like pruners, loppers also come in anvil, bypass and ratchet styles.

Pruning Saw or Hand Saw

The pruning saw is a manually operated hand tool for trimming of tree branches or cutting lumber, which are beyond the capacity of the secateurs or tree pruners. It is like a carpenter saw. It essentially consists of a serrated blade and a handle. The saw has large teeth and a thick blade. The blade has a longer pitch to avoid clogging during cutting of green branches. The blade either straight or curved (Figure 15.10 left) is made from tool steel having a carbon content of more than 0.7 per cent. The cutting teeth are made sharp and hardened to 45-48 HRC. The handle is made from good quality wood joined to the blade with the help of rivets. Repeated movement of the cutting blade over the branch helps to cut the same. The blades are designed to cut on the pull stroke, creating a clean and smooth cut. These saws can be used by one person or by two people together. The two-person saws have a handle at either end and are also sometimes called a lumberjack saw.

Power Driven Chainsaw

A chainsaw is a light and portable machine. Power driven saws are available in many sizes and styles. These are capable of cutting branches from 4-12 cm in diameter. A chainsaw operates with the use of a linked chain that rotates around a piece of steel and has specially designed teeth (Figure 15.10 right). It is capable of cutting through wood at high speed. The chainsaw is driven either by a small two- stroke petrol engine or an electric motor. Chainsaw powered by battery or electric cables is better for longer jobs as they will induce less fatigue on the user. The roller type chain has left and right-hand cutters spaced alternately along the length. A depth gauge controls the depth of cut made by the cutter. The chainsaw can be operated by one person. It is mainly used to trim dead or diseased wood from trees, to remove inconvenient branches or managing fallen trees. Heavier chainsaws discussed in Chapter 16 are used in forestry.

Figure 15.10. A Pruning Saw with Curved Blade (Left) and a Chainsaw (Right).

Pole Pruner

A pole or tree pruner is capable of cutting branches 3 to 3.5 cm in diameter. These pole pruners can reach to a height of 2.5 m or more (Figure 15.11), eliminating the need for a ladder in many cases. Electric pole pruners are now available that have made the task of tree top lopping easier.

Hedge Cutters

It is also called hedge trimmer. Portable hedge cutters are operated by a small petrol engine, electric motor or 12 V batteries. Cutting takes place between two blades, one of which reciprocates in close contact with a stationary one at a rate 33-66 strokes/min. Hedge trimmers may also have a circular cutting head. Some hedge cutters have an extension like a small saw on the end of a moving blade, which can be used to cut branches that are too thick not amenable to cutting by blades.

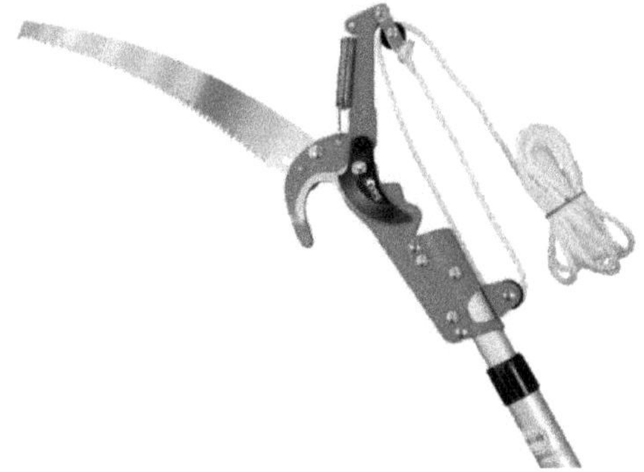

Figure 15.11. A 2.5 m Long Pole Pruner.

The Flower Scissors

It is a very simple hand tool, which resembles a surgical scissor. It is used to cut flowers with stems and other soft vegetative materials. It consists of two short blades with handles (Figure 15.12). The scissors blades are made from cast iron, mild steel, and stainless steel. Hard chrome plating is done on the scissors made from cast iron and steel. The handles have a grip to accommodate the fingers for actuating the blades. The blades are joined together with a rivet and perform a close cutting clearance operation. During operation, the tool is held in one hand and the material to be cut is brought in between the blades. The movement of the fingers closes the blades for performing cutting operation. The scissors are specified according to their lengths, which vary from 100-200 mm.

Figure 15.12. A Flower Scissor.

Draining and Tapping Knife

Commonly used in rubber gardens, it has a blade attached to a wooden handle and is designed to have a V-shaped cutting edge to make narrow channels in the bark of rubber trees. It opens the latex cells and facilitates easy tapping and draining of latex from the tree.

GRAFTING AND BUDDING TOOLS

Grafting and budding are asexual or vegetative methods of plant propagation. These are horticultural techniques used to join parts from two or more plants so that they appear to grow as a single plant. The new plant that grows from the scion or bud will be exactly like the plant it came from. *Grafting is a method of asexual plant propagation widely used in horticulture in which the upper part (scion) of one plant grows on the root system (rootstock) of another plant.* The best quality scion wood usually comes from shoots grown the previous season. Scions are severed with sharp, clean shears or knives. These are immediately placed in moistened burlap or plastic bags. The scion contains the desired quality to be duplicated in the stock/scion plant. Stem grafting is a common method of grafting used in plant science. In this method, a shoot of a selected desired plant cultivar is grafted onto the stock of another type. The vascular cambium tissues of the stock and scion plants are placed in contact with each other and are kept alive until the graft takes place. It may take a period of few weeks. Successful grafting only requires a vascular connection between the two tissues. Thus, a physically weak point may still exist at the graft, because the structural tissue of the two plants, such as wood may not fuse. The grafted plant derives certain characteristic of the rootstock as well such as hardiness, drought tolerance, or disease resistance. Budding is another grafting technique. *In this technique a single bud from the desired scion is used rather than an entire scion, which may contain many buds.* Most budding operations are performed just before or during the growing season. It should be made sure that the scion and rootstocks are compatible, that the scion has mature buds and that the cambia of the scion and rootstock match. When the bud has fused successfully, the stem above the new bud is cut so that the new bud grows. Grafting and budding can be performed only at very specific times. Both the weather conditions and the physiological stage of plant growth should be optimum at this time. The actual timing depends on the species and the technique used.

Benefits of Grafting and Budding

Grafting and budding is practiced for the following reasons.

- ☆ Changing the varieties or cultivars or produce certain plant forms
- ☆ To achieve higher yields and better quality products
- ☆ Optimization of cross-pollination and pollination
- ☆ Take advantage of rootstock's characteristics such as its drought tolerance, hardiness or tolerance to abiotic stresses
- ☆ Disease tolerance
- ☆ Benefit from inter stocks

☆ Perpetuate clones

☆ Repair damaged plants

☆ Increase the growth rate of seedlings

☆ As a practice for ornamental value additions

There may be few disadvantages of grafting and budding as well. One of them is higher prices of the seedlings.

Tools used for budding and grafting are dibber, budding and grafting knives, grafting tools, grafting tape, and pruning and lopping shears (Dixit and Singh, 2020). Budding and grafting knives are pliers-like tools that help to make accurate cuts with specific blades for different types of grafting techniques. These make budding and grafting easy. Metal parts of these tools are coated for protection against rust.

Grafting Tool

Grafting tool is designed for making the cleft graft. It is used when the rootstock's diameter is more than 2.5 cm. The wedge-shaped blade is used to split the stock, and the flat pick opens the cleft so that scions can be easily inserted. Once in place, the flat pick is removed and the cleft comes together that holds the scions in position (Figure 15.13 right).

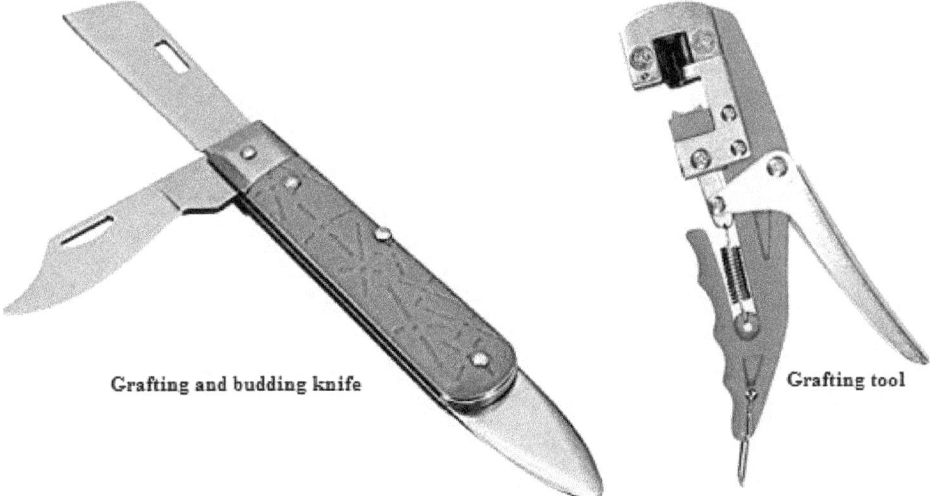

Figure 15.13. Pictures of Grafting-cum-Budding Knife (Left) and Grafting Tool (Right).

A grafting tool must have the following characteristics.

☆ Easy to use for quickly creating precise grafts

☆ Prepare stock for grafting

☆ Grafting blades should be replaceable to make sure that one always gets a clean cut. The clean cuts ensure perfect fit and improves the chances of the graft 'taking', resulting in much less wastage

☆ Made of materials that provide longer lasting strength and service

☆ It should contain both a pruner and grafting guillotine. It first allows preparing the plants to be grafted by trimming away excess leaves and twigs. Then it makes a perfect cut into the grafting stock using special grafting guillotine

☆ Ergonomic handles for ease of operation

Budding cum Grafting Knife

The budding-cum-grafting knife is a multipurpose knife to accomplish both the budding and grafting operations. The razor-sharp knives are foldable into the handle, thus making it as simple as a sturdy pocket knife, or a special grafting knife. Functionally, it consists of two blades each for budding and grafting, which are either joined to a common hinge or are fixed to the two ends of the handle. Both the knives are made of high carbon or alloy steel. The knives are 6.5–7.5 cm long and 1.5 cm wide (Figure 15.13 left). Since budding and grafting knives are designed specifically for these purposes, they should not be used for carving and whittling wood, as it will blunt the cutting edge.

Grafting Wax, Tape and Budding Rubbers

The slender pieces of the scion and the rootstock may dry out and are susceptible to diseases. Therefore, grafting wax either alone or in combination with grafting tape or string is used. It prevents the surfaces from drying out, keeps the scion and rootstock surfaces pressed tightly until they grow together, and keeps out water and air. The grafting wax is usually a paraffin-based wax also known as cold wax or hand wax. Another kind of wax is an asphalt-based tree wound dressing. It is also known as hot grafting wax, which needs reheating to use in grafting. It can be applied to the graft area with a small paint brush.

Grafting tape is used as a cover/protecting bandage when using a splint for broken/cracked branches/twigs. It is also used to hold the soil and roots for the root over rock design of Bonsai style. Electrical tape can be used, which can be cut later to ensure it falls off before girdling the branch. Cloth-backed grafting tapes that decompose on their own with time are also available.

Budding rubbers are elastic bands, typically 20 cm long, that do a fine job of maintaining adequate pressure.

OTHER EQUIPMENT AND ACCESSORIES

Sprayers

It is used for spraying insecticides, fungicides, herbicides, fertilizers and various other chemicals. A variety of sprayers are available in the market to suit different requirements of gardening. For potted plants and small areas manually operated sprayers can be used while for large areas power operated sprayers of different sizes should be preferred (See chapter 13 of this book).

Lawn Mower

Lawn mowers, used for cutting grass in lawns and fields, are available as manually operated and power operated (Figure 15.14). It consists of a cylindrical reel on the surface of which cutting blades are mounted in spiral fashion, anvil stationery blade, handle, drive wheel, drive pawl, grass gatherer (bucket) and roller assembly. The cutting blades are made of carbon steel and fastened to flanges mounted on a center shaft. Because of their spiral mounting, cutting blades while rotating result in progressive cutting action across the anvil blade. The anvil blade is a flat blade sharpened at the edges and can be adjusted. The grass is trapped between the rotating and anvil blades and cutting takes place due to shearing action. The machine has a front roller for adjustment of height of the cut and a grass box at the rear to collect the cut grass while the machine is in operation

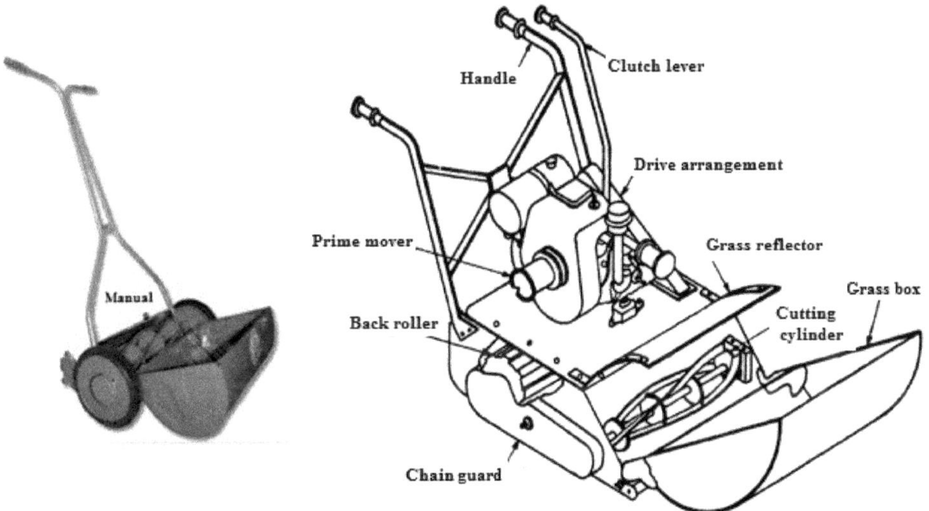

Figure 15.14. Lawn Mowers: Manual (Left) and Power Operated (Right).

Auger Bore Diggers

These are used to dig holes for tree plantations (See Chapter 16). Both power tiller operated auger digger and post hole tractor operated diggers are available. The former consists of a small frame with the provision to lower and raise the working element. Pits up to a depth of 45 to 60 cm can be dug. The diameter of the posthole is around 30 cm. Deeper and higher diameter bores can be dug using a tractor post hole digger.

PRECAUTIONS IN USE OF GARDEN TOOLS

☆ Handle each tool and equipment with care and according to the instructions given in the manual provided by the manufacturer

☆ Ensure that all equipment are functional

☆ Clean the equipment before and after use in the field

☆ During the spraying of insecticides, pesticides and fungicides, put on the mask, gloves, etc., besides ensuring that no person is around who might inhale the spray

CARE AND MAINTENANCE OF GARDEN TOOLS

☆ Store all machinery and equipment in a dry place

☆ Drain the tank and flush it with clean water, wash the pump nozzle before and after the use of a sprayer

☆ Regularly overhaul the tools/equipment/machines and replace the worn out parts

☆ Grease and oil all moving parts of the machinery as per the requirement

☆ Always keep all spare parts in the tool kit

☆ Regularly sharpen the blades of cutters

FRUIT HARVESTERS

Orchard management operations like harvesting, pruning, and other canopy management practices in tall fruit trees such as mango, citrus and sapota are difficult and labour-intensive. All these operations in India are carried out manually using ladders, poles, *'dhoti'*, chippers, *etc.* For example, workers use a pole with a hook at the end to detach the fruit. Later on, fruits are collected manually for final packing and transport to destinations. Mechanical harvesting machines can significantly ease the fruits harvesting operation. These machines are broadly classified as contact machines and mass-removal machines. The contact machines consist of the positioning mechanism and the picking hand or arm. The mass-removal machines operate by applying external force, shaking the limb or tree trunk mechanically or applying force in the form of a jet of water or air to vibrate limbs, foliage, and twigs. In these machines, fruit either drops on padded catch frames or on ploughed ground.

Contact Machines

The contact machines are based on the principle of selective picking. Two most commonly used contact machines are:

☆ Fixed height type with wooden head and metal cutting blade

☆ Adjustable height type with metal head and metal cutting blade

Fixed Height Harvester

This type of harvester is made of a light bamboo pole of about 12 mm in diameter and 3.5 m long (Figure 15.15). The other components include a wooden head, a metal blade, a spring, a fruit collecting net, and a hand lever and nylon rope. The cutting mechanism consists of the wooden head, the cutting blade and the retaining spring. The wooden head is stationary and acts as a support against the cutting blade when the fruit stem comes between the metal blade and the head. The blade is properly sharpened and heat treated so that it does not wear fast. The blade always remains in open position by means of a soft spring of 64 mm long and 90 mm in diameter.

It moves against the tension of the spring during its cutting action. A hand lever is fitted to the pole at about 380 mm from the bottom end. The metal cutting blade on the top and the hand lever are connected by a nylon rope of appropriate length. By pushing the hand lever down, the blade moves against the head to cut the fruit. The blade, immediately after the cutting operation, is pulled back to its original open position by the spring connected between the blade and the pole. A nylon net is fixed around a ring of 300 mm diameter made of 8 mm steel rod just below the cutting mechanism. The net is around 270 mm deep. The cut fruit is collected in the net to avoid any damage to the fruit.

Figure 15.15. A View of Fixed Height Fruit Plucker/Harvester.

Adjustable Height Harvester

The cutting mechanism in this kind of harvester is essentially similar to the fixed height type harvester. The variations are only in its head and the pole length. The cutting head, fabricated from an angle iron, is arranged at the top of a 12 mm pipe by means of a coupling. This helps in dismantling and assembly of the cutting mechanism from the pole. The pole is also composed of two telescopic galvanized iron pipes of 12 mm and 18 mm sizes to achieve adjustable height up to 5.4 m. A clamp is used to fix the poles to the desired height. This arrangement helps in varying the height to which the equipment is able to harvest the fruits. A base plate at the bottom of the pole with U-clamp arrangement helps to freely move the pole in any direction as per the fruit orientation. The only limitation is its weight, being heavier than the fixed type harvester discussed in the previous section.

Manually Operated Sapota Harvester

The harvester is commonly used to harvest sapota but can be used to harvest any other small fruits like lemon, *etc.* It consists of a main body made of PVC having cylindrical shape (Figure 15.16). The upper end of the body is closed while the bottom end is open to which a nylon net is tied for collecting the fruits (Figure 15.16). The other end of the net is closed with the help of a string. Two fingers with small sharp blades cut in V-shape are provided at the closed end of the body of the harvester. A metal holder fixed on the lower surface of the body holds the bamboo/pole of required length. During the harvesting operation, the fingers hold the selected fruit from the bunch. By pulling the harvester, fruit gets detached from the bunch. A gate is provided on the body for entry of the harvested fruits initially in the body, which then rolls into the net. The stretched string at the closed end of the net is loosened to unload the harvested fruits.

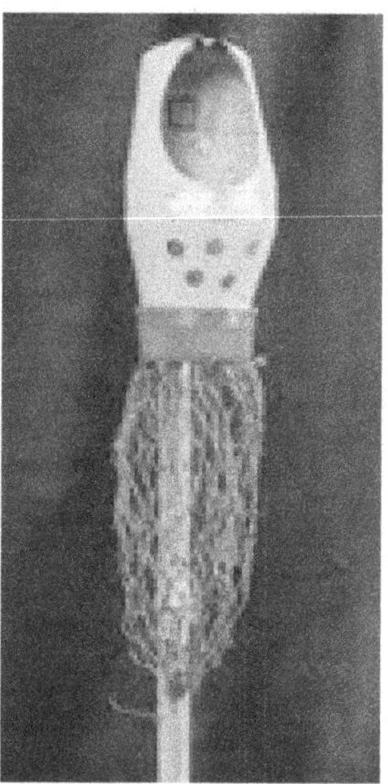

Figure 15.16. A Pictorial view of Sapota Harvester.

Platform Type Fruits Harvesting System

Tractor or other vehicle operated or self-propelled hydraulic multi-purpose systems are now available. Many new ones are being designed and developed at the research Institutions. CIAE, Bhopal has developed a self-propelled machine hydraulically powered by 8.7 kW petrol engine. It is suitable for medium height fruit trees having the vertical reach of 6 m. It has a load carrying capacity of 200 kg. It can be operated at maximum ground speed of 3 km/hr. This machine is designed with a 2.20 x 6.32 x 1.89 m platform to reach to the fruits for easy picking. Lifting and lowering of the platform, forward and backward movement, and steering of the machine are controlled by the operator on the platform. Most modern machines now use mechanical fingers, which are quite flexible and imitate human fingers. Most advanced versions of these machines use robots or mechanical arms and the torsion method of detachment.

Mechanical Harvesting of Fruits

Very little efforts have been made to mechanize harvesting operation on fruit and vegetables in India. It is mainly due to abundance of labour and lack of organized large-scale fruit or vegetable farming. Although mechanical harvesting can sort out many problems yet it often causes a reduction in harvested crop value per unit area. It is mainly because of the following factors:

☆ Crop do not mature uniformly

☆ Fruit and vegetable damage

☆ Higher field losses

☆ Reduced produt quality

In spite of this and considering the future need of mechanical harvesters, it is high time that field applications and design and development efforts are initiated in right earnest. Mainly, two types of mechanical harvesters are used in harvesting operation of fruits. These are continuous canopy shake and trunk shake systems.

Continuous Canopy Shake System

A continuous canopy shake system is a self-propelled unit that shakes the tree canopy. It causes the fruit to fall from the tree onto a catch frame. It is why, it is also known as shake and catch system. The core unit of this mechanical harvesting system consists of a series of whirls stacked horizontally. Each whirl in turn consists of a series of approximately 1.8 m long, 4-5 cm diameter rods called tines. These are mounted to the whirls and are connected to a central drum. The drum is mounted on a controlling unit. The interaction of the harvesting tines with the canopy begins at about 1 m from the soil surface. Since the drum can be elevated up to 6 m above the soil surface, the machine is able to harvest up to 6 m of canopy height during one pass. The tines penetrate into the tree canopy, and shake it horizontally to remove the fruits. The shaking frequency of the tines can be adjusted depending upon the force required to remove the fruit. In some models, even the provision to adjust the angle of the tine penetration to the angle of a hedged tree has been provided. The chances of damage to the fruit are minimal as the harvested fruits fall on a padded catch system. If such a catch system is not provided/used, land is ploughed before the harvesting operation begins. Self-propelled harvesting units often work in pairs, one unit harvesting one side of the tree. The tractor drawn canopy shake and catch units work similarly but without the catch frame. Additional labour may be required to harvest the left out fruits on the tree as well as to collect fallen fruit that misses the catch frame. Studies have shown that there are no adverse effects of such systems on tree health, productivity, or tree longevity. Following practices help to maximize fruit harvest without any damage to the fruits.

☆ Hedging of trees insures maximum tine penetration into the tree canopy, thereby increasing fruit harvest

☆ Topping of trees to a maximum of 5-6 m height minimizes fruit splitting from impact with the ground or catch frame

Trunk Shake System

These machines work on the principle of accelerating each fruit so that inertia force developed exceeds the bonding force between the fruit and the tree. A trunk shake machine simply shakes the tree so that fruits fall on the ground. The harvested fruits are later on picked-up by the labours. Tractor mounted cable shakers, fixed stroke boom shakers and boom type impact knockers are some of the machines designed for this purpose. These trunk shaking machines so far have been quite commonly used in the nut industry for fruits like walnut, almonds *etc.* Several design modifications are required before these can be used to harvest other fruits. As an example, nut shakers in general use low frequency - high amplitude shaking. On the other hand, harvesting of other fruits may require high frequency – low amplitude shaking with quick wind up and wind down to avoid damage to the tree trunk. Such changes in design will help to shake large range of tree sizes say trunks over 6 cm to fully grown trees. Besides, it will help in gentler fruit removal with lesser damage. It will also improve capability of the machine to harvest table fruits. Another issue that confronts the use of such machines in harvesting fruits arises from the fact that the fruit removal in the trunk shaker type machines occurs simultaneously over the entire tree and the falling fruits are distributed over the entire catching space. It may require considerable amount of power and may not be suitable for very large trees.

QUESTIONS (THEORY)

1. Name some of the manually operated general purpose tools and describe at least 3 of them.

2. Write short notes on hand saw and power driven chainsaw.

3. Differentiate between grafting and budding. List the benefits of grafting and budding.

4. What do you understand by a grafting tool? List the various characteristics of a grafting tool.

5. Discuss the role of grafting wax and tape in grafting.

6. List some care and maintenance activities in respect of garden tools for their long life.

7. Describe fixed and adjustable height fruit harvesters.

8. Write a descriptive note on mechanical fruit harvesting giving brief introduction to the equipment used for this purpose.

9. Differentiate between pruning and lopping, spade and a fork spade, anvil and bypass pruner.

10. Write short notes on axe, shovel, secateurs, pole pruner and lawn mower.

Chapter 16

Tools and Machines for Forestry

Forest utilization deals with felling, extraction, on site/landing processing, loading of tree logs or other tree parts on trucks and disposal of the forest produce. Various kinds of machines are used to perform these operations. For example, trees are felled manually using axes, and manually operated or power saws. Some delimbing is done using axes and machetes. Many kinds of multi-functional equipment such as forest harvesters have also come in use. Four broad categories of equipment are: animal power and logging, small sized machinery and ancillary equipment, tractor-based equipment and cable systems. The focus of this chapter is mainly on power sources used in forestry followed by an introduction to the manual and simple mechanical equipment used in forest utilization such as tree felling cutters and harvesters, skidding, loading, log transport and chipping.

POWER SOURCES IN FORESTRY

Hydraulic (Water) Power

Hydraulic power was commonly used in forwarding operation of logs to industrial mills. In that sense early logging operations were mainly restricted to the banks along rivers and large streams. Lumberjacks manually felled trees, trimmed off branches, and cut them into sawn logs. Animals dragged these logs to the river's edge where these were thrown into the stream. The river divers rode these logs to keep them moving and to prevent jams. The logs were taken out at the destination by the divers and labour.

Animal Power

Like agriculture, animals were the main source of power used to extract timber throughout the world until the advent of the farm tractor. It was only during the 1940s to 1960s, the predominance of animal extraction in forestry shifted to mechanized extraction. Small woodland owners used horses, mules, and oxen for timber extraction. Elephants were the main power source in many countries notably, Sri Lanka, Thailand and Myanmar. Even today, elephants are used on some large scale in Myanmar. Horses are highly maneuverable and have the flexibility to handle a variety of timber sizes and lengths. On the other hand, animal power logging is a physically demanding operation. Coupled with the unavailability of skilled operators especially *mahouts*, the elephant riders, trainers and keepers has severely limited the use of animal power. Still it can complement the work alongside forwarders, processors and skylines especially in areas too small to be harvested using conventional methods.

Advantages of Animal Power

- ☆ Low overhead and investment costs
- ☆ Lower operational and move in costs
- ☆ Damage to residual stand is relatively lighter than with mechanical power
- ☆ Stands do not have to be thinned as heavily as in case of machinery
- ☆ Soil disturbance is low
- ☆ Easy to pull down hung-up trees
- ☆ Acceptability with the public is quite high

Disadvantages

- ☆ Takes longer to log an area
- ☆ Horses have limited power to work only with smaller logs
- ☆ Limited availability
- ☆ Limited extraction distances
- ☆ Daily attention required for the animals even if they are not engaged to do work

Mechanical Power

It is mainly derived from gasoline, diesel engines and ranges from farm tractors with or without modification to all terrain vehicles to self-propelled multi-purpose vehicles.

Farm tractor

The agricultural tractor with pneumatic wheels with or without certain modifications is the most widely used vehicle in traditional logging for the extraction of wood. A 4-wheel power drive tractor with 50-70 kW power is capable of doing a good job. It can easily be equipped at the back with a winch or trailer with hydraulic knuckle-boom loader. Since forestry work can be more demanding than agricultural operations, basic modifications such as window and radiator protection screens/grills, belly pans, deflector and cab protection bars, valve stem protectors and front weights may be required. When PTO-driven attachments are used, 3-point hitch may be sufficient for some equipment, but for heavier equipment it may have to be modified to a four point hitch (Nilsson and Forshed, 2017). A major problem in using a farm tractor in forestry is due to the orientation of the operator's seat. Almost all tractors are designed exclusively for a forward facing operating position. On the other hand, many forestry implements such as grapple–loader trailers, skidding and forwarding grapples, winch processors require the operator to face the rear of the tractor. Reversible seats and controls and increased space at the rear of the cab can overcome this difficulty.

All-Terrain Vehicles (ATVs)

ATVs developed mainly as recreational vehicles have found widespread use in forestry due to their versatility and low price tag. ATVs are 3- or 4-wheeled motorized vehicles, which have been designed primarily for off road use. They have handlebars like a motorcycle, and the rider straddles the body of the vehicle. With large soft tyres, ATVs have a relatively high centre of gravity. ATVs are approximately 1 m wide and typically weigh from 200 to 300 kg. In forestry, ATVs are primarily used for transporting individuals or materials such as plants, fertilizers, tools *etc.* ATVs can also function as prime movers in the extraction of timber using appropriate attachments. ATVs used for logging related activities should have at least 300 cc engine capacity and 4-wheel drive. Desirable modifications in the recreational ATVs may include the addition of tracks or traction chains, wide tyres and placing counterweights on the front of the vehicle. Track conversion can greatly improve ATV traction and stability. Other equipment that may be added to ATVs include a front bumper, a protective belly pan under the engine, foot guards and a recovery winch. One must take into account the limitations of ATVs while deciding to perform a job or using equipment. Generally, the total load hauled should not exceed the weight of the ATV. Many accessories and attachments have been developed for use with ATVs and some of these are widely used.

Self-Propelled Machines

Large number of self-propelled forestry machines such as fellers, bunchers, delimbers, forwarders, log loaders, skidders, processors, harvesters, mulcher, stump remover and multi-function versions of these machine types are available. In these machines power source is integrated with the machines. Electronically controlled operations of many of these machines provide hundred percent accuracy and quality without much supervision.

OPERATIONS

While different methods may be used for wood harvesting, all of them involve similar type of operations. The sequence of operations may however vary from one site to another site depending upon the forest type, the kind of product desired and the equipment/technology available. The common operations are:

Tree felling: Cutting the tree from the stump and bringing it down.

Topping and debranching (delimbing): Cutting off the unusable tree crown and the branches.

Debarking: Removing the bark from the stem. Note that this operation is usually done at the processing plant and may not be needed at all in fuel wood harvesting.

Extraction: Moving the stems or logs from the stump to a place close to a forest road where they are temporarily stored before being transported.

Bucking: Also known as log making/cross-cutting the trees in lengths required for the intended use of the log.

Scaling: Determining the quantity of logs produced, usually by measuring volume or weight.

Sorting, piling and temporary storage: Logs being of variable dimensions and quality are classified into assortments as per their potential use as pulpwood, saw logs and so on. The cleared area where these operations, as well as scaling and loading take place is called landing (site).

Loading: Moving the logs onto the transport medium, typically a truck, and attaching the load.

All these operations require tools that may differ in size and capacity varying from hand tools to sophisticated computer controlled self propelled harvesters. We will look at few of them under various heads.

HAND TOOLS AND ANCILLARY EQUIPMENT

Main implements used for felling are: axe, saws operated either by hand or power. The same tools can also be used for conversion. Some of these implements are also used in gardening. These have been introduced in Chapter 15 but are being discussed in a greater detail in this chapter.

Axes

Most felling axes can be used for felling, trimming, splitting and grubbing. Felling axes vary in size and shape depending upon the kind of job. An axe has two parts namely the metal head and the handle.

Axe head: The axe head has a cutting edge and an eye or hole to fit the handle. The best type of axe head is either a solid piece of iron with a steel wedge weld into it, or two tapering pieces of iron between which a steel wedge is fixed. The edge is carefully tempered and made strong enough to withstand the stresses and strains. It easily penetrates the wood and is slightly curved to protect the corners from breaking. The sides of the blade are slightly convex so that it enters the wood easily and emerges clear. The weight of the head is generally in the range of 1.5 to 2.0 kg depending on the type of wood to be cut.

Axe handle: Generally round handles are used in India as these are easy to fit and can be easily replaced. However, round handles are liable to slide around the eye. The oval shaped handles are now used to minimize the slip. The length of the handle varies from 70 to 120 cm

Hooks

Cant-hook: It is mostly used as a lever to roll, stop and turn logs (Figure 16.1 left).

Log hook: It is used for dragging, lifting and rolling of logs (Figure 16.1 right).

Figure 16.1. A Cant Hook (Left) and a Steel Log Hook (Right).

Hand Sappie or Nursery *Kutla*

It is used for hoeing and weeding in forest nursery. Hand sappie, depending upon the size, is of three types namely Type A - Large hand sappie (Large *Kutla*), Type B - Medium hand sappie (Medium *Kutla*) and Type C - Small hand sappie (Small *Kutla*). The blade of the hand sappie is made of carbon steel. The hardness of sappie blade is 38 ± 3 HRC. Handle is made of any hard wood.

Debarking Spud

A debarking spud is used to debark the logs. It has a blade and a handle along with a blade support and supporting block. The blade is made of hard carbon steel having hardness of 45±3 HRC. It can be operated either by a single or by both the hands (Figure 16.2). The mass of single hand barking spud does not exceed 600 gm including the wooden handle (BIS, 1992b).

Tree Felling Lever

Felling lever is used for felling the trees. It consists of a plate made of carbon steel and a lever made of steel tube fitted with a rubber handle grip for greater safety (Figure 16.3 left). Small teeth on the front section of the lift plate dig into the

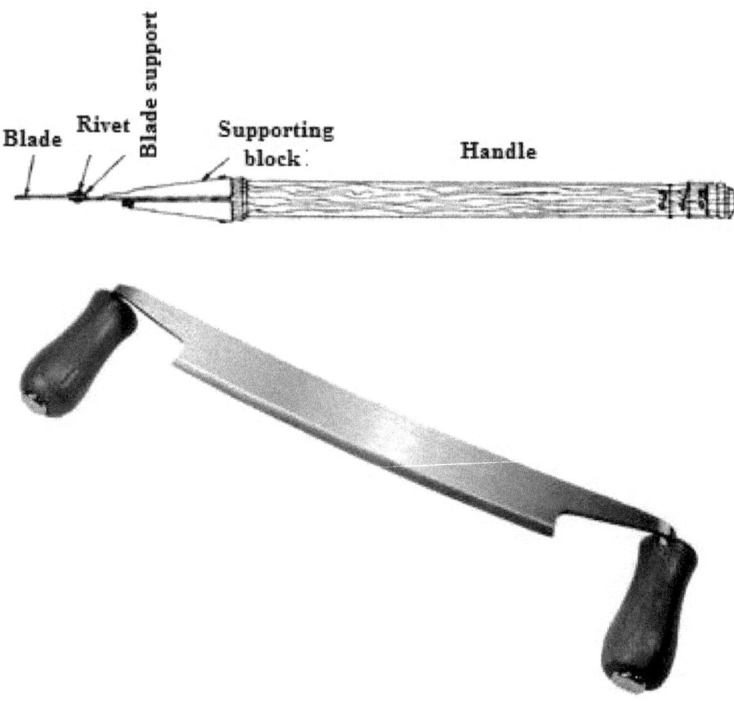

Figure 16.2. Hand Operated Debarking Spuds.

stump and stop the lever from slipping out of the saw kerf. Some models have a cant hook to help with turning a stem or dislodging a hung-up tree. The hardness of plate is 38 ± 2 HRC. The mass of the felling lever does not exceed 1500 gm. Steel models weigh about 2500 gm.

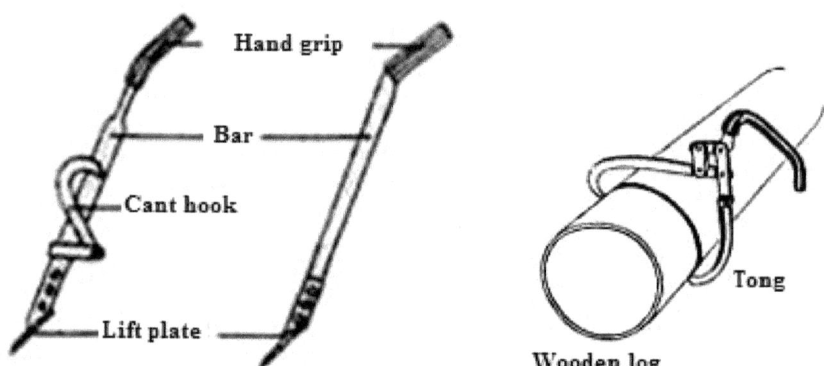

Figure 16.3. Tree Felling Lever (Left) and a Tong (Right).

Tongs

Tongs are used for grasping, holding or lifting wood. Regular sized tongs can be used to lift wood of diameter up to 20-25 cm (Figure 16.3 right). Several forest workers now use two hooks instead of tongs. Two men log tongs are used for medium size logs (normally up to 250 kg mass). A wooden pole of 60 mm diameter and 2000 mm length is supplied along with the tong at the option of the purchaser.

Pickaroon

A pickaroon is a metal-topped log handling tool with sharply bent tip having a wooden handle. It is a short pole, 85-100 cm long, with a curved pike or hook for drawing or pulling small logs. Wood handle is partly painted with orange color making the tool visible in the forest (Figure 16.4).

Figure 16.4. A Photograph of a Pickaroon.

Stalk Puller

Stalk puller is used for uprooting congested bushes such as lantana, *etc.* It consists of tongue, wheel, knurling plate and handle. Whereas tongue is made of mild steel other components respectively are made of rubber, high carbon steel and steel tube respectively. A stalk puller freely and smoothly moves on the wheels. The jaws have a minimum opening of 100 mm (BIS, 1998b).

Measuring Stick

Fitted with marking ends, a measuring stick is used to measure the log length.

Stem Tightener

Its function is to prevent the stems from splitting at butt ends. It consists of 13 mm wire rope having a steel core. It is laid round the stem just above the felling cut and tightened with the help of a lever mechanism. The wire rope is held fast with the help of a clamping device which consists of a guide groove for the rope, a movable support and a wedge.

SAWS

Saws are used for felling, cross-cutting, ripping logs into scantlings, sleepers and other converted materials. Saws are broadly categorized as hand saws and power saws.

Hand Saws

These are cheap, easy to use and maintain. Although their output is low compared to power saws, yet these are commonly used because of the earlier defined characteristics. Saws are also characterized as one-man or two-man, according to whether they require one man or two men to operate. A saw essentially comprises a broad blade or plate of steel of small thickness, one edge of which is toothed. Some of the saws are discussed in the following paragraphs.

A crosscut saw is a general term for any saw blade that cuts the wood perpendicular to the wood grain. Although smaller crosscut saws are used for fine work like wood-working, large saws are used in forestry for operations like felling of trees and conversion into logs. Length, breadth, thickness and tooth form are the main characteristics of a saw blade that needs to be chosen properly to suit a particular job in hand.

Length: It is guided by the diameter of the tree and the ease of arm movement. Suitable stroke length lies in the range 80 to 100 cm. An allowance of 5 to 15 cm should be given at both ends for fixing handles and protection of arms. General purpose crosscut saws of following lengths are commonly used in forestry operations.

Diameter of Stems (cm)	Length of Saw (cm)
30	140-150
30-70	165-170
80 and over	180-200

Width/Breadth: Narrow saws give less friction during cuts than the broad ones. The saws which are either 85 mm broad along the whole length, or 90 mm at the ends and 120 mm at the middle, have proved to be the best.

Thickness: Saw blade should not be very thin as guiding and controlling the saw becomes difficult. Moreover, a thick saw produces broad kerf (width of saw cut) and consume higher energy to saw. For 80 mm wide and 165 cm long saw, a thickness of about 1.82 mm has proved quite suitable.

Teeth: Saw teeth perform the following three functions:

- ☆ Cut through the fibers
- ☆ Break loose the cut fibers
- ☆ Remove the loose fibers (sawdust) from the kerf

Mostly saw with peg tooth and raker tooth are used (Figure 16.5). In peg-tooth saws, all the three functions are done by one kind of tooth. In raker-tooth saws the first of the three actions is done by a group of cutters cutting on alternate sides of the kerf. The second and third actions are done by the raker tooth following the group of cutters. Peg-tooth type saws are easy to maintain. Raker-tooth type saws are preferred by professional workers. They cut faster but require more skill in maintenance.

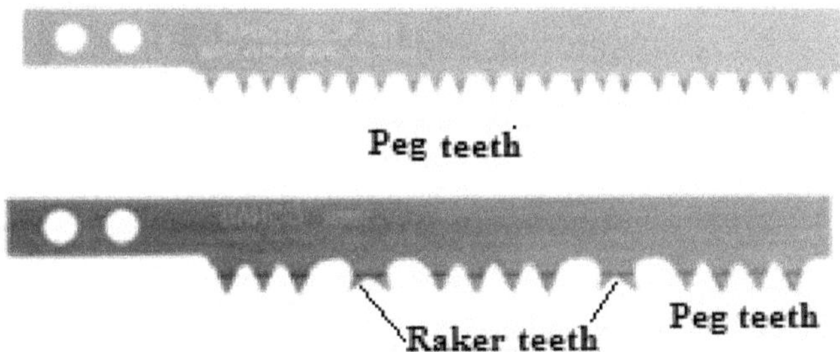

Figure 16.5. Saw Blades with Peg Teeth and Peg and Raker Teeth.

Bow Saw

Bow saw or one man saw is used for felling of small trees and poles or cutting other small sized products. It carries a frame with a thin narrow saw blade. The oval section of the tubing makes a comfortable grip. The frame may be a solid frame which uses standard length of the blade or may consists of two halves, one sliding into the other. The frame is adjustable and can be used with blades of different length. The blades of the bow saw are generally 90-105 cm long.

Power Saws

Power saws are now commonly used for logging operation. The power saws are either chain, circular or drag type, and are used mainly for felling and cross-cutting. One man power saw is chainsaw type. In many countries, chainsaw is the only choice for cutting of small and medium sized plants. It is used in activities such as tree felling, logging, pruning *etc.* A chainsaw is a portable, mechanical saw which cuts with a set of teeth attached to a rotating chain that runs along a guide bar. Power saws are generally powered by a small gasoline engine attached with the machine. Advantages and disadvantages of power saws are listed as follows:

Advantages

☆ Trees can be felled to low stumps

☆ Production rate per hour is high

☆ Light weight and therefore easily portable

☆ Higher precision and less loss of wood

☆ Can be used for sawing horizontally and vertically

Disadvantages

☆ Moving power saws from one tree to another is sometimes cumbersome and time consuming especially in rugged and sloping terrain and in places with heavy undergrowth

☆ Operation of power saws requires skilled manpower

☆ Maintenance and replacement of parts may pose problems in interior locations

Chainsaw

Chainsaws are widely used in small-scale tree harvesting in most countries. A chainsaw is mainly composed of a two-stroke internal-combustion engine. A centrifugal clutch moves a sprocket wheel to which the chain with cutting links is hooked. A separate oil tank is used to release oil between the grooved bar and the serrated chain to reduce friction. The chainsaws are normally classified based on engine power, bar length and weight. One amongst the several options is selected as per the requirement of operation in hand. In the felling operation, chainsaw is used to make a notch that decides the direction of the fall of the tree. It is produced with two cuts, one obliquely and the other horizontally. The two cuts must meet perfectly without overlapping, forming an angle of about 40-45°. The depth of the horizontal cut has to fall between 1/4 and 1/3 of the diameter of the tree in the cutting zone. A second cut is performed in the opposite side (back cut) that releases the tree and causes its fall. Felling is normally carried out by one person. In special cases, another person may be needed when the trees lean in the opposite direction to their fall. He inserts aluminium or plastic wedges using a sledgehammer or the eye of a hatchet in the back cut to avoid the tree to sit on the bar of the chainsaw. The wedges also cause the tree to tilt so that it falls to the ground in the chosen direction.

Limbing or delimbing as it is sometimes called is the systematic removal of branches from a felled tree. Limbing takes approximately 50 per cent of the time a worker needs to harvest a tree. Delimbing should be performed making cuts as close as possible to the stem so that no stumps are left. The operation begins with the removal of smaller branches that allows good access to the large ones. Thick branches can even be cut into sections. For this purpose begin from the outside and work inward toward the tree trunk until the entire limb is removed. For thick limbs, it is useful to cut from both sides of the limb. It is always useful to use a lighter and more compact chainsaw for smaller branches. If the tree is lying directly on the ground, tree is rolled over to access branches on the underside.

Cutting to size operation consists of cross-cutting the stem into pieces of a set length, based on customer requirement and final use. If the trees are perfectly supported all along their length on the ground, a single cut perpendicular to the longitudinal axis of the stem is made so as to cover the whole diameter. This operation has to be carefully planned. Operator should never stand below the tree trunks lying on a slope. The trunk may roll over and cause injury to the operator. Besides what is stated in previous paragraphs, chainsaws continue to be single most dangerous piece of machinery in forestry in spite of many improvements made over the years. Many serious accidents and health problems are associated with their use. Therefore, it is imperative that operators are well trained and equipped. They should have a clear understanding of how trees are felled, trimmed and cross cut.

Self-Propelled Forest Harvester

The first fully mobile timber harvester was introduced in 1973 by a Finnish engineer. Harvesters can be employed effectively in level to moderately steep terrain for clear cutting forest areas. Harvesters are routinely used to cut trees up to 900 mm in diameter. The harvesters are built on a robust all terrain wheeled, or track type vehicle. A diesel engine powers both the vehicle and the harvesting mechanism through hydraulic drive. A boom, similar to that on an excavator, reaches out from the vehicle and carries the harvester head. The felling head is composed of a grapple with three or four metal arms that hold the tree during and after felling. A proper cutting instrument such as a clipper, a disc or a cutting chain such as a chainsaw is placed at the base. The chainsaw is much stronger than any manually operated saw. It is equipped with two or more curved delimbing knives which reach around the trunk to remove branches. Two feed rollers are used to grasp the tree. The rollers are driven in rotation to force the cut tree stem through the delimbing knives. In more sophisticated harvesters, diameter sensors are used to calculate the volume of timber harvested. A measuring wheel is also provided to measure the length of the stem as it is fed through the head. The operator handles all the operations from inside the vehicle cabin. A computer not only helps to simplify mechanical movements but also keeps records of the length and diameter of cut trees. Combi-machines are now available which combine the felling capability of a harvester with the load-carrying capability of a forwarder, allowing a single operator and machine to fell, process and transport trees. The main limitations are that such equipment require highly trained and skilled operators besides being very costly.

EXTRACTION

Extraction as already stated involves moving the stems or logs from the stump to a landing or roadside where they can be processed or piled into assortments. The commonly recognized major types of extraction systems are:

Ground-skidding: In this case stems or logs are dragged on the ground by draught animals, humans or machines

Forwarders: The stems or logs are carried on a machine/trailer

Cable systems: The logs are conveyed from the stump to the landing by one or more suspended cables

Aerial systems: Helicopters or balloons are used to airlift the logs

SKIDDING

Skidding is lumbering operation that consists in the collection of trees from which the branches have been removed and hauling them from the cutting area to loading areas on logging routes. In the past, manual and animal power was used for this activity. Carrying small logs in wire slings, by crews of four to eight men, two per sling, is still common for handling in the log yard or for loading on railway cars, trucks and boats. Currently mechanical power comprising skidding tractors called skidders and winches is used. Log skidding with farm tractors is one of traditional extraction systems in flat and on mountain slopes with a grade not exceeding 22°.

Skidding Attachments

Several kinds of attachments are used to efficiently skid the wood with the help of tractors (Russell and Mortimer, 2005). The selection and use of the attachment is decided on the basis of horsepower of the tractor. Normally the attachments are grouped under two categories *i.e.* for tractors less than 30 HP and tractors of higher than 30 HP.

Cable Skidders

These skidders pull logs behind them. Steel ropes are bound around each log to pull the logs. One end of the tree, trunk, or log is attached to the tractor, and the other end drags it on the ground. In the case of trees, the butt usually rests on the tractor, and in the case of trunks, the top.

Winch

A skidding winch is the most widely used tractor-mounted attachment. Skidding winches use a cable and choker to pull one or more trees by the tractor in semi-suspended or suspended hauling. Mostly used in mountains and in swampy areas, it can winch logs to the tractor from about 30 m in the woods and is also useful for other tasks on the woodlot. The skidding winch is normally attached to the 3-point hitch and receives its power from the tractor PTO. The three point hitch allows the tractor to raise and lower the winch for winching or skidding operations. Power winches are available to suit a very wide range of tractors varying from 30 to 100 HP (Figure 16.6 top).

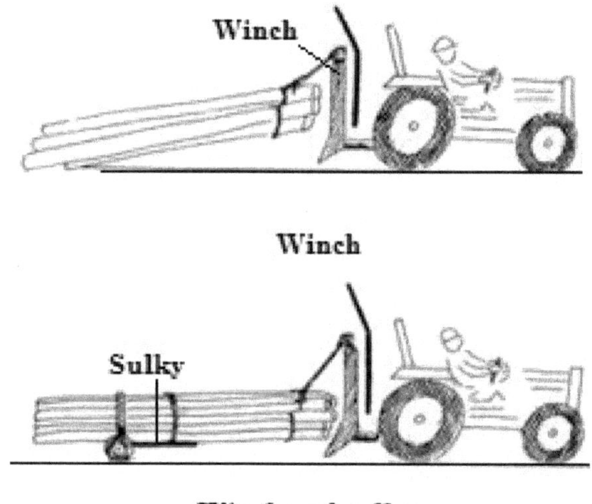

Figure 16.6. Application of a Winch (Top) and Winch and Sulky for Skidding using a Tractor (Bottom).

Advantages

- ☆ Low to medium cost
- ☆ Suited to a wide range of tractors and sites
- ☆ When using winches in difficult terrain one can drop the load until tractor moves to a more favourable terrain and again winch the log from a distance

Disadvantages

- ✩ Limited application in thinning
- ✩ Skidding often produces dirty logs, which can cause difficulties at the processing stage
- ✩ Contribute to both soil and tree damage

Sulkies

Sulkies are typically used with an ATV though they can also be used with smaller tractors (<30 HP) or horses. A sulky consists of a cradle two-wheeled unit that can be pulled behind the vehicle. It can also have a winch and shaft. One end of the log (s) is winched into the cradle and the other end skids behind. It can be converted into a trailer for forwarding by attaching a pole to some rear wheels with a cradle. A jack or a manual winch may be used to lift the log into the cradle. The addition of a sulky to the winch makes a good working pair (Figure 16.6 bottom).

Skidding Bar and Butt Plate

A notched skidding bar is a heavy metal bar with slots along the top attached to the back of the tractor by the two draft arms of the three point hitch (Figure 16.7 left). A butt plate is similar but higher (Figure 16.7 middle). These are used for skidding logs which are choked and attached on the bar. The chain is attached to the log, put through the slot and then the log is lifted up. This keeps the logs cleaner compared to pulling the logs along the ground. It is very simple equipment and does not require much modification to the agricultural tractor except to add front-end weights. Most wire cranes and skid winches are equipped with skidding bars. Power requirement is about 30 to 50 HP. The advantages lie in its low cost and being suitable for a wide range of tractors. The disadvantages include that the productivity is low. Other disadvantages include their unsuitability in more difficult and wet sites and limited application in thinning.

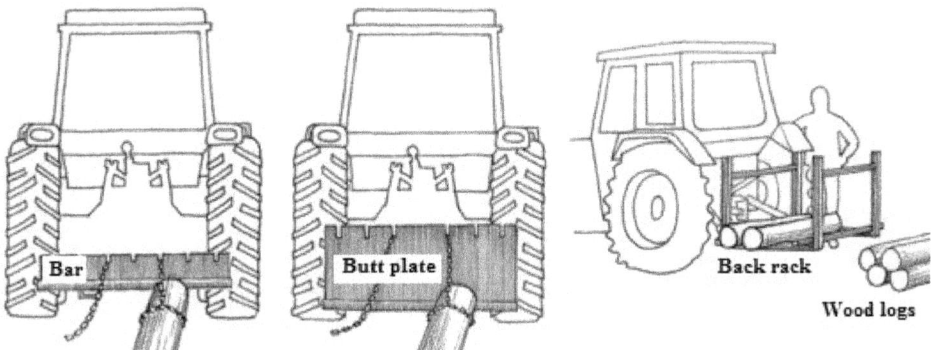

Figure 16.7. Skidding Attachments: a Bar (Left), a Butt Plate (Middle) and a Back Rack (Right).

Back Fork/Back Rack

A back fork or back rack (Figure 16.7 right) is useful for carrying cut lengths of wood up to 2.5 m. These are low cost extraction implements mounted on the 3-point hitch to allow logs to be forwarded from the site. The logs are piled in the rack manually and a chain is used to secure them. The load is then lifted off the ground to a height needed to clear obstacles on the way. The back rack is inexpensive and useful for small amounts of wood. It can also be used to carry other materials around the woodlot. The wood stays clean as the wood lot is held off the ground. The disadvantages include that it requires manual loading and works well only on fairly even sites to ensure load stability. Tractors used with this attachment need good front to rear weight ratio. Therefore, 4-wheel drive tractors with compensatory front weight attachments are needed for the use of this attachment.

Grapple Skidders

These skidders scoop up a bunch of logs with the help of hydraulic arms mounted on the back of skidders.

Skidding and Forwarding Grapple

Large hydraulic grapples mounted on the 3-point hitch can be used for transporting cut-to-length logs or full pole length timber. The operator reverses up to the logs or timber stack and 'grapples' the load, which is then hydraulically lifted for transportation. Tractors need good front to rear weight ratio and are best suited to 4-wheel drive tractors with compensatory front weight attachments. The actual size of the tractor depends on the size of the implement and the weight of the wood to be carried/skidded but minimum size requirement is about 55 HP. The major advantages are that the equipment is relatively inexpensive, wood can be extracted clean and operator can work from his cabin. Major disadvantage is that it does not have the flexibility and versatility of skidders.

The Skidder

This is a 60-90 kW tractor having good maneuverability. At the back a hydraulic grapple is mounted on the tractor's hydraulic lift that ensures both horizontal and vertical movement of the grapple. Alternatively or in some cases together with the grapple there is a winch with one or two drums. Fifty percent or even more weight of the machine is distributed at the front to avoid rearing of the machine during skidding. The speed of this kind of skidders can go up to 30 km/h.

Forwarders

A forwarder differs from a skidder in the sense that large felled logs are moved from the stump to a roadside landing by lifting logs at least two feet above the ground. These are truck- like vehicles that carry logs without dragging them. It reduces soil impact. Moreover, these can be used in muddy conditions. Forwarders are categorized on the basis of their load carrying capabilities. The smallest towed by all-terrain vehicles can carry loads of 1-3 tonnes. On the other hand, forwarders towed by farm tractors can carry between 12 to 15 tonnes. Medium-sized forwarders used in clear-cutting are able to carry 12-16 tonnes.

Steel Cables

Logs are carried aloft by steel cables driven by traction ropes in tough terrain. These cables are hung across canyons and valleys and are driven by traction ropes. A carrier moves on a carrier cable. On the rehaul, the carrier brings cable chokers to the felling area, and on the return pass it pulls the logs, trunks, or trees to the loading area.

LOADERS AND TRAILERS

Manual loading is often practiced in developing countries. Simple hand tools like hooks, levers, sappies, pulleys and so on discussed in the section on hand tools are used. Human and animal power or their combination may be involved. Currently, loading is mechanized, usually with swing-boom, knuckle-boom or front-end loaders. Swing-boom and knuckle-boom loaders may be mounted on wheeled or tracked carriers or on trucks. They are usually equipped with grapples.

Wire Crane Loader

Wire crane loaders can be used in conjunction with horse drawn forwarding trailers or any other forestry trailer for winching, loading and forwarding wood. The wire crane comprises a PTO powered winch with a high A-frame and stabilizer legs. The operator walks up to each bundle, puts a wire around it and follows the bundle as it is winched back to the tractor.

Advantages

- ☆ Suitable for thinning with widely spaced racks
- ☆ Short wood can be extracted clean
- ☆ Wire-crane loaders are the least expensive mechanical loading systems

Disadvantages

- ☆ It requires well trained operators for safety considerations
- ☆ Substantial wood handling and walking is required
- ☆ Productivity tends to be lower than the grapple loaders

Grapple Loader

A grapple loader is like a hydraulic crane with a grapple used to pick up single trees, or bunches of logs for loading or unloading a trailer. They can be mounted on the tractor itself, its 3-point hitch or on the trailer.

Advantages

- ☆ Allows fast efficient loading and unloading of logs eliminating manual handling
- ☆ Large payloads can be moved than by skidding systems
- ☆ Tractor stability remains unaffected even if mounted on the trailer

The only disadvantage is that it is quite costly.

Forestry Trailers

A wide range of timber trailers are available for use with animal power to more sophisticated power driven trailers. A trailer is fitted with one or two powered axles, a strong chassis to resist stress during transport and systems that increase its maneuverability. It consists of a chassis, which is formed by steel axles, connected to the tractor by a drawbar with a hook with revolving eye. The smallest capacity trailers may have only 2 wheels but usually most forest models have a 4-wheel construction. It has a low centre of gravity and wheels have wide tyres. Most popular tractor drawn model may have a pay loads ranging from 3 to 8 m^3 with 5 tonne models. These can also be used in agricultural applications. The investment for a tractor and trailer is substantially low compared to a conventional forwarder.

Advantages

☆ High load capacity to quickly remove large volumes

☆ Short wood can be extracted clean

☆ Can be used in agricultural applications

Disadvantages

☆ Detailed planning and site layout is required especially in thinning

☆ It is expensive especially the more sophisticated ones

CHIPPERS

Until now chips have been a less valuable product. In recent years, the growing demand of chips in alternate energy sources, paper manufacturing and as mulch has expanded the use of chips and so its demand has increased manifold. Moreover, production of chips makes it possible to recover on an average 20-30 per cent of biomass that would otherwise be wasted while clearing the site. Advantages of chipping are listed as under.

☆ It is possible to produce fuel chip from unmarketable timber

☆ Leaves sites neat and tidy

☆ Allows unwanted material to be transported efficiently off-site

Some disadvantages are:

☆ The process is expensive

☆ Feeding chippers by hand is hazardous

☆ Its use and scale in forest plantations is subject to policy decision in relation to renewable energy

A chipping/cutting machine that reduces tops, branches or whole trees into small chips is known as chipper. Most chippers are typically made of a hopper with a collar, the chipper mechanism itself, and an optional collection bin for the chips. A tree limb is inserted into the hopper and then into the chipping mechanism. The chips exit through a chute and can be directed onto the ground or directly into a truck-mounted container. Typical output of chips is of the order of 2.5–5 cm across sizes. Depending upon the scale of operation, chipping is classified as small-scale chipping, in-forest chipping and large-scale chipping. The machine is selected on the basis of the scale of operation. Depending upon the cutting instrument used, chippers have been classified as disc, drum, and endless types. The chipping operation is either performed directly on the field or at the landing site depending on the available machinery. Small portable units, typically powered by gasoline engines, can be moved to the work location manually or with the help of a garden tractor. Tractor mounted PTO-driven units are powered by the PTO of a tractor and mounted on a three-point hitch.

Disc Chipper

The basic design of the disc chipper developed in 1889 is known as the Wigger chipper. The functional components of a disc chipper schematically shown in Figure 16.8a include a flywheel and knives mounted radially on the chipper disc like spokes on a bicycle wheel. The cutting speed is in the range of 20 to 40 m/s. The chip length is normally set to about 20- 25 mm (in the grain direction). The average chip thickness of a disc chipper is about 3-5 mm. The wood chippers rely on energy stored in the heavy flywheel, which is accelerated by an electric motor or internal combustion engine or a tractor. A 45 to 60 HP tractor can chip logs up to 15 cm in

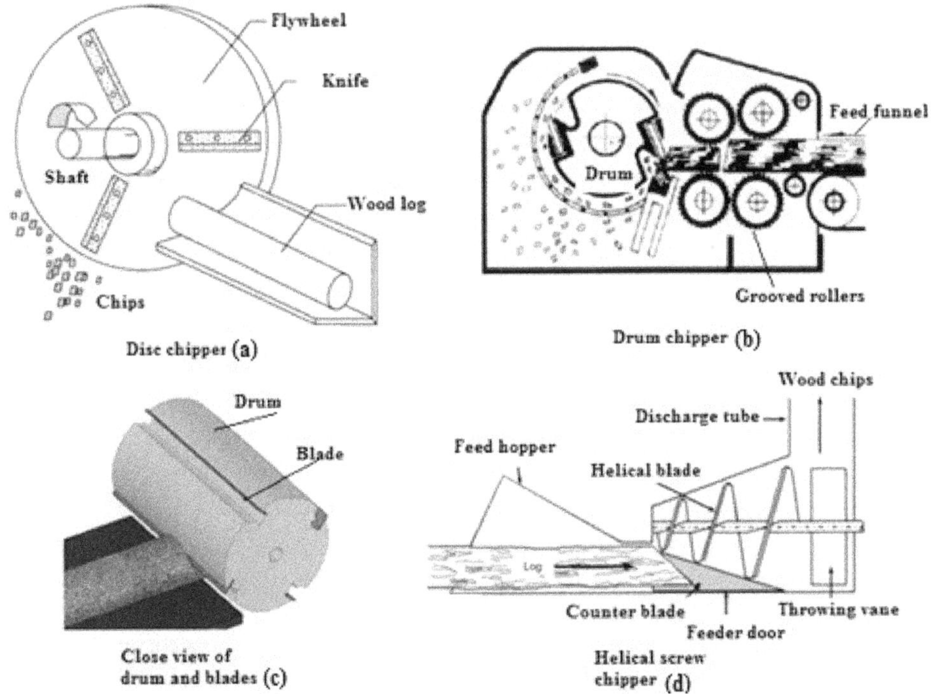

Figure 16.8. Various kinds of Chippers used in Forestry.

diameter. Commercial grade disk-style chippers can chip logs up to 45 cm diameter (Figure 16.9). The power requirement in that case is quite high. A high-speed airflow fan is used to blow out the wood chips out of the cutting chamber. The chips made from disk chippers can be used in the manufacturing of particle board.

Drum Chipper

Drum chippers employ a large steel drum powered by a motor (Figure 16.8b). The drum is mounted parallel to the hopper and spins toward the chute. Blades mounted on the outer surface of the drum cut the material into chips and propel the chips into the discharge chute (Figure 16.8c). Chipper blades are made from high grade steel and usually contain a minimum of 8 per cent chromium for hardness. Commercial grade drum-style chippers can chip log diameter as large as 60 cm. Large wood chippers are also equipped with grooved rollers in the throat of their feed funnels (Figure 16.8b). Once a branch is gripped in the rollers, rolling rollers move it towards the chipping blades at a steady rate. These chippers use a set of hydraulically powered wheels to regulate the rate of feed material into the chipper drum.

Figure 16.9. A Photographic View of a Tractor Operated Disc Chipper.

Other Machines

Some of the machines may also have different kinds of blades such as helical blades (Figure 16.8d). Much larger machines for wood processing are called whole tree chippers and recyclers. These can typically handle log diameters of 60–180 cm and may employ drum, disk, or a combination of both kinds of chippers. These machines usually require power in the range of 200 to 1000 HP. Being heavy, they require a semi-trailer truck for transport. Smaller models can be towed by a medium duty truck.

The general precautions for operating a chipper or shredder are as follows:

☆ Operate the unit only on stable, uncluttered ground to avoid potential slips and falls. Avoid their operation on slopes

☆ Do not leave the unit unattended

☆ Make sure the chute or hood is securely attached and chips exit in a safe direction

☆ Keep the hands and feet away from the discharge chute or any other moving parts

☆ Make sure that the PTO shaft is properly covered to avoid entanglement

☆ Make sure the material to be chipped is free from stones, metal and foreign objects

☆ Bystanders should not be nearer than the safe distance of 15 m

☆ Do not operate the equipment when fatigued or under the influence of alcohol

☆ Make adequate arrangement of light

Stump Grinders

Stump grinders vary in size, being as small as a lawn mower or as large as a truck. Most comprise a cutter wheel with fixed carbide teeth. The high-speed disk with teeth grinds the stump and roots into small chips. Cutter wheels receive power from hydraulic cylinders that push the cutter head through the stump. Some stump grinders can be attached to tractors or excavators. They can either completely remove the roots of the trees or recover the central part of the roots.

QUESTIONS (THEORY)

1. Name and describe the various power sources used in forestry.

2. Define skidding. Name some the attachments for skidding with tractors. Discuss winch and sulky, the two attachments for skidding with a tractor.

3. Write descriptive note on a power chainsaw and discuss its various functions.

4. What is a chipper? What are the benefits of using a chipper in forestry? List the necessary precautions during operation of a chipper.

5. Write short notes on: all terrain vehicles, debarking spud, tree felling lever, hand saw, self-propelled forest harvester and stump grinder.

6. Differentiate between cant hook and log hook, skidder and forwarder, disc and drum chipper

Chapter 17

Miscellaneous Equipment

Farmers routinely perform many activities on their farms. Some of these activities and related equipment have been discussed in the previous chapters of this book. Still many activities and the related equipment have been left out. Many of these activities also cause drudgery as well as take a toll on the farmer's time and money. Therefore, several other tools and machines have been designed and developed to easily accomplish these tasks. This chapter includes few common activities of general nature and few crop specific tools for sugarcane crop in particular to illustrate the plethora of tools/equipment/machines used at the Indian farms.

CHAFF CUTTER

A chaff cutter, also known as *tokka* in Punjab and Haryana, is a mechanical device to cut chaff, grass, green and dry fodder crops and paddy straw into bits and pieces. The material is then fed to the animals sometimes alone but mostly after mixing with other forages and/or concentrates. Chaff cutters have evolved from the basic hand operated machines into tractor driven chaff cutters and commercial machines. In the latter type, both single and three-phase motors operated models are available in the market (Figure 17.1). These are driven at various speeds and can cut the chaff to various sizes depending upon the animal preference. A tractor driven chaff cutter is a portable version such that the machine can be directly operated in the field. The hand operated machine in general comprises a fly wheel (roller), frame stand, cutting wheel, feeding trough (*Parnala*), and wooden handle (Figure 17.1). The cutter is a steel welded structure of convex shape in the design of word U.

Hand Operated Chaff Cutter

Two kinds of hand operated chaff cutters are used on Indian farms. The first kind that has almost become obsolete consists of a feeding tray, a curved blade fixed on to a spring-loaded lever, an anvil which also acts as a cutting blade and a suitable framework. The cutting action in this cutter is similar to the commonly used shearing machines in the workshops. The blades, made from medium carbon steel or low alloy steel, are hardened and tempered to about 45 HRC (Hardness Rockwell C scale). Fodder is fed in the tray and pushed by one hand. The other hand and a leg actuate the curved cutting blade, which progressively shears the fodder into small pieces.

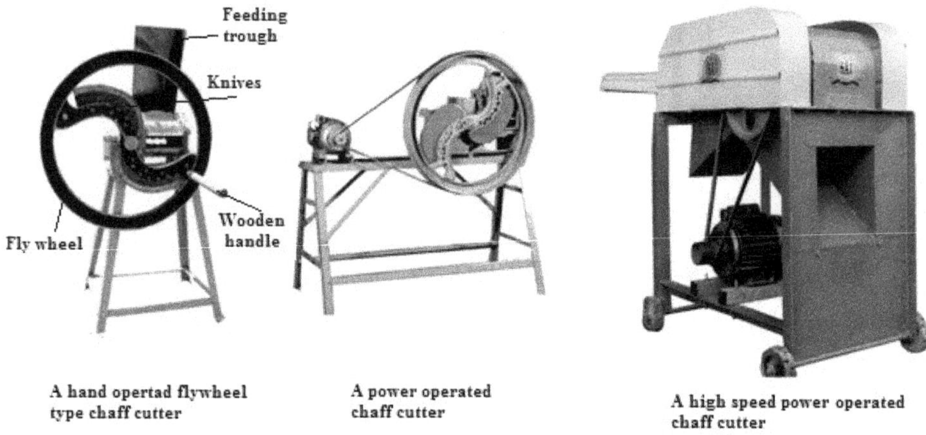

A hand opertad flywheel type chaff cutter

A power operated chaff cutter

A high speed power operated chaff cutter

Figure 17.1. Hand and Power Operated Chaff Cutters.

Flywheel type chaff cutter is a more commonly used machine. It consists of a cast iron flywheel and radially mounted knives (Figure 17.1 left). The knives are fastened on the flywheel by means of two or three counter sunk bolts. While manual chaff cutters normally have two curved blades, the power chaff cutters have two to six straight knives depending upon the type of chaff cutter and size of chaff required. The average working speed of a hand chaff cutter with two knives is about 35 rpm, while it is higher at about 50 rpm for chaff cutters having one knife. Hand operated machines are provided a feed-table or merely a feed box in which feeding is done by hand. Feed rollers of the hand-operated machine have variable spacing and are operated by the worm mounted on the flywheel shaft. For proper cutting of silage, the knife should operate close to the shear plate without striking. A set of screws is provided to adjust this clearance.

The bullock operated chaff cutter is quite similar to the hand operated one. The flywheel of this chaff cutter is relatively smaller in diameter with two or three knives mounted on it. The machine is operated at about 250 rpm. Gears and universal joints are used to increase the slow speed of the bullocks to the desired rpm. The capacity of the animal drawn chaff cutters is 4 to 6 times higher than that of a hand driven machine.

The feeding mechanism of a power operated chaff cutter consists of an apron generally provided with metal or wooden slates to give a positive feed. The apron moves on a table of about 2 m long to accommodate tall grasses or fodder crops. Besides, it has two spring-loaded corrugated or toothed feed rollers that compress and feed the uncut material to the cutting head. The working speed of power operated chaff cutters ranges from 600 to 1000 rpm.

Care and Adjustments of Hand Chaff Cutter

A hand operated chaff cutter is a very simple and useful machine. It is why; it is very popular amongst the Indian farmers. Moreover, it is easy to convert hand chaff cutters into power chaff cutters using 1 HP electric motor or small engine. The following general procedure of maintenance and lubrication should be adopted to get long efficient service from the machine.

- It must be properly installed on a rigid foundation
- Before the machine is put into operation, all loose nuts, bolts, and screws should be tightened and its gears and bearings including other moving parts should be properly lubricated
- The blunt knives should be regularly sharpened by hand filing. If the cutting edge is too blunt, it may be sharpened on a grinder to the proper bevel only on one side
- The clearance between the knives and the shear plate should be properly adjusted for effective cutting
- After the day's work is over, wipe off the moisture from the knives and dirt from the gears and bearing parts
- The flywheel of the machine should be kept in locked position when idle, so that it is not unintentionally operated especially by the children
- Most parts of the machine particularly the frame and the flywheel made of cast iron are quite susceptible to breakage by hammering. Such actions should always be avoided for the long-term safety of the machine

HOISTS

A hoist is a device used for vertically lifting or lowering a load by means of a drum or lift-wheel. The load is attached to the hoist by means of a lifting hook, which uses chain, fiber or wire rope as its lifting medium. Many orchard owners use manually operated while others use sophisticated tractor operated (Figure 17.2), electrically or pneumatically driven hoists. Besides hoist, cranes and forklifts are also used to perform similar tasks.

A tractor mounted aerial access hoist to manage coconut and tall trees has been designed and developed at TNAU, Coimbatore in collaboration with M/s Vanjax, Chennai. A full length chassis is extended from the front of the tractor to the rear and bolted to the tractor chassis. This forms the support for the hoist and for mounting the stabilizers. The total weight of the hoist and moments are transmitted through the chassis to the stabilizers without transferring to the tractor chassis. The hoist is able to go up to 15 m height. It is possible to access four coconut trees from a single setting. The time required for relocating the unit and operating stabilizers is around 1 min and positioning the hoist against a tree of 10 m height is 2 min. The positioning of the operator platform can be done by the operator using electro hydraulic controls (https://icar.org.in/hi/content/harvesting-and-threshing-equipment-aicrp-farm-implements-and-machinery).

Figure 17.2. A Tractor Operated Hoist. Note the Post Hole Auger at the Rear.

Tractor Mounted Telescopic Hoist

The tractor-mounted hoist is used in orchards and plantation crops for trimming, pruning, plant protection and harvesting operations. It consists of two aluminium ladders, each made of U-section as frames and round hollow pipes as cross members. In order to facilitate the sliding of one ladder over the other, U-sections of the outer ladder are made to face inward and that of the inner ladder to face outward. Two wire ropes driven with a hydraulic motor are used during operation. The motor while running in clockwise direction lifts the inner ladder and lowers it down when running in anti-clockwise direction. A support frame fitted to the platform on the top end of the inner ladder acts as a safety frame. During transport, the hoist can be folded and tilted in a horizontal position over the tractor canopy.

GREEN MANURE TRAMPLER

This implement is used to trample and press the green manure crops such as *dhaincha* (*Sesbania aculeate*), sun hemp and cowpea *etc.* in the field. Two types of tramplers in common use are slat type (Figure 17.3) and disc type (BIS, 1965). The main parts of the trampler are:

☆ Frame

☆ Discs/slats

☆ Axle and bearing

☆ Handle and handle support

☆ Foot board

Figure 17.3. A Slat type Green Manure Trampler.

In slat type trampler, long radical slats of flats are fitted to a central axle through supporting discs. The slats have minimum width and length as 50 mm and 230 mm. Minimum thickness is 5 mm. The number depends upon the size of the trampler. The slats are riveted to supporting discs. In disc type trampler, discs with minimum diameter of 230 mm made of a plate of minimum thickness of 6 mm are used. The flat discs are fitted to a central axle. The distance between the discs is at least 200 mm but should not exceed 275 mm. The size of the trampler is given by its maximum working width. Weight of the green manure trampler (without beam) is in the range of 30 to 40 kg.

MANURE SPREADER

A manure or muck spreader or honey wagon is used to distribute manure over a field. A typical manure spreader consists of a trailer (open-topped box, a hopper) mounted on wheels, with an attachment point that can be hooked to a tractor. The floor of the hopper has a conveyor belt mechanism driven by the tractor's power take off (PTO). For operation, a manure spreader is filled with manure. The conveyor belt mechanism pushes the manure towards the back, where it is pressed into rotating corkscrews that ejects the manure while spreading it on to the field (Figure 17.4). The working area of a manure spreader is in the shape of an arc, the sides of the arc extending outwards by more than 60° to either side of the machine. The widest part of the arc is equal to the working width of the manure spreader, and may range from 6.0 to 15.0 m.

Figure 17.4. A Pictorial view of a Manure Spreader.

DRAINAGE DITCHERS OR TRENCHERS

Trenchers are used for making irrigation/drainage ditches. These may range in size from walk-behind models, to attachments to a tractor, to very heavy tracked equipment. Different digging implements are used depending on the required width and depth of the trench and the hardness of the soil. Trenchers mostly use either a chain or wheel for the digging operation (Figure 17.5). Both have teeth that cut into the ground. As they work, trenchers remove the soil, leaving it in a pile next to the machine's path. Operators use the machine's control panel to raise, lower, start, and stop the wheel or chain.

Figure 17.5. A Chain type (Left) and Wheel type (Middle and Right) Ditchers.

A chain trencher uses a digging chain around a rounded metal frame, or boom to cut the soil to form a trench. It resembles a giant chainsaw. This type of trencher can be effectively used to form channels in soils that are too hard to cut with a bucket-type excavator. These trenchers can cut narrow and deep trenches. Thus, trenches can be made with minimal damage to lawns and gardens. It is possible to adjust the boom angle that controls the depth of the cut. To cut a trench, the boom is held at a fixed angle while the machine moves slowly. Walk-behind trenchers rely on the operator to walk behind the machine and pull it backward. Since these models are labour-intensive, they work best for small jobs and shallow trenches. They are quite suitable to dig trenches 30 to 75 cm deep. The power requirement is in the range of 12 to 31 HP.

A wheel trencher comprises a toothed metal wheel. A wide range of these ditchers suitable for excavation and maintenance of ditches and drains is available. They are mostly single-wheel or double-wheel ditchers although other models are also available. The single or double wheel ditchers are used under hard excavation conditions, such as construction of the ditches from scratch or when the excavated soil needs to be directed towards one side only. Vertical ditchers are suitable for digging vertical ditches. They are one of the most emerging solutions for water control and management, as well as drainage and irrigation purposes (https://www.cosmecosrl.com/news-fairs/the-ditchers-for-water-drainage). A pull-type ditcher-terracer is a multi-purpose machine designed to make surface drainage ditches, to construct and clean waterways, and to build and maintain terraces. It can be operated with a farm tractor. Depth of cut may be up to 25 cm in a single pass and up to 75 cm in multiple passes. Micro-trenchers can be small or large, but all use a wheel to make small, precise trenches. These are cheaper to operate and maintain than chain-type trenchers. Trencher may be combined with a drainage pipe or geo-textile feeder unit and back filler, so that pipe and textile envelope may be placed and the trench filled in one pass. These machines are commonly used in subsurface drainage of agricultural lands.

TREE PLANTING

Tree planting is the process of transplanting tree seedlings of fruit, ornamental and forest trees *etc.* in gardening, forestry, land reclamation, soil conservation or landscaping activities. Properly planted tree/plant/shrub will require much less management than an improperly planted tree/plant/shrub. Moreover, the properly planted tree is able to easily overcome adverse soil conditions. Some of the factors that may decide the appropriate planting technique are: the way the plant was grown in the nursery, the plant's drainage requirement, the soil type and its drainage characteristics, and the availability of irrigation water and its quality. The plant should be appropriate to the site conditions, or the site should be amended to specifically fit the plant requirements especially under abiotic stresses such as acid, alkali or saline soils. There are numbers of planting techniques for raising tree plantations. Depending upon the topography and resources, one of the following techniques can be successfully used.

- ✰ Pit method (In normal soils plant nutrients and FYM is added to the dugout soil while refilling (Figure 17.6 top left). In alkali soils one needs to add some kind of amendment such as gypsum besides the nutrients and FYM (Figure 17.6 top right). In acid soils, appropriate amounts of lime, nutrients and FYM are added
- ✰ Auger hole method (Figure 17.6 bottom left)
- ✰ Combination of pit and auger hole method (Figure 17.6 bottom right)

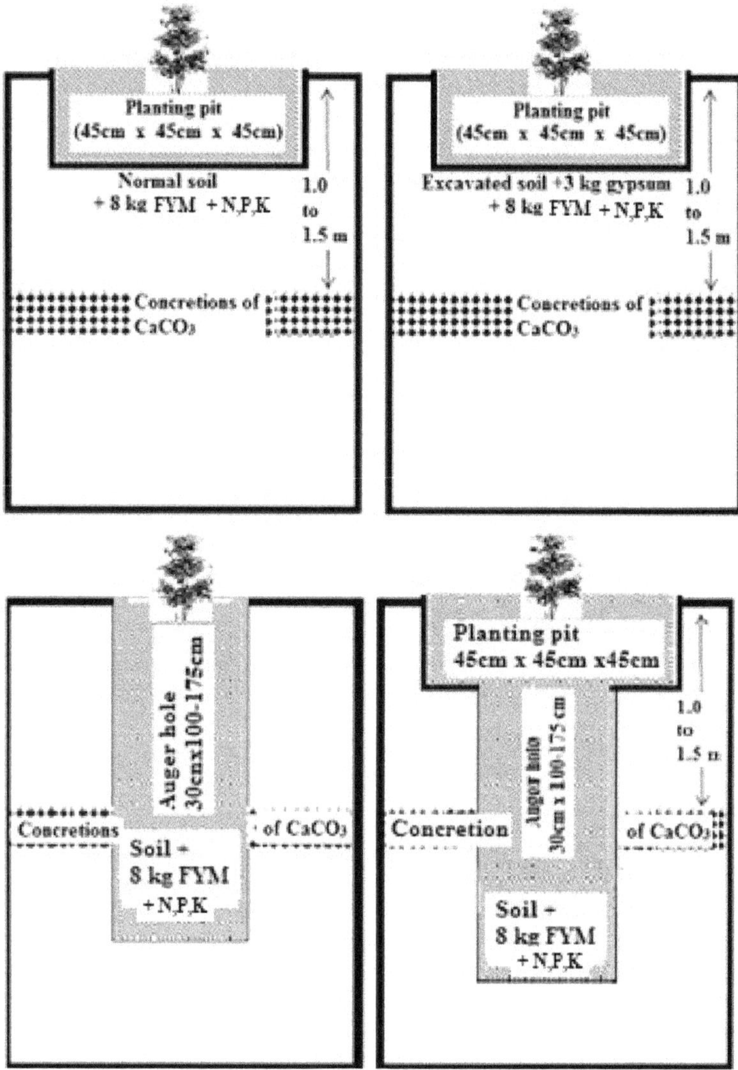

Figure 17.6. Method of Planting: Pit Method in Normal (Top left) and Alkali Soils (Top right), Auger Hole Method in Alkali Soils (Bottom left) and Pit-cum-Auger Hole Method in Alkali Soils (Bottom right).

☆ Ridge plantation

☆ Trench plantation

The depth of the pits or auger holes is so adjusted that it pierces any impermeable layer that may interfere in root proliferation. This adjustment allows the tree roots to have unrestricted movement beyond the impermeable layer.

Both traditional and improved mechanical equipment are used for tree planting. *Kassi* is the most commonly used agriculture tool for digging the planting pits supported by small tools such as *fawda*, *khurpi*, pick axe or shovel (Chapter 15). Ridges can also be formed manually with a *kassi* or a ridger can be used for this purpose. Tractor or engine operated powered auger are widely used to plant forest plantations (Figure 17.7 left, right).

Tractor Operated Post Hole Auger Digger

Post hole augers are available to fit most landscaping, farming, or ranching applications such as drill holes for footings, fencing, sign posts, and trees plantations. The auger drive mechanism either uses planetary drive or hydraulic drive, although the latter is most common. It consists of a hitch frame, an auger, yoke tubing, gear box drive, and power shaft, *etc.* The post hole auger is connected to the tractor and moved to places where required. The drive to the gear box is obtained from PTO of the tractor (Figure 17.7 left). Holes of diameter 15 to 100 cm and depth 90 to 175 cm can be made by using appropriate augers.

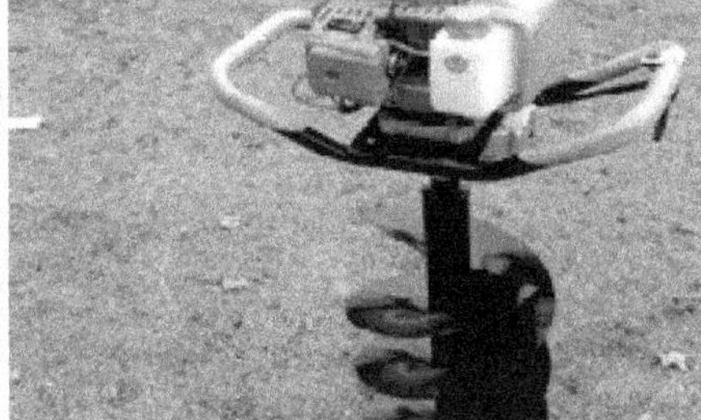

Figure 17.7. A Tractor Operated Auger (Left) and an Engine Operated Auger (Right).

Engine Operated Post Hole Auger Digger

Auger is powered by a 4 cycle, 5 HP engine. The auger can make holes from 5 to 20 cm in diameter and depth up to 75 cm (Figure 17.7 right). An optional 45 cm extension is supplied by several manufacturers. Two men are required to operate the machine.

TRAILERS

Trailers are used for transportation activities at the farm or to transport the inputs and produce from and to the market. The trailers are pulled by a tractor. Trailers may be classified according to the number of wheels as one-wheel, two-wheel, and four-wheel types (Smith, 1937). The two-wheel heavy-duty industrial trailers equipped with dual tyres are sometimes considered as a special class. One-wheel trailers have now become obsolete. According to BIS (2000), agricultural trailers have been categorized as semi-trailers (single axle, two wheel, Figure 17.8 top) and balanced trailers (double axel, four wheel, Figure 17.8 bottom). Both types of trailers may be fitted with fixed or tipping platforms. A properly hitched two-wheel trailer has no tendency to whip and moderately high speeds can be maintained. A well built four-wheel trailer is more standard being stable and can be used to transport heavy loads. The wheels of such trailers are equipped with roller bearings, demountable rims, and pressure fittings for good lubrication. The rigid construction prevents whipping. The capacity of a trailer shall be its gross load. Both the gross load and the pay load are declared by the manufacturer. The capacity of the trailers is in the range of 2 to 10 tonnes, but the capacity of a semi-trailer is limited to 5 tonnes. The trailer shall be provided with suitable spring suspension for recommended load carrying capacity. Two-wheel trailers suitable for farm use in India are mostly manufactured by local blacksmith. It is why many standard fittings and design features recommended by BIS (2000) are compromised. Besides the lack of such features, trailers are overloaded and improperly hitched resulting in their overturning and accidents. Farmers should always prefer trailers manufactured by standard companies having ISI mark.

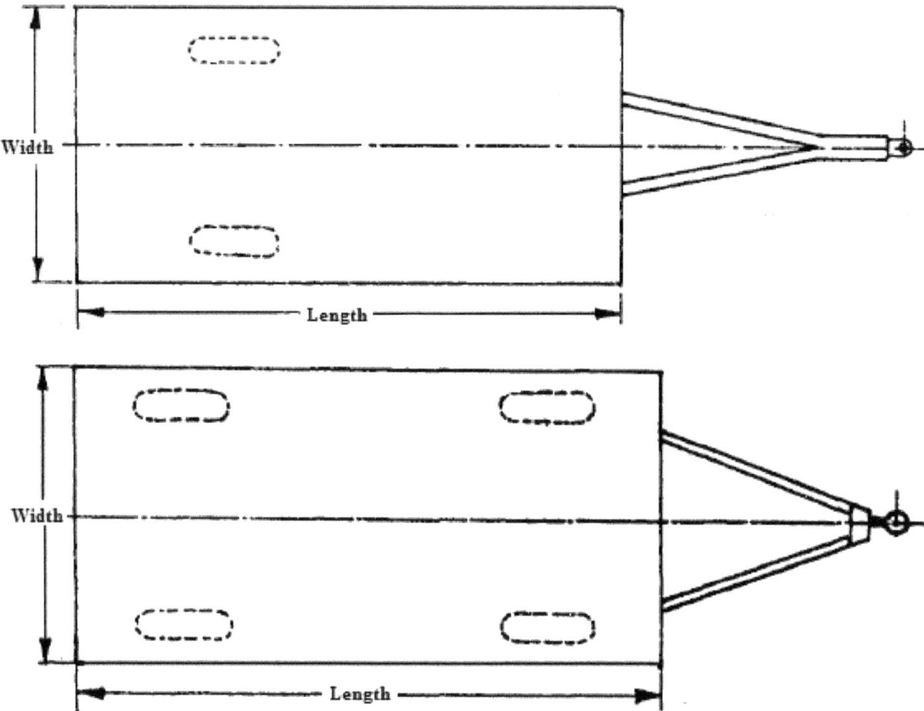

Figure 17.8. A Two Wheel Semi-trailer (Top) and a Four Wheel Stable Trailer (Bottom).

MECHANIZATION IN SUGARCANE CULTIVATION

Sugarcane is one of the most important cash crops. Total acreage under sugarcane in the country was around 5.50 M ha in 2018-19. With an average productivity of 78.3 t/ha, the total sugarcane production was 400.4 million tons during the year. Cultivation of sugarcane is a strenuous job involving major operations of sett cutting, planting, hoeing, earthing-up, cutting, de-husking, shredding, trashing, ratoon management, bundling of cut cane and loading for transportation. All these operations in India and many other developing countries are mostly performed manually. The challenges in sustainable sugarcane agriculture can be met effectively by adopting appropriate mechanical alternatives not only for planting and harvesting but also for other operations. Besides increasing productivity, it will help to increase cost efficiency in sugarcane production system as a whole.

Two Row Pit Digger

Sugarcane productivity can be increased by adopting pit planting or ring-pit method of planting. Nearly 10,000 pits/ha are required for pit planting of sugarcane. A tractor operated two rows pit-digger has been developed for making 25-30 cm deep, circular pits of 75 cm diameter at 30 cm spacing for planting sugarcane (Figure 17.9). It consists of a frame, 3 point hitch system, two rotating auger blades powered by tractor PTO. Sugarcane setts are planted manually in these pits. It can dig 150 pits/hr giving a field capacity of 0.017 ha/hr. It saves 400 man-days/ha and saves about 70 per cent cost on digging pits compared to manual digging.

Figure 17.9. A Two Row Pit Digger for Sugarcane Cultivation

Power Operated Sugarcane Sett Cutting Machine

Use of a sharp knife has been a common practice to manually cut setts of varying lengths and varying number of buds. Indian Institute of Sugarcane Research (IISR), Lucknow has developed a power operated sugarcane sett cutting machine for planting. Several such machines with location specific modifications have been developed at other places in the country. These machines can be operated by an electric motor, a diesel engine or even by a tractor. All these machines have circular saw blades with sharp and fine teeth. These are capable of cutting setts of varying lengths. Some of these machines have a provision to simultaneously perform the fungicidal treatment of cut setts.

Manual Bud-chipping Machine

To reduce the amount of sugarcane used in seeding, only buds of sugarcane are used for seeding while the remaining sugarcane is sent for milling. Bud chippers have been developed to ease the operation. A manually operated floor-mounted bud chipper is equipped with a knife with a semi-circular edge. It has an ergonomic spring-loaded handle, and can be operated by a person sitting on the ground. While sugarcane is continuously fed with the left hand, swinging action of the right hand in a smooth arc helps to cut the sugarcane buds. Pedal operated and pneumatic bud-chipping machines have also been designed and developed.

Animal Drawn Sugarcane Planter

A bullock drawn semi-automatic sugarcane planter has been designed for flat planting of sugarcane in light to medium soils. The machine has furrow openers to open the furrows. The machine uses pre-cut sugarcane setts, which are manually dropped in the furrows. Fertilizer and insecticide are applied and the planted setts are covered with the soil followed by compacting the soil cover. All the operations are completed in one go.

Tractor Operated Sugarcane Planter

The machine used for planting sugarcane setts consists of two furrow openers, a cane hopper, two rotating distributor discs, two fertilizer hoppers, pesticide tank with a distribution valve. All the components are mounted on a frame and two wheels. Adjustments can be made to maintain desired row to row spacing. Seed distributor and applicator are powered from ground wheel through a set of roller chains and gears. The fertilizer and chemical pesticides are applied simultaneously. Two persons are needed to manually drop the cane sets stored in the hopper to the seed rotor. Machine mounted on a tractor has an output of about 0.6 ha/hr. It requires 4-6 men to complete the operations.

Tractor Drawn Sugarcane Cutter Planter

This implement has been developed by IISR, Lucknow for ridge and furrow system of planting. It is a semi- automatic planter consisting of ridger body attached to frame to create furrows, sett cutting unit, fertilizer application unit, chemical application unit, sett covering unit and seed hopper. The planter is PTO driven and can be mounted on a three point linkage system of the tractor. The machine is designed to operate with whole cane replacing the need to prepare cane setts beforehand. Horizontal blades provided in the machine cuts the sugarcane into setts of 350 mm length each having 2-3 buds. Setts are treated with insecticides at the cutting point. The setts are conveyed through a chute behind the two ridger bodies and placed in the furrows providing an overlapping of up to 30 per cent. Fertilizer is metered through fluted roller mechanism provided at the bottom of fertilizer boxes. The pesticide is also sprayed at the ends of sett in the furrows (Srivastava, 2010). The setts placed in the furrow are covered immediately after the treatment. The furrow is then closed and rows are leveled by the leveler provided in the machine. The field capacity of the machine is 0.20 ha/hr. A mechanical sugarcane planter developed by Vasantdada Sugar Institute is shown in Figure 17.10. It helps in better germination and saves planting cost.

Figure 17.10. Sugarcane Planter Developed by Vasantdada Sugar Institute, Haveli.

Animal Drawn Earthing-up Hoe

Earthing-up or ridging is the technique in agriculture and horticulture of piling-up soil around the base of a plant. Earthing-up operation helps to suppress the growth of excess tillers, facilitate irrigation and economize water, drains out the excess water from the field, helps to control weed infestation and protects the crop from lodging. Animal drawn sugarcane earthing-up hoe loosens the soil between the rows and moves it towards the standing plant for earthing-up. It consists of three iron tines for loosening the soil and scraper with adjustable wings for earthing-up. It is suitable for earthing-up operation of sugarcane crop having row spacing of 900 to 1200 mm. Field capacity of the machine is 0.14 ha/hr at a operational speed of 2.0 km/hr.

HARVESTING EQUIPMENT

Sugarcane harvesting involves de-topping, base cutting of the crop, de-trashing, bundling, loading and transportation. De-topping and de-trashing of crop itself takes about two-third of manpower required for harvesting. Some commonly used hand tools for cutting, stripping, peeling and trashing are shown in Figure 17.11. Although some improvements in these tools have been made yet a major breakthrough is required to mechanize sugarcane harvesting. Shukla (2017) made a comprehensive review of the emerging innovation and challenges in sugarcane mechanization.

Soldier-Type Cane Harvester

It is a de-topping and base cutting machine. Both these operations are performed simultaneously on one row of green cane at a time. It consists of a de-topper with a gathering chain. This unit removes the tops from the standing crop and drops them towards the right of the row being harvested. The base cutting unit consists of two sets of pick-up chains arranged in a V and two cutting discs. The chains pick and feed the cane to the base cutter. A cane conveying system conveys the harvested cane to the windrow. The machine does not uproot the cane. Its capacity is about 0.4 ha/hr. Further cleaning and loading is done either manually or mechanically.

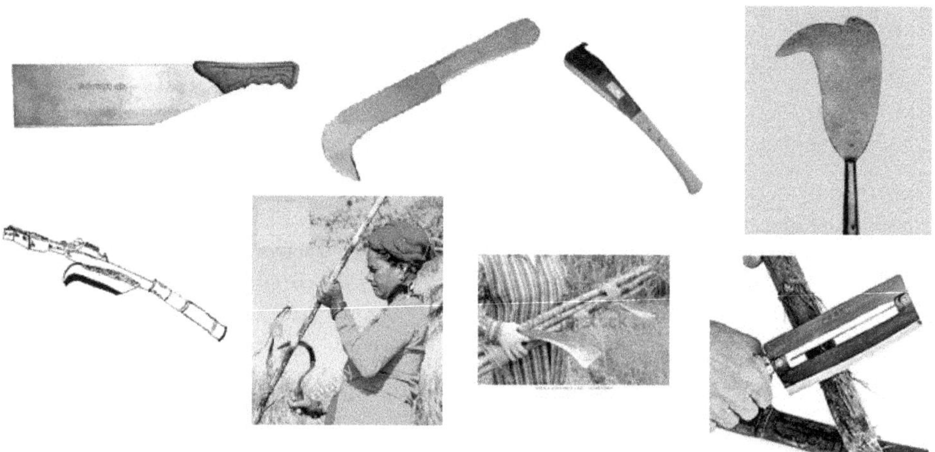

Figure 17.11. Various kinds of Sugarcane Cutting, Striping, Trashing and Peeling Knives.

Sugarcane Detrasher

The detrasher developed by IISR, Lucknow is a tractor operated PTO driven machine for removal of green tops as well as dry trash from the harvested sugarcane stalks (Rajesh Kumar, 2016). It consists of mechanisms for cane feeding, detrashing and delivery. It separates the top from the cane by breaking it at the natural weak point, the joint of the immature top with mature cane stalk. One can feed 2-3 cane stalks at a time. The output of the detrasher is 2.4 t/ha with trash removal efficiency of 77.5 to 94.5 per cent. It can be transported to the site on three point linkage of the tractor. The machine can also be operated by an electric motor or diesel engine.

Heera Sugarcane Harvester

Heera Sugar Industries in District Belgaum has developed a single row whole cane sugarcane harvester operated by 60 HP tractor. The machine cuts the crop, detops, de-trashes, collects the whole cane on the rear and de-loads at one place. The machine is suitable for sugarcane crop with row to row spacing of 0.9 m. The machine has since been patented.

Tractor Operated Sugarcane Trash Shredder

A tractor operated sugarcane trash shredder has been developed at TNAU, Coimbatore. The main component of the unit is the rotating member with two rotary units. The spacing between them can be varied as per the row spacing of the crop. Swinging type curved blades are fitted on the rotary member in a staggered manner. The rotating speed of the rotary unit is around 2000 rpm. Its field capacity is 2 ha/day. A similar machine has also been developed at MPKV, Rahuri. The unit consists of shredding unit, trash compressing unit, and power transmission unit. The width of operation is 60 cm. It can be operated with a 35 HP tractor having the field capacity of 1.5 ha/day.

HARVESTER COMBINE

The basic components of a sugarcane harvester combine comprise (Abdel-Mawla, 2014):

☆ Gathering mechanism

☆ Topping mechanism

☆ Base cutter

☆ Feed conveyor

☆ Choppers

☆ Elevator

☆ Cleaning by air blast

These basic components are mounted on a frame and a vehicle, which constitutes the cane combine (Figure 17.12).

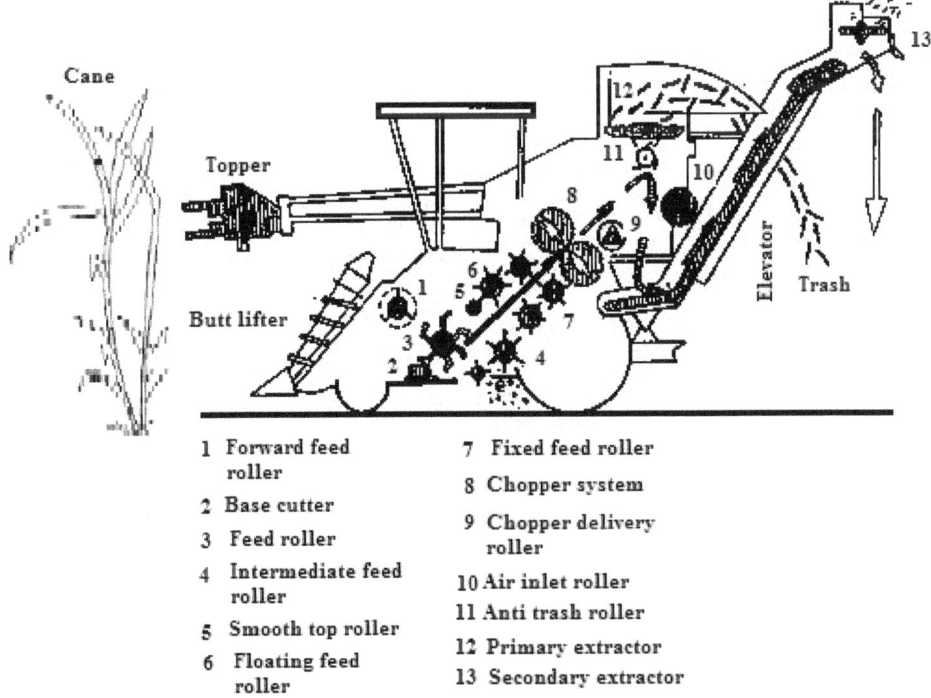

Figure 17.12. Line Diagram of Sugarcane Combine Harvester.

Gathering mechanism: As the name suggest, its function is to separate sprawled cane and align the row to be harvested. They are made of revolving scrolls fitted on two gathering triangular walls. The walls are approximately 140 cm apart at the tips converging to the throat width of the machine just forward of the base cutters. The tips of each wall are fitted with ground engaging points which move under the stalks lying on the ground to lift them.

Topping mechanism: It gathers cuts and discards the non-productive tops. The gathering operation is performed by gathering chains while cutting is done by a horizontally rotating disc fitted with mower blades which operate against a fixed anvil.

Base cutter: Its function is to cut the stalk at or just below the ground level. While some machines use single diameter blades for base cutting the cane, others use twin contra-rotating discs fitted with a number of replaceable knife blades.

Feed conveyor: It is made of endless chain slat conveyor, or a series of rollers or augers. It conveys the whole stalks of cane from the base cutter to the choppers.

Choppers: Their function is to receive the whole stalks from the feed conveyor and chop them into short uniform billets. It is a pair of parallel shafts each with paddle shaped blades, which during rotation come together in the plane containing the shafts to give a flying or travelling cut. Although other chopping mechanisms are also used but the flying cut mechanism has the advantage of being aggressively self-feeding.

Elevator: It consists of an inclined chain and slat conveyor, which receives the billets from the choppers and conveys them into a receptacle for transporting to the mill. In most machines, the elevator can be rotated 180° to facilitate harvesting in either direction.

Air-blast cleaning: One of the biggest problems with mechanically harvested cane is the foreign matter consisting of leaves, tops, dirt, stones and many other materials picked-up from the field. Therefore, one or in many cases two fans are used to extract leaves and dirt from the cane.

RATOON MANAGEMENT DEVICE

A ratoon management system is essentially required because ratooning is an integral component of sugarcane production system. *Ratooning is a method of propagation in sugarcane in which subterranean buds on cane left underground after harvesting give rise to a new crop stand.* It is also known as stubble crop as opposed to plant crop raised from sugarcane setts. Thus, a ratoon management device executes most of the operations involved in ratoon management such as stubble shaving, deep tilling, off-barring at initiation of ratoon, placing manure, fertilizer/ bio-agents, chemicals in liquid form and earthing-up operations in a single pass of operation. A tractor drawn implement mounted on a 3 point linkage driven by the PTO is capable of accomplishing many of these operation notably stubble shaving, off-barring and inter-culturing (deep tilling), fertilizer as well as manure dispensing and earthing-up in ratoon crop. It shaves the stubble evenly and the off-barring helps to trim the roots. Harrowing promotes aeration and water penetration, both helping to induce and rejuvenate the growth of the sugarcane. Its capacity is around 0.40 ha/hr. Note that ratooning is also practiced in banana and few other crops although ratoon practices and management may vary from one crop to another.

QUESTIONS (THEORY)

1. Write a descriptive note on chaff cutter.

2. What is a hoist? Write a note on tractor operated hoist.

3. What do you understand by sugarcane mechanization? Name some of the operations and machines used in mechanization of sugarcane cultivation.

4. Write a descriptive note on the sugarcane harvester combine.

5. Write short notes on: green manure trampler, manure spreader, post hole auger hole digger and trailer.

List of Select BIS Standards of Farm Power and Machinery for Engineers and Industries

AGRICULTURAL TRACTORS AND POWER TILLERS

Standards Published

1. IS 789 (Part 8):1991 Feb 2012/ISO Agricultural tractors - Test procedure: Part 8 Engine air cleaner

2. IS 4468 (Part 1):1997 Agricultural wheeled tractors - Rear-mounted three-point linkage, Feb 2012: Part 1 Categories 1, 2 3 and 4 (fourth revision)

3. IS 4468 (Part 2):1993 Mar 2009/ISO 730-2:1979 Agricultural wheeled tractors - Three-point linkage: Part 2 Category 1 N (Narrow hitch) (third revision)

4. IS 4931:1995 Agricultural tractors - Rear mounted power take off Mar 2009, Types 1, 2 and 3 (third revision)

5. IS 5994:1998 Agricultural tractor - Test code (third revision), Mar 2009

6. IS 6283 (Part 1):2006 Mar 2009/ISO 3767-1:1991 Tractors and machinery for agriculture and forestry, powered lawn and garden equipment - Symbols for operator controls and other displays Part 1: Common symbols (second revision)

7. IS 6283 (Part 2):2007 Mar 2009/ISO 3767-2:1991 Tractors and machinery for agriculture and forestry, powered lawn and garden equipment - Symbols for operator controls and other displays: Part 2 Symbols for agricultural tractors and machinery (first revision)

8. IS 6483:2004 Tractors and machinery for agriculture and forestry - Linch pins and spring pins - Dimensions and requirements (second revision), Feb 2012

9. IS 6840:1972 Code of practice for preventive maintenance of agricultural wheeled tractor, Feb 2012

10. IS 6847:1995 Code of practice for installation of agricultural wheeled tractor (first revision), Mar 2009

11. IS 8132:1999 Mar 2009 | ISO 3600:1996 Tractors and machinery for agriculture and forestry - Powered lawn and garden equipment - Operators manuals - Content and presentation (second revision)

12. IS 8133:1983 Mar 2009 2 | ISO 3789:1982 Guidelines for location and operation of operator controls on agricultural tractors and machinery (first revision)

13. IS 8213:2000 Agricultural tractor trailers – Specifications, Mar 2011

14. IS 8265:1996 Agricultural tractors - Guards for power take-off (PTO) drive-shafts (second revision), Feb 2012

15. IS 9253:2001 Guideline for field performance and haulage tests of agricultural wheeled tractors (second revision), Feb 2012

16. IS 9864:1981 Guidelines for stocking spare parts of agricultural tractors, Feb 2012

17. IS 9869:1981 Technical requirements for power take-off pulley assembly for agricultural tractors with type 1 PTO shaft, Feb 2012

18. IS 9935:2002 Power tiller - Test code (second revision), Feb 2012

19. IS 9939:1981 Glossary of terms relating to agricultural tractors and power tillers, Feb 2012

20. IS 9980:1988 Guidelines for field performance and haulage tests of power tillers (first revision), Mar 2009

21. IS 10273:1987 Guidelines for declaration of power and specific fuel consumption and labeling of agricultural tractors (first revision), Mar 2009

22. IS 10274:1993 Agricultural wheeled tractors - Maximum travel speed - Method of determination (first revision), Mar 2009

23. IS 10282:1982 Cage wheel for power tillers, Mar 2009

24. IS 10318:2002 Feb 2012/ISO 5673:1993 Agricultural tractors and machinery - Power take-off drive shafts and position of power-input connection (first revision)

25. IS 10703:1992 Feb 2012 1 | ISO/TR 3778:1987 Agricultural tractors - Maximum actuating forces required to operate controls (first revision)

26. IS 10740:1983 Recommendation for operating requirement for power take-off driven implements, Mar 2009

27. IS 10743:1983 Mar 2009 | ISO 789-6:1982 Method for determination of centre of gravity of agricultural tractors

28. IS 10747:1983 Dimensions of hitch for power tillers, Mar 2009

29. IS 11081:1993 Agricultural tractors - Half cage wheel – Specifications, Mar 2009 (first revision)

30. IS 11082:1984 Technical requirements of agricultural tractors for wet land cultivation, Feb 2012

31. IS 11442:1996 Feb 2012/ISO 5721:1989 Agricultural Tractors - Operator's field of vision - Test procedures (first revision)

32. IS 11821(Part 1):1992, Feb 2012/ISO 3463:1989 Method of test and acceptance conditions for protective structures of agricultural tractors: Part 1 Dynamic test (first revision)

33. IS 11821(Part 2):1992 Feb 2012/ISO 5700:1989 Method of test and acceptance conditions for protective structures of agricultural tractors: Part 2 Static test (first revision)

34. IS 11822:1986 Methods of tests for spark arrester of agricultural tractors and power tillers, Feb 2012

35. IS 11858:1986 Handle grip for power tiller, Feb 2012

36. IS 11859:2004 Agricultural tractors - Turning and clearance diameters - Methods of test, Feb 2012

37. IS 12036:1995 Agricultural tractors - Test procedures - Power tests for power take-off (first revision), Mar 2009

38. IS 12061:1994 Agricultural tractors - Braking performance - Method of test (first revision), Mar 2009

39. IS 12062:1987 Method for measurement of exhaust smoke emitted by agricultural tractors, Mar 2009

40. IS 12180 (Part 1):2000 Tractors and machinery for agriculture and forestry - Noise measurement - Method of test: Part 1 Noise at the operator position - Survey method (first revision), Feb 2012

41. IS 12180 (Part 2):2000 Tractors and machinery for agriculture and forestry - Noise measurement - Method of test: Part 2 Noise emitted when in motion (first revision), Feb 2012

42. IS 12207:2008 Agricultural tractors - Recommendations on selected performance characteristics of tractors (second revision)

43. IS 12224:1987 Method of test for hydraulic power and lifting capacity of agricultural tractors, Mar 2009

44. IS 12226:1995 Agricultural tractors - Power tests for drawbar - Test procedure (first revision), Mar 2009

45. IS 12239 (Part 1):1996 Feb 2012/ISO 4254-1:1989 Guide for safety and comfort of operator of agricultural tractors and power tillers: Part 1 General requirements (first revision)

46. IS 12239 (Part 2):1999 Tractors and machinery for agriculture and forestry - Mar 2009 Technical means for ensuring safety Part 2: Tractors (first revision)

47. IS 12239 (Part 3):1988 Guide for safety and comfort of operator of agricultural tractors and power tillers: Part 3 Requirements relating to power tillers, Mar 2009

48. IS 12343:1998 Agricultural tractors - Operators seat technical requirements (first revision), Mar 2009

49. IS 12362 (Part 1):2007 Feb 2012/ISO 6489-1:2001 Agricultural vehicles - Mechanical connections between towed and towing vehicles: Part 1 Dimensions of hitch hooks (second revision)

50. IS 12362 (Part 3):1994 Mar 2009/ISO 6489-3:1992 Agricultural vehicles - Mechanical connections on towing vehicles - Part 3: Tractor drawbar (first revision)

51. IS 12363:1988 Recommendation on track width of agricultural tractors, Mar 2009

52. IS 12953:1990 Drawbar for agricultural tractors - Link type -Specification, Feb 2012

53. IS 13064:1991 Power tillers - Installation and preventive maintenance – Guidelines, Feb 2012

54. IS 13539:2008 Power tillers - Recommendations on selected performance characteristics (first revision)

55. IS 13548:1992 Feb 2012/ISO 5008:1979 Agricultural wheeled tractors and field machinery - Measurement of whole body vibration of the operator

56. IS 13549:1992 Feb 2012/ISO 5676:1983 Tractors and machinery for agriculture and forestry - Hydraulic coupling - Braking circuits

57. IS 13581:1993 Mar 2009/ISO 5007:1990 Agricultural wheeled tractors - Operator's seat - Laboratory measurement of transmitted vibration

58. IS 13642:1993 Agricultural tractors - Silencers - Technical requirements, Mar 2009

59. IS 13732:1993 Mar 2009/ISO 5675:1992 Agricultural tractors and machinery - General purpose quick-action hydraulic couplers

60. IS 14414:1996 Feb 2012/ISO 789-7:1991 Agricultural tractors - Axle power determination - Test procedures

61. IS 14683:1999 Agricultural tractors and machinery - Lighting device for travel on public roads, Mar 2009

62. IS 14980 (Part 1):2001 Feb 2012/ISO 8759-1:1998 Agricultural wheeled tractors - Front mounted equipment: Part 1 Power take-off and three point linkage

63. IS 14980 (Part 2):2001 Feb 2012/ISO 8759-2:1998 Agricultural wheeled tractors - Front-mounted equipment: Part 2 Stationary equipment connections

Finalized Drafts under Print

1. DOC.FAD 11(2328) Draft Indian Standard- Agricultural tractors-Recommendations on selected performance characteristics (Third revision)

2. DOC.FAD 11(2336) Draft Amendment No. 2 to IS 10273:1987 Guidelines for declaration of power and specific fuel consumption and labeling of agricultural tractors (first revision)

3. DOC.FAD 11(2129) Draft Indian Standard - Code of practice for preventive maintenance of agricultural wheeled tractor (first revision of IS 6840)

4. DOC.FAD 11(2130) Draft Indian Standard - Guideline for field performance and haulage tests of agricultural wheeled tractors (third revision of IS 9253)

Draft Standards Issued for Wide Circulation

1. DOC.FAD 11(2478) Draft Indian Standard-Agricultural tractors and machinery — Power Take-Off drive shafts and power input connection-Part 1: General manufacturing and safety requirements (Second revision of IS 10318) (Adoption of ISO 5673-1:2005)

2. DOC.FAD 11(2479) Draft Indian Standard- Agricultural tractors and machinery — Power Take-Off drive shafts and power-input connection-Part 2: Specifications for use of PTO drive shafts, and position and clearance of PTO drive line for various attachments

3. DOC.FAD 11(2480) Draft Indian Standard- Tractors for agriculture and forestry-Roll over protective structures (ROPS) Part 1-Dynamic test method and acceptance conditions [Second revision of IS 11821(part 1)] (Adoption of ISO 3463:2006)

4. DOC.FAD 11(2483) Draft Indian Standard-Agricultural vehicle's mechanical connections between towed and towing vehicles- Part 3 Tractor drawbar [(Second revision of IS 12362 (Part 3)] (Adoption of ISO 6489-3:2004)

5. DOC.FAD 11(2484) Draft Indian Standard- Agricultural wheeled tractors and field machinery- Measurement of whole – body vibration of the operator (First revision of IS 13548) (Adoption of ISO 5008:2002)

6. DOC.FAD 11(2485) Draft Indian Standard- Agricultural wheeled tractors — Operator seat- Laboratory measurement of transmitted vibration (First revision of IS 13581) (Adoption of ISO 5007:2003)

7. DOC.FAD 11(2486) Draft Indian Standard-Agricultural tractors and machinery -General purpose quick-action hydraulic couplers (First Revision of IS 13732) (Adoption of ISO 5675:2008)

AGRICULTURAL MACHINERY

1. IS 619: 1979 Pruning and slashing knives hooked and curved (second revision), Dec 99

2. IS 1970: 1995 Hand operated compression knapsack sprayer (fifth revision), Dec 99

3. IS 1971:1996 Hand operated stirrup-type sprayer (fifth revision)

4. IS 1976:1976 Rotary paddy weeder, manually operated (second revision), Dec 99

5. IS 2192:2000 Soil working equipment-Animal drawn mould-board plough, fixed type-Specification (second revision)

6. IS 2559:1978 Garden rake (first revision), Dec 99

7. IS 2563:1978 Hedge shears, straight-edge type (first revision), Dec 99

8. IS 2565:1979 Ridger, Animal drawn (first revision), Jan 01

9. IS 3062:1995 Rocker sprayer (fourth revision), Jan 01

10. IS 3092:1982 Rubber draining and tapping knife (first revision), Dec 99

11. IS 3093:1981 *Dah, jungle* cutting (first revision), Dec 99

12. IS 3094:1982 Bill-hook (first revision), Dec 99

13. IS 3122:1982 Budding and grafting knife combined (first revision), Mar 97

14. IS 3327:1982 Pedal-operated paddy thresher (first revision), Dec 99

15. IS 3342:1998 Soil working equipment-cultivators, Animal drawn-Specification (second revision)

16. IS 3369:1965 Puddler, Animal drawn, Mar 97

17. IS 3372:1965 Bund former, Animal drawn, Dec 99

18. IS 3394:1978 Pruning secateurs (first revision), Dec 99

19. IS 3606:1998 Soil working equipment, Disc harrow, Animal drawn-Specification (second revision)

20. IS 3652:1995 Foot sprayer (fourth revision), Jan 01

21. IS 3906:1995 Hand-operated knapsack sprayer (fourth revision), Jan 01

22. IS 4358:1996 Sickles (first revision)

23. IS 4366(PT-1)1985 Agricultural tillage discs: Part 1 concave type (second revision), Jan 01

24. IS4366(PT-2)1985 Agricultural tillage discs: Part-2: Flat type (second revision), Jan 01

25. IS 4930:1985 Axle assembly for pneumatic wheeled animal drawn vehicles (first revision), Dec 99

26. IS 5135(PT-1):1994 Hand rotary duster: Part 1 Belly mounted type (second revision), Dec 99

27. IS 5135(PT-2):1994 Hand rotary duster: Part 2 Shoulder mounted type (first revision), Dec 99

28. IS 6024:1983 Guards for harvesting machines (first revision), Dec 99

29. IS 6025:1982 Knife sections for harvesting machines (first revision), Dec 99

30. IS 6284:1985 Test code for power thresher for cereals (second revision), Dec 99

31. IS 6288:1971 Test code for mould-board ploughs, Dec 99

32. IS 6316:1993 Seed-cum-fertilizer drill-Test code (first revision), Feb 99

33. IS 6320:1985 Power thresher, hammer mill type (first revision), Nov 95

34. IS 6635:1972 Tractor operated disc harrows, Jan 01

35. IS 6638:1972 Tractor-mounted spring loaded cultivators, Jan 01

36. IS 6690:1981 Blades for rotavator for power tillers (first revision), Feb 96

37. IS 6813:2000 Sowing equipment-Seed-cum-fertilizer-drill specifications (second revision)

38. IS 7201(PT-1):1987 Method of sampling for agricultural machinery and equipment Part-1 Hand tools and hand operated/animal drawn equipment (first revision), Dec 99

39. IS 7230:1974 Plain spool for tractor operated disc harrows, Jan 01

40. IS 7353:1974 Blade for tractor operated terrace, Jan 01

41. IS 7565(PT-1):1975 Tines for tractor operated cultivators Part-1 Rigid tines, Dec 99

42. IS 7565(PT-2):1988 Tines for tractor operated cultivators Part-2 S. type tines, Dec 99

43. IS 7593(PT-1):1986 Power operated pneumatic sprayer-cum-duster Part 1 Knapsack type (first revision), Feb 96

44. IS 7640:1975 Test code for disc harrows, Dec 99

45. IS7825(PT-1):1975 Cylinder type hand lawn mower: Part 1 Wheel type, Feb 99

46. IS 7825(PT-2):1975 Cylinder type hand lawn mower: Part 2 Roller type, Feb 99

47. IS 7927:1975 Method of field testing for manually operated paddy weeder, Dec 99

48. IS 8122(PT-1):1994 Test code for combine harvester thresher: Part 1 Terminology (first revision), Dec 99

49. IS 8122(PT-2):2000 Combine harvester-thresher-Test code: Part 2 Performance tests (first revision), Nov 99

50. IS 8213:2000 Agricultural tractor trailer-Specification (third revision)

51. IS 8480:1996 Glossary of terms relating to crop protection equipment (first revision)

52. IS 8548:1977 Test code for power operated hydraulic sprayer, Dec 99

53. IS 9019:1979 Code of practice for installation, operation and preventive maintenance of power thresher, Jan 01

54. IS 9020:1979 Safety requirements for power threshers, Nov 95

55. IS 9129:1979 Technical requirements for safe feeding systems for power threshers, Nov 95

56. IS 9164:1979 Guide for estimating cost of farm machinery operation, Feb 96

57. IS 9217:1979 Test code for agricultural discs, Jan 01

58. IS 9575:1980 Power lawn mower, pedestrian-controlled cylinder (reel) type, Feb 99

59. IS 9581:1980 Safety and operational requirements for pedestrian controlled cylinder (reel) power lawn mowers, Feb 99

60. IS 9632:1980 Code of practice for operation and preventive maintenance of crop protection equipment, Feb 96

61. IS 9813:1981 Tractor-mounted blade terracers, Dec 99

62. IS 9818(PT-1):1981 Glossary of terms relating to tillage and inter cultivation equipment Part 1 General terms, Dec 99

63. IS 9818(PT-2):1981 Glossary of terms relating to tillage and inter cultivation equipment Part 2 Terms relating to equipment, Dec 99

64. IS 9821:1981 Glossary of terms relating to farm transport equipment, Dec 99

65. IS 9826:1981 Glossary of terms relating to harvesting and threshing equipment, Mar 97

66. IS 9855:1981 Glossary of terms relating to sowing, planting, fertilizers and manure application equipment, Dec 99

67. IS 9856:1981 Test code for potato planters

68. IS 9877:1981 Code of practice for installation, operation and preventive maintenance of grain combine, Jan 01

69. IS 10134:1994 Methods of tests for manually operated sprayers (first revision), Dec 99

70. IS 10225:1982 Bearing spools for tractor-operated disc harrows, Dec 99

71. IS 10233:1982 Tractor-operated disc ploughs, Dec 99

72. IS 10239:1982 Blade for tractor-operated scraper, Dec 99

73. IS 10254:1982 Share for animal-drawn ridger, Dec 99

74. IS 10378:1982 Knife back for harvesting machines, Jan 01

75. IS 10618:1983 Pictorial representation for cautionary notices for power threshers, Jan 01

76. IS 10683:1983 *Khurpi*, Dec 99

77. IS 10684:1983 Tree pruner, Jan 01

78. IS 10691:1983 Share for tractor-operated mould-board ploughs, Jan 01

79. IS 10806:1984 Trailers for power tillers, Jan 01

80. IS 10807:1984 Axles with brakes of trailer for power tillers, Jan 01

81. IS 11033:1984 Animal drawn potato digger, ridger types, Jan 01

82. IS 11204:1985 Sugarcane harvesting knife, Jan 01

83. IS 11234:1985 Test code for power thresher for groundnut, Jan 01

84. IS 11235:1985 Test code for groundnut digger, animal drawn, Jan 01

85. IS 11269:1985 Axle assembly for conventional animal drawn vehicles, Feb 99

86. IS 11270:1985 Technical requirements for ring-type hitches for agricultural trailers, Feb 99
87. IS 11271:1985 Groundnut planter, Jan 01
88. IS 11272:1985 Glossary of terms relating to horticultural equipment, Jan 01
89. IS 11313:1985 Hydraulic power sprayer, Dec 99
90. IS 11429:1985 Methods for calibration of sprayers, Dec 99
91. IS 11467:1985 Test code for cereal - harvesting machines, Jan 01
92. IS 11531:1985 Test code for puddler, Jan 01
93. IS 11580:1986 Glossary of terms relating to forestry equipment, Feb 99
94. IS 11691:1986 Specification for power thresher, spike tooth type, Mar 97
95. IS 11893:1986 Specification for potato planter- semi - automatic, Mar 97
96. IS 11905:1986 Specification for shaft assembly for rotavator for power tiller, Jan 01
97. IS 11976:1986 Specification for sugarcane planter, semi - automatic, Mar 97
98. IS 12161:1987 Test code for animal carts, Feb 99
99. IS 12162:1987 Specification for lifting hook, Dec 99
100. IS 12163:1987 Specification for cant hook, Dec 99
101. IS 12164:1987 Specification for log tongs, Dec 99
102. IS 12191:1987 Specification for felling wedges, Dec 99
103. IS 12199:1987 Pneumatic wheeled animal drawn vehicle, Dec 99
104. IS 12334:1988 Tractor mounted bund former, Dec 99
105. IS 12337:1988 Manually operated fertilizer broadcaster, Dec 99
106. IS 12371:1988/ISO 5670:1984 Technical requirements for single acting telescopic tipping cylinders for agricultural trailers, Dec 99
107. IS 12409:1988 Garden sword, Dec 99
108. IS 12482:1988 Methods of test for manually operated dusters, Aug 97
109. IS 12653:1989 Rubber tapping tools-coagulation tray, Dec 99
110. IS 12698:1989 Forestry tools-tree marking digit set, Dec 99
111. IS 12703:1989 Forestry tools-tree marking hammers, Dec 99
112. IS 12917(PT 1):1990 Farm transport equipment-Conventional animal drawn vehicle Part 1 Cattle, donkey and mule, Nov 95
113. IS 12917(PT 2):1991 Farm transport equipment-Conventional animal drawn vehicle Part 2 Camel, Nov 95
114. IS 13151:1991 Forestry tools-Two-men log tongs, Jan 01
115. IS 13373:1992 Grass shear, Aug 97
116. IS 13374:1992 Forestry tools-Hand sappie, nursery (*kutla*), Aug 97
117. IS 13376:1992 Forestry tools-Debarking spud, Aug 97

118. IS 13377:1992 Forestry tools-Shrub cutting knife, Aug 97
119. IS 13378:1992 Forestry tools-Felling lever, Aug 97
120. IS 13483(PT 2):1998 Forestry tools-Stalk puller Part 2 Heavy duty
121. IS 13485(PT 1):1992 Forestry tools-Stalk puller Part 1 light duty, Aug 97
122. IS 13733:1993/ISO 5669:1982 Agricultural trailers and trailed equipment-Braking cylinders, Feb 99
123. IS 13818:1993 Harvesting equipment-Tractor operated potato digger shakers-Test code, Feb 99
124. IS 13821:2000 Forestry tools-Wooden filling vice, Feb 99
125. IS 14459:1997 Battery operated rotary atomizer disc type ULV sprayer
126. IS 14523:1998 Forestry tools-Dibbling board
127. IS 14524:1998 Forestry tools-Dibble iron
128. IS 14525:1998 Forestry tools-Seed plucker
129. IS 14526:1998 Forestry tool-Plant carrier
130. IS 14540:1998 Inter cultivation equipment-Hand hoes specification
131. IS 14541:1998 Forestry and plantation tools-Hoes and forks specification (Amalgamated revision of IS 621,13375, 13486 (Pt 1), 4315
132. IS 14855(PT 1):2000 Fogging machines-specification Part 1:Pulse-jet type thermal fogger

Draft Standards Finalized but not yet Under Print

1. DOC:FAD 59(684) Amalgamated revision of IS 6320, 1691 and Doc 59(2866)
2. DOC:FAD 59(685) Amalgamated revision of IS 4930,11269,12199,12917(1 and 2)
3. DOC:FAD 59(2975) Post hole digger
4. DOC:FAD 59(686) Amalgamated revision of IS 6635,10225
5. DOC:FAD 59(990) Fogging machine-Specification Part 2: Air blower type

Draft Standards Issued for Wide Circulation

1. DOC:FAD 59(692) Amalgamated revision of IS 7825 (1 and 2)
2. DOC:FAD 59(694) Amalgamated revision of IS 10806, IS 10807

Draft Standards under Preparation

1. DOC:FAD 59(687) Amalgamated revision of IS 2565,10254,11035
2. DOC:FAD 59(688) Amalgamated revision of IS 7353, 9813
3. DOC:FAD 59(689) Amalgamated revision of IS 6638, 7565 (1 and 2)
4. DOC:FAD 59(690) Amalgamated revision of IS 4366 (1 and 2), 9217
5. DOC:FAD 59(691) Amalgamated revision of IS 9575, 9581

List of Farm Equipments for Assistance under Sub-Mission on Agricultural Mechanization (SMAM), 2020-21

Tractors

Tractor 2-Wheel Drive (WD) (08-20 PTO HP), Tractor 4-WD (08-20 PTO HP), Tractor 2-WD (20 to 40 PTO HP), Tractor 4-WD (20 to 40 PTO HP), Tractor 2-WD (40 to70 PTO HP), Tractor 4-WD (40 to70 PTO HP)

Power Tillers

Power tiller (below 8 BHP), Power tiller (8 BHP and above)

Rice Transplanter

Self propelled rice transplanter (4 rows), Self propelled rice transplanter 4 to 8 rows and 8 to 16 rows

Self Propelled Machinery

Reaper, Reaper cum binder (3 wheel), Reaper cum binder (4 wheel), Post hole digger/auger, Pneumatic or other Planters

Self Propelled Horticultural Machinery

Track trolley, Nursery media filling machine, Multi-purpose hydraulic system, Power operated horticulture tools for pruning, budding, grafting, shearing *etc.*

All manual/animal drawn equipment/implements/Tools

MB plough, Disc plough, Cultivator, Harrow, Leveler blade, Furrow opener, Ridger, Puddler

Paddy planter, Seed cum fertilizer drill, Raised bed planter, Planter, Dibbler, Equipment for raising paddy nursery, Marker for SRI, Seed treating drum, Rice-wheat seeder, Drum seeder (below 4 rows), Drum seeder (above 4 rows)

Groundnut pod stripper, Thresher, Winnowing fan, Tree climber, Horticulture hand tools, Maize sheller, Feed block machine, Spiral grader, Chaff cutter (up to 3'),

Grass weed slasher, Weeder, Cono-weeder, Garden hand tools

Manual horticultural equipments such as Aluminium ladder, Aluminium pole, Plucker

Tractor (20- 35 BHP) driven equipment

MB plough, Disc plough, Cultivator, Harrow, Leveler blade, Cage wheel, Furrow opener, Ridger, Weed slasher

Roto-puddler, Furrow opener, Bund former, Crust breaker, Roto-cultivator, Power harrow, Rotavator 5 feet, Chisel plough, Reversible hydraulic plough (2 bottom), Reversible mechanical plough (2 bottom), Laser land leveler

Post hole digger, Potato planter, Potato digger, Groundnut digger, Tractor drawn reaper, Onion harvester, Raised bed planter, Sugar cane cutter/stripper, Multi-crop planter, Ridge furrow planter

Grass weed slasher, Power weeder (engine operated above 2 BHP)

Sugarcane thrash cutter, Coconut frond chopper, Rake (small capacity), Round balers (below 14 kg per bale), Straw reaper, Feed block machine (100- 200 kg/hr), Stubble shaver, Straw chopper/Shreder/Mulcher mounted type 5 ft, Trailer/Trolley (up to 3 tonnes capacity)

Groundnut pod stripper, Thresher, Multi-crop threshers, Paddy thresher, Brush cutter, Maize sheller, Mower, Flail harvester, Mower shredder (all purpose/ all crops), Reaper-cum-binder (tractor drawn), Chaff cutter (operated by engine/ electric motor, power tiller, and tractor of below 35 BHP tractor)

Tractor (above 35 BHP) driven equipments

MB plough, Disc plough, Cultivator, Harrow, Leveler blade, Cage wheel, Furrow opener, Ridger, Weed slasher, Laser land leveler, Rotavator (5 to 8 ft), Roto-puddler, Reversible hydraulic plough (2 and 3 bottom), Reversible mechanical plough (2 and 3 bottom), Sub-soiler, Trench makers (PTO operated), *Bund* former (PTO operated), Backhoe loader dozer (tractor operated), Power harrow (PTO operated), Furrow opener, *Bund* former, Crust breaker, Roto-cultivator, Power harrow (PTO operated)

Raised bed planter, Seed drill/zero till seed drill (9 tines), Potato digger, Tractor drawn reaper, Onion harvester, Seed-cum-fertilizer drill, Zero till seed-cum-fertilizer drill (9,11,13 and 15 tines), Direct rice seeder (DRS), Post hole digger, Potato planter (automatic), Groundnut digger, Sugar cane cutter/stripper/planter, Multi-crop planter, Zero–till multi-crop planter, Ridge furrow planter, Happy/Turbo seeder (9, 10 and 11 tines), Pneumatic planter, Pneumatic vegetable transplanter, Pneumatic vegetable seeder, Cassava planter, Manure spreader, Fertilizer spreader – PTO operated, Plastic mulch laying machine, Automatic rice nursery sowingmachinery, Aqua-ferti-seed drill, Raised bed planter with inclined plate planter and shaper attachment, Grass/Weed slasher, Weeder (engine operated above 5 BHP), Weeder (PTO operated)

Groundnut pod stripper, Thresher/Multi-crop threshers up to 4 tonnes/hr capacity, Paddy thresher, Chaff cutter, Forage harvester, Maize sheller, Crop reaper-cum-binder (tractor drawn), Combine harvester (self propelled, up to 14 feet cutter bar), Combine harvester (tractor operated (without tractor up to 10 feet cutter bar), Combine harvester (track, 6-8 feet cutter bar), Combine harvester (track, below 6 feet cutter bar), Thresher/Multi-crop threshers above 4 tonnes/hr capacity, Mower, Flail harvester, Mower shredder (all purpose/all crops), Sugarcane thrash cutter, Coconut frond chopper, Hay rake, Baler (round, 14-16 kg per bale), Baler (round, above 16-25 kg per bale), Balers (round, 180-200 kg per bale), Baler (rectangular, 18-20 kg per bale), Wood chippers, Sugarcane ratoon manager, Cotton stalk uprooter, Straw reaper, Feed block machine (above 200 kg/hr), Stubble shaver, Straw chopper/shreder/mulcher (mounted type 5,6,7 and 8 ft, trailed type, combo type), Super Straw Management System (Super SMS), Shrub master/Cutter-cum-spreader, Rotary straw slasher, Briquette making machine (500-1000 kg/hr capacity)

Chain saw, Wheel barrow, Mango grader/planter and other suitable self propelledmachineries and equipments for horticulture crops

Establishment of PHT units for transfer of primary processing technology, value addition, low cost scientific storage, packaging units and technologies for by-product management in the production catchments

Mini rice mill, Mini *dal* mill, Millet mill, Oil mill with filter press (for all types of horticulture/food grain/oilseed crop), Extractor (for all types of horticulture/food grain/oilseed crops), Pomegranate aril extractor, Custard apple pulper (for all types of horticulture/food grain/oilseed crops), Dehydration unit/Pricking machine/Humidifier (for all types of horticulture/food grain/oilseed crop), Packing machines (for all types of horticulture/food grain/oilseed crops), All types of power driven dehuskar/sheller/threshers/harvesters/de-spiking/deconing machine/peeler/splitter/stripper (for all type of horticulture/food grain/oilseed crops), All types of boiler/steamer/drier (for all types of horticulture/foodgrain/oilseed crop), All types of solar driers (for all type of horticulture/food grain/oilseed crops) with floor area of about 400 to 1000 sq. feet, All types of washing machines (for all types of horticulture/food grain/oilseed crops), All types of grinder/pulverizer/polisher (for all types of horticulture/food grain/oilseed crops), All types of cleaner-cum-grader/gradient separator/specific gravity separator (for all types of horticulture/food grain/oilseed crop)

Plant Protection Equipment

Manual sprayer: Knapsack/foot operated sprayer, Powered knapsack sprayer/power operated sprayer (capacity 8 - 12 L), Powered knapsack sprayer/Power operated sprayer (capacity above 12-16 L), Powered knapsack sprayer/power operated sprayer (capacity above 16 L), Tractor operated sprayer (air carrier/

assisted), Tractor operated sprayer (boom type), Eco-friendly light trap, Tractor operated electrostatic sprayer, Bird scarer

Specialized Agricultural Machinery

Solar operated/electric operated animal deterrent bioacoustics equipment (with solar panel), Solar operated/electric operated animal deterrent bioacoustics equipment (without solar panel), Solar operated/electric operated hydroponic machine for raising nursery of crops

For SC, ST, small and marginal farmers, women and NE States beneficiary the subsidy is 50 per cent except for post harvest technology equipment where it is 60 per cent. For other beneficiaries subsidy is 40 per cent. For all the beneficiaries these per cent ages are subject to a maximum value. Source: Ministry of Agriculture and Farmer's Welfare (2020).

Glossary of Terms

A

Aeration (Improvement): Any method of loosening soil or improving its physical properties to allow air to circulate.

Agitator: A device to operate a mechanism that agitates fertilizer in a hopper to prevent bridging. Hydraulic, mechanical and pneumatic agitators are also used in sprayers/dusters for similar purposes.

Agri-business: A way of farming combining agriculture and business and usually involves large amounts of land, animals, and expensive technology.

Agricultural tractor: A self-propelled wheeled vehicle having two axles, or a track-laying or semi-track-laying machine, designed to pull, push, carry and operate implements and machines used in agricultural operations including forestry works.

Air filter: A device that filters incoming air fed to the engine.

Air-fuel ratio: Ratio of mass of air to mass of fuel input into engine. Sometimes it is also reported as fuel-air ratio.

Alternating current: The electric power supplied by an AC generator and distributed in one, two, and three phase form.

Ampere: Standard unit of measurement for electric current. 1 Ampere current is defined as flow of 1 coulomb charge through a conductor in 1 s.

Anti-drip device: A device which is part or fitted within the nozzle to prevent any further flow or dripping from the nozzle after the spray boom has been shut-off.

Armature: The laminated iron core with wire wound around it in which electromotive force is produced by magnetic induction in a motor or generator. It is the rotating portion of the magnetic structure of a DC or universal motor.

Asynchronous motor: An AC motor which does not run at synchronous speed. The ordinary induction motor is an asynchronous motor. Also called non-synchronous motor.

Auxiliary cylinder: Rotating component, used in head-feed combines, for rethreshing and separating.

Axial flow pumps: In these pumps, fluid enters and exits along the same direction parallel to the rotating shaft.

Axle load (trailer): The technically permissible axle load for each axle as stated by the manufacturer.

B

Back furrow: A raised ridge left at the center of the strip of land during ploughing from center to side.

Balanced trailer: A double or multiple axle trailer whose total load is supported by its wheels when detached from the tractor.

Basic machine: It is a device with few or no moving parts that are used to modify motion and the magnitude of a force in order to perform work.

Bedding or ridging: A tillage operation which places soil into a specific configuration of ridges and furrows.

Bed planting: Planting on elevated level beds separated by furrows.

Belt power: Power transmitted through a belt with the governor control in the position recommended by the tractor manufacturer for belt pulley work.

Belt pulley: A pulley driven by a power unit to transmit power to another machine by means of a belt.

Biomass: A renewable energy source that comes from recently living plants and animals, such as wood, crops, manure and even garbage.

Block (engine): Body of engine made of cast iron or aluminium, which contains the cylinders. In some older engines, the valves and valve ports were also mounted in the block. Whereas the block of water-cooled engines includes a water jacket cast around the cylinders, the exterior surface of the block has cooling fins in air-cooled engines.

Bore: Diameter of the cylinder or diameter of the piston face. Both are nearly the same as clearance between the two is very small.

Bottom-dead-center (BDC): Piston position at the point closest to the crankshaft at which it stops during its reciprocating motion. Some sources also call this crank-end-dead-center (CEDC) because it is not always at the bottom of the engine.

Brake horsepower: It is the horsepower available on the crankshaft and is measured by a suitable dynamometer.

Brake linings: High-friction, heat-resistant material attached to the brake shoes in a rear drum brake system.

Brake master cylinder: A device that stores brake fluid which is hydraulically forced to the brakes upon pressing of brake pedal.

Brake pads: High-friction material attached to a metal braking plate.

Broadcasting: The process of scattering of agricultural inputs, such as seed, fertilizer and manure on the surface of the soil either by hand or mechanically.

Broadcast tillage: Coverage of an entire area as contrasted to a partial coverage as in bands or strips. It is also known as overall tillage.

Brush: A conductor usually composed of some element of carbon that serves to maintain an electrical connection between stationary and moving parts of a machine.

Budding: It is kind of grafting technique in which a single bud from the desired scion is used rather than an entire scion, which may contain many buds.

Budding and grafting knife: A knife used for budding and grafting purposes.

Bulldozer: A heavy duty powerful vehicle moving on metal tracks (crawler type) or big pneumatic wheels having a large curved piece of metal at the front used to move soil and rocks or pushover trees and other structures.

Bypass: A device that allows all or part of the fluid delivered by the pump back to the sprayer tank.

C

Cage wheel: A wheel or an attachment to a wheel with space crossbars for reducing ground pressure and/or improving traction of a tractor or power tiller. Generally used in wetland seedbed preparation.

Cage wheel, half: A cage wheel, which is used in conjunction with pneumatic wheel mostly for wetland seedbed preparation.

Calibration: Operation of adjusting and checking the application appliances to give the desired application rate *e.g.* calibration of seed drills, sprayers *etc.*

Calorific value: The heat liberated by combustion of a fuel is known as calorific value or heat value of the fuel.

Camshaft: Rotating shaft used to push open valves at the proper time in the engine cycle, either directly or through mechanical or hydraulic linkages.

Capacitance: The value in microfarads of a capacitor or condenser.

Capacitor: A device, which when connected in an alternating current circuit causes the current to lead the voltage in time phase. The peak of the current wave is reached ahead of the voltage wave.

Carburetor: A device used on petrol engines mounted on the engines intake manifold that meters the proper amount of fuel into the air flow by means of a pressure differential.

Cetane number: The percentage of cetane in a mixture of cetane ($C_{16}H_{34}$) and alphamethyl naphthelene ($C_{11}H_{16}$) that produces the same knocking effect as the fuel under test is called cetane number of the fuel.

Chaffer: The upper sieve of the combine harvester on which grain and chaff mixture-falls from stepped grain bed for initial cleaning

Check-row planting: The process of planting in which row to row and plant to plant distances are uniform and plants across the rows are also in line.

Chiseling: A tillage operation used to break-up hard pan in the soil. It is usually performed using a narrow tool at depths greater than the normal ploughing depth.

Clearance volume: Minimum volume in the combustion chamber with piston at top dead center (TDC).

Clods: A lump or mass especially of earth or clay. Tillage tools cut, shear or break loose the clods during tillage operation.

Cloud point: The lowest temperature at which wax and other substance in the oil crystallize and separate out from oil or oil becomes cloudy is called cloud point.

Combined tillage operations: Two or more operations performed simultaneously utilizing two or more different types of tillage implements say sub-soiler-lister, lister-planter or plough-planter combinations to simplify control or reduce the number of operations over a field.

Combine harvester: A machine designed to harvest different grain crops by combining three traditionally separate harvesting operations namely reaping, threshing, and winnowing into a single mechanical process.

Combine height: The vertical distance from the horizontal plane on which the combine is standing to the highest point on the combine expressed in cm.

Combustion chamber: The end of the cylinder between the head and the piston face where combustion occurs. The size of the combustion chamber continuously changes from a minimum volume when the piston is at TDC to a maximum when the piston is at BDC.

Commutator: A cylindrical device mounted on the armature shaft and consisting of a number of wedge-shaped copper segments arranged around the shaft (insulated from it and each other). The motor brushes ride on the periphery of the commutator and electrically connect and switch the armature coils to the power source.

Compression ignition engine: An engine in which the combustion process starts when the air-fuel mixture self ignites due to high temperature in the combustion chamber caused by high compression. It is commonly known as diesel engine.

Compression ratio: The ratio of the volume of an engine cylinder with its piston at bottom dead center to the volume of the same cylinder with its piston at top dead center.

Cone (forestry): This is a cone shaped metal accessory attached to the end of one or more logs during skidding to help them slide.

Connecting rod: Rod connecting the piston with the rotating crankshaft.

Conservation tillage: It refers to a tillage practice or system of practices that leaves at least 30 per cent plant residues on the soil surface for erosion control and moisture conservation.

Contour ploughing: The method of ploughing in which soil is broken and turned along contours.

Conventional tillage: The combined primary and secondary tillage operations normally performed in preparing a seedbed.

Converter (electricity): The devices for changing AC to DC and DC to AC.

Coolant: Also called antifreeze is a mixture of water and ethylene glycol or other additives that has both a higher boiling point and a lower freezing point than plain water.

Cooling fins: Metal fins on the outside surfaces of cylinders and head of an air-cooled engine that helps to cool the cylinders by conduction and convection.

Coupling: The mechanical connector joining the motor shaft to the equipment to be driven.

Covering device: A device used to refill a furrow after seeds have been placed in it.

Crankcase: Part of the engine block surrounding the rotating crankshaft. In many engines, the oil pan makes up part of the crankcase housing.

Crankshaft: Rotating shaft through which engine work output is supplied to external systems. The crankshaft is connected to the engine block with the main bearings. It is rotated by the reciprocating pistons through connecting rods connected to the crankshaft, offset from the axis of rotation.

Crown: The top of the piston

Current: The movement of electrons (usually from positive to negative) through conducting materials. It is measured in amperes.

Cut-fill ratio: Volume of cut to volume of fill. It should always be more than 1.0 in land leveling/grading.

Cut-off device: A hand-operated mechanism located between the delivery hose and the spray lance to control the flow of the liquid from the sprayer.

Cutter bar: The assembly comprising guard (finger) bar, guards (fingers), knife guides, wearing plate, outer shoe and main shoe excepting the reciprocating part of the cutting mechanism.

Cutting: A vegetative method of plant propagation whereby a piece of plant leaf, stem, root or bud is cut from a parent plant. It is then inserted into a growing medium to form roots, thus developing a new plant.

Cylinders: The circular cylinders in the engine block in which pistons reciprocate back and forth.

D

DC (direct current): A current that flows only in one direction in an electric circuit. It may be continuous or discontinuous and it may be constant or varying.

DC motor - compound wound: Type of DC motor having both shunt and series field connections. This motor has good speed regulation and starting torque.

DC motor - permanent magnet: Type of DC motor where the field poles and the armature poles are electromagnets. The only current used by the motor is that of the armature. It has high starting torque, good speed regulation and a definite maximum speed.

DC motor - series wound: Type of DC motor which has its field winding connected in series with the armature. This motor has very high starting torque, but has a tendency to 'run-away' when lightly loaded or unloaded, having poor speed regulation.

DC motor - shunt wound: Type of DC motor which has its armature winding and field winding done in parallel circuits. The motor has very good speed regulation.

Dead furrow: An open trench left in between two adjacent strips of land after finishing the ploughing operation.

Deep tillage: A primary tillage operation that manipulates soil to a greater depth than normal ploughing.

Depth of cut: The maximum depth of the penetration of the tool measured with reference to the initial soil surface.

Dibbler: A stick for making holes for seedlings to be transplanted.

Dibbling: The process of placing seeds in the holes made in a seedbed and covering them.

Differential unit: A differential unit is a special arrangement of gears in the driving axle that makes one of the rear wheels of the tractor to rotate slower or faster than the other.

Direct drive: Drives that send power directly to an application rather than going through gears or another method of power transfer.

Diesel engine: An engine that uses diesel fuel and compression ignition. See compression ignition engine.

Differential: A device that enables two wheels driven from a single shaft to rotate at different speeds.

Direct injection system: A fuel injection system in which the injection nozzles are located inside the combustion chamber of each piston.

Direct sowing: To plant seeds directly in the soil where the plant is to grow. The term has now become quite popular in rice cultivation where rice seeding is done directly rather than through transplanting.

Disbudding: Removing surplus buds or shoots so that those remaining grow larger or stronger.

Disc angle: It is the angle at which the plane of the cutting edge of the disc is inclined to the direction of travel.

Displacement volume: Volume displaced by the piston as it travels through one stroke. Displacement can be given for one cylinder or for the entire engine (one cylinder times number of cylinders).

Disks: Rolling circular blades that have straight or fluted edges and are intended to cut residues, pulverize soil structure, and/or level the soil surface. Disks are often mounted in gangs of parallel blades.

Dozer: Like a bulldozer but mostly operated by a tractor that uses metal tracks (crawler type) or pneumatic wheels. It also has a large curved piece of metal at its front used for moving soil and rocks and pushing over trees and other structures.

Draft: The horizontal component of the pull.

Drawbar: A device in a tractor by which the pulling power of the tractor is transmitted to the trailing implements.

Drawbar horsepower: It is the power of a tractor measured at the end of the drawbar. It is the power available to pull loads.

Drawbar power: Power obtained at the drawbar with the governor control in the position recommended by the tractor manufacturer for drawbar work and the tractor moving on a horizontal surface, with the drawbar pull applied horizontally.

Drift: Fraction of the applied pesticide which is not deposited within the target area.

Drilling: It is the process of placing the seeds in rows at uniform rate and at controlled depth with or without covering them with soil.

Drummy-type thresher: A hammer mill type thresher without separation and cleaning systems. Usually a centrifugal blower is provided for partial separation and cleaning of grains.

Dust: Finally divided particles of an inert solid substance carrying the active ingredient and ready for application using a duster.

Duster: An appliance used for dusting.

E

Earthing-up: Drawing up the soil around a plant's stem to blanch it, to cover tubers and prevent greening, or to support it.

Earth moving: Tillage action and transport operations utilized to loosen, load, carry, and unload soil.

Earth work: It involves total volume of shifting of soil in a field (plot) through cut and fill operations in land leveling operation.

Effective field capacity: The actual area covered by the implement based on its total time consumed and its width.

Efficiency (Electric motors): The efficiency of a motor is the ratio of mechanical output to electrical input. It represents the effectiveness with which motor converts electrical energy into mechanical energy at the output shaft.

Efficiency (general): Ratio of output to input power.

Efficiency (pump): Pump efficiency is defined as the ratio of water horsepower output from the pump to the brake horsepower input for the pump.

Electrical motor: It is an electromechanical machine that converts electrical energy to mechanical energy.

Electromotive force (EMF): A synonym for voltage usually restricted to generated voltage.

Energy: The capacity for doing work; usable power (as heat or electricity); the resources for producing such power.

Engine oil: A substance that lubricates and cools the moving parts of the engine and reduces corrosion and the formation of rust.

Engine power: power measured at the crank shaft of the engine with the governor control lever in the position recommended by the manufacturer.

Engine size: A vehicle engine's displacement, in liters as per manufacturer.

F

Feed metering mechanism: The mechanism of a seed drill, planter or fertilizer drill which delivers seeds or fertilizers from the hopper at pre-decided rates.

Fertilizer: Any natural or synthetic material that is applied to soil or to plant tissues to supply one or more nutrients essential for the growth of plants.

Fertilizer broadcaster: A fertilizer distributor with a spreading width substantially greater than the width of the machine.

Fertilizer drill: A machine to deposit fertilizer in soil at regulated and selected rates and at pre-determined depth.

Field capacity (machines): The field capacity of a farm machine is the rate at which it performs its primary function.

Field efficiency: The ratio of effective field capacity to theoretical field capacity expressed in percentage.

Final drive: A device provided for additional reduction of speed between the drive shaft from the transmission and the axle connecting the drive member.

Fire point: The temperature at which vapours are released from the oil continuously so that the flame persists for a longer period than a momentarily flash is known as fire point.

Firing order: The order or sequence in which the firing in different cylinders of a multi-cylinder engine takes place.

Flash point Flash point is the lowest temperature at which oil is heated until sufficient inflammable vapours come off, which when brought to flame produces a momentary flash.

Flail mower: A grass-cutting machine which utilizes a power source to rotate a horizontal shaft with blade. Cutting action is accomplished by impact of rotating blades.

Flywheel: A heavy rotating wheel having a large moment of inertia connected to the crankshaft of the engine. Its main purpose is to store energy and furnish a large angular momentum that keeps the engine rotating between power strokes and smoothens out engine operation. It is also used in several other tools and machines notably the chaff cutter and the disc chipper.

Fog (fogging machine): When droplets smaller than 15 μm fill a volume of air to such an extent that visibility is reduced, it is termed as fog.

Foliar application: A technique of feeding plants by applying liquid fertilizer directly to plant leaves.

Force: It is an invisible agent that always tries to change the state of the body *i.e.* a change in speed, direction or shape.

Forwarding: It is process of transporting logs with both ends off the ground.

Fossil fuels: Fuels formed in the Earth's crust from the remains of ancient plants and animals buried underground *e.g.* petroleum (oil), coal, and natural gas.

Frequency: The rate at which alternating current reverses its direction of flow. Measured in hertz (Hz); 1 Hz = 1 cycle per second.

Fuel capacity: The amount of fuel that a fuel tank of a vehicle can hold.

Fuel injector: A pressurized nozzle that sprays fuel into the incoming air on SI engines or into the cylinder on CI engines.

Fuel pump: Electrically or mechanically driven pump to supply fuel from the fuel tank (reservoir) to the engine.

Full load speed: The speed that the output shaft of the drive motor attains with rated load connected and with the drive's controller adjusted to deliver rated output at rated speed.

Fungicides: Compounds used to prevent the spread of fungi in gardens and crops, which can cause serious damage to plants.

Furrow: The trench formed by any tool in the soil during operation.

Furrow crown: The peak of the turned furrow slice.

Furrow planting: Planting in the bottom of furrows.

Furrow opener: A part of seed drill for opening a furrow and assisting in placing the seeds in the furrow.

Furrow slice: The soil mass cut and turned by a tool.

Furrow sole: The bottom surface of the furrow.

Furrow wall: The undisturbed side of furrow.

G

Garden tool: Any tool that may be a hand tool or power tool that makes gardening easy.

Gear ratio: In gearboxes and gear motors, the gear ratio is derived by dividing the input speed by the output speed.

Governor: It is a mechanical device designed to control the speed of the engine within specified limits.

Governor hunting: A governor is said to be hunting, if the speed of the engine fluctuates continuously above and below the mean speed.

Grafting: An asexual method of plant propagation in which the upper part (scion) of one plant grows on the root system (rootstock) of another plant.

Grafting knife: A knife to cut the wood from scion and to insert the same in the stock.

Grain elevator: The device in a combine harvester that carries the grains from grain auger to grain tank or bin.

Granules (pesticides): Particles, within a defined size range, of an inert substance containing or carrying the active ingredients.

Granules applicator: A device used to apply materials in the form of granules.

Grapple: A grapple is a set of tongs attached to the end of a cable to skid or load a log.

Green manure: A crop that is grown and then incorporated into the soil to increase soil fertility or organic matter content.

Gross load (trailer): The sum of pay load and the unladen mass of the trailer.

Ground clearance: The height of the lowest point of the trailer from a level supporting surface when the trailer is loaded to its pay load and tyres inflated as per the recommended pressure.

H

Hand feed tuber planter: A tuber planter, the planting element of which is fed by hand.

Harrow: A secondary tillage implement used to break up and smooth out the surface soil. Harrowing often follows coarser ploughing with the purpose of breaking up large lumps of soil so as to provide a better tilth or sometimes to remove weeds or to cover seed after sowing.

Harrowing: A secondary tillage operation which pulverizes, smoothens and packs the soil in seedbed preparation and also to control weeds.

Harvesting: Harvesting is the operation of cutting, picking, plucking or digging, or a combination of these operations to remove the crop from under or above the ground. Also removing the useful part or fruit from the plants.

Head: The piece which closes the end of the cylinders, usually containing part of the clearance volume of the combustion chamber.

Header: The portion of the combine comprising the mechanism for gathering, cutting, stripping or picking the crop and deliver it to the cylinder.

Header loss: The loss of grains and ear heads being shed and left over on the ground as a result of operations of cutter bar and header unit in a combine.

Headland: A wide strip of land at each end of a planted field used for turning or maneuvering large farm machinery such as ploughs.

Heat engine: It is defined as an engine that converts the chemical energy of the fuel into thermal energy for performing some useful work.

Hill dropping: The process of placing seeds in small groups at regular intervals along straight parallel furrows and covering the seeds with soil.

Hoist: A hoist is a device used for vertically lifting or lowering a load by means of a drum or lift-wheel.

Horsepower: It is that amount of force which is capable of displacing 75 kg force through a distance of one meter in one second. A unit of measurement for engine power originally developed as a way to express the output of steam locomotives in terms of the strength of draft horses.

Hydraulic power lift: A mechanism driven by a tractor's power unit to raise, hold or lower mounted or semi-mounted equipment by hydraulic means.

Hydropower: Electricity generated using energy of the flowing water to turn a generator. Also called hydro energy.

I

Idle speed: The RPM of the engine free from load (all accessories off) at normal operating temperature and in neutral.

Indicated horsepower: The power actually developed in the cylinder is called indicated horsepower. The amount of power measured on the flywheel is always less than the power generated in the engine on account of expansion of the combusted fuel.

Indicator diagram: Indicator diagram is a chart used to measure the thermal or cylinder performance of reciprocating steam or IC engines. It is used to calculate the work done and power produced in the engine. It is also known as pressure-volume (PV) diagram.

Induction motor: An AC electric motor in which the electric current in the rotor needed to produce torque is obtained by electromagnetic induction from the magnetic field of the stator winding. Also called an asynchronous motor.

Insecticide: A kind of pesticide that kills insects.

Intake manifold: Piping system which delivers incoming air to the cylinders usually made of cast metal, plastic, or composite material.

Inter-cultivation: Soil cultivation performed in standing crop.

Internal combustion engine: An engine that obtains its power from heat and pressure produced by the combustion of a fuel-and-air mixture inside a closed chamber or cylinder.

Inversion tillage: Inversion tillage performed by traditional tillage equipment comprise flip over a soil layer (often 15-30 cm), burying surface residues (and associated weed seeds, spores, and insect larva and eggs) in the process. The result is a surface with minimal residues for easy management, but susceptible to erosion.

Inverter: An electronic component that turns power supplied to a motor from DC voltage into AC voltage.

Irrigation: The application of controlled amounts of water to plants at needed intervals for the purposes of growing agricultural crops, maintaining landscapes, or revegetating disturbed or drought-affected soils. It may also be used as a means of protecting crops from frost, suppressing the growth of weeds, and preventing soil consolidation.

K

Kilowatt: One thousand watts, a watt being a unit of measure of power, or how fast energy is used. It is typically used to describe electrical power usage.

Kilowatt hour (kWh): A unit of measure for energy normally applied to electricity usage. It is equal to the amount of energy used at a rate of 1000 watts over an hour.

Knife: The reciprocating parts of the cutting mechanism comprising of knife head, knife back and knife sections.

Knife registration: The alignment of centre line of knife section with the centre line of guard.

L

Land forming: Tillage operation that moves soil to alter land configuration for orderly movement of water, reduces runoff and conserves soil moisture or creates desired soil configurations.

Land grading: An operation of land leveling that reshapes the field surface to a planned grade.

Land leveling: It is defined as a process by which the surface relief of a field is modified to a desired grade and to certain specifications that provide a more suitable surface for efficient application of irrigation water and for improved drainage conditions.

Lawn mower: A lawn mower is a grass cutting machine that consists of one or more rotatory blades used to cut or trim grass and lawn vegetation to maintain the grass length short and tidy.

Leveling: 1) The tillage operation in which soil is moved to establish a desired slope. 2) The surveying process of determining the difference in elevation between two or more points by measuring the vertical distance between the points; the determination of elevation of points above a datum.

Listing: A tillage and land-forming operation using a tool which splits the soil and turns two furrows laterally in opposite directions, thereby providing a ridge-and-furrow soil configuration.

Lopping: It is 'basically the trimming of branches of trees in order to reduce its size.

M

Maximum drawbar pull: The maximum horizontal drawbar pull at a drawbar height recommended by the manufacturer which a tractor is able to sustain in the line of its longitudinal axis.

Maximum horsepower: The maximum horsepower is measured at the engine flywheel without any of the power consuming accessories attached.

Mean effective pressure: The mean effective pressure (MEP) is the average pressure during the power stroke, minus the average pressure during the other three strokes. The MEP is the pressure that actually forces the piston down during the power stroke.

Mechanical advantage (machine): It is the ratio of the force delivered by the machine to the force applied.

Mechanized agriculture: It is defined as the process of using agricultural machinery to mechanize agricultural operations to increase the productivity of farm workers.

Metric ton: Also referred to as a tonne. It is a measure of mass equal to 1,000 kilograms, or the mass of one cubic meter of water. This is different from the short ton, a unit used in the United States, which is equal to 2,000 lbs.

Micro-nutrients: The mineral elements such as Fe, Mn, Zn *etc.* needed by plants in very small quantities. Also called trace elements.

Minimum tillage: A type of conservation tillage performed to conserve soil quality by minimizing the amount of soil manipulation necessary for successful crop production. It is mostly achieved through completely avoiding primary tillage and practicing only minimal secondary tillage.

Mixed flow pumps: The flow in these pumps is a compromise between radial and axial flow pumps. The fluid experiencing both radial acceleration and lift and exits the impeller somewhere between $0\text{-}90^{\circ}$ from the axial direction.

Motor: A device that takes electrical energy and converts it into mechanical energy to turn a shaft.

Motor winding: Used to generate force in an electric motor. Power is produced through an interaction between the electrical current in the wire winding and the magnetic field of the engine.

Mould-board plough: A traditional primary tillage tool comprising mould-board that lifts and rolls the soil, bringing about soil inversion.

Mower: A machine to cut herbage crops and leave them in swath.

Muffler: Silencer used to reduce engine noise.

Mulch tillage: Preparation of soil in such a way that plant residues or other mulching materials are specially left on or near the surface.

Multi-crop thresher: Equipment used for threshing more than one crop with or without minor adjustments.

N

Net horsepower: Net HP is measured at the engine flywheel in the same manner as the maximum HP, but after equipping with accessories. Net HP is the basis for rating the HP of industrial and farm tractors.

Non-renewable resource: A resource that is not replaceable once it has been used. Fossil fuels such as coal and natural gas are non-renewable resources.

Non-return valve: An automatic device which permits the flow of a liquid in one direction only.

Notch (Forestry): A notch made of two cuts as low as possible at the base of the tree and toward the direction of fall.

No-tillage planting: Direct planting in essentially unprepared seedbeds.

No-till farming: Any method of growing crops or maintaining pasture without disturbing the soil through tillage.

Nozzle (spray nozzle): A part or an assembly of parts having orifice(s) which transforms the fluid being ejected under pressure into a spray.

Nuclear energy: Energy stored in the nucleus of an atom. It is released when atoms join together (fusion) or split (fission). The fusion reaction in the sun provides warmth and light, while the fission reaction at a nuclear power plant creates enough energy to power large cities.

O

Octane number: The percentage of iso-octane (C_8H_{18}) in the reference fuel consisting of a mixture of iso-octane and normal heptane (C_7H_{16}), when it produces the same knocking effect as the fuel under test, is called octane number of the fuel.

Oil filter: A cartridge-filled canister placed in an engine's lubricating system to strain dirt and abrasive materials out of the oil.

Oiliness: The ability of the lubricating oil to adhere to the surface is known as oiliness.

Oil pan: Oil reservoir usually bolted to the bottom of the engine block, making up part of the crankcase. It acts as an oil sump for most engines.

Oil pump: Pump used to distribute oil from the oil sump to various lubrication points.

Overhead cam: Camshaft mounted in engine head giving more direct control of valves in the engine head.

Overhead valves: Valves mounted in engine head.

P

Peg drum: A cylinder having rows of spikes or pegs.

Photovoltaic: The process of creating energy through the conversion of sunlight into electricity through a photovoltaic (PV) cell, also called a solar cell.

Piston: The cylindrical-shaped device that reciprocates back and forth in the cylinder, transmitting the pressure forces in the combustion chamber to the rotating crankshaft.

Piston rings: Metal rings that fit into circumferential grooves around the piston and form a sliding surface against the cylinder walls. At the top of the piston are two or more compression rings while below the compression rings on the piston is at least one oil ring, which assists in lubricating the cylinder walls and scrapes away excess oil to reduce oil consumption.

Pitman: A connecting rod to transmit reciprocating motion to the knife head.

Planter: A machine used for precision drilling, hill dropping or check-row planting.

Plough: A primary tillage farm implement used to loosen or overturn soil while preparing for sowing seed or transplanting.

Ploughing: A primary tillage operation performed with the help of a plough that cuts, breaks and partially or completely inverts the soil.

Plough pan: A hard layer immediately below the depth of regular tillage by mould-board ploughs, disks, and rotary tillers.

Ploughshare: The large metal blade that is the leading edge of the mould-board of a plough, used to cut through large amounts of soil to the bottom of the furrow.

Ply rating: Identification of a given tyre with its maximum recommended load when used in a specific type of service. It is an index of the strength and does not necessarily represent the number of cord plies in the tyre.

Pour point: The lowest temperature at which oil is observed to flow by gravity in a specified laboratory test is known as the pour point.

Power: It is the rate at which work is done. It is also defined as the rate at which energy is consumed or utilized. The rate at which the engine can do work is measured in horsepower (HP).

Power outlet: Any outlet which transmits the engine power to the tractor in order to make it functional, such as PTO, belt pulley and drawbar.

Power take off horsepower: It is the power delivered by a tractor through its PTO shaft. It is approximately the same as the belt horsepower.

Power tiller: A prime mover in which direction of travel and its control for field operation is performed by the operator walking behind it. It is also known as hand or walking type tractor.

Precision: It denotes relative or apparent nearness to the design values.

Precision seeding: A method of seeding that involves placing seed with precise spacing and depth, either manually or mechanically, as opposed to broadcast seeding.

Pressure bar: A spring or weight loaded bar which assists in penetration of the furrow opener.

Pressure regulator: An automatic device to control the pressure of a fluid or gas within a pre-defined range.

Press wheel: The wheel that compacts the soil and covers seeds in the furrow.

Primary tillage: The operations performed to open up any cultivable land with a view to prepare a seedbed for growing crops is known as primary tillage. More often it is also referred as ploughing, the process of cutting, breaking and inverting the soil either partially or completely.

Primary winding: The winding of a motor, transformer or other electrical device which is connected to the power source.

Priming: The operation in which suction pipe, casing of the pump and a portion of delivery pipe is filled by the fluid from outside is known as priming.

Pruning: The selective removal of certain unwanted plant parts or tissues, such as branches, buds, or roots, from crops or landscape plants during cultivation for any of a variety of reasons, including controlling or redirecting growth, improving or sustaining the plant's health or appearance, reducing risk from falling branches, preparing juvenile plants for transplanting, and increasing the yield or quality of harvestable flowers and fruits.

PTO power: Power obtained at the main PTO with the governor control in position recommended by the tractor manufacturer for PTO work, and the tractor being stationary.

Puddling: The tillage operation designed to disrupt aggregates and disperse clay, creating an impermeable layer that helps to control deep percolation losses.

Pull: The total force required to pull an implement.

Pulley power: Power measured by coupling the pulley shaft directly to the dynamometer with the governor control in the position as recommended by the tractor manufacturer for pulley work.

Pump (irrigation): A hydraulic pump is any device that uses input force to create pressure or kinetic energy, which in turn creates flow.

Push rods: Mechanical linkage between the camshaft and valves on overhead valve engines with the camshaft in the crankcase.

R

Radial flow pumps: The fluid in these pumps enters along the axial plane and is accelerated by the impeller to exit at right angle to the shaft (radially).

Radiator: Liquid-to-air heat exchanger of honeycomb construction used to remove heat from the engine coolant.

Ratooning: The practice of harvesting most of the above ground portion of the plant with roots and shoots left intact. It permits the plant to recover and produce a fresh crop in a subsequent growing season.

Reaper: A machine to cut grain crops.

Reaper-binder: A reaper that not only cuts the crops but makes them into neat and uniform sheaves/bundles.

Reconnaissance survey: An exploratory study of the area to identify problems, degree and extent of the problem and even possible solutions although only qualitatively.

Rectifier: An electronic circuit which converts alternating current into direct current.

Reduced-till: Reduced-till leaves 15-30 per cent residue cover after planting throughout the critical wind erosion period. Sometimes it is also referred as minimum till.

Reduced tillage: A tillage system in which primary tillage operation is performed in conjunction with special procedures to reduce or eliminate secondary tillage operations.

Reel: A part of the combine harvester with revolving slats or arms with battens arranged parallel to the cutter bar. It holds the crop being cut by the knife and to push and guide it to a conveyor platform or feeder conveyor auger.

Relay: An electrically controlled device that causes electrical contacts to change status. Open contacts will close and closed contacts will open when rated voltage is applied to the coil of the relay.

Renewable resources: Energy sources that are replaced naturally such as energy from the sun, wind and water.

Repeat leveling: The repetition of land leveling operation in terms of number of years after the previous land leveling operation. Sometimes it is given in terms of number of seasons.

Returns: The process of recirculating incompletely threshed or completely unthreshed grain for further processing in a combine harvester. Also earnings from investments.

Ridge planting: Planting on ridges.

Ripper: A trailer or mounted tined device with elements for attachment to the prime mover and a hydraulic drive mostly used for stripping work.

Root crop harvesters: A machine to expose the root crops from the soil for picking.

Rootstock: A root system and stem onto which another plant is grafted.

Root zone: The part of soil profile exploited by the roots of plants.

Rotary tillage: Tillage operation employing rotary action to cut, break and mix soil.

Rotor: The rotating member of an induction motor with a shaft. Current is normally induced in the rotor which reacts with the magnetic field produced by the stator to produce torque and rotation.

Rough leveling: Removal of abrupt irregularities such as mounds, dunes or filling of pits, depressions and gullies.

S

Scarifying: Using a spring-tooth rake to pull out moss and dead vegetation from a lawn.

Scavenging: The process of removal of burnt gases from the engine cylinder is known as scavenging.

Secateur: A scissor-like tool used for pruning the plants.

Secondary tillage: Tillage operations performed to create proper soil tilth for seeding and planting following primary tillage are known as secondary tillage.

Seedbed: The zone where seeds are seeded.

Seed-cum-fertilizer drill: A machine which drills seeds and fertilizers simultaneously in the same or different rows.

Seed drill: A machine to place seeds at uniform rate at selected depth and row spacing.

Semi-trailer: The trailer which while in use transfers a part of its load on the towing tractor and rest on its axle(s).

Separating: It is the process of isolating the detached grain, small debris, incompletely threshed and completely unthreshed grains from the bulk of straw, stem or stalk.

Shanks: Stiff tines are generally referred to as shanks. See tines.

Short-circuit: A defect in a winding which causes part of the normal electrical circuit to be bypassed. This frequently results in reducing the resistance or impedance to such an extent as to cause overheating of the winding, and subsequent burnout.

Shrub: A perennial plant with a number of persistent woody stems.

Side delivery reaper: A reaper which delivers the harvested crop on the side to clear the space for the next run.

Side draft: The horizontal component of the pull perpendicular to the direction of motion. This is developed if the centre of resistance is not directly behind the centre of pull.

Single phase motor: Simplest electric motor with minimal power output – usually only around one horsepower (HP). Electric motors also come in two phase and three phase designs.

Skidding: Skidding is a lumbering operation that consists of collection of trees from which the branches have been removed and hauling them from the cutting area to loading areas on logging routes. Also transport of logs when one or both ends of the log are on the ground.

Skirt: Sides of the piston are called the skirt.

Smoothening: An operation required nearly every cropping season, particularly where cultivation and harvesting processes significantly disrupts the field surface. It is also a final process in the land leveling to smoothen the topographic variations that remains following the operation.

Soil compaction: Reduction in the specific volume of soil by means of mechanical manipulation.

Soil cultivation: Tillage operations performed to create soil conditions conducive to improved aeration, infiltration, compaction and moisture conservation as well as to control weeds.

Solar panels: Solar panels are made up of photovoltaic cells, which are used to generate electricity from the sun.

Sowing: The process of placing seeds in seedbed.

Spark ignition engine: It is also called a petrol engine, the engine in which the combustion process in each cycle is started with the help of a spark plug.

Spark plug: Electrical device used to initiate combustion in an SI engine by creating a high-voltage discharge across an electrode gap.

Special purpose motor: A motor with special operating characteristics, special mechanical construction, or both, designed for a particular application and not falling within the definition of a general purpose or definite purpose motor as defined by NEMA.

Specific fuel consumption: Specific fuel consumption of an engine may be defined as the amount of fuel consumed per unit work done *i.e.* SFC = Amount of fuel consumed (in kg/gm/lbs/)/Amount of work done (in kgm/Nm/Joules/BHP. hr). In electrical engine, it is the amount of electricity (in KWh) consumed per unit work done.

Split phase start: Motors, which employ a main winding and an auxiliary winding called the starting winding, which are unlike and thereby "split" the single phase of the power supply by causing a phase displacement between the currents of the two windings thus producing a rotating field.

Spray boom: A device on which the nozzles are mounted and which may form or support one or more pipelines which are carrying the liquid to the nozzle.

Spray lance: A hand-held tube through which the liquid after being released from cut-off device reaches to nozzle.

Spray overlap: Amount by which the spray from adjacent nozzles overlaps, as measured at the target surface level.

Starting torque: The torque exerted by the motor during its starting period. It is a function of speed and slip.

Stator: That part of an induction motor's magnetic structure which does not rotate. It usually contains the primary winding.

Stover: Material left behind after a plant has been harvested such as leaves and stalks.

Straw: An agricultural byproduct consisting of the dry stalks of cereal plants after the grain and chaff have been removed.

Straw walker: The assembly of two or more racks which agitates the straw and separates the remaining grains from straw.

Strip tillage: A tillage system in which only isolated bands of soil are tilled.

Stroke: The distance between the centerline of an engine's crankshaft and the centerline of its connecting rod journal. It may also be defined as the distance travelled by the piston from one extreme position to the other: TDC to BDC or BDC to TDC.

Submersible motor/pump: A hermetically sealed motor close-coupled to the pump body with whole assembly being submerged in the fluid to be pumped.

Sub-soiler: A tractor-mounted farm implement used for tilling soil at depths much below the levels normally worked by ploughs and harrows. It is used to break the hard pan created by ploughs and harrows over the years.

Sub-soiling: Also called ripping, chiseling or aerating is the process of breaking up of the dense horizons below the plough layer and to loosen up the soil.

Supercharger: Compressor that forces increased oxygen into the cylinders of an internal combustion engine.

Swather/Windrower: A machine that cuts hay or small grain crops and forms them into a windrow, mainly to decrease the time required for drying the crop to moisture content suitable for harvesting and storage.

Swept volume: See displacement volume.

Synchronous motor: A motor which operates at a constant speed up to full load. The rotor speed is equal to the speed of the rotating magnetic field of the stator – there is no slip. Also an AC motor that rotates in sync with the supply current's frequency.

Synchronous speed: The speed of an AC induction motor's rotating magnetic field. It is determined by the frequency applied to the stator and the number of magnetic poles present in each phase of the stator windings. Mathematically it is expressed as Speed (RPM) = 120 x Applied frequency (Hz)/Number of poles per phase.

Syndicator-type thresher: A thresher, the threshing unit of which consists of a corrugated fly wheel with serrated chopping knives and a closed cylinder casing and concave. It is also known as chaff cutter type thresher.

T

Tailing auger: A device in a combine harvester that carries tailings from pan to elevator.

Theoretical field capacity: The rate of field coverage of an implement based on 100 per cent of time at the rated speed and covering 100 per cent of its rated width.

Thermal (heat) energy: Thermal energy (also called heat energy) is produced as a result of rise in temperature causing atoms and molecules to move faster and collide with each other.

Thermal protector (inherent): An inherent overheating protective device which is responsive to motor temperature and when properly applied to a motor it protects the motor against dangerous overheating due to overload or failure to start.

Thermal overload relay: A thermal overload relay functions (trips) by means of a thermally responsive system.

Three point linkage: A combination of one upper link and two lower links each articulated to the tractor and the implement at their ends to connect the implement to the tractor.

Threshing: Threshing is the process of loosening the edible part of the cereal grain from the scaly, inedible chaff that surrounds it. In other words, it is the operation of detaching the grains from the spikes, panicles, ear heads, cobs or pods without removing the bran.

Throttle system: The components used to control the volume of air to the engine.

Throw: The movement of soil in any direction as a result of kinetic energy imparted to the soil by tillage tool. Also the distance reached by the jet or spray.

Tillage: The preparation of agricultural soil by any of various types of machines, whether human-powered, animal-powered, or mechanized, such as ploughing, harrowing, digging, hoeing and raking.

Tillage depth: Vertical distance from the initial soil surface to a specific point of penetration of the tool.

Tilt angle: It is the angle at which the plane of the cutting edge of the disc is inclined to a vertical line.

Tilth: Describes the general health of the soil. Soil that is healthy and has good physical qualities such as texture, structure, and general condition is in good tilth. Parameters such as moisture content, aeration, soil aggregate stability, rate of water infiltration, and drainage *etc.* indicate the general quality of tilth.

Tine: A slender prong of a fork or similar tool, straight or curved, stiff or flexible, and varying with respect to angle of soil contact.

Tractor: A type of vehicle designed specifically to deliver high tractive effort or torque at slow speeds for the purpose of hauling various kinds of machinery or a trailer. From agricultural point of view, it provides the power and traction to mechanize agricultural tasks.

Transplanter: A machine which sets seedlings into the ground in rows.

Transplanting: The process of planting seedlings in prepared seedbeds especially from one growing medium to another.

Top-dead-center (TDC): Position of the piston when it stops at the farthest point away from the crankshaft.

Torque: A measurement of an engine's power that indicates how forcefully it can rotate the crankshaft at a given engine speed. The unit of torque is kg-m.

U

Unit draft: Draft per unit cross-sectional area of the furrow

V

Velocity ratio (VR): It is defined as the distance moved by the effort to the distance moved by load.

Volumetric efficiency (engine): Volumetric efficiency is the ratio of actual air taken into the cylinder divided by the swept volume.

Volumetric efficiency (sprayers): The ratio of the actual volume of the spray discharged in one cycle to the piston displacement volume in the same cycle.

W

Water pump: A device that circulates coolant through a vehicles cooling system.

Watt: A unit of measure of power, or how fast energy is used. One watt of power is equal to one ampere (a measure of electric current) moving at one volt (a measure of electrical force).

Wettability: The ability of a liquid to maintain contact with a solid surface is known as its wettability.

Wheel base (trailers): The horizontal distance between front and rear wheels measured at the center of ground contact.

Winch: A winch is a cable system used to winch in logs. It can be powered by hand, hydraulics, the PTO or by electric.

Winching: The process wherein logs are attached to a cable and then moved along the ground by a machine that pulls the logs.

Wind farm: An area of land containing few or large number of wind turbines that generate electricity.

Windrow: A row of material formed by combining two or more swaths.

Windrower: A machine that cuts crops and delivers them in a uniform manner in a row.

Windrowing attachment: A series of bars attached to the cutter bar, curved upward at the rear end, to roll the swath into a windrow.

Winnower fan: A machine with one or two sieves and fan using air steam across falling grain. They may be manually or power-operated types.

Winnowing: Winnowing is the process of separation of heavier components of a mixture such as grains from lighter substances such as chaff with the help of wind either manually or mechanically.

Work: The component of the force in the direction of the displacement times the magnitude of the displacement in the direction of force.

Objective Type Questions and Answers

QUESTION 1: FILL IN THE BLANKS

Chapter 1

1. The horsepower developed by a bullock is nearly _________HP.
2. The horsepower developed by a man is _________HP.
3. 1 horsepower is equal to _________ Watts
4. Agricultural productivity has been observed to _________ with increasing power used in agricultural.
5. Biomass energy is a kind of _________ source of energy
6. Velocity ratio, VR in a jack is given as: $VR = 2\pi L/$_________.
7. For an ideal inclined plane, the mechanical advantage is given as $1/$_________.
8. The velocity ratio in case of wheel and axle is given as: $VR = R/r$ such that R is the radius of _________ and r is the radius of _________.
9. In the first class lever the _________is in the middle of the effort and the load.
10. The ratio of the force delivered by the machine to the force applied is known as _________ advantage of a machine is.

Chapter 2

11. Based on cycle of combustion, IC engines are categorized in _________ groups.

12. Compared to external combustion engines, IC engines have high power output per unit _________.

13. Cylindrical wall of the piston is known as piston _________.

14. Temperature of the compressed air in a compression ignition engine rises in the range of 600 to _________ °C.

15. Thermal efficiency of a petrol engine is _________ than the diesel engine.

16. Direction of the piston motion is reversed at the _________ centers.

17. 1 standard atmosphere is approximately equal to _________ kg/cm².

18. Indicator diagram is a plot between pressure and _________.

19. In a four stroke engine, the camshaft speed is _________ of the crankshaft speed.

20. Compression ratio of the _________ engine is in the range of 14:1 to 22:1.

21. Compression ratio of the _________ engine is in the range of 5:1 to 10:1.

22. For the same size of the engine, theoretical power produced by the two stroke engine is _________ that of a four stroke engine.

23. In a horizontal engine BDC is also known as _________ dead center.

24. Pressure – volume diagram is known as _________ diagram.

25. The equipment used to measure BHP is known as a _________

Chapter 3

26. The word tractor has its origin from the Latin word _________ meaning drawing.

27. _________ paper air filter is kind of a dry air filter.

28. Calorific value of petrol is about _________ kcal/kg.

29. Fuel injection pump supplies fuel to the injector in the pressure range of _________ kg/cm² to _________ kg/cm².

30. Both the flash and fire points of lubricant oil should be _________ than the temperature at which engine operates.

31. A good lubricant to give good service at low temperatures should have _________ pour point and cloud point.

32. The ability of lubricant oil to adhere to surfaces is known as _________.

33. The ability of lubricant oil to maintain contact with the solid surfaces is known as _________.

34. Thin or diluted oil can result in _________ oil pressure in the lubrication system.

35. In the thermo-siphon system of cooling, water circulation occurs due to _________ differences between the hot and cold water.

36. The complete path of power from the engine to the wheel is called as power _________.

37. In the quantitative governing, air fuel ratio of the mixture remains __________.

38. A governor is said to be _________, if the speed of the engine fluctuates above and below the mean speed.

39. A governor with a high degree of precision or stability is known as _________ deal governor.

40. Lay shaft in a sliding mesh gearbox is also known as _________ shaft.

Chapter 4

41. Tractors with _________ wheels are called wheeled tractors.

42. For an irrigated land an area of about _________ ha can be managed per 1 HP of tractor.

43. The strength of tyres is indicated by its _________ rating.

44. Belt pulley is generally made from _________ iron.

45. Crawler tractor exerts a much _________ force per unit area of the ground surface.

46. _________ wheels on the tractor are used at the time of puddling operation.

47. Slackened _________ repeatedly jump off the Sprockets.

48. To change the direction of rotation of the driven pulley, one needs to use _________ belt drive system.

49. The grip in a _________ belt drive system is greater than in the _________ belt drive system of pulleys.

50. The _________ are toothed wheels over which an endless chain is fitted.

Chapter 5

51. An electrical motor converts electrical energy into _________ energy.

52. A DC motor can be run with AC using a _________.

53. In a synchronous motor, rotation of shaft at steady state is synchronous with the _________ of the supply current.

54. Speed of an asynchronous motor is _________ than that of synchronous motor.

55. The line voltage is reduced to $1/$_________ value on each phase when connected in star equivalent design.

56. Fuse wire melts as soon as the current in the circuit exceeds the rated _________ of the fuse wire.

57. MCCB stands for molded case _________ breaker.

58. MCB stands for _________ circuit breaker.

59. Submersible motors used to pump water are _________ cooled.

60. Star-delta connection can only be used in motors where _________ motor terminals can be accessed.

Chapter 6

61. Groundwater pumping is one of highly _________ intensive farm activity.

62. On an average productivity increase due to farm mechanization can be to the tune of _________ per cent.

63. One of the factor for increasing farm productivity through farm mechanization is _________ completion of farm operations.

64. SMAM stands for sub-mission on _________ mechanization

65. Straw management systems on combine harvesters can help to overcome the environmental problems arising from _________of straw.

Chapter 7

66. Tillage operation carried out in a standing crop is known as _________ tillage.

67. Hoeing is an _________ tillage operation.

68. Land preparation operation for transplanting semi-aquatic crops such as rice is known as _________ tillage.

69. Tillage that covers the entire field in such a way that no living plant is left undisturbed is known as _________ tillage.

70. _________ tillage is completely avoided in no-till system.

71. A tillage system in which only isolated bands of soil are tilled is known as _________ tillage.

72. A raised ridge is formed at the center of the field during _________ type of ploughing.

73. In a country plough raising or lowering the beam helps to increase or decrease the _________ of ploughing.

74. Part of the plough bottom to which other components of the plough bottom are bolted is known as _________.

75. Size of the mould-board plough is expressed by the _________ of cut it is designed to cut.

76. Tilt angle of a good plough varies in the range of $15°$ to _________.

77. Depth measured at the center of the disc after the concave side of the disc is placed on a flat surface is known as _________ of the disc.

78. Low plough penetration may result from _________ tilt angle

79. Uneven furrows may result if _________ angles are not uniform.

80. Draft per unit cross-section area of the furrow is known as the _________ draft of plough.

Chapter 8

81. There is not much soil inversion in _________ tillage.

82. The function of a scraper is to scrape the soil from the _________ side of the disc.

83. Spools are normally made of __________ iron.

84. In a single action disc harrow gangs throw the soil in the __________ direction to each other.

85. In a double action disc harrow front and back gangs through the soil in the __________ direction to each other.

86. The angle between the axis of the gang and a line perpendicular to the direction of travel is known as __________ angle.

87. __________ tooth harrows are suitable for working in hard and stony soils.

88. Hoeing operation helps to improve soil __________ besides removing weeds.

89. A manually operated pull type hoe is known as __________.

90. Cage wheels are used to improve the __________ of tractors.

91. The flanged tube, mounted on the gang axle between every two discs to retain them at fixed position laterally on the shaft is called __________.

Chapter 9

92. Field efficiency is defined as the ratio of __________ field capacity to the __________ field capacity expressed in percentage.

93. Fill in the blank for the equation of material capacity, MC of a machine MC = __________ field capacity x crop yield.

94. Draft force is measured using some kind of __________.

95. Ownership cost of machinery increases almost __________ with size of the machine.

96. Operating cost of the machine __________ with increasing size of the machine.

Chapter 10

97. Fairly level lands are amenable to land __________ rather than the costly land leveling.

98. A tractor unit having a front mounted blade is known as __________.

99. A __________ can also be used for earthing-up operation.

100. Receiver of laser land leveling is a light __________ mounted on a mast on the land grading implement.

101. The rotating laser beam __________ creates a plane of laser light above the field, which acts as the leveling reference.

Chapter 11

102. Random scattering of seeds on the seedbed is known as __________.

103. A bamboo tube with a funnel shaped mouth used for sowing is known as __________.

104. A circular disc with V shaped serrations is known as __________ disc commonly used to meter fertilizer application.

105. A wheel made of aluminium with 6-12 spurs is known as ________wheel that is used to meter fertilizer application.

106. The disc in a single disc type furrow opener is set at a tilt angle of about ________ degrees.

107. The included angle between the discs in a double disc type furrow opener is about ________ degrees.

108. ________ chain is used as a covering device in seed drills.

109. T-type rather than a tine type furrow opener is used in ________ till drill.

110. Turbo ________ seeder is used to sow crops in fields laden with heavy surface residue loads.

111. Pointed________ type furrow openers are used to form a narrow slit in heavy soils.

112. A ________ is used to maintain seed to seed distance in a row.

Chapter 12

113. A turbo-machine that adds energy to the fluid is known as a ________.

114. Rotary pumps are grouped in ________displacement pumps.

115. Working head in a high head pump is more than ________ m.

116. In an________ flow pump, fluid enters and exits along the same direction parallel to the rotating shaft.

117. Head versus discharge curve rises with ________discharge.

118. The head developed by the pump at zero discharge is known as ________ head.

119. The operation in which suction pipe, casing of the pump and a portion of delivery pipe is filled by the fluid from outside is known as ________.

120. Locating the pump near to the water resource helps to reduce ________ lift.

121. The direction of rotation of the pump impeller is marked on the pump ________.

122. The ________ valve should be closed before stopping the operation of the pump.

123. 1 Newton-m/s is equal to ________ Watt.

124. In a turbine pump, line shafting transmits torque from the driver to ________ pump assembly.

125. In a ________ pump, both the pump and motor are submerged in water.

126. PM-KUSUM stands for *Pradhan-Mantri Kisan* ________ *Suraksha evam Utthaan Mahabhiyaan.*

127. Submersible pumps are a kind of ________ pumps.

Chapter 13

128. In gm ai/ha ai stands for _________ ingredient.
129. A _________ agent is used in foam spraying.
130. In ultra low volume spraying, spraying rate is less than _________ L/ha.
131. The pressure range in aircraft spraying/dusting is _________ than the pressure range in engine operated sprayers/dusters.
132. EC in pesticides formulations stands for _________ concentrate.
133. For all other conditions remaining the same, required spray volume will be _________ for coarse droplets compared to fine droplets.
134. Low pressure below _________ kg/cm^2 is undesirable as nozzles do not perform well at such low pressures.
135. Exit velocity of a nozzle is directly proportional to the square root of operating _________.
136. Paddle tip speed of agitators in power sprayers should not exceed _________ m/s to avoid foaming.
137. In a fogger, droplets smaller than _________ µm fill a volume of air reducing visibility.
138. Solid cone nozzle is also known as flood _________ nozzle.
139. Bigger size droplets are required during _________ application.

Chapter 14

140. Forged end of the blade to fix the handle of the sickle is called a _________.
141. _________ used to manually harvest fodder crops is operated in standing posture.
142. Assembly and working together of 2 or more cylinder mowers is known as _________ mower.
143. Cylinder mower is also known as _________ mower.
144. A conventional mower performs the cutting operation by impact and _________.
145. A properly registered knife of a conventional mower/reaper stops at the _________ of the guard on each stroke.
146. The outer end of the cutter bar is given a lead of _________ cm per meter length of the cutter bar.
147. One of the crop factors affecting threshing performance is the _________ content of the crop material.
148. In the operational mode, the crank pin, _________ head and the outer end of the knife in a cuter bar should be in a straight line.
149. Olpad thresher has its origin from village _________ in Gujarat.
150. Separation of heavier components of a mixture from its lighter substances with the help of wind is known as _________.

Chapter 15

151. *Khurpi* is a smaller version of _________.

152. A garden line makes it easy to create perfectly _________ rows and edges.

153. A garden sword is used to cut _________.

154. Pruning and slashing knife is also known as _________.

155. The blades of hand saw are designed to cut on the _________ stroke.

156. Electric pole pruners make the task of _________ of tree tops easy.

157. Draining and tapping knife is used for tapping and draining latex from the _________ trees.

158. Low frequency - high amplitude shaking is used for harvesting _________ fruits.

159. Grafting is a _________ method of plant propagation.

160. The handle in an axe is fixed in the _________ of the axe.

Chapter 16

161. The _________ is a hand tool used for grasping, holding and lifting wood logs.

162. Raker teeth saw cuts _________ than peg teeth saw.

163. Abbreviation ATV stands for All _________ Vehicle.

164. Drum _________ has a large steel drum powered by a motor.

165. The basic design of the disc chipper developed in 1889 is known as the _________ chipper.

166. There is no need of _________ operation in fuel wood harvesting

167. A stalk _________ is used for uprooting congested bushes.

168. The horizontal cut for the notch that decides the direction of the fall of the tree is made up to 1/4 to _________ of the diameter of the tree.

169. Production of chips makes it possible to recover on an average _________ to _________ per cent of biomass that would otherwise be wasted.

170. A skidding bar and a butt plate are almost similar but differ in their _________.

Chapter 17

171. Manually operated chaff cuter normally has _________ curved blades.

172. A _________ is a machine used to _________ lift or lower the loads.

173. The capacity of a trailer is given by the _________ load in tonnes.

174. The process to cut and discard the non-productive tops of sugarcane is known as _________.

175. _________ is a method of plant propagation in sugarcane in which cane cut above ground is left to continue growing.

176. The working area of a manure _________ is in the shape of an arc.

177. As per BIS standard, the capacity of a semi-trailer should be limited to _________ tonnes.

178. The productivity of sugarcane is observed to increase with _________ planting method.

179. Gathering mechanism is provided in sugarcane _________ harvester.

180. Piling-up of soil around the base of a plants in agriculture and horticulture is known as _________.

ANSWERS TO QUESTION 1

1. 0.5
2. 0.1
3. 747
4. Increase
5. Renewable
6. Pitch
7. Sin Θ
8. Wheel, Axle
9. Fulcrum
10. Mechanical
11. 3
12. Weight
13. Skirt
14. 900
15. Less
16. Dead
17. 1.03
18. Volume
19. Half
20. Diesel
21. Petrol
22. Twice
23. Outer
24. Indicator
25. Dynamometer
26. Tractus
27. Pleated
28. 11100
29. 120, 300
30. Above/more
31. Low
32. Oiliness
33. Wettability
34. Low
35. Density
36. Train
37. Constant
38. Hunting
39. Dead
40. Counter
41. Pneumatic
42. 1.5
43. Ply
44. Cast
45. Lower/lesser
46. Cage
47. Chains
48. Crossed
49. Crossed, Open
50. Sprockets
51. Mechanical
52. Rectifier
53. Frequency
54. Less
55. $\sqrt{3}$
56. Current
57. Circuit
58. Miniature
59. Water
60. Six
61. Power
62. 15-20
63. Timely
64. Agricultural
65. Burning
66. Inter
67. Inter
68. Wet
69. Clean

70. Primary
71. Strip
72. Gathering
73. Depth
74. Frog
75. Width
76. 25°
77. Concavity
78. High
79. Disc
80. Unit
81. Secondary
82. Concave
83. Cast
84. Opposite
85. Opposite
86. Gang
87. Spring
88. Aeration
89. Grubber
90. Traction
91. Spool
92. Effective, Theoretical
93. Effective
94. Dynamometer
95. Linearly
96. Decreases
97. Smoothening
98. Dozer
99. Ridger
100. Sensor
101. Emitter
102. Broadcasting
103. *Pora*
104. Serrated
105. Spur
106. 5
107. 10
108. Drag
109. Zero/no
110. Happy
111. Bar
112. Planter
113. Pump
114. Positive
115. 40
116. Axial
117. Decreasing
118. Shut-off
119. Priming
120. Suction
121. Casing
122. Delivery
123. 1
124. Bowl
125. Submersible
126. *Urja*
127. Rotodynamic/hydrodynamic
128. Active
129. Foaming
130. 5
131. Less
132. Emulsifiable
133. More
134. 1.5
135. Pressure
136. 2.5
137. 15
138. Jet
139. Herbicides
140. Tang
141. Scythe
142. Gang
143. Reel

144. Shear
145. Center
146. 2
147. Moisture
148. Knife
149. Olpad
150. Winnowing
151. *Khurpa*
152. Straight
153. Grass
154. Billhook
155. Pull
156. Lopping
157. Rubber
158. Nut
159. Asexual/vegetative
160. Eye
161. Tong
162. Faster
163. Terrain
164. Chipper
165. Wigger
166. Debarking
167. Puller
168. 1/3
169. 20, 30 per cent
170. Height
171. 2
172. Hoist, Vertically
173. Gross
174. Detopping
175. Ratooning
176. Spreader
177. 5
178. Pit
179. Combine
180. Earthing-up

QUESTION 2: MATCH THE FOLLOWING
(PUT SR. NUMBER FROM COLUMN B IN THE _______ (BLANK) OF COLUMN A)

Chapter 1

	Column A			Column B
1	Acceleration due to gravity		A	Watt
2	Joules		B	Wedge
3	Potential energy		C	Derived from movement of electrons
4	Electrical energy		D	Due to position of the body
5	1 Horsepower		E	9.81 m/s^2
6	1 Calorie		F	A process for biomass conversion
7	Force		G	Screw
8	Chisel		H	4.2 joules
9	Jack		I	Newton, N
10	Pyrolysis		J	0.746 kW

Chapter 2

	Column A			Column B
1	Air fuel mixture		A	Gudgeon pin
2	Clean air		B	Oil ring
3	Combustion is outside the engine		C	Stroke length
4	Piston pin		D	Mechanical efficiency
5	Brake horsepower/ Indicated horsepower		E	Petrol engine
6	Grooved ring in the piston		F	Diesel engine
7	Ports		G	Engine capacity
8	The distance between the top and bottom dead centers		H	Frictional horsepower
9	Swept volume times the numbers of cylinders in the engine		I	Two stroke engine
10	Indicated horsepower – Brake horsepower		J	External combustion engine

Chapter 3

	Column A			Column B
1	Carburetion		A	Priming pump
2	Air venting		B	Silencer
3	Turbocharger		C	Battery is not used
4	Muffler		D	Large dead bands
5	Clutch		E	High degree of precision or stability
6	Magnet ignition system		F	Supply air to the engine under pressure
7	Hit and miss governing		G	Rear axle
8	Mechanical governors		H	Uneven turning moment
9	Dead deal governor		I	Preparation of air fuel mixture
10	Emergency brakes		J	Located between the engine and gearbox

Chapter 4

	Column A			Column B
1	Power tiller		A	Weight not supported by the tractor
2	Track type tractor		B	Depreciation
3	Dash board		C	Helps start the engine
4	Decompression lever		D	Crawler tractor
5	Part of ownership cost		E	*Krishi*
6	Trailed type implement		F	Instrument panel
7	Power tiller		G	Walking or walk behind tractor

Chapter 5

	Column A			Column B
1	Synchronous motor		A	Brushless motors
2	Direct on-line starters		B	Starting winding 90° out of phase with the running winding
3	Submersible pumps		C	Low and medium power motors
4	Capacitor motor		D	Suction head is beyond the reach of centrifugal pumps
5	DC motors		E	AC motor

Chapter 6

	Column A			Column B
1	One of the causes of burning of crop residue		A	Early threshing machines
2	Loss of limbs		B	Cooperative and contract farming
3	To overcome problems related to small farm size		C	An impediment to farm mechanization
4	Socio-economic condition		D	Early model of custom hiring
5	Use of threshers on share basis		E	Combine harvesting

Chapter 7

	Column A			Column B
1	Tilian/Teolian		A	Summer tillage
2	Off-season tillage		B	Looks like a miniature plough
3	Secondary tillage		C	Inter tillage
4	Earthing up		D	Zero tillage
5	Wet tillage		E	Horizontal component of the pull
6	No till		F	Turns a furrow slice
7	Unploughed land on each end of field		G	Tillage
8	Tail piece		H	Harrow
9	Jointer		I	Head land
10	Side draft		J	Puddling

Chapter 8

	Column A			Column B
1	Hoof shovel		A	*Bakhar*
2	Blade harrow		B	Knife harrow
3	Acme harrow		C	Clod crusher
4	Cono-weeder		D	Goose foot
5	Cage roller		E	Used in line planted rice fields

Chapter 9

	Column A			Column B
1	Timeliness cost		A	Kilowatt
2	Draft force		B	Ratio of actual to the theoretical field capacity
3	_____ x 1.34 = Horsepower (HP)		C	Loss of forward motion by the tractor
4	Field efficiency		D	Delay in farm operations
5	Wheel slip		E	Load cells

Chapter 10

	Column A			Column B
1	Ripper		A	Maximum horizontal distance between the farming boards
2	*Patella* harrow		B	Battery operated
3	Size of the *bund* former		C	Wooden plank normally of *Sal* wood with curved steel hooks
4	Laser emitter		D	Double mould-board plough
5	Ridger		E	Used to rip and loosen the soil, break old pavements

Chapter 11

	Column A			Column B
1	Check row planting		A	Picker wheel mechanism
2	Hill dropping		B	Residue cutting
3	Transplanting		C	Furrow closer
4	Dropping of seeds and fertilizers		D	Related to rice and some vegetables
5	Vertical plate with radially projected arms		E	Direct seeded rice
6	Covering device		F	Checker board
7	Zero till drill		G	Used in paddy transplanter
8	Coulter		H	No till
9	Mat type rice seedlings		I	Seed dropping at fixed spacing
10	Helps to save water and fuel		J	Seed-cum-fertilizer drill

Chapter 12

	Column A			Column B
1	Related to cost of electrical energy		A	Lack of prime
2	Radial flow pump		B	kW-h
3	Vortex casing		C	0.746 kW
4	Foot valve		D	Improper alignment of the pump and driving unit
5	Water power		E	Photovoltaic
6	1 horsepower		F	Fluid exits at right angle to the shaft
7	Water is not delivered by the pump		G	Pump bowl assembly
8	Pump is noisy		H	Theoretical power to lift fluid
9	Turbine pump		I	A circular chamber between casing and impeller
10	Solar panels		J	Non-return valve

Chapter 13

	Column A			Column B
1	Spray nozzle		A	It is provided in reciprocating pumps
2	Agitator		B	Spray drift problem
3	Air chamber		C	Derives power from land wheels
4	Stirrup sprayer		D	mg/L
5	Air craft spraying		E	Breaks chemical into fine droplets
6	Adjustable nozzle		F	Suited to dry land crops
7	Traction dusters		G	Located near the bottom of the container
8	Wet duster		H	Rotary duster
9	Parts per million		I	Triple action nozzle
10	Belly mounted		J	Bucket sprayer

Chapter 14

	Column A			Column B
1	Shoes of conventional mower/reaper		A	Root crop
2	Drummy		B	Ratio of threshed grain from all outlets divided by grain input expressed in per cent
3	Tapioca		C	Traditional methods of threshing
4	Threshing efficiency		D	Breaking of knives
5	Syndicator type		E	Height of cut
6	Low quality of product		F	Stores/releases energy uniformly to maintain steady speed
7	Groundnut		G	Cassava
8	A major problem of mower/reaper		H	Chaff cutter type
9	Flywheel		I	Swath
10	Path made by the width of scythe		J	No separation or cleaning systems

Chapter 15

	Column A			Column B
1	Spade		A	Operated by two persons on either side
2	Wheelbarrow		B	Nut industry
3	Dibber		C	Make soil into fine tilth
4	Lumberjack saw		D	Improve tolerance to abiotic stresses
5	Fixed height fruit harvester		E	Nylon rope
6	Grafting		F	To make holes in the seedbed for planting
7	Loppers		G	A blade fixed on a wooden handle
8	Traditional trunk shaking fruit harvesters		H	Light bamboo pole
9	Garden line		I	Requires use of both hands
10	Garden rake		J	Used to move heavy materials

Chapter 16

	Column A			Column B
1	Elephant		A	Stem tightener
2	Splitting at butt end		B	Pickaroon
3	Hand sappie		C	Used for mulching
4	Log handling tool with sharply bent tip		D	Hoeing and weeding in forest nursery
5	Wood chips		E	Power source in forestry

Chapter 17

	Column A			Column B
1	Ratooning		A	Manure spreader
2	A sugarcane planting technique		B	*Dhaincha*
3	Handle operated chaff cutter		C	Developed to reduce the quantity of sugarcane required for seeding
4	Honey wagon		D	Gears and universal joints to increase speed
5	Green manure crop		E	Stubble crop
6	Chain trencher		F	A tree planting technique
7	Combination of pit and auger hole method		G	Fly wheel
8	Bud-chipping machine		H	A component of sugarcane combine harvester
9	Gathering mechanism		I	Narrow and deep channels
10	Bullock drawn chaff cutter		J	Ring-pit method

ANSWERS TO QUESTION 2

Chapter 1		Chapter 2		Chapter 3	
1	E	1	E	1	I
2	A	2	F	2	A
3	D	3	J	3	F
4	C	4	A	4	B
5	J	5	D	5	J
6	H	6	B	6	C
7	I	7	I	7	H
8	B	8	C	8	D
9	G	9	G	9	E
10	F	10	H	10	G

Chapter 4		Chapter 5		Chapter 6	
1	G or E	1	E	1	E
2	D	2	C	2	A
3	F	3	D	3	B
4	C	4	B	4	C
5	B	5	A	5	D
6	A				
7	E or G				

Chapter 7		Chapter 8		Chapter 9	
1	G	1	D	1	D
2	A	2	A	2	E
3	H	3	B	3	A
4	C	4	E	4	B
5	J	5	C	5	C
6	D				
7	I				
8	F				
9	B				
10	E				

Chapter 10		Chapter 11		Chapter 12	
1	E	1	F	1	B
2	C	2	I	2	F
3	A	3	D	3	I
4	B	4	J	4	J
5	D	5	A	5	H
		6	C	6	C
		7	H	7	A
		8	B	8	D
		9	G	9	G
		10	E	10	E

Chapter 13		Chapter 14		Chapter 15	
1	E	1	E	1	G
2	G	2	J	2	J
3	A	3	G	3	F
4	J	4	B	4	A
5	B	5	H	5	H
6	I	6	C	6	D
7	C	7	A	7	I
8	F	8	D	8	B
9	D	9	F	9	E
10	H	10	I	10	C

Chapter 16		Chapter 17	
1	E	1	E
2	A	2	J
3	D	3	G
4	B	4	A
5	C	5	B
		6	I
		7	F
		8	C
		9	H
		10	D

QUESTION 3: SELECT THE CORRECT OR MOST APPROPRIATE ANSWER (A, B, C OR D) FROM THE FOLLOWING

Chapter 1

1. **In an ideal basic machine**
   (A) Mechanical advantage is equal to velocity ratio
   (B) Mechanical advantage is greater than velocity ratio
   (C) Mechanical advantage is less than velocity ratio
   (D) None of these

2. **Kinetic energy is given as (m is the mass and v the velocity)**
   (A) mV^2 (B) $0.5\,mV^2$
   (C) $0.5\,m/V^2$ (D) None of these

3. **In a wedge the mechanical advantage increases with**
   (A) Increasing vertex angle   (B) Decreasing vertex angle
   (C) Vertex angle has no effect   (D) None of these

4. **Which of the following machines is a wedge?**
   (A) Jack   (B) Hammer
   (C) Bolt   (D) Axe

5. **Power is expressed as**
   (A) Rate of doing work
   (B) Amount of energy consumed per unit time
   (C) Any of (A) and (B)
   (D) None of these

Chapter 2

6. **Based on the cycle of combustion, IC engines are**
   (A) Otto cycle engines   (B) Diesel cycle engines
   (C) Dual cycle engines   (D) All of these

7. **Compression rings are placed in grooves**
   (A) Nearest to the piston head
   (B) Lowest groove above the piston pin
   (C) Below the gudgeon pin
   (D) None of these

8. **The functions of a flywheel**
   (A) Stores energy during power stroke to return it during idle strokes
   (B) Serves as one of the surfaces for the clutch plate
   (C) Can serve the purpose of pulley to transfer power
   (D) All of these

9. **Inlet valve is open during**
   (A) Power stroke
   (B) Compression stroke
   (C) Suction stroke
   (D) Exhaust stroke

10. **Injection pumps are provided in**
    (A) Otto cycle engines
    (B) Diesel cycle engines
    (C) Both (A) and (B)
    (D) None of these

11. **One power stroke in each revolution of the crankshaft**
    (A) Two stroke engine
    (B) Four stroke engine
    (C) Both (A) and (B)
    (D) None of these

12. **Which of the two sequences of events does not have separate strokes in the two stroke engines?**
    (A) Suction and exhaust
    (B) Suction and compression
    (C) Compression and power
    (D) Power and exhaust

13. **Clearance and swept volume of the engine are 3000 and 1000 cm^3 respectively. What is the compression ratio?**
    (A) 3
    (B) 1.33
    (C) 4
    (D) None of these

14. **In horizontal engines TDC is also known as**
    (A) Outer dead center
    (B) Inner dead center
    (C) Middle dead center
    (D) None of these

15. **Power generated in the engine is**
    (A) Indicated horsepower
    (B) Brake horsepower
    (C) PTO horsepower
    (D) Drawbar horsepower

Chapter 3

16. **Lubrication system besides lubrication helps in**
    (A) Acts as a seal
    (B) Cooling of heated parts
    (B) Prevents corrosion
    (D) All of these

17. **A gear on the main shaft is called as**
    (A) Countershaft gear
    (B) Main shaft gear
    (C) Primary gear
    (D) None of these

18. **PTO stands for**

(A) Power take off

(B) Power train on

(C) Polish take off

(D) None of these

19. **PTO shaft is designed to rotate at**

(A) 540 RPM

(B) 1000 RPM

(C) Both (A) and (B)

(D) None of these

20. **Which statement well describes the characteristic of disc brake over drum brakes**

(A) Efficient

(B) Better stopping power

(C) Work better in wet condition

(D) All of these

21. **Which amongst the following has highest calorific value?**

(A) Light speed diesel

(B) Petrol

(C) High speed diesel

(D) All have equal value

22. **Which kind of filter has now become obsolete?**

(A) Dry type

(B) Oil wetted mesh type

(C) Water bath type

(D) Oil bath type

23. **The chemical formula for iso-octane is**

(A) C_8H_{18}

(B) C_7H_{16}

(C) $C_{16}H_{34}$

(D) $C_{11}H_{16}$

24. **Governor regulation is defined as (S_0 is the speed at no load and S_1 the speed at maximum load)**

(A) $100(S_0 - S_1)/(S_0/2)$

(B) $100(S_0 - S_1)/[(S_0 + S_1)/2]$

(C) $100\,S_0/[(S_0 + S_1)/2]$

(D) $100(S_0 - S_1)/(S_0 + S_1)$

25. **The lowest temperature at which oil flows under gravity in a specified laboratory test is known as**

(A) Flash point

(B) Cloud point

(C) Pour point

(D) None of these

Chapter 4

26. **A tyre size of 12-38 means that**

(A) Sectional diameter of tyre is 12″

(B) Sectional diameter of tyre is 38″

(C) Sectional diameter of tyre is 26″

(D) None of these

27. Ownership cost of the tractor comprises

(A) Depreciation

(B) Interest on investment

(C) Taxes, insurance and housing

(D) All of these

28. Average cost of the tractor over its life span is given as (A is the purchase price and S the salvage value)

(A) A+S

(B) (A-S)/2

(C) (A+S)/2

(D) A-S

29. Average diesel consumption, A in L/hr of a tractor can be estimated from the formula (B is rated power in kW)

(A) A= 0.25 B

(B) A= 0.15 B

(C) A= 0.20 B

(D) None of these

30. Average petrol consumption, A in L/hr of a tractor can be estimated from A= 0.25 B such that B is given as

(A) Rated power in kW

(B) Rated power in HP

(C) Rated power as A or B

(D) None of these

31. Depreciation cost of the tractor is attributed to

(A) Wear and tear

(B) Obsolescence

(C) Age of the tractor

(D) All of these

32. Dry mass is the mass of the tractor that includes

(A) Operator's weight

(B) All components necessary for its operation

(C) Water in the radiator

(D) All of these

33. In a chain drive, following expression can be used to calculate the speed of the driven sprocket (T_1 and T_2 are the teeth of the driving and driven sprockets and S_1 and S_2 are their speeds respectively).

(A) $T_1/T_2 = S_2/S_1$

(B) $T_1/T_2 = S_1/S_2$

(C) $T_2 = S_1/S_2$

(D) None of these

34. Which is not required to calculate the ownership cost of tractor?

(A) Initial cost

(B) Salvage value

(C) Fuel consumption

(D) Life of the machine

35. The production of indigenous tractors in India started in

(A) 1947-48

(B) 1950-51

(C) 1989-90

(D) 1960-61

Chapter 5

36. **A split phase motor is**
    (A) Capacitor start motor
    (B) Resistance start motor
    (C) Induction start motor
    (D) None of these

37. **Which kind of protection is provide by direct on line starters**
    (A) Over current protection
    (B) Overload protection
    (C) Both (A) and (B)
    (D) None of these

38. **A relay protects the motor from**
    (A) Overloads
    (B) Phase loss/failure
    (C) Phase imbalance
    (D) All of these

39. **Which one of them is AC motor?**
    (A) Series wound
    (B) Shunt wound
    (C) Induction
    (D) Permanent magnet

40. **Rotor of a split-phase motor consists of**
    (A) Laminated core
    (B) The shaft
    (C) Squirrel cage winding
    (D) All of these

41. **Safety of motor is ensured through**
    (A) MCBs
    (B) MCCBs
    (C) Overload relays
    (D) Any or more than one of these

42. **A submersible motor used to pump groundwater is**
    (A) Water cooled
    (B) Air cooled
    (C) Both (A) and (B)
    (D) None of these

43. **With a star-delta starter motor is always started in**
    (A) Delta configuration
    (B) Star configuration
    (C) Any of (A) and (B)
    (D) None of these

44. **Current flowing through the motor in star configuration is**
    (A) More than delta configuration
    (B) Equal to delta configuration
    (C) Less than delta configuration
    (D) None of these

45. **Characteristics of an overloaded motor**
    (A) Draws excessive current
    (B) Gets overheated
    (C) Reduced life of winding
    (D) All of these

Chapter 6

46. Farm mechanization reduces

(A) Corruption

(B) Drudgery

(C) Population

(D) None of these

47. RKVY stands for

(A) *Rashtriya Krishi Vikas Yojana*

(B) Regional *Krishi Vikas Yojana*

(C) *Rashtriya Kisan Vikas Yojana*

(D) None of these

48. SMAM envisages to increase availability of average farm power by 2022 to

(A) 5.0 kW/ha

(B) 1.5 kW/ha

(C) 2.5 kW/ha

(D) None of these

49. Constraints in farm mechanization of small and marginal farmers

(A) Lack of resources

(B) Fear and risk involved

(C) Lack of knowledge

(D) All of these

50. Timely harvest of crops helps in

(A) Reducing losses

(B) Good quality of produce

(C) Reduced vagaries of nature

(D) All of these

Chapter 7

51. Puddling helps to

(A) Reduce deep percolation losses

(B) Incorporate green manure

(C) Control weeds

(D) All of these

52. Puddling is known to result in

(A) Destroy soil structure

(B) Formation of hard pans

(C) Both (A) and (B)

(D) None of these

53. Reduced tillage leaves a residue cover of ____ per cent on the soil surface

(A) 15-30 per cent

(B) <15 per cent

(C) >30 per cent

(D) None of these

54. Which function(s) are performed by the coulter?

(A) Cut vertical furrow slice

(B) Cuts the trash

(C) Makes ploughing smoother

(D) All of these

55. Which one of them is a special kind of plough?

(A) Mould-board plough

(B) Rotavator

(C) Disc plough

(D) Vertical disc plough

56. Conventional tillage leaves a residue cover of _____ per cent on the soil surface
    (A) <15 per cent
    (B) 15-30 per cent
    (C) >30 per cent
    (D) None of these

57. Which is known as wheat land plough?
    (A) Vertical disc plough
    (B) Mould-board plough
    (C) Standard disc plough
    (D) None of these

58. Relatively speaking a sub-soiler requires a _____ HP tractor than a chiseler
    (A) Lower
    (B) Higher
    (C) Equal in both
    (D) None of these

59. The angle at which the plane of the cutting edge of the disc is inclined to the vertical plane is
    (A) Disc angle
    (B) Vertical suction
    (C) Tilt angle
    (D) None of these

60. Basin lister is used to form
    (A) Ridges
    (B) Broken furrows with small dams and basins
    (C) Contour cultivation
    (D) None of these

61. Which plough accessory is used to turn over a small ribbon like furrow slice directly in front of the main plough bottom?
    (A) Coulter
    (B) Gauge wheel
    (C) Jointer
    (D) Land wheel

62. What is the disc angle of a good standard disc plough?
    (A) 35°-39°
    (B) 42°-45°
    (C) 23°-27°
    (D) 59°-63°

Chapter 8

63. Disc penetration can be increased by
    (A) Increasing the gang angle
    (B) Decreasing the disc angle
    (C) Raising the hitch point
    (D) All of these

64. Disc penetration can be increased by
    (A) Increasing the gang angle
    (B) Increasing the disc angle
    (C) Lowering the hitch point
    (D) All of these

65. Blades of a rotavator can be
    (A) L type
    (B) Twisted
    (C) Straight
    (D) Any of these

66. **Which is not the kind of puddler**

    (A) *Patela*

    (B) ANGRAU

    (C) Cage wheel

    (D) Peg type

67. **Which is not a category of cultivator?**

    (A) Disc type

    (B) Blade type

    (C) Tine type

    (D) Rotary

68. **The purpose of harrowing is**

    (A) Pulverize soil

    (B) Destroy weeds

    (C) Break clods

    (D) All of these

69. **Acme harrow is also known as**

    (A) Knife harrow

    (B) *Guntaka*

    (C) *Bakhar*

    (D) None of these

70. **Which is not a category of roller?**

    (A) Corrugated roller

    (B) Land packer

    (C) *Patela*

    (D) Cambridge type

71. **Which disc harrow has two gangs in tandem and fitted one behind the other?**

    (A) Tractor drawn

    (B) Off-set

    (C) Tandem

    (D) Double action

72. **What is the range of the gang angle?**

    (A) $30°$-$40°$

    (B) $36°$-$56°$

    (C) $0°$-$27°$

    (D) $90°$-$100°$

Chapter 9

73. **Effective operating time per hectare is given as**

    (A) Effective field capacity

    (B) 1/Effective field capacity

    (C) Theoretical field capacity

    (D) 1/Theoretical field capacity

74. **Tractive and transmission coefficient is**

    (A) Drawbar power/PTO power

    (B) PTO power/ Drawbar power

    (C) Engine power/Drawbar power

    (D) Engine power/PTO power

75. If the cost of operation of a tractor and machine is Rs. 550 per hour and the effective capacity of the machine is 0.8 ha/hr, then the cost of operation in Rs. per ha nearly is

    (A) 551   (B) 1.4

    (C) 440   (D) 687

76. Material capacity of the machine is given as (C is the effective field capacity of the machine)

    (A) C× crop yield   (B) Crop yield/C

    (C) Crop yield   (D) None of these

77. Theoretical field capacity of a 5×25 cm double action disc harrow operating at a speed of 6 km/hr is

    (A) 7.5   (B) 75

    (C) 0.75   (D) None of these

Chapter 10

78. Which statement well describes the phases of land levelling?

    (A) Rough leveling or land clearing

    (B) Land leveling

    (C) Smoothening

    (D) All or any combination of the A, B and C

79. Ridger cannot be used for

    (A) To form ridges for sowing seeds and plants of row crops in well tilled soil

    (B) Grading the land

    (C) Earthing-up and similar other operations

    (D) To make field furrows or channels

80. Which is not a part of laser land leveler

    (A) Laser emitter   (B) Laser receiver

    (C) *Patella*   (D) Control box

81. Animal drawn bulk scraper consists of

    (A) Vertical board   (B) Tail board

    (C) Handle and hitch   (D) All of these

82. Which is not used in initial rough levelling or land clearing operation

    (A) Laser leveler   (B) Bulldozer

    (C) Dozer   (D) Ripper

Chapter 11

83. **Which is not a part of the seed drill**
    (A) Seed rate adjustment lever (B) Furrow opener
    (C) Dibbler (D) Frame

84. **Give the most appropriate response in respect of covering device**
    (A) Drag chain (B) Press wheel
    (C) Disc hiller (D) Any of these

85. **In a turbo-happy seeder it clears the residue in front of the furrow opener.**
    (A) Flail (B) Drive wheel
    (C) Slit opener (D) None of these

86. **Functional processes of seed drills include**
    (A) Storing the seeds
    (B) Place the seeds accurately and uniformly
    (C) Meter the seeds
    (D) All of these

87. **When the seeds behind the plough are dropped by hand, the method is known as**
    (A) *Kera* (B) *Pora*
    (C) Both (A) and (B) (D) None of these

88. **Sticky ball method is used**
    (A) For calibration of seed drill
    (B) To test uniformity of seed placement
    (C) To assess the breakage of the seeds
    (D) For calibration of a planter

89. **Knock out arrangement is used in a**
    (A) Planter (B) Seed drill
    (C) Seed-cum-fertilizer drill (D) None of these

90. **Which of the following is not a seed metering device**
    (A) Fluted roller feed type (B) Internal double run type
    (C) Shovel type (D) Cell feed mechanism type

91. **In a single disc furrow opener, disc is set at a tilt angle of**
    (A) 10^0 (B) 5^0
    (C) 15^0 (D) None of these

92. **Drawbacks of direct seeding of rice**

    (A) Excessive weed infestation

    (B) Requirement of proper land leveling, which may need additional investment

    (C) Relatively high seed rate

    (D) All of these

Chapter 12

93. **According to casing design, pumps are characterized as**

    (A) Volute type

    (B) Vortex type

    (C) Diffuser type

    (D) Any of these

94. **Which of the following pumps has highest specific speed**

    (A) Radial flow pumps

    (B) Propeller pumps

    (C) Mixed flow pumps

    (D) All have the same specific speed

95. **According to disposition of shaft, pumps are**

    (A) Horizontal shaft pumps

    (B) Vertical shaft pumps

    (C) Any of (A) and (B)

    (D) None of these

96. **This impeller is preferred when fluids have solid particles**

    (A) Open

    (B) Semi-open

    (C) Closed

    (D) All of these are preferred

97. **A reflux valve in a pump is used**

    (A) Only on suction side

    (B) Only on delivery side

    (C) On any or both sides

    (D) None of these

98. **In the water horsepower equation, WHP = Q H/75, if H the head is in meters then Q, the pump discharge is in**

    (A) m^3/hr

    (B) L/s

    (C) m^3/s

    (D) None of these

99. **The water horsepower equation, WHP = Q H/75 is valid for (H is the head in meters then Q, the pump discharge is in L/s)**

    (A) Only water

    (B) Any fluid including gases

    (C) All kinds of liquids

    (D) All of these

100. **For direct driven pumps**

    (A) BHP > SHP

    (B) BHP = SHP

    (C) BHP < SHP

    (D) Any of these

101. Pump discharge is low

    (A) Air leak in stuffing box   (B) Speed too low

    (C) Suction lift high   (D) All of these

102. Pump leaks excessively at stuffing box

    (A) Worn out shaft   (B) Worn out packing material

    (C) Both (A) and (B)   (D) None of these

103. Energy consumed by a motor in kilowatts is

    (A) IHP x 746   (B) IHP x 0.746

    (C) IHP/0.746   (D) None of these

104. If a pump creates perfect vacuum, the water can be theoretically sucked up to _____ m at sea level

    (A) 10.34 m   (B) 8.34 m

    (C) 5.34 m   (D) None of these

105. In a submersible pump

    (A) Motor is under water

    (B) Pump is under water

    (C) Both motor and pump are under water

    (D) Both motor and pump are above the water

106. If pump is working but no liquid is being delivered, then it could be due to

    (A) Pump not primed

    (B) Strainer or foot valve on the suction line clogged

    (C) End of suction line is not submerged in water

    (D) Any or all of these

Chapter 13

107. A duster is used to apply pesticides

    (A) Solid powder form

    (B) Aerosols

    (C) Liquid form

    (D) Any or a combination of these

108. In modern agriculture chemicals are also used for

    (A) Defoliation   (B) Growth regulation

    (C) Prevent fruit dropping   (D) All of these

109. Discharge from the nozzle is related to the operating pressure as

(A) Directly proportional

(B) Proportional to the square root of the pressure

(C) Proportional to the cube root of the pressure

(D) Proportional to the square of the pressure

110. Which of the sprayer has a bean shaped tank?

(A) Hand atomizer   (B) Hand compression sprayer

(C) Knapsack sprayer   (D) Foot or pedal sprayer

111. This type of nozzle is a miniature type of disc nozzle.

(A) Solid stream nozzle   (B) Cap nozzle

(C) Vermorel nozzle   (D) Modified vermorel nozzle

112. The non-toxic inert material added to the dry dust formulations of pesticides are

(A) Talc

(B) Attapulgite, a naturally mined clay

(C) Soapstone

(D) Any of these

113. 8 L of spray was used to cover an area of 250 m². What is the application rate of the pesticide?

(A) 320 L/ha   (B) 800 L/ha

(C) 250 L/ha   (D) None of these

114. Volumetric efficiency of a sprayer is given as (A is the actual and B the theoretical suction capacity respectively)

(A) 100 B/A   (B) 100 A/B

(C) 100 (B –A)/A   (D) None of these

115. A prime mover is used in

(A) Hand atomizer   (B) Manually operated sprayers

(C) Power operated sprayers   (D) None of these

116. The pesticide application by any sprayer is regulated by

(A) Nozzle discharge rate   (B) Swath width

(C) Walking speed of operator   (D) All of these

Chapter 14

117. Which action(s) help the harvesting of the crops?

    (A) Slicing action

    (B) Tearing action

    (C) Scissor type action

    (D) Any or a combination of these

118. A sickle is categorized as

    (A) Plain sickle   (B) Serrated sickle

    (C) Both (A) and (B)   (D) None of these

119. Functions of a pitman in a conventional mower/reaper

    (A) To transmit reciprocation motion to knife head

    (B) Support weight

    (C) Absorb vibrations

    (D) All of these

120. Breaking of knife is caused by

    (A) Play in bearings   (B) Worn knife head holder

    (C) Poor alignment   (D) Any or all of these

121. Controlling factors to properly align a cutter bar

    (A) Cutter bar lead   (B) Registration of knives

    (C) Angle of pitman   (D) All of these

122. Blades of a super straw management system rotate at

    (A) 550 RPM   (B) 1000 RPM

    (C) > 1500 RPM   (D) None of these

123. Threshing is achieved by

    (A) Rubbing

    (B) Impact

    (C) Stripping

    (D) Any or a combination of them

124. Which one is not the root crop

    (A) Potato   (B) Pearl millet

    (C) Tapioca   (D) Groundnut

125. **Full form of SSMS is**

    (A) Super stubble management system

    (B) Super straw management system

    (C) Stubble straw management system

    (D) None of these

126. **Spacing between cylinder and concave in case of rice and wheat is in the range of**

    (A) 5-13 cm

    (B) 1-5 cm

    (C) 15-20 cm

    (D) None of these

Chapter 15

127. **A shovel based on the shapes of the blade and its tip can be**

    (A) Round

    (B) Square

    (C) Scoop

    (D) Any of these

128. **Which is not a kind of dibber**

    (A) Straight dibber

    (B) M-shaped dibber

    (C) T-handled dibber

    (D) L-shaped dibber

129. **Tree pruning is a process of carefully removing the ______ parts of a plant**

    (A) Unhealthy parts

    (B) Healthy parts

    (C) Both (A) and (B)

    (D) None of these

130. **Mechanical harvesting may result in reduced harvested crop value per unit area because of**

    (A) Crops do not mature uniformly

    (B) Reduced quality of produce

    (C) Damaged product

    (D) All of these

131. **A garden rake is used to perform**

    (A) Breaking up the soil surface into a fine tilth

    (B) Collect weeds and stones

    (C) Remove dead grass from the lawns

    (D) All of these

Chapter 16

132. **Normally the total load hauled by an ATV should not exceed**

    (A) Weight of ATV

    (B) Twice the weight of ATV

    (C) Thrice the weight of ATV

    (D) No restriction

133. **A hand tool for debarking of wood is**
    - (A) Debarking puller
    - (B) Debarking spud
    - (C) Debarking lumber
    - (D) None of these

134. **The process of log making/cross-cutting the trees in lengths required for the intended use of the log is known as**
    - (A) Extraction
    - (B) Lumbering
    - (C) Bucking
    - (D) Topping

135. **Which is not a type of hand sappie**
    - (A) Type A
    - (B) Type B
    - (C) Type C
    - (D) Type D

136. **The saw teeth help to perform**
    - (A) Cut through the fiber
    - (B) Break loose the cut fiber
    - (C) Remove the loose fiber from the kerb
    - (D) All of these

137. **Raker teeth perform**
    - (A) Break loose the cut fiber
    - (B) Remove the loose fiber from the kerb
    - (C) Both (A) and (B)
    - (D) Cut through the fiber

Chapter 17

138. **The working speed (rpm) of power operated chaff cutter is in the range of**
    - (A) 35-50
    - (B) 200-300
    - (C) 600-1000
    - (D) >2000

139. **A hoist is a machine used to lift the loads**
    - (A) Vertically
    - (B) Horizontally
    - (C) Both (A) and (B)
    - (D) None of these

140. **The methods of tree planting**
    - (A) Pit method
    - (B) Auger hole method
    - (C) Combination of A and B
    - (D) All of these

141. **The capacity of a tractor trailer is given by**
    - (A) The pay load
    - (B) Gross load
    - (C) Pay load + Gross load
    - (D) None of these

142. **Ratooning is practiced in**
    - (A) Only in banana
    - (B) Only in sugarcane
    - (C) Few other crops including banana and sugarcane
    - (D) None of these

ANSWERS TO QUESTION 3

1. (A)	30. (A)	59. (C)	88. (B)	117. (D)
2. (B)	31. (D)	60. (B)	89. (A)	118. (C)
3. (B)	32. (B)	61. (C)	90. (C)	119. (D)
4. (D)	33. (A)	62. (B)	91. (B)	120. (D)
5. (C)	34. (C)	63. (A)	92. (D)	121. (D)
6. (D)	35. (D)	64. (D)	93. (D)	122. (C)
7. (A)	36. (B)	65. (D)	94. (B)	123. (D)
8. (D)	37. (C)	66. (C)	95. (C)	124. (B)
9. (C)	38. (D)	67. (B)	96. (A)	125. (B)
10. (B)	39. (C)	68. (D)	97. (C)	126. (A)
11. (A)	40. (D)	69. (A)	98. (B)	127. (D)
12. (A)	41. (D)	70. (C)	99. (A)	128. (B)
13. (C)	42. (A)	71. (B)	100. (B)	129. (C)
14. (B)	43. (B)	72. (C)	101. (D)	130. (D)
15. (A)	44. (C)	73. (B)	102. (C)	131. (D)
16. (D)	45. (D)	74. (D)	103. (B)	132. (A)
17. (B)	46. (B)	75. (D)	104. (A)	133. (B)
18. (A)	47. (A)	76. (C)	105. (C)	134. (C)
19. (C)	48. (C)	77. (C)	106. (D)	135. (D)
20. (D)	49. (D)	78. (D)	107. (A)	136. (D)
21. (B)	50. (D)	79. (B)	108. (D)	137. (C)
22. (C)	51. (D)	80. (C)	109. (B)	138. (C)
23. (A)	52. (C)	81. (D)	110. (C)	139. (A)
24. (B)	53. (A)	82. (A)	111. (B)	140. (D)
25. (C)	54. (D)	83. (C)	112. (D)	141. (B)
26. (A)	55. (B)	84. (D)	113. (A)	142. (C).
27. (D)	56. (A)	85. (A)	114. (B)	
28. (C)	57. (A)	86. (D)	115. (C)	
29. (B)	58. (B)	87. (A)	116. (D)	

References

Abdel-Mawla, H. A. 2014. State of the art: Sugarcane mechanical harvesting-discussion of efforts in Egypt. International Journal of Engineering and Technical Research (IJETR). 2: 57-68.

Abu-Aligah, M. 2011. Design of photovoltaic water pumping system and compare it with diesel powered pump. Jordan Journal of Mechanical and Industrial Engineering (JJMIE). 5: 273–280.

Adegunloye, F. O. 2009. Introduction to Farm Mechanization. National Open University of Nigeria. 68 p.

AICRP (Farm Implements and Machinery). 2008. Success Stories: Tractor Operated Straw Baler. All India Coordinated Research Project on Farm Implements and Machinery. Central Institute of Agricultural Engineering, Bhopal. 9 p.

AICRP (Farm Implements and Machinery). 2016. Success Stories-2016: Tractor Mounted Vertical Belt Paired Row Potato Planter. Extension Bulletin No. CIAE/FIM/2016/191. Central Institute of Agricultural Engineering, Bhopal.

Ambast, S. K., Gupta, S. K. and Sharma, D. K. 2015. Laser Land Leveling. ICAR-Central Soil Salinity Research Institute, Karnal, Technical Bulletin: CSSRI/Karnal/2015/01. 53 p.

Amponsah, S. K., Berchie, J. N., Manu-Aduening, J., Danquah, E. O., Adu, J. O., Agyeman, A. and Bessah, E. 2017. Performance of an improved manual cassava harvesting tool as influenced by planting position and cassava variety. 12: 309-319.

Beard, F. R. and Hill, R. W. 2000. Maintaining Electric Motors Used for Irrigation. ENGR/BIE/WM/06. Utah State University - Extension Service, Logan, UT. 5 p.

Bello, S. R. 2102. Farm Tractor Systems: Maintenance and Operations. Federal College of Agriculture, Ishiagu, Nigeria

BIS. 1965. Indian Standard: Specification for Green Manure Trampler, Animal Drawn. IS 3301-1965 (Reprint August 1982). Indian Standard Institution, New Delhi. 11p.

BIS. 1978. Indian Standard: Specifications for Pruning Secateur (Reaffirmed 1999). IS 3494-1978. Bureau of Indian Standards, New Delhi. 5 p.

BIS. 1979. Indian Standard: Guide for Estimating Cost of Farm Machinery Operation. IS: 9164. Bureau of Indian Standard, New Delhi. 12 p.

BIS. 1981. Indian Standard: Glossary of Terms Relating to Harvesting and Threshing Equipment. IS: 9826 – 1981. Bureau of Indian Standards, New Delhi. 12 p.

BIS. 1992a. Indian Standard: Code of Practice for Installation and Maintenance of Induction Motors (Revision 2003). IS 900:1992. Bureau of Indian Standards, New Delhi. 33 p.

BIS. 1992b. Indian Standard: Forestry Tools-Debarking Spud Specifications. IS 13376. Bureau of Indian Standard, New Delhi. 2 p.

BIS. 1996. Indian Standard: Three-Phase Induction Motors - Specification (Revision 1997). IS 325:1996. Bureau of Indian Standards, New Delhi. 11 p.

BIS. 1998a. Indian Standard: Soil Working Equipment-Animal Drawn Mould-board Plough, Fixed Type Specification (Second Revision 2009). Bureau of Indian Standards, New Delhi. 5 p.

BIS. 1998b. Indian Standard: Forestry Tools-Stalk Puller Specifications. IS 13485 (Part II). Bureau of Indian Standard, New Delhi. 3 p.

BIS. 2000. Indian Standard: Agricultural Tractor Trailer—Specification (Third Revision). Bureau of Indian Standards, New Delhi. 9 p.

BIS. 2004. Indian Standard: Rotating Electrical Machines- Part 1: Rating and Performance. IS/IEC 60034-1:2004. Bureau of Indian Standards, New Delhi. 63 p.

Cherian, A., Punnan, J. S. and Varghese, A. 2016. A review on power tiller attachments. Kerala Technological Congress (KETCON-2016 Technology for Sustainability). DOI: 10.13140/RG.2.1.4843.3528. 1-4.

Coates, W. E. 2002. Agricultural Machinery Management. Extension Note of the University of Arizona, Tucson, Arizona, USA.

Davidson, J. B. and Chase, L. W. 1908. Farm Machinery and Farm Motors. Agricultural and Biosystems Engineering Books. 2. Orange Judd Company, London. 513 p.

Desai, S. R. and Sivakumar, S. S. 2017. Farm Power and Machinery (Horticultural). www.agrimoon.com. 154 p.

Dixit, A. and Singh, S. 2020. Farm Machinery and Equipment II. www.agrimoon. com. 197 p.

Dogra, R. and Singh, K. K. 2020. Farm Machinery and Equipment I. www.agrimoon. com. 175 p.

Edwards, W. 2017. Farm Machinery Selection. Iowa State University Extension, Ames, Iowa https://image1.slideserve.com/3430420/bowers-86-method2-n.jpg.

Escorts Ltd. Service and Maintenance Manual of Tractor. Escorts Ltd., Faridabad https://www.escortsgroup.com/agri-machinery/international/templates/agriintl_home/ images/pdf/FT-6045.pdf

FASAR, Yes Bank. 2016. Farm Mechanization in India: The Custom Hiring Perspective. *FASAR,* Mumbai and German Asia-Pacific Business Association (GAA), Hamburg. 83 p.

Goodwin, W. H. 1912. Spraying Machinery Accessories. Bulletin of the Ohio Agricultural Experiment Station Number 248. 32 p.

Gulich, J. F. 2007. Centrifugal Pumps. Springer, New York, USA.

Gupta, A. and Narain, S. S. 2015. Effects of turbo-charging of spark ignition engines. HIDRAULICA. 4: 62-67.

Gupta, S. K. 2020. Fundamentals of Soil and Water Conservation Engineering. Astral Publications, New Delhi. 671 p.

Gupta, S. K. 2021. A Textbook of Fluid Mechanics. Astral International Publishing, New Delhi.

Hancock, J. N., Swetnam, L. D. and Benson, F. J. 1991. Calculating Farm Machinery Field Capacities. University of Kentucky. https://uknowledge.uky.edu/aen_reports/20 Pub. 5-1991. Opened on 12.11.2020.

Hartung, H. and Pluschke, L. 2018. The Benefits and Risks of Solar-Powered Irrigation - A Global Overview. The Food and Agriculture Organization of the United Nations and Deutsche Gesellschaft für Internationale Zusammenarbeit. Rome. 67 p.

http://www.eagri.org/eagri50/FMP211/pdf/lec11.pdf. Opened on 26.12.2020

http://ecoursesonline.iasri.res.in/course/view.php?id=59. Dogra, B. and Sharma, D. N. Field Operation and Maintenance of Tractors and Farm Machinery II. Opened on 26.12.2020.

http://www.ikisan.com/sprayers.html Opened on 06.01.2021

http://www.smallpelletmachines.com/product/auxiliary-equipment/drum-chipper.html opened on 20.12.2021.

https://beeindia.gov.in/sites/default/files/3Ch2.pdf. Electric Motors. Bureau of Energy Efficiency, New Delhi. 25-44.

https://icar.org.in/hi/content/harvesting-and-threshing-equipment-aicrp-farm-implements-and- machinery

https://niphm.gov.in/Recruitments/ASO-PHE-Manual-NIPHM-03102013.pdf. ASO-PHE-Manual -NIPHM-03102013.doc. Plant Health Engineering Division, NIPHM.

https://www.ceew.in/sites/default/files/CEEW-Solar-for-Irrigation-Deployment-Report-17Jan18_0.pdf. Opened on 10.12.2020.

https://www.cosmecosrl.com/news-fairs/the-ditchers-for-water-drainage. Opened on 10.02.2021.

https://www.extension.iastate.edu/agdm/crops/html/a3-28.html Opened on 06.01.2021

https://www.Progressivegardening.com/agricultural-engineering-2/converting-tractor-power-ratings.html Opened on 10.12.2020.

https//www.sameng.co.za/full-service-pump-manufacturers/centrifugal-pumps-troubleshooting-guide. Opened on 10.12.2020.

Huber, M. G. 1948. The Mower: How to Repair and Adjust. Extension Bulletin 686. Cooperative Extension Service- Oregon State College, Corvallis. 28 p.

ICID. 2019. Solar Powered Irrigation Systems in India: Lessons for Africa through a FAO Study Tour. Draft Report. International Commission on Irrigation and Drainage (ICID), New Delhi. 136 p.

Jat, M. L., Kapil, Kamboj, B. R., Sidhu, H. S., Singh, M., Bana, A., Bishnoi, D., Gathala, M., Saharawat, Y. S., Kumar, V., Kumar, A., Jat, H. S., Sharma, R. K., Sharma, P. C., Singh, R. K., Rajbir, Sapkota, T. B., Malik, R. K. and Gupta, Raj. 2013. Operational Manual for Turbo Happy Seeder- Technology for Managing Crop Residues with Environmental Stewardship. International Maize and Wheat Improvement Center (CIMMYT), Indian Council of Agricultural Research (ICAR), New Delhi, India.

Macmillan, R. H. 2002. The Mechanics of Tractor - Implement Performance: Theory and Worked Examples. University of Melbourne, Victoria, Australia.170 p.

Mehta, C. R., Chandel, N. S., Jena, P. C. and Jha, A. 2019. Indian agriculture counting on farm mechanization. Agricultural Mechanization in Asia, Africa and Latin America. 50: 84-89.

Miller, J. E. 1995. The Reciprocating Pump: Theory, Design and Use. Krieger Publishing Company; 2nd revised edition. Florida, USA.

MoAFW (Ministry of Agriculture and Farmers Welfare). 2017. Report of the Expert Committee on Power Tillers. Department of Agriculture, Cooperation and Farmers Welfare, New Delhi. 48 p.

MoAFW (Ministry of Agriculture and Farmers Welfare). 2018. Open Invitation of Proposals for Empanelment of Manufacturers for Supply of Machinery and Equipment for *in-Situ* Management of Crop Residue in the States of Punjab, Haryana, Uttar Pradesh and NCT of Delhi during the Financial Year 2018-19. F. No. 13-1/2018-M and T (I and P). Ministry of Agriculture and Farmers Welfare, New Delhi.

MoAFW (Ministry of Agriculture and Farmers Welfare). 2019. Agricultural Statistics at a Glance 2018. Department of Agriculture, Cooperation and Farmers Welfare, New Delhi. 468 p

MoAFW (Ministry of Agriculture and Farmers Welfare). 2020. Sub-Mission on Agricultural Mechanization: Operational Guidelines. Department of Agriculture, Cooperation and Farmers Welfare (Mechanization and Technology Division) Government of India, Krishi Bhawan, New Delhi. 85 p.

Nilsson, M. and Forshed, N. 2017. The Farm Tractor in the Forest. https://www.maine.gov/dacf/mfs/publications/general_publications/farm_tractor_in_the_forest.pdf

Ojha, T. P and Michael, A. M. 2006. Principles of Agricultural Engineering. Reliance Industries (P) Ltd, New Delhi.

Pioneer Pumps. 2018. Pioneer Self Priming Series: PT Series. Operation and Maintenance Manual. Manual #2501. Pioneer Pumps Canby, OR. 22 p.

PWC and FICCI. 2019. Farm Mechanization: Ensuring a Sustainable Rise in Farm Productivity and Income. PWC and FICCI, New Delhi. 53 p

Rajesh Kumar, Chaudhary, S. K. and Zubair, A. 2016. Forty Five Years of AICRP on Sugarcane. Indian Institute of Sugarcane Research, Lucknow. 77 p.

Richard, N. 2014. The Role of Farm Power in Agriculture. Lecture delivered at College of Agricultural Engineering and Technology, JAU, Junagarh.

Russell, F. and Mortimer, D. 2005. A Review of Small-scale Harvesting Systems in use Worldwide and their Potential Application in Irish Forestry. COFORD, National Council for Forest Research and Development, Dublin-Ireland. 48 p.

Sahay, J. 2006. Elements of Agricultural Engineering. Standard Publishers and Distributors, New Delhi.

Sahay, J. 2019. Elements of Agricultural Engineering. Standard Publishers Distributors, Delhi. 542 p.

Sahu, G. K. 2017. Pumps: Theory, Design and Applications. New Age International Publishers, Lucknow.

Sharma, D. N. and Mukesh, S. 2004. Design of Agricultural Tractor (Principles and Problems). Jain Brothers, New Delhi.

Shukla, P., Mehta, C. R., Agrawal, K. N. and Potdar, R. R. 2021. Studies on operational frequencies of controls on self-propelled combine harvesters in India. Journal of Agricultural Engineering. 58: 101-111.

Shukla, S. K., Sharma, L., Awasthi, S. K. and Pathak, A. D. 2017. Sugarcane in India (Package of Practices for Different Agro-climatic Zones). AICRP (S) Technical Bulletin No. 1. ICAR-Indian Institute of Sugarcane Research, Lucknow. 53 p.

Singh, A. K. 2014. More crop per drop of water. Lecture delivered in Brainstorming session on Water for Agriculture. 7-8 October, 2014. ICAR Research Complex for North Eastern Region. Barapani, Shillong

Singh, R. S., Singh, S. and Singh, S. P. 2015. Farm power and machinery availability on Indian farms. Agricultural Engineering Today. 39: 45–56.

Singh, S., Kingra, H. S. and Sangeet. 2013. Custom hiring services of farm machinery in Punjab: Impact and policies. Indian Research Journal of Extension Education. 13: 45-50.

Smith, H. P. 1937. Farm Machinery and Equipment. McGraw-Hill Book Company, Inc. New York. 460 p.

Srinivasan, K., Narayanan, V. V., Singh, S. K., Lakshmi, L. G. 2011. Tractors and Agricultural Machinery. New India Publishing Agency. Pitam Pura, New Delhi.

Srivastava, A. C. 2010. Mechanical aids for raising sugarcane ratoon crop. In: Mechanization of Sugarcane Cultivation (Singh, J. *et al.*, Eds.). Indian Institute of Sugarcane Research, Lucknow: IISR. 18-31.

The Working Group Report. 2018. Demand and Supply Projections Towards 2033: Crops, Livestock, Fisheries and Agricultural Inputs. NITI AAYOG, New Delhi. 209 p.

Wadhwa, D. S., Dhingra, H. S. and Singh, S. 2003. Field Operation and Maintenance of Tractor and Farm Machinery: Laboratory Manual. Department of Farm Machinery and Power Engineering, PAU Ludhiana.

Waterman, B. 2013. Solar Water Pumping Basics. http://newfarmerproject. wordpress.com/2012/06/14/solar-water-pumping-basics/Opened on 23.12.2020.

www.AgriMoon.com. 2016. ICAR e-Course: Farm Power and Machinery. 168 p. Opened on 20.07.2020.

Yadav, R., Patel, M., Shukla, S. P. and Pund, S. 2014. Ergonomic evaluation of manually operated six-row paddy transplanter. International Agricultural Engineering Journal. 16: 147-157.

Zakiuddin, K. S., Sondawale, H. V., Modak, J. P. and Ceccarelli, M. 2012. History of human powered threshing machines: a literature review. In: Explorations in the His. of Machines and Mech., HMMS 15 (Koetsier and Ceccarelli, Eds.). Springer Science + Business Media, Dordrecht. 431–445.

Index